Multidimensional Signal and Color Image Processing Using Lattices

Multidimensional Signal and Color Image Processing Using Lattices

Eric Dubois
University of Ottawa

WILEY

This edition first published 2019

Registered Office(s)
John Wiley & Sons, Inc., 111 River Street, Hoboken, NJ 07030, USA
John Wiley & Sons Ltd, The Atrium, Southern Gate, Chichester, West Sussex, PO19 8SQ, UK

Editorial Office
The Atrium, Southern Gate, Chichester, West Sussex, PO19 8SQ, UK

For details of our global editorial offices, customer services, and more information about Wiley products visit us at www.wiley.com.

Wiley also publishes its books in a variety of electronic formats and by print-on-demand. Some content that appears in standard print versions of this book may not be available in other formats.

Library of Congress Cataloging-in-Publication Data

Names: Dubois, E. (Eric), author.
Title: Multidimensional signal and color image processing using lattices / Eric Dubois.
Description: Hoboken, NJ : John Wiley & Sons, 2019. | Includes bibliographical references and index. |
Identifiers: LCCN 2018048220 (print) | LCCN 2019002308 (ebook) | ISBN 9781119111757 (Adobe PDF) | ISBN 9781119111764 (ePub) | ISBN 9781119111740 (hardcover)
Subjects: LCSH: Signal processing. | Lattice theory. | Image processing–Mathematics.
Classification: LCC TK5102.9 (ebook) | LCC TK5102.9 .D83 2019 (print) | DDC 621.382/201511332–dc23
LC record available at https://lccn.loc.gov/2018048220

Cover image: Courtesy of Eric Dubois
Cover design by Wiley

Set in 10/12pt WarnockPro by SPi Global, Chennai, India

Printed in Singapore by C.O.S. Printers Pte Ltd

10 9 8 7 6 5 4 3 2 1

To Sheila

Contents

About the Companion Website

The supplementary material is available on an accompanying website

www.wiley.com/go/Dubois/multiSP.

1

Introduction

This book presents the theory of multidimensional (multiD) signals and systems, primarily in the context of image and video processing. MultiD signals are considered to be functions defined on some domain $\mathcal{D}$ of dimension two or higher with values belonging to a set $\mathcal{R}$, the range. These values may represent the brightness or color of an image or some other type of measurement at each point in the domain. With this interpretation, a multiD signal f is represented as

$$f : \mathcal{D} \to \mathcal{R} : \mathbf{x} \mapsto f(\mathbf{x}), \tag{1.1}$$

i.e. each element $\mathbf{x}$ of the domain $\mathcal{D}$ is mapped to the value $f(\mathbf{x})$ belonging to the range $\mathcal{R}$. In conventional continuous-time one-dimensional (1D) signals and systems theory [Oppenheim and Willsky (1997)], $\mathcal{D}$ and $\mathcal{R}$ are both the set of real numbers $\mathbb{R}$, and so a real 1D signal would be written $f(t)$, $t \in \mathbb{R}$. MultiD signals arise when the domain $\mathcal{D}$ is a space with two or more dimensions. The domain can be continuous, as in the case of real-world still and time-varying images, or discrete, as in the case of sampled images. In addition, the range $\mathcal{R}$ can also be a higher-dimensional space, for example the three-dimensional color space of human vision.

In this book, we are mainly concerned with examples from conventional still and time-varying images, although the theory has broader applicability. A conventional planar image is written $f(x, y)$, where $\mathbf{x} = (x, y)$ lies in a planar region associated with the Euclidean space $\mathbb{R}^2$. Here, x denotes the horizontal spatial position and y is the vertical spatial position while f denotes image brightness or color. The domain $\mathcal{D}$ can be $\mathbb{R}^2$ itself, or a discrete subset in the case of sampled images. Similarly, a conventional time-varying image is written $f(x, y, t)$, where $\mathbf{x} = (x, y, t)$ lies in a subset (possibly discrete) of $\mathbb{R}^2 \times \mathbb{R}$, which may also be written $\mathbb{R}^3$. Here, x and y are as above, and t represents time. Higher-dimensional cases also exist, for example time-varying volumetric images with $\mathcal{D} = \mathbb{R}^3 \times \mathbb{R}$ and where f denotes some measurement taken at location $\mathbf{x}$ at time t. The domain $\mathcal{D}$ can also be a more complicated manifold such as a cylinder or a sphere, as in panoramic imaging.

MultiD signal processing has been an active area of study for over fifty years. Early work was in optics and the continuous domain. Papoulis's classic text on *Systems and Transforms with Applications in Optics* appeared in 1968 [Papoulis (1968)]. Soon after, work on two-dimensional digital filtering started to appear, for example [Hu and Rabiner (1972)]. Over the years, there have been several books devoted to multiD digital signal processing and numerous books on image and video processing. The present book is distinguished from these works in a number of aspects. The book is mainly concerned with

Multidimensional Signal and Color Image Processing Using Lattices, First Edition. Eric Dubois.

Companion website: www.wiley.com/go/Dubois/multiSP

the theory of discrete-domain processing of real- or vector-valued multiD signals. The application examples are drawn from grayscale and color image processing and video processing. In particular, the book is not intended to present the state-of-the-art algorithms for particular image processing tasks.

Most previous books on multiD signals considered rectangularly sampled signals for the main development and presented non-rectangular sampling on a lattice as a subsidiary extension. A lattice, as in crystal lattice, is a mathematical structure from which we can construct more general sampling structures. In this book, the theory is developed on lattices from the beginning, and rectangular sampling is considered a special case. Another difference is that most books use normalized representations for non-rectangular sampling that are dependent on the lattice basis. Although this may be convenient for certain manipulations, this approach obscures much of the geometrical and scale information of the signal. We prefer to use basis-independent representations as much as possible, and introduce the basis where needed to perform calculations or implementations. Thus, we do not use such normalized representations but rather use the actual units of space-time and frequency.

Another distinguishing feature of this book is the treatment of color. Color signals are viewed as multiD signals with values in a vector space, in this case the vector space of human color vision, and color signal processing is viewed as vector-valued signal processing. Most multiD signal processing books deal mainly with scalar signals, representing a grayscale or brightness value. If color models are introduced, color signal processing generally involves separate processing of three color channels. Here we present the theory of multiD signal processing where the input and output are vector-valued signals, further developing the theory introduced in Dubois (2010).

In general, multiD signals in the real world, such as still and time-varying images, are functions of the continuous space and time variables. Consider for example a light signal falling on a camera sensor or emanating from a motion-picture screen. These multiD signals are converted to discrete-domain signals for digital processing, storage, and transmission. They may eventually be converted back to continuous-domain signals, for example for viewing on a display device. Thus, we begin with an overview of scalar-valued continuous-domain multiD signals and systems, i.e. the domain is $\mathbb{R}^D$ for some integer D. In particular we introduce the concepts of signal space, linear shift-invariant systems and the continuous-domain multiD Fourier transform, develop properties of the Fourier transform and present some examples. Continuous-domain signal spaces and transforms involve advanced mathematical analysis to provide a general theory for arbitrary signal spaces. We do not attempt to provide a rigorous analysis. We assume that signals belong to a suitable signal space for which transforms are well defined and the properties hold. For example, a space of tempered distributions would be satisfactory. However, we do not develop the theory of distributions and take an informal approach to the Dirac delta and related singularities. We refer the reader to references for a rigorous analysis, e.g., [Stein and Weiss (1971), Richards and Youn (1990)].

There are many possible domains for multiD signals, generally subsets of $\mathbb{R}^D$ for some D. These domains can be continuous or discrete, or a hybrid that is continuous in some dimensions and discrete in others, like in analog TV scanning. The domain can also correspond to one period of a periodic signal, whether continuous or discrete. Among the possible domains, certain of them allow for the possibility of linear shift-invariant filtering. These domains have the algebraic structure of a locally-compact Abelian (LCA) group. While we cannot go into the detail of such structures, their main feature is that the concept of shift is well defined and commutative. $\mathbb{R}^D$ is an example, as is any lattice in $\mathbb{R}^D$. The LCA group is the classical setting for abstract harmonic analysis, e.g., as presented in Rudin (1962). An early work on signal processing in this setting is the Ph.D. thesis of Rudolph Seviora [Seviora (1971)]), which considered generalized digital filtering on LCA groups. More recently Cariolaro has developed signal processing on LCA groups in a comprehensive book [Cariolaro (2011)].

In this book, we have elected to provide a separate development for the cases of continuous-domain aperiodic signals, discrete-domain aperiodic signals, discrete-domain periodic signals, and continuous-domain periodic signals (Chapters 2–5). Each case has its own sphere of application, and while the development may be redundant from an abstract mathematical perspective, the concrete details are sufficiently different to warrant their own presentation. Each of these chapters follows a similar roadmap, presenting concepts of signal space, linear shift-invariant (LSI) systems, Fourier transforms and their properties. For discrete-domain signals, we use lattices to describe the sampling structure. For periodic signals, we use lattices to describe the periodicity. Since lattices form an underlying tool used throughout the book, we have chosen to gather all definitions and results about lattices that we need for this work in Chapter 13, which may be consulted any time as needed. We prefer not to interrupt the flow of the book at the beginning with this material, and we wish to give it a higher status than an appendix. This is why we have chosen to include it as the last chapter in the book.

In Chapter 6 we see the relationship between the four representations. Discrete-domain aperiodic and periodic signals can be obtained for the corresponding continuous-domain signals by a sampling operation. This is shown to induce a periodization in the frequency domain. In another view, discrete and continuous-domain periodic signals can be obtained by periodization of corresponding aperiodic signals, resulting in sampling in the frequency domain. These results are all explored in Chapter 6 and various sampling theorems are presented. We do not explicitly explore hybrid signals, which may correspond to a different one of the above types in different dimensions. This extension is usually straightforward; many examples are given in Cariolaro (2011).

Having developed the theory of processing of multiD scalar signals, we address the nature of the signal range in Chapter 7, specifically for color image signals. Here we take up the vector-space view of color spaces as presented in Dubois (2010), where colors are viewed as equivalence classes of light spectral densities that give the same response to a viewer. This viewer can be a typical human with a three-dimensional color space or maybe a camera, which could have a higher-dimensional color space. With this representation, we develop signal processing theory for color signals in Chapter 8, again taking up and extending the presentation of Dubois (2010). Such a unified theory of color signal processing has not yet been widely adopted in the literature. One unique aspect is that color signals can have different sampling structures for different subspaces of the color space, as in the Bayer color filter array (CFA) of digital cameras, for example. This aspect is developed in Chapter 8.

MultiD signals can often be usefully considered to be realizations of a random process. This concept has been used extensively in multiD signal processing applications using many types of random field models. In Chapter 9, we present some basic concepts on multiD random fields, limiting the presentation to wide-sense stationary random fields characterized by correlation functions and spectral densities. One novel aspect is the development of vector-valued random field models for color signals. Most books on random processes deal mainly with scalar signals, and may just mention the case of vector signals (e.g., Leon-Garcia (2008), Stark and Woods (2002)). However, some classic texts (e.g., Brillinger (2001), Priestley (1981)) consider vector-valued random processes in detail and we follow this approach for correlation functions and spectral densities of color or other vector-valued signals. However, we do not go into more elaborate random field models such as Gibbs–Markov random fields, which have been very successful. These are covered in several books, including for example Fieguth (2011).

In Chapter 10, we present some multiD filter design methods and examples, and then consider the specific example of change of sampling structure of an image in Chapter 11. These developments are all carried out in the general context of lattices. We do not delve into general multiD multiresolution signal processing such as filter banks in this book. This topic can also be developed using a general lattice formulation, such as in our previous work [Coulombe and Dubois (1999)], or in the works by Suter (1998) and Do and Lu (2011).

We then present a novel development of symmetry-invariant signals and systems in Chapter 12. This extends the concept of linear shift-invariant systems to systems invariant to other transformations of Euclidean space, such as rotations and reflections. This uses the theory of symmetry groups, widely used in fields such as crystallography. One outcome of this development is the generalization of the widely used separable discrete cosine transform (DCT) to arbitrary lattices.

As mentioned above, Chapter 13 gathers all the material and results on lattices used throughout the book into a single chapter, which can stand alone. This is followed by some brief appendices describing a few aspects of equivalence relations, groups and vector spaces. One should consult suitable textbooks on algebra, such as the ones cited, for a complete presentation of these topics. Throughout the book, all results requiring a proof have been called theorems regardless of their importance, rather than trying to categorize them as theorem, proposition, lemma, etc.

Most of the material in this book (other than Chapter 12) has been used for many years in graduate and undergraduate courses in image processing. For such courses, it is usually supplemented with additional material on certain applications such as image and video compression or image restoration for which there are numerous textbooks. I would like to acknowledge the numerous students, both in my courses as well as thesis students and postdoctoral fellows, who have contributed to my understanding and development of this theory over the years. I would particularly like to thank Stéphane Coulombe for his many helpful comments on the manuscript and Luis Gurrieri for his input and help with some of the figures.

2

Continuous-Domain Signals and Systems

2.1 Introduction

This chapter presents the relevant theory of continuous-domain signals and systems, mainly as it applies to still and time-varying images. This is a classical topic, well covered in many texts such as Papoulis's treatise on *Systems and Transforms with Applications in Optics* [Papoulis (1968)] and the encyclopedic *Foundations of Image Science* [Barrett and Myers (2004)]. The goal of this chapter is to present the necessary material to understand image acquisition and reconstruction systems, and the relation to discrete-domain signals and systems. Fine points of the theory and vastly more material can be found in the cited references.

A continuous-domain planar time-varying image $f(x, y, t)$ is a function of two spatial dimensions x and y, and time t, usually observed in a rectangular spatial window $\mathcal{W}$ over some time interval $\mathcal{T}$. In the case of a still image, $f(x, y, t)$ has a constant value for each (x, y), independently of t. In this case, we usually suppress the time variable, and write $f(x, y)$. We use a vector notation $f(\mathbf{x})$ to simplify the notation and handle two and three-dimensional (and higher-dimensional) cases simultaneously. Thus $\mathbf{x}$ is understood to mean (x, y) in the two-dimensional case and (x, y, t) in the three-dimensional case. We will denote $\mathbb{R}^2 = \{(x, y) \mid x, y \in \mathbb{R}\}$ and $\mathbb{R}^3 = \{(x, y, t) \mid x, y, t \in \mathbb{R}\}$, where $\mathbb{R}$ is the set of real numbers. To cover both cases, we write $\mathbb{R}^D$, where normally $D = 2$ or $D = 3$; also the one-dimensional case is covered with $D = 1$ and most results apply for dimensions higher than 3. For example, the domain for time-varying volumetric images is $\mathbb{R}^4 = \{(x, y, z, t) \mid x, y, z, t \in \mathbb{R}\}$. It is often convenient to express the independent variables as a column matrix, i.e.

$$\mathbf{x} = \begin{bmatrix} x \\ y \end{bmatrix} \quad \text{or} \quad \mathbf{x} = \begin{bmatrix} x \\ y \\ t \end{bmatrix}.$$

Since there is no essential difference between $\mathbb{R}^D$ and the space of $D \times 1$ column matrices, we do not distinguish between these different representations. We will often abbreviate two-dimensional as 2D and three-dimensional as 3D.

The spatial window $\mathcal{W}$ is of dimensions $\mathrm{ph} \times \mathrm{pw}$ where pw is the picture width and ph is the picture height. Since the absolute physical size of an image depends on the sensor or display device used, we often choose to adopt the ph as the basic unit of spatial

Multidimensional Signal and Color Image Processing Using Lattices, First Edition. Eric Dubois.

Companion website: www.wiley.com/go/Dubois/multiSP

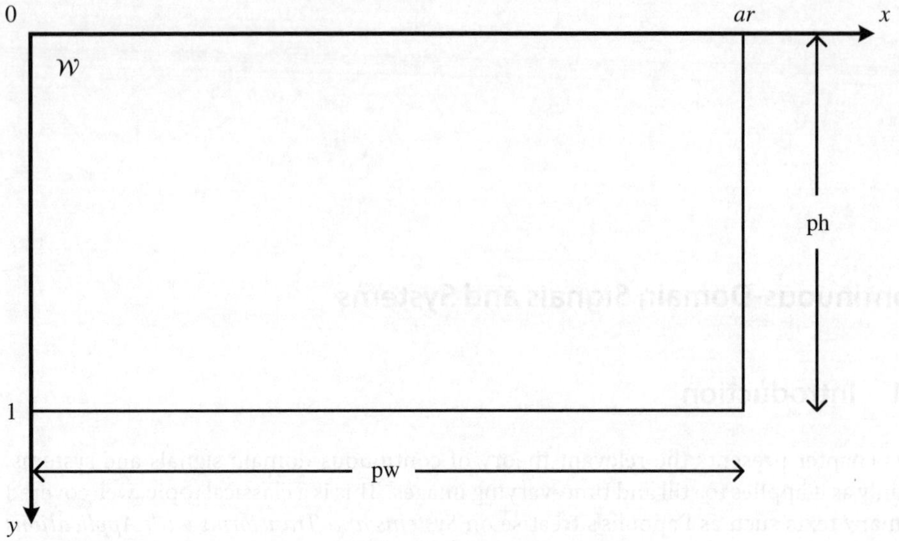

Figure 2.1 Illustration of image window $\mathcal{W}$ with aspect ratio ar : 1 pw = ar ph.

distance, as has long been common in the broadcast video industry. However, we are free to choose any convenient unit of length in a given application, for example, the size of a sensor or display element, or an absolute measure of distance such as the meter or micron. The ratio $ar = \text{pw}/\text{ph}$ is called the aspect ratio, the most common values being 4/3 for standard TV and 16/9 for HDTV. With this notation, 1 pw = ar ph (see Figure 2.1). Time is measured in seconds, denoted s. Examples of continuous-domain space-time images include the illumination on the sensor of a video camera, or the luminance of the light reflected by a cinema screen or emitted by a television display.

Since the image is undefined outside the spatial window $\mathcal{W}$, we are free to extend it outside the window as we see fit to include all of $\mathbb{R}^D$ as the domain. Some possibilities are to set the image to zero outside $\mathcal{W}$, to periodically repeat the image, or to extrapolate it in some way. Which of these is chosen depends on the application.

There are two common ways to attach an xy coordinate system to the image window, involving the location of the origin and the orientation of the x and y axes, as shown in Figure 2.2. The standard orientation used in mathematics to graph functions would place the origin at the lower left corner of the image with the y-axis pointing upward. However,

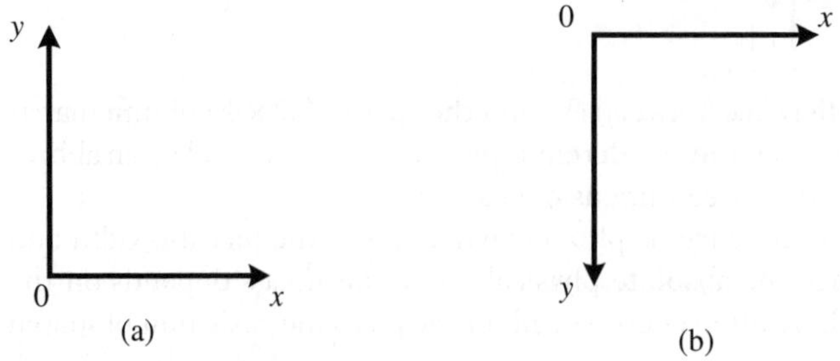

Figure 2.2 Orientation of xy-axes. (a) Common bottom-to-top orientation in mathematics. (b) Scanning-based top-to-bottom orientation.

because traditionally images have been scanned from top to bottom, most image file formats store the image line-by-line, with the top line first, and line numbers increasing from top to bottom of the image. This makes the orientation shown in Figure 2.2(b) more convenient, with the origin in the upper left corner of the image and the y-axis pointing downward. For this reason, we will generally use the orientation of Figure 2.2(b).

2.2 Multidimensional Signals

A multiD signal can be considered to be a function from the domain, here $\mathbb{R}^D$, to the range. In this and the next few chapters, we consider only real and complex valued signals, which we call scalar signals. In later chapters, where we consider color signals, we will take the range to be a suitable vector space. In addition to naturally occurring continuous space-time images, many analytically defined D-dimensional functions are useful in image processing theory. A few of these are introduced here.

2.2.1 Zero–One Functions

Let $\mathcal{A}$ be a region in the D-dimensional space, $\mathcal{A} \subset \mathbb{R}^D$. We define the *zero–one function* $p_{\mathcal{A}}(\mathbf{x})$ as illustrated in Figure 2.3(a) by

$$p_{\mathcal{A}}(\mathbf{x}) = \begin{cases} 1, & \text{if } \mathbf{x} \in \mathcal{A}; \\ 0, & \text{otherwise.} \end{cases} \tag{2.1}$$

Sometimes, $p_{\mathcal{A}}(\mathbf{x})$ is called the indicator function of the region $\mathcal{A}$. Different functions are obtained with different choices of the region $\mathcal{A}$. Such functions arise frequently in modeling sensor elements or display elements (sub-pixels). The most commonly used ones in image processing are the rect and the circ functions in two dimensions. Specifically, for a unit-square region $\mathcal{A}$ we obtain (Figure 2.3(b))

$$\mathrm{rect}(x, y) = \begin{cases} 1, & \text{if } |x| \le 0.5 \text{ and } |y| \le 0.5; \\ 0, & \text{otherwise.} \end{cases} \tag{2.2}$$

For a circular region of unit radius we have (Figure 2.3(c))

$$\mathrm{circ}(x, y) = \begin{cases} 1, & \text{if } x^2 + y^2 \le 1; \\ 0, & \text{otherwise.} \end{cases} \tag{2.3}$$

These definitions can be extended to the three-dimensional case (where the region $\mathcal{A}$ is a cube or a sphere) or to higher dimensions in a straightforward fashion, and the single notation $\mathrm{rect}(\mathbf{x})$ or $\mathrm{circ}(\mathbf{x})$ can be used to cover all cases. We will see later how these basic signals can be shifted, scaled, rotated or otherwise transformed to generate a much richer set of zero–one functions. Other zero–one functions that we will encounter correspond to various polygonal regions such as triangles, hexagons, octagons, etc.

2.2.2 Sinusoidal Signals

As in one-dimensional signals and systems, real and complex-exponential sinusoidal signals play an important role in the analysis of image processing systems. There are several reasons for this but a principal one is that if a complex-exponential sinusoidal signal is applied as input to a linear shift-invariant system (to be introduced shortly), the output

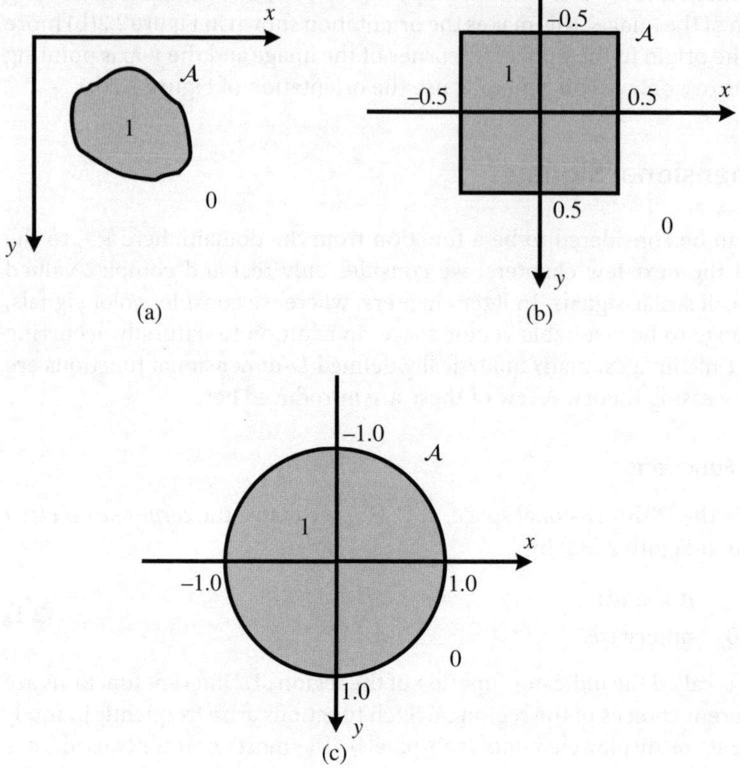

Figure 2.3 Zero–one functions. (a) General zero–one function. (b) Rect function. (c) Circ function.

is equal to the input multiplied by a complex scalar. A general, real two-dimensional sinusoid with spatial frequency (u, v) is given by

$$f_R(x, y) = A\cos(2\pi(ux + vy) + \phi) \tag{2.4}$$

where x and y are spatial coordinates in ph (say), u and v are *fixed* spatial frequencies in units of c/ph (cycles per picture height) and ϕ is the phase. In general, for a given unit of spatial distance (e.g., μm), spatial frequencies are measured in units of cycles per said unit (e.g., c/μm).

Figure 2.4 illustrates a sinusoidal signal with horizontal frequency 1.5 c/ph and vertical frequency 2.5 c/ph in a square image window of size 1 ph by 1 ph. The actual signal displayed is $f(x, y) = 0.5\cos(2\pi(1.5x + 2.5y)) + 0.5$, which has a range from 0 (black) to 1 (white). From this figure, we can identify a number of features of the spatial sinusoidal signal. The sinusoidal signal is periodic in both the horizontal and vertical directions, with horizontal period $1/|u|$ and vertical period $1/|v|$. The signal $f(x, y)$ is constant if $ux + vy$ is constant, i.e. along lines parallel to the line $ux + vy = 0$.

The one-dimensional signal along any line through the origin is a sinusoidal function of distance along the line. The maximum frequency along any such line is $\sqrt{u^2 + v^2}$, along the line $y = (v/u)x$, as illustrated in Figure 2.4. The proof is left as an exercise.

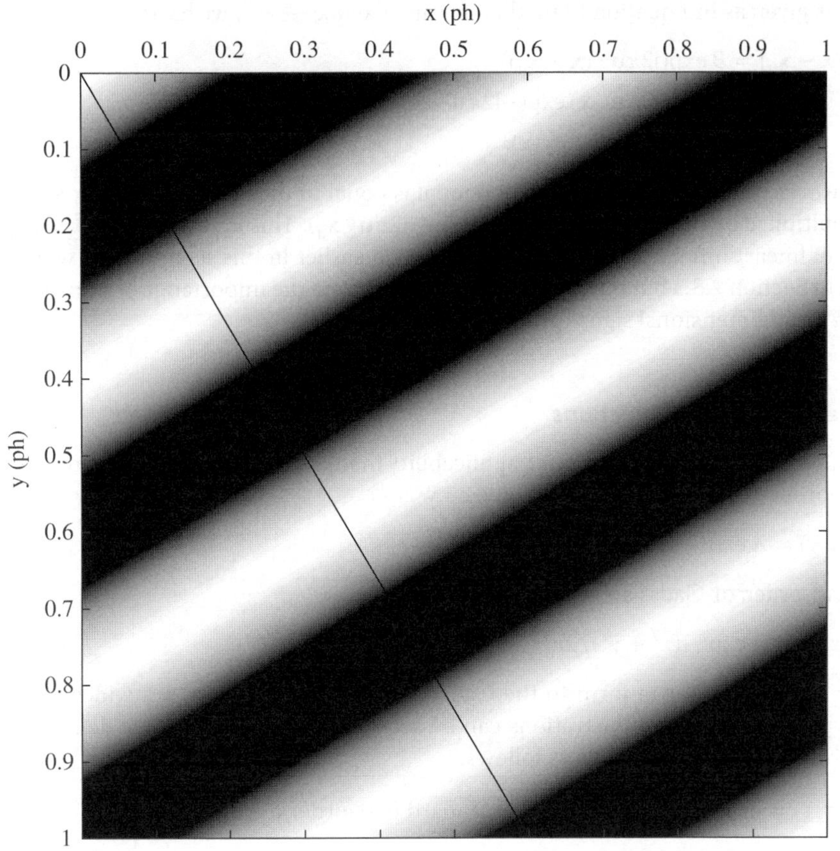

Figure 2.4 Sinusoidal signal with $u = 1.5$ c/ph and $v = 2.5$ c/ph. The horizontal period is (2/3) ph and the vertical period is 0.4 ph. The frequency along the line $y = \frac{5}{3}x$ is 2.9 c/ph, corresponding to a period of 0.34 ph.

As in one dimension, the complex exponential sinusoidal signals play an important role, e.g., in Fourier analysis. The complex exponential corresponding to the real sinusoid of Equation (2.4) is

$$f_C(x, y) = B\exp(\mathrm{j}2\pi(ux + vy)) \tag{2.5}$$

where $B = A\mathrm{e}^{\mathrm{j}\phi}$ can be complex. In this book, we use j to denote $\sqrt{-1}$. We often will adopt the vector notation

$$f_C(\mathbf{x}) = B\exp(\mathrm{j}2\pi\mathbf{u}\cdot\mathbf{x}) \tag{2.6}$$

where in the two-dimensional case $\mathbf{x}$ denotes (x, y) as before, $\mathbf{u}$ denotes (u, v) and $\mathbf{u}\cdot\mathbf{x}$ denotes $ux + vy$. When using the column matrix notation, then $\mathbf{u} = [u\ v]^T$ and we can write $f(\mathbf{x}) = B\exp(\mathrm{j}2\pi\mathbf{u}^T\mathbf{x})$. This is extended to any number of dimensions D in a straightforward manner. Using Euler's formula, the complex exponential can be written in terms of real sinusoidal signals

$$f_C(x, y) = A\cos(2\pi(ux + vy) + \phi) + \mathrm{j}A\sin(2\pi(ux + vy) + \phi). \tag{2.7}$$

If $f_C(\mathbf{x})$ is given as in Equation (2.6), then for any fixed $\mathbf{x}_0 \in \mathbb{R}^D$, we have

$$\begin{aligned} f_C(\mathbf{x}-\mathbf{x}_0) &= B\exp(\mathrm{j}2\pi\mathbf{u}\cdot(\mathbf{x}-\mathbf{x}_0)) \\ &= B\exp(\mathrm{j}2\pi\mathbf{u}\cdot\mathbf{x})\exp(-\mathrm{j}2\pi\mathbf{u}\cdot\mathbf{x}_0) \\ &= \exp(-\mathrm{j}2\pi\mathbf{u}\cdot\mathbf{x}_0)f_C(\mathbf{x}). \end{aligned} \tag{2.8}$$

In other words, the shifted complex exponential is equal to the original complex exponential multiplied by the complex constant $\exp(-\mathrm{j}2\pi\mathbf{u}\cdot\mathbf{x}_0)$. This in turn leads to the key property of linear shift-invariant systems mentioned earlier in this section. This will be analyzed in Section 2.5.5 but is mentioned here to motivate the importance of sinusoidal signals in multidimensional signal processing.

2.2.3 Real Exponential Functions

Real exponential signals also have wide applicability in multidimensional signal processing. First-order exponential signals are given by

$$f(x,y) = A\exp(-(a|x|+b|y|)) \tag{2.9}$$

and second-order, or Gaussian, signals by

$$f(x,y) = A\exp(-(x^2+y^2)/2r_0^2). \tag{2.10}$$

Gaussian signals, similar in form to the Gaussian probability density, are widely used in image processing. Some illustrations can be seen later, in Figure 2.6. We define the standard versions of these signals as follows:

$$f_E(x,y) = \exp(-2\pi\sqrt{x^2+y^2}), \qquad \text{standard first-order exponential signal;} \tag{2.11}$$

$$f_G(x,y) = \exp(-(x^2+y^2)/2), \qquad \text{standard Gaussian signal.} \tag{2.12}$$

2.2.4 Zone Plate

A very useful two-dimensional function that is often used as a test pattern in imaging systems is the *zone plate*, or Fresnel zone plate. The sinusoidal zone plate is based on the function

$$f(x,y) = \cos(\pi(x^2+y^2)/r_0^2) \tag{2.13}$$

where r_0 is a parameter that sets the scale. Figure 2.5 illustrates the zone plate $0.5 + 0.5\cos(\pi(x^2+y^2)/r_0^2)$ with $r_0 = 1/250$ ph. It is often convenient to consider the zone plate as the real part of the complex exponential function $\exp(\mathrm{j}\pi(x^2+y^2)/r_0^2)$.

Examining Figure 2.5, it can be seen that locally (say, within a small square window) the function is a sinusoidal signal with a horizontal frequency that increases with horizontal distance from the origin, and similarly a vertical frequency that increases with vertical distance from the origin. To make this concept of local frequency more precise, consider a conventional two-dimensional sinusoidal signal

$$\cos(2\pi(ux+vy)) = \cos(\phi(x,y)) \tag{2.14}$$

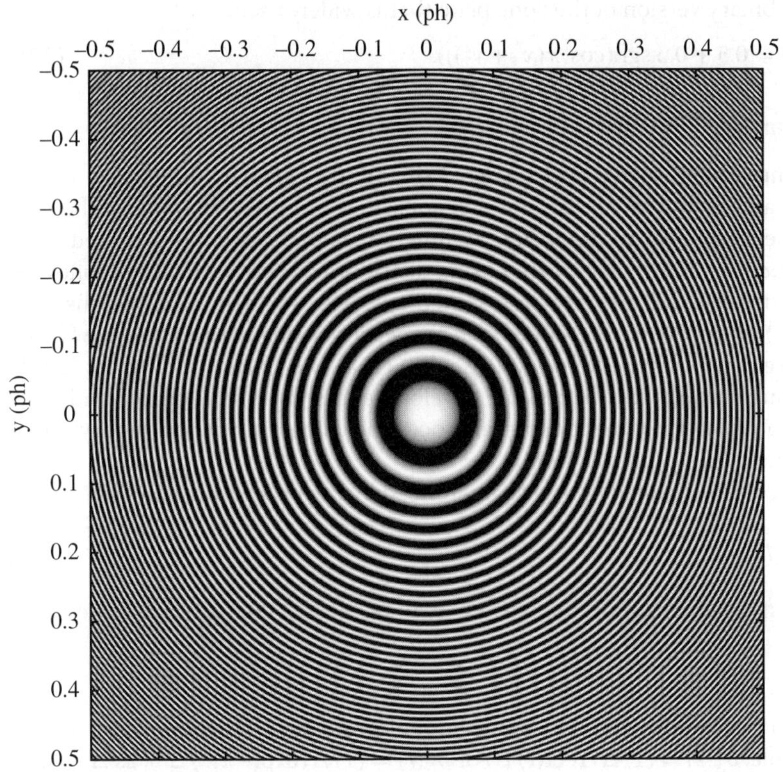

Figure 2.5 Illustration of a zone plate with parameter $r_0^2 = (1/250)$ ph. Local horizontal and vertical frequencies range from 0 to 125 c/ph.

where $\phi(x, y) = 2\pi(ux + vy)$. In this case we see that the horizontal and vertical frequencies are given by

$$u = \frac{1}{2\pi}\frac{\partial\phi}{\partial x};$$
$$v = \frac{1}{2\pi}\frac{\partial\phi}{\partial y}.$$

We use these definitions to define the *local frequency* of a generalized sinusoidal signal $\cos(\phi(x, y))$. For the zone plate, we have $\phi(x, y) = \pi(x^2 + y^2)/r_0^2$ and so obtain

$$u(x, y) = \frac{x}{r_0^2} \tag{2.15}$$

$$v(x, y) = \frac{y}{r_0^2} \tag{2.16}$$

which confirms that local horizontal and vertical frequencies vary linearly with horizontal and vertical position respectively.

We define standard versions of the real and complex zone plates by

$$f_{\text{ZR}}(x, y) = \cos(\pi(x^2 + y^2)) \tag{2.17}$$

$$f_{\text{ZC}}(x, y) = \exp(\text{j}\pi(x^2 + y^2)) \tag{2.18}$$

There is also a binary version of the zone plate that is widely used:

$$f_{ZB}(x, y) = 0.5 + 0.5 \operatorname{sgn}(\cos(\pi(x^2 + y^2))). \tag{2.19}$$

2.2.5 Singularities

As in one-dimensional signals and systems, singularities play an important role in multiD signal and system analysis. These singularities are not functions in the conventional sense; they can be described as functionals and are often referred to as *generalized functions*, rigorously treated using distribution theory. However, following common practice, we will nevertheless sometimes refer to them as functions (e.g., delta functions). We present here some basic properties of singularities suitable for our purposes. More details can be found in Papoulis (1968) and Barrett and Myers (2004). A more rigorous but relatively accessible development is given in Richards and Youn (1990), and a careful development of signal processing using distribution theory can be found in Gasquet and Witomski (1999).

The 1D Dirac delta, denoted $\delta(t)$, is characterized by the property that

$$\int_{-\infty}^{\infty} \delta(t) f(t)\, \mathrm{d}t = f(0) \tag{2.20}$$

for any function $f(t)$ that is continuous at $t = 0$. In particular, taking $f(t) = 1$, we have

$$\int_{-\infty}^{\infty} \delta(t)\, \mathrm{d}t = 1. \tag{2.21}$$

The Dirac delta can be considered to be the limit of a sequence of narrow pulses of unit area, for example, $\delta_\Delta(t) = (1/\Delta)\operatorname{rect}(t/\Delta)$ or $\delta_\Delta(t) = (1/\Delta)\exp(-\pi t^2/\Delta^2)$, as $\Delta \to 0$.

The 2D Dirac delta is defined in a similar fashion by the requirement

$$\int_{-\infty}^{\infty}\int_{-\infty}^{\infty} \delta(x, y) f(x, y)\, \mathrm{d}x\, \mathrm{d}y = f(0, 0) \tag{2.22}$$

for any function $f(x, y)$ that is continuous at $(x, y) = (0, 0)$. Again, the 2D Dirac delta can be considered to be the limiting case of sequences of narrow 2D pulses of unit volume, e.g.,

$$\delta_\Delta(x, y) = \frac{1}{\Delta^2}\operatorname{rect}(x/\Delta, y/\Delta),$$

$$\delta_\Delta(x, y) = \frac{1}{\Delta^2}\exp(-\pi(x^2 + y^2)/\Delta^2),$$

$$\delta_\Delta(x, y) = \frac{1}{\pi\Delta^2}\operatorname{circ}(x/\Delta, y/\Delta).$$

The 2D Dirac delta satisfies the scaling property

$$\delta(ax, by) = \frac{1}{|ab|}\delta(x, y). \tag{2.23}$$

Other important properties of Dirac deltas will emerge as we investigate their role in multiD system analysis.

The Dirac delta can be extended to the multiD case in an obvious fashion, with the notation $\delta(\mathbf{x})$ covering all cases. The conditions (2.20) and (2.22) are written

$$\int_{\mathbb{R}^D} \delta(\mathbf{x}) f(\mathbf{x})\, \mathrm{d}\mathbf{x} = f(\mathbf{0}) \tag{2.24}$$

in the general case, where $d\mathbf{x}$ is understood to mean dt, $dx\,dy$ or $dx\,dy\,dt$ according to context. As a consequence of (2.23) in the general case,

$$\delta(-\mathbf{x}) = \delta(\mathbf{x}). \tag{2.25}$$

In addition to the point singularities defined above, we can have singularities on lines or curves in two or three dimensions, or on surfaces in three dimensions. See Papoulis (1968) for a discussion of singularities on a curve in two dimensions.

2.2.6 Separable and Isotropic Functions

A two-dimensional function $f(x, y)$ is said to be separable if it can be expressed as the product of one-dimensional functions,

$$f(x, y) = f_1(x)f_2(y). \tag{2.26}$$

Several of the 2D functions we have seen are separable, including $\text{rect}(x, y) = \text{rect}(x)$ $\text{rect}(y)$, the complex exponential $\exp(j2\pi(ux + vy)) = \exp(j2\pi ux) \cdot \exp(j2\pi vy)$, and the exponential functions. Also, the 2D Dirac delta $\delta(x, y)$ can be considered to be separable: $\delta(x, y) = \delta(x)\delta(y)$. The extension to higher-dimensional separable signals is evident,

$$f(x_1, x_2, \dots, x_D) = f_1(x_1)f_2(x_2)\cdots f_D(x_D). \tag{2.27}$$

Separability is a convenient way to generate multiD signals from 1D signals. Note that signals can also be separable in other variables than the standard orthogonal axes $x_1, \dots, x_D$.

A 2D signal is said to be isotropic (circularly symmetric) if it is only a function of the distance r from the origin,

$$f(x, y) = f_1(r) = f_1(\sqrt{x^2 + y^2}). \tag{2.28}$$

Examples of isotropic signals that we have seen are $\text{circ}(x, y)$, the Gaussian signal, and the zone plate. Again, the extension to multiD signals is evident: $f(\mathbf{x}) = f_1(\|\mathbf{x}\|)$. We may also call such a signal *rotation invariant*, since it is invariant to a rotation of the domain $\mathbb{R}^D$ about the origin.

2.3 Visualization of Two-Dimensional Signals

It is easy to visualize a 1D signal $f(t)$ by drawing its graph. If the graph is drawn to scale, we can derive numerical information by reading the graph, e.g., $f(2.5) = 5.2$. There are various ways that we can visualize a 2D signal. The three main visualization techniques are:

1) Intensity image: the signal range is transformed to the dynamic range of the display device and viewed as an image.
2) Contour plot: this shows curves of equal value of the function to be displayed. The levels to be shown must be selected to get the most informative visualization, and they should be labeled if possible.
3) Perspective plot: this shows a wireframe mesh of the surface as seen from a particular point of view. The density of the mesh and the point of view must be chosen for the best effect. Various ways of shading or coloring the perspective view can also be used.

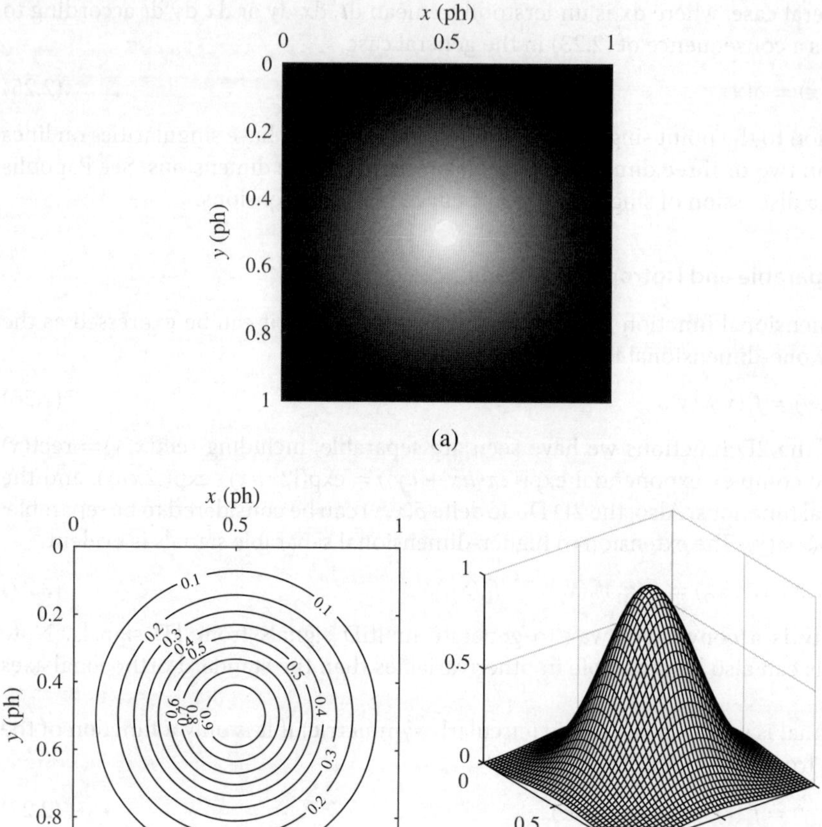

Figure 2.6 Visualization of a two-dimensional Gaussian function centered at (0.5,0.5) and scaled with $r_0 = 0.2$ ph. (a) Intensity plot. (b) Contour plot. (c) Perspective view.

Of course, method 1 is probably the most appropriate method to visualize a 2D signal that represents an image in the usual sense, but it may be useful in other cases as well. Contour and perspective plots are mainly useful for special 2D functions like the ones described in Section 2.2. Figure 2.6 illustrates the three visualization methods for a 2D Gaussian function. MATLAB provides all the necessary tools to generate such figures. See for example chapter 25 of Hanselman and Littlefield (2012) for a good description of how to generate such graphics in MATLAB.

2.4 Signal Spaces and Systems

A signal space S is a collection of multiD signals defined on a specific domain D and satisfying certain well-defined properties. For example, we can consider a space of all two-dimensional signals $f(x, y)$ defined for $-\infty < x < \infty, -\infty < y < \infty$, or we can define

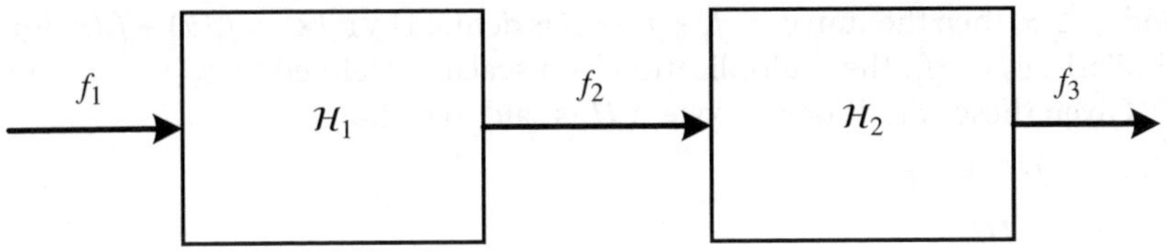

Figure 2.7 Cascade of two systems: $\mathcal{H}_3 = \mathcal{H}_2\mathcal{H}_1$.

a space of 2D signals defined on the spatial window $\mathcal{W}$ of Figure 2.1. We can also impose additional constraints, such as boundedness, finite energy, or whatever constraints are appropriate for a given situation. An example of a specific signal space is

$$\mathcal{S} = \{f : \mathcal{W} \to \mathbb{R} \mid |f(\mathbf{x})| < \infty \text{ for all } \mathbf{x} \in \mathcal{W}\}. \tag{2.29}$$

As we will see later, we can also consider spaces of signals defined on discrete sets, corresponding to sampled images. We denote the fact that a given signal belongs to a signal space by $f \in \mathcal{S}$. The object f as a single entity is understood to encompass all the signal values $f(x, y)$ as (x, y) ranges over the specified domain of the signal space $\mathcal{S}$.

A system $\mathcal{H}$ transforms elements of a signal space $\mathcal{S}_1$ into elements of a second signal space $\mathcal{S}_2$ according to some well-defined rule. We write

$$\mathcal{H} : \mathcal{S}_1 \to \mathcal{S}_2 : g = \mathcal{H}f \tag{2.30}$$

where $f \in \mathcal{S}_1$ and $g \in \mathcal{S}_2$. This equation can be read "the system $\mathcal{H}$ maps elements of $\mathcal{S}_1$ into elements of $\mathcal{S}_2$, where the element $f \in \mathcal{S}_1$ is mapped to $g = \mathcal{H}f \in \mathcal{S}_2$." If $g = \mathcal{H}f$, then we write $g(\mathbf{x}) = (\mathcal{H}f)(\mathbf{x})$ for the value of $\mathcal{H}f$ at location $\mathbf{x}$. In most cases, $\mathcal{S}_1$ and $\mathcal{S}_2$ are the same, but sometimes they are different, e.g., for a system that samples a continuous-space image to produce a discrete-space image.

If we have two systems $\mathcal{H}_1 : \mathcal{S}_1 \to \mathcal{S}_2$ and $\mathcal{H}_2 : \mathcal{S}_2 \to \mathcal{S}_3$, then we can apply $\mathcal{H}_1$ and $\mathcal{H}_2$ successively to obtain a new system $\mathcal{H}_3 : \mathcal{S}_1 \to \mathcal{S}_3$ called the cascade of $\mathcal{H}_1$ and $\mathcal{H}_2$. As shown in Figure 2.7, $f_2 = \mathcal{H}_1 f_1$ and $f_3 = \mathcal{H}_2 f_2 = \mathcal{H}_2\mathcal{H}_1 f_1$. Thus, we write $\mathcal{H}_3 = \mathcal{H}_2\mathcal{H}_1$.

Alternatively, if the two systems have the same domain and range, $\mathcal{H}_1 : \mathcal{S}_1 \to \mathcal{S}_2$ and $\mathcal{H}_2 : \mathcal{S}_1 \to \mathcal{S}_2$, they can be connected in parallel to give the new system $\mathcal{H}_4 : \mathcal{S}_1 \to \mathcal{S}_2$ defined by $((\mathcal{H}_1 + \mathcal{H}_2)f)(\mathbf{x}) = (\mathcal{H}_1 f)(\mathbf{x}) + (\mathcal{H}_2 f)(\mathbf{x})$. In this case, we write $\mathcal{H}_4 = \mathcal{H}_1 + \mathcal{H}_2$.

2.5 Continuous-Domain Linear Systems

Let $\mathcal{H} : \mathcal{S} \to \mathcal{S}$ be a system defined on a space $\mathcal{S}$ of continuous-domain multiD signals with some domain $\mathcal{D}$. Many systems of practical interest satisfy the key property of *linearity*. This is convenient since linear systems are normally simpler to analyze than general nonlinear systems.

2.5.1 Linear Systems

We first make the assumption that the signal space $\mathcal{S}$ has the properties of a real or complex vector space. This ensures that elements of the signal space can be added together, and can be multiplied by a scalar constant, and that in each case the result also lies in the given signal space. This holds for most signal spaces of interest.

If $f_1 \in S$ and $f_2 \in S$, then the sum $g_1 = f_1 + f_2 \in S$ is defined by $g_1(\mathbf{x}) = f_1(\mathbf{x}) + f_2(\mathbf{x})$ for all $\mathbf{x} \in D$. Similarly, $g_2 = \alpha f_1$, the multiplication by a scalar, is defined by $g_2(\mathbf{x}) = \alpha f_1(\mathbf{x})$ for all $\mathbf{x} \in D$. Given these conditions, a system $\mathcal{H}$ is said to be linear if

$$\mathcal{H}(f_1 + f_2) = \mathcal{H}f_1 + \mathcal{H}f_2, \tag{2.31}$$

$$\mathcal{H}(\alpha f_1) = \alpha(\mathcal{H}f_1) \tag{2.32}$$

for all $f_1, f_2 \in S$ and for all $\alpha \in \mathbb{R}$ (or for all $\alpha \in \mathbb{C}$ for complex signal spaces). The definition of a linear system extends in an obvious fashion if S_1 and S_2 in Equation (2.30) are different vector spaces. The most basic example of a linear system is $(\mathcal{H}f)(\mathbf{x}) = cf(\mathbf{x})$, which is easily seen to satisfy the definition. A system that does *not* satisfy the conditions of a linear system is said to be a *nonlinear* system. A simple and common example of a nonlinear system is given by $(\mathcal{H}f)(\mathbf{x}) = (f(\mathbf{x}))^2$.

Linear systems are of particular interest because if we know the response of the system to a number of basic signals $f_1, f_2, \ldots, f_K$, namely $g_i = \mathcal{H}f_i$, then we can determine the response to any linear combination of the f_i:

$$\mathcal{H}\left(\sum_{i=1}^{K} \alpha_i f_i\right) = \sum_{i=1}^{K} \alpha_i \mathcal{H}f_i = \sum_{i=1}^{K} \alpha_i g_i. \tag{2.33}$$

As a simple example of a linear system, consider the shift (or translation) operator $\mathcal{T}_\mathbf{d}$ for some *fixed* shift vector $\mathbf{d}$. If $g = \mathcal{T}_\mathbf{d} f$, then $g(\mathbf{x}) = f(\mathbf{x} - \mathbf{d})$. For this operation to be well defined, the domain of the signal space must be all of $\mathbb{R}^D$. In two dimensions, we would have $\mathbf{d} = [d_x, d_y]^T$, and $g(x, y) = f(x - d_x, y - d_y)$. It can easily be verified from the definitions that $\mathcal{T}_\mathbf{d}$ is a linear system. Figure 2.8 illustrates the shift operator with $\mathbf{d} = [.25, -.25]^T$.

Another important class of linear systems consists of systems induced by an arbitrary nonsingular linear transformation of the domain $D = \mathbb{R}^D$ for some D. Let $\mathcal{A} : D \to D : \mathbf{x} \mapsto \mathbf{Ax}$ be such a transformation, where $\mathbf{A}$ is a $D \times D$ nonsingular matrix. The induced system $\mathcal{M}_\mathbf{A}$ is defined by

$$\mathcal{M}_\mathbf{A} : S \to S : g = \mathcal{M}_\mathbf{A} f : g(\mathbf{x}) = f(\mathbf{Ax}). \tag{2.34}$$

Again, it is easily verified from the definitions that $\mathcal{M}_\mathbf{A}$ is a linear system. We will mainly use this category of systems for scaling and rotating basic signals such as

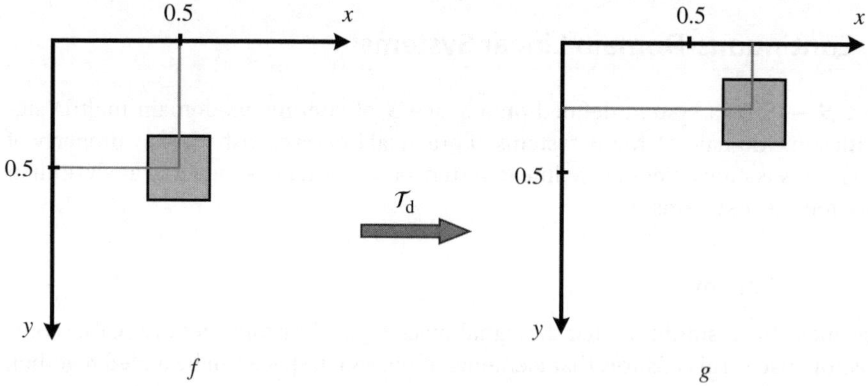

Figure 2.8 Shift operator $g = \mathcal{T}_\mathbf{d} f$ with $\mathbf{d} = [0.25, -0.25]^T$ giving $g(x, y) = f(x - 0.25, y + 0, 25)$.

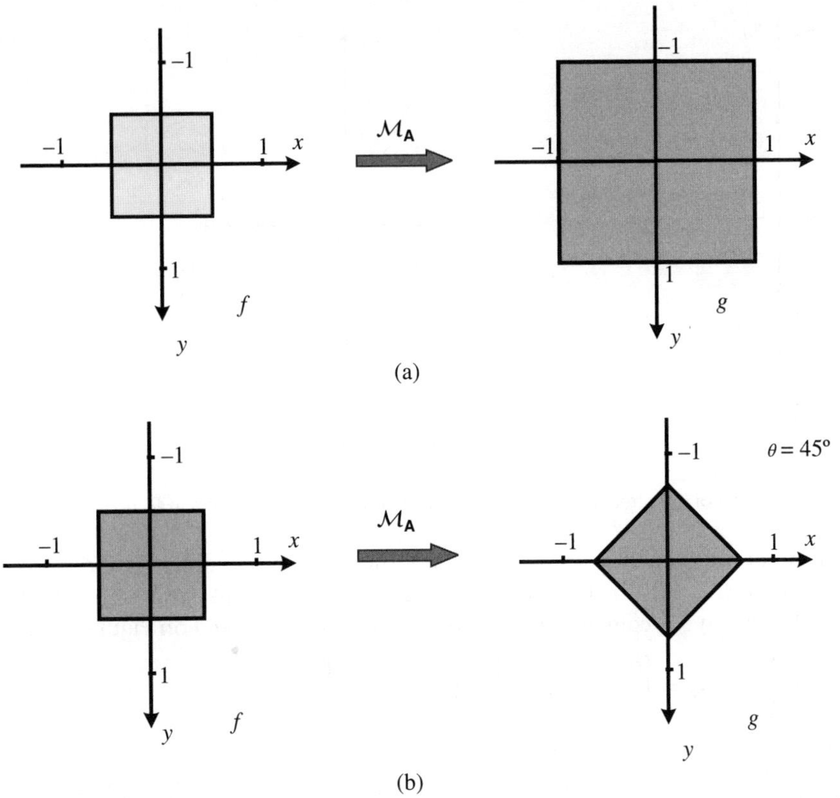

Figure 2.9 Transformations $g = \mathcal{M}_\mathbf{A} f$. (a) Scale operator with $\mathbf{A} = \mathrm{diag}(0.5, 0.5)$. (b) Rotation operator with $\theta = \pi/4$.

those of Section 2.2, but more general cases are widely used as well. If $\mathbf{A}$ is a diagonal matrix, the transformation $\mathcal{M}_\mathbf{A}$ carries out a separate scaling along each of the axes, as illustrated in two dimensions in Figure 2.9(a). In this example, $\mathbf{A} = \mathrm{diag}(0.5, 0.5)$, which has the effect of magnifying the image by a factor of two in each dimension. In general, if $\mathbf{A} = \mathrm{diag}(a_1, a_2)$, the system $\mathcal{M}_\mathbf{A}$ will scale the image by a factor of $1/a_1$ in the horizontal dimension and by $1/a_2$ in the vertical dimension.

If $\mathbf{A}$ is a two-dimensional rotation matrix,

$$\mathbf{A}_\theta = \begin{bmatrix} \cos\theta & \sin\theta \\ -\sin\theta & \cos\theta \end{bmatrix}$$

the transformation $\mathcal{M}_{\mathbf{A}_\theta}$ rotates the signal in the clockwise direction, as illustrated in Figure 2.9(b). The rotation of the domain $\mathbb{R}^2$ is given by

$$\mathcal{A}_\theta : \mathbb{R}^2 \to \mathbb{R}^2 : \mathbf{x} \mapsto \mathbf{s} = \mathbf{A}_\theta \mathbf{x} \tag{2.35}$$

where explicitly

$$\begin{bmatrix} s \\ t \end{bmatrix} = \begin{bmatrix} \cos\theta & \sin\theta \\ -\sin\theta & \cos\theta \end{bmatrix} \begin{bmatrix} x \\ y \end{bmatrix} = \begin{bmatrix} x\cos\theta + y\sin\theta \\ -x\sin\theta + y\cos\theta \end{bmatrix}. \tag{2.36}$$

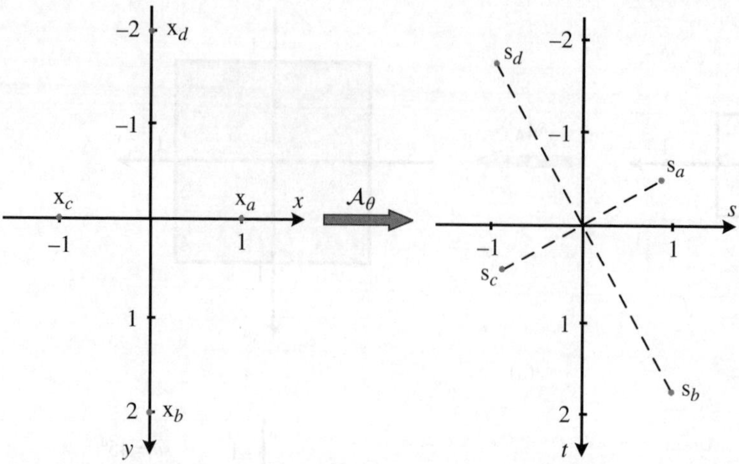

Figure 2.10 Illustration of the effect of a rotation operator with $\theta = 30°$ on points in $\mathbb{R}^2$.

The rotation of the domain and the rotation of the signal are in opposite directions. Since this sometimes leads to confusion, we present a specific illustration. To demonstrate how $\mathcal{A}_\theta$ acts on $\mathbb{R}^2$, consider the following four points, marked on Figure 2.10.

$$\mathbf{x}_a = \begin{bmatrix} 1 \\ 0 \end{bmatrix} \quad \mathbf{x}_b = \begin{bmatrix} 0 \\ 2 \end{bmatrix} \quad \mathbf{x}_c = \begin{bmatrix} -1 \\ 0 \end{bmatrix} \quad \mathbf{x}_d = \begin{bmatrix} 0 \\ -2 \end{bmatrix}$$

These points are mapped by $\mathcal{A}_\theta$ to

$$\mathbf{s}_a = \begin{bmatrix} \cos\theta \\ -\sin\theta \end{bmatrix} \quad \mathbf{s}_b = \begin{bmatrix} 2\sin\theta \\ 2\cos\theta \end{bmatrix} \quad \mathbf{s}_c = \begin{bmatrix} -\cos\theta \\ \sin\theta \end{bmatrix} \quad \mathbf{s}_d = \begin{bmatrix} -2\sin\theta \\ -2\cos\theta \end{bmatrix}$$

These are marked on the right of Figure 2.10 for $\theta = 30°$, where $\cos\theta = \sqrt{3}/2$ and $\sin\theta = 1/2$. We see that this transformation has rotated points in $\mathbb{R}^2$ *counterclockwise* by θ.

Now let us consider what happens when a two-dimensional signal is mapped through the linear system given by Equation (2.34) for this transformation. Thus,

$$\mathcal{M}_{\mathbf{A}_\theta} : \mathcal{S} \rightarrow \mathcal{S} : g = \mathcal{M}_{\mathbf{A}_\theta} f : g(\mathbf{x}) = f(\mathbf{A}_\theta \mathbf{x}). \tag{2.37}$$

Thus, explicitly, $g(x, y) = f(x\cos\theta + y\sin\theta, -x\sin\theta + y\cos\theta)$. For example,

$$g(\cos\theta, \sin\theta) = f(\cos^2\theta + \sin^2\theta, -\sin\theta\cos\theta + \cos\theta\sin\theta) = f(1, 0).$$

This is illustrated in Figure 2.11, which shows the effect of applying this linear system to a diamond-shaped zero-one function. From the figure, we see that the linear system $\mathcal{M}_{\mathbf{A}_\theta}$ has rotated the two-dimensional signal *clockwise* by θ, which is the opposite direction to that in which $\mathcal{A}_\theta$ has rotated the points of $\mathbb{R}^2$.

Another more general class of linear systems involves an affine transformation of the independent variables. One way to express this is

$$\mathcal{Q}_{\mathbf{A},\mathbf{d}} : g = \mathcal{Q}_{\mathbf{A},\mathbf{d}} f : g(\mathbf{x}) = f(\mathbf{A}(\mathbf{x} - \mathbf{d})), \tag{2.38}$$

where $\mathbf{A}$ is a nonsingular $D \times D$ matrix. The affine transformation can be expressed as a cascade of the two preceding types of linear systems in two ways:

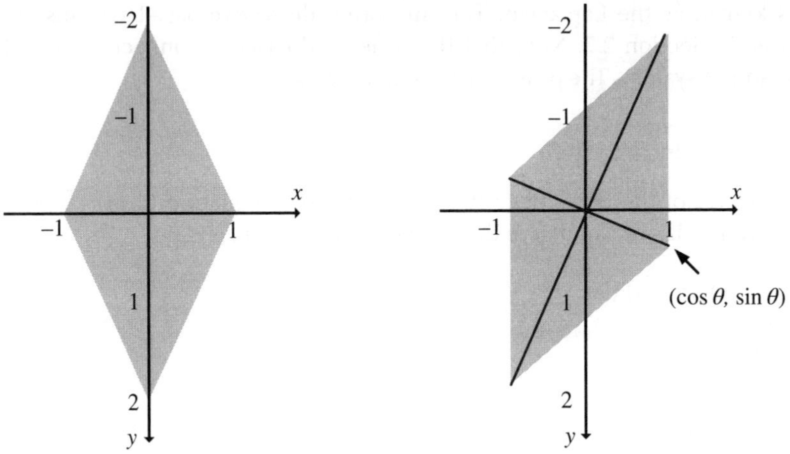

Figure 2.11 Illustration of the effect of a rotation operator with $\theta = 30^\circ$ on a zero-one function with a diamond-shaped region of support.

$\mathcal{Q}_{\mathbf{A},\mathbf{d}} = \mathcal{T}_{\mathbf{d}}\mathcal{M}_{\mathbf{A}} = \mathcal{M}_{\mathbf{A}}\mathcal{T}_{\mathbf{Ad}}$. Generally the representation $\mathcal{T}_{\mathbf{d}}\mathcal{M}_{\mathbf{A}}$ is more convenient. The signal is first rotated, scaled and perhaps sheared with respect to the origin, then centered at point $\mathbf{d}$. It can be shown that the cascade of any linear systems is also a linear system, and thus the system induced by an affine transformation of the domain is a linear system. As an example of an affine transformation, the Gaussian image of Figure 2.6 can be obtained from the unit variance, zero-centered Gaussian $f(x,y) = \exp[-(x^2 + y^2)/2]$ using the affine transformation with

$$\mathbf{A} = \begin{bmatrix} 5 & 0 \\ 0 & 5 \end{bmatrix} \qquad \mathbf{d} = \begin{bmatrix} 0.5 \\ 0.5 \end{bmatrix}.$$

2.5.2 Linear Shift-Invariant Systems

An important subclass of linear systems is the class of *linear shift-invariant* (LSI) systems. In such a system, if the response to an input f is g, then if f is shifted by any amount $\mathbf{d}$, the resulting output is equal to g shifted by the same amount $\mathbf{d}$. Using the above terminology, if $\mathcal{H}f = g$, then $\mathcal{H}(\mathcal{T}_{\mathbf{d}}f) = \mathcal{T}_{\mathbf{d}}(\mathcal{H}f) = \mathcal{T}_{\mathbf{d}}g$ for any $\mathbf{d}$. For an LSI system $\mathcal{H}$, $\mathcal{H}\mathcal{T}_{\mathbf{d}} = \mathcal{T}_{\mathbf{d}}\mathcal{H}$ for any $\mathbf{d}$. We say that the two systems $\mathcal{H}$ and $\mathcal{T}_{\mathbf{d}}$ *commute*. The shift system $\mathcal{T}_{\mathbf{d}}$ is itself an LSI system, since $\mathcal{T}_{\mathbf{d}}\mathcal{T}_{\mathbf{s}} = \mathcal{T}_{\mathbf{d+s}} = \mathcal{T}_{\mathbf{s+d}} = \mathcal{T}_{\mathbf{s}}\mathcal{T}_{\mathbf{d}}$ for any $\mathbf{s}$. However, the general affine transformation system $\mathcal{Q}_{\mathbf{A},\mathbf{d}}$ is *not* an LSI system if $\mathbf{A} \neq \mathbf{I}$, since $\mathcal{Q}_{\mathbf{A},\mathbf{d}}\mathcal{T}_{\mathbf{s}} = \mathcal{M}_{\mathbf{A}}\mathcal{T}_{\mathbf{Ad+s}}$ but $\mathcal{T}_{\mathbf{s}}\mathcal{Q}_{\mathbf{A},\mathbf{d}} = \mathcal{T}_{\mathbf{s+d}}\mathcal{M}_{\mathbf{A}} = \mathcal{M}_{\mathbf{A}}\mathcal{T}_{\mathbf{Ad+As}}$.

Another important class of LSI systems are those that involve partial derivatives of the input function with respect to the independent variables. As the simplest example, consider the 2D system ∂_x defined by $(\partial_x f)(x,y) = \frac{\partial}{\partial x} f(x,y)$. Such systems are often connected in series or parallel, for example the system $\mathcal{H}_L = \partial_x\partial_x + \partial_y\partial_y$, where the output is given by

$$(\mathcal{H}_L f)(x,y) = \frac{\partial^2 f}{\partial x^2}(x,y) + \frac{\partial^2 f}{\partial y^2}(x,y). \tag{2.39}$$

This system is known as the Laplacian. This and other derivative-based systems are further discussed in Section 2.7. Note that the series and cascade connection of LSI systems is also an LSI system. The proof is left as an exercise.

2.5.3 Response of a Linear System

The defining property of the Dirac delta function is given in Equation (2.24): $\int_{\mathbb{R}^D} \delta(\mathbf{x}) f(\mathbf{x})\, d\mathbf{x} = f(\mathbf{0})$. From this, we can derive the so-called *sifting property*

$$f(\mathbf{x}) = \int_{\mathbb{R}^D} f(\mathbf{s})\delta(\mathbf{x} - \mathbf{s})\, d\mathbf{s}. \tag{2.40}$$

This follows from

$$\begin{aligned}
\int_{\mathbb{R}^D} f(\mathbf{s})\delta(\mathbf{x} - \mathbf{s})\, d\mathbf{s} &= \int_{\mathbb{R}^D} f(\mathbf{s})\delta(\mathbf{s} - \mathbf{x})\, d\mathbf{s} && \text{(from Equation (2.25))} \\
&= \int_{\mathbb{R}^D} f(\mathbf{t} + \mathbf{x})\delta(\mathbf{t})\, d\mathbf{t} && \text{(letting } \mathbf{t} = \mathbf{s} - \mathbf{x}) \\
&= f(\mathbf{x}) && \text{(from Equation (2.24)).}
\end{aligned}$$

The sifting property (2.40) can be interpreted as the *synthesis* of the signal f by the superposition of shifted Dirac delta functions $\mathcal{T}_{\mathbf{s}}\delta$ with weights $f(\mathbf{s})d\mathbf{s}$:

$$f = \int_{\mathbb{R}^D} (f(\mathbf{s})\, d\mathbf{s})\mathcal{T}_{\mathbf{s}}\delta. \tag{2.41}$$

For a linear system $\mathcal{H}$, we can conclude that

$$g = \mathcal{H}f = \int_{\mathbb{R}^D} (f(\mathbf{s})\, d\mathbf{s})\mathcal{H}(\mathcal{T}_{\mathbf{s}}\delta) \tag{2.42}$$

where $\mathcal{H}(\mathcal{T}_{\mathbf{s}}\delta)$ is the response of the system to a Dirac delta function located at position $\mathbf{s}$. If we denote this impulse response $h_{\mathbf{s}}$, we obtain

$$g(\mathbf{x}) = \int_{\mathbb{R}^D} f(\mathbf{s})h_{\mathbf{s}}(\mathbf{x})\, d\mathbf{s}. \tag{2.43}$$

It is important to recognize that while the above result is broadly valid and applies to essentially all cases of interest to us, the development is informal and there are many unstated assumptions. For example, the existence of $\mathcal{H}\delta$ (or more generally of $\mathcal{H}\mathcal{T}_{\mathbf{s}}\delta$) and the applicability of the linearity condition to an integral relation are assumed. The development can be treated more rigorously in several ways, such as assuming a specific signal space with a given metric, and assuming that the linear system is continuous. Then, an arbitrary signal can be approximated as a superposition of a finite number of pulse functions of the form $\delta_\Delta(\mathbf{x} - \mathbf{x}_i)$, and the output to this approximation determined. Taking the limit as $\Delta \to 0$ and the extent tending to infinity yields the desired result. More details of this approach can be found in section 7.2 of Barrett and Myers (2004).

2.5.4 Response of a Linear Shift-Invariant System

Equation (2.43) describes the response of a general space-variant linear system. Most optical systems are indeed space variant, with the response to an impulse in the corner of the image being different than the response to an impulse in the center of the image, for example. However, the design goal is usually to have a system that is as close to being shift invariant as possible. Thus, shift-invariant systems are an important class. In this case,

$$\mathcal{H}(\mathcal{T}_{\mathbf{s}}\delta) = \mathcal{T}_{\mathbf{s}}(\mathcal{H}\delta) = \mathcal{T}_{\mathbf{s}}h \tag{2.44}$$

where h is the response of the LSI system to an impulse at the origin, and so

$$g = \int_{\mathbb{R}^D} (f(\mathbf{s})\,\mathrm{d}\mathbf{s})\mathcal{T}_{\mathbf{s}}h. \tag{2.45}$$

Evaluating at position $\mathbf{x}$ gives

$$g(\mathbf{x}) = \int_{\mathbb{R}^D} f(\mathbf{s})h(\mathbf{x}-\mathbf{s})\,\mathrm{d}\mathbf{s} \tag{2.46}$$

which is called the *convolution integral*, and is denoted

$$g = f * h. \tag{2.47}$$

By a simple change of variables in the integral (2.46), we can show that $f * h = h * f$, i.e. convolution is commutative.

Example 2.1 A common model for blurring due to a camera being out of focus is *uniform out-of-focus blur*, in which a point source anywhere in the scene is imaged as a small disk of radius R [Lagendijk and Biemond (2009)]. In this case the impulse response, or point spread function, of the system is given by $h(x,y) = \frac{1}{\pi R^2}\mathrm{circ}(\frac{x}{R},\frac{y}{R})$. Suppose that we image a pure step function in the horizontal direction with this camera:

$$f(x,y) = u_0(x-0.5) = \begin{cases} 0 & \text{if } x \le 0.5, \\ 1 & \text{if } x > 0.5. \end{cases}$$

The output is computed as

$$\begin{aligned} g(x,y) &= \int_{-\infty}^{\infty}\int_{-\infty}^{\infty} h(s_1,s_2)f(x-s_1,y-s_2)\,\mathrm{d}s_1\,\mathrm{d}s_2 \\ &= \frac{1}{\pi R^2}\int_{-\infty}^{\infty}\int_{-\infty}^{\infty} \mathrm{circ}\left(\frac{s_1}{R},\frac{s_2}{R}\right) u_0(x-s_1-0.5)\,\mathrm{d}s_1\,\mathrm{d}s_2 \\ &= \frac{1}{\pi R^2}\int_{-\infty}^{\infty}\int_{-\infty}^{x-0.5} \mathrm{circ}\left(\frac{s_1}{R},\frac{s_2}{R}\right)\,\mathrm{d}s_1\,\mathrm{d}s_2. \end{aligned}$$

This amounts to computing the area of the part of the circle overlapping the step that has been shifted horizontally by x. Clearly this area is zero if $x < 0.5 - R$ and it is one if $x > 0.5 + R$. Otherwise, using simple geometry we can determine the area to obtain the resulting image

$$g(x,y) = \begin{cases} 0 & x \le 0.5 - R \\ \frac{1}{\pi R^2}\left[R^2\cos^{-1}\left(\frac{0.5-x}{R}\right) - (0.5-x)\sqrt{R^2-(0.5-x)^2}\right] & 0.5 - R < x \le 0.5 \\ 1 - \frac{1}{\pi R^2}\left[R^2\cos^{-1}\left(\frac{x-0.5}{R}\right) - (x-0.5)\sqrt{R^2-(x-0.5)^2}\right] & 0.5 < x \le 0.5 + R \\ 1 & x > 0.5 + R. \end{cases}$$

The filtered step is now spread out over a distance $2R$.

2.5.5 Frequency Response of an LSI System

Suppose that the input to an LSI system is the complex sinusoidal function $f(\mathbf{x}) = \exp(\mathrm{j}2\pi\mathbf{u}\cdot\mathbf{x})$. According to Equation (2.46), the corresponding output is

$$\begin{aligned} g(\mathbf{x}) &= \int_{\mathbb{R}^D} \exp(\mathrm{j}2\pi\mathbf{u}\cdot\mathbf{s})h(\mathbf{x}-\mathbf{s})\,\mathrm{d}\mathbf{s} \\ &= \int_{\mathbb{R}^D} \exp(\mathrm{j}2\pi\mathbf{u}\cdot(\mathbf{x}-\mathbf{t}))h(\mathbf{t})\,\mathrm{d}\mathbf{t} \qquad (\text{setting } \mathbf{t}=\mathbf{x}-\mathbf{s}) \\ &= \left(\int_{\mathbb{R}^D} h(\mathbf{t})\exp(-\mathrm{j}2\pi\mathbf{u}\cdot\mathbf{t})\,\mathrm{d}\mathbf{t}\right)\exp(\mathrm{j}2\pi\mathbf{u}\cdot\mathbf{x}) \\ &= H(\mathbf{u})\exp(\mathrm{j}2\pi\mathbf{u}\cdot\mathbf{x}) \end{aligned} \tag{2.48}$$

where $H(\mathbf{u})$ is a complex scalar (assuming the integral converges). Thus, exactly as in one dimension, if the input to an LSI system is a complex sinusoidal signal with frequency vector $\mathbf{u}$, then the output is that same complex sinusoidal signal multiplied by the complex scalar $H(\mathbf{u})$. Taken as a function of the two or three-dimensional frequency vector, $H(\mathbf{u})$ is referred to as the frequency response of the LSI system. According to this observation, $\exp(\mathrm{j}2\pi\mathbf{u}\cdot\mathbf{x})$ is called an *eigenfunction* of the linear system $\mathcal{H}$ with corresponding *eigenvalue* $H(\mathbf{u})$.

Multiplication by $H(\mathbf{u}) = |H(\mathbf{u})|\exp(\mathrm{j}\angle H(\mathbf{u}))$ amounts to multiplying the magnitude of $\exp(\mathrm{j}2\pi\mathbf{u}\cdot\mathbf{x})$ by $|H(\mathbf{u})|$ and introducing a phase shift of $\angle H(\mathbf{u})$, i.e.

$$g(\mathbf{x}) = |H(\mathbf{u})|\exp(\mathrm{j}(2\pi\mathbf{u}\cdot\mathbf{x} + \angle H(\mathbf{u}))). \tag{2.49}$$

2.6 The Multidimensional Fourier Transform

From Equation (2.48) we identify

$$H(\mathbf{u}) = \int_{\mathbb{R}^D} h(\mathbf{x})\exp(-\mathrm{j}2\pi\mathbf{u}\cdot\mathbf{x})\,\mathrm{d}\mathbf{x} \tag{2.50}$$

as the multiD extension of the continuous-time Fourier transform. The multiD Fourier transform has properties that are completely analogous to the familiar properties of the 1D Fourier transform, as shown in Table 2.1. In particular, the inverse Fourier transform is given by

$$h(\mathbf{x}) = \int_{\mathbb{R}^D} H(\mathbf{u})\exp(\mathrm{j}2\pi\mathbf{u}\cdot\mathbf{x})\,\mathrm{d}\mathbf{u}. \tag{2.51}$$

The reason for this will be given in Chapter 6. The Fourier transform can be applied to any signals in the signal space, not just the impulse response, as long as it converges. We denote that $f(\mathbf{x})$ and $F(\mathbf{u})$ form a multidimensional Fourier transform pair by $f(\mathbf{x}) \xleftrightarrow{\text{CDFT}} F(\mathbf{u})$, where CDFT denotes continuous-domain Fourier transform.

The property that makes the Fourier transform so valuable in linear system analysis is the *convolution property* (Property 2.4): the Fourier transform of $f * h$ is $F(\mathbf{u})H(\mathbf{u})$. Thus, if the input f to an LSI system with frequency response $H(\mathbf{u})$ has Fourier transform $F(\mathbf{u})$, the output g has Fourier transform $G(\mathbf{u}) = F(\mathbf{u})H(\mathbf{u})$.

Table 2.1 Multidimensional Fourier transform properties.

	$f(\mathbf{x}) = \int_{\mathbb{R}^D} F(\mathbf{u}) \exp(j2\pi\mathbf{u}\cdot\mathbf{x})\, d\mathbf{u}$	$F(\mathbf{u}) = \int_{\mathbb{R}^D} f(\mathbf{x}) \exp(-j2\pi\mathbf{u}\cdot\mathbf{x})\, d\mathbf{x}$
(2.1)	$Af(\mathbf{x}) + Bg(\mathbf{x})$	$AF(\mathbf{u}) + BG(\mathbf{u})$
(2.2)	$f(\mathbf{x} - \mathbf{x}_0)$	$F(\mathbf{u}) \exp(-j2\pi\mathbf{u}\cdot\mathbf{x}_0)$
(2.3)	$f(\mathbf{x}) \exp(j2\pi\mathbf{u}_0\cdot\mathbf{x})$	$F(\mathbf{u} - \mathbf{u}_0)$
(2.4)	$f(\mathbf{x}) * g(\mathbf{x})$	$F(\mathbf{u})G(\mathbf{u})$
(2.5)	$f(\mathbf{x})g(\mathbf{x})$	$F(\mathbf{u}) * G(\mathbf{u})$
(2.6)	$f(\mathbf{A}\mathbf{x})$	$\frac{1}{\lvert\det \mathbf{A}\rvert} F(\mathbf{A}^{-T}\mathbf{u})$
(2.7)	$\nabla_{\mathbf{x}} f(\mathbf{x})$	$j2\pi\mathbf{u}F(\mathbf{u})$
(2.8)	$\mathbf{x}f(\mathbf{x})$	$\frac{j}{2\pi}\nabla_{\mathbf{u}}F(\mathbf{u})$
(2.9)	$f^*(\mathbf{x})$	$F^*(-\mathbf{u})$
(2.10)	$F(\mathbf{x})$	$f(-\mathbf{u})$
(2.11)	$f_1(x_1)\cdots f_D(x_D)$	$F_1(u_1)\cdots F_D(u_D)$
(2.12)	$\int_{\mathbb{R}^D} f(\mathbf{x})g^*(\mathbf{x})\, d\mathbf{x} = \int_{\mathbb{R}^D} F(\mathbf{u})G^*(\mathbf{u})\, d\mathbf{u}$	

2.6.1 Fourier Transform Properties

The proofs of the properties in Table 2.1 are straightforward (aside from convergence issues) and similar to analogous proofs for the one-dimensional Fourier transform, as given in many standard texts such as Bracewell (2000), Gray and Goodman (1995). They are presented briefly here. Again, the proofs are informal and assume that the relevant integrals converge, as would be the case for example if all the functions involved are absolutely integrable. Note that Property 2.6 relating to a linear transformation of the domain $\mathbb{R}^D$ is a more complex generalization from the 1D case. In some proofs, we assume the validity of the inverse Fourier transform of Equation (2.51), although it has not been proved at this point.

Property 2.1 *Linearity*: $Af(\mathbf{x}) + Bg(\mathbf{x}) \xleftrightarrow{\text{CDFT}} AF(\mathbf{u}) + BG(\mathbf{u})$.

Proof: Let $q(\mathbf{x}) = Af(\mathbf{x}) + Bg(\mathbf{x})$. Then

$$\begin{aligned} Q(\mathbf{u}) &= \int_{\mathbb{R}^D} (Af(\mathbf{x}) + Bg(\mathbf{x})) \exp(-j2\pi\mathbf{u}\cdot\mathbf{x})\, d\mathbf{x} \\ &= A\int_{\mathbb{R}^D} f(\mathbf{x}) \exp(-j2\pi\mathbf{u}\cdot\mathbf{x})\, d\mathbf{x} + B\int_{\mathbb{R}^D} g(\mathbf{x}) \exp(-j2\pi\mathbf{u}\cdot\mathbf{x})\, d\mathbf{x} \\ &= AF(\mathbf{u}) + BG(\mathbf{u}). \end{aligned} \tag{2.52}$$

□

Property 2.2 *Shift*: $f(\mathbf{x}-\mathbf{x}_0) \stackrel{\text{CDFT}}{\longleftrightarrow} F(\mathbf{u})\exp(-\text{j}2\pi\mathbf{u}\cdot\mathbf{x}_0)$.

Proof: Let $g(\mathbf{x}) = f(\mathbf{x}-\mathbf{x}_0)$ for some $\mathbf{x}_0 \in \mathbb{R}^D$. Then

$$\begin{aligned} G(\mathbf{u}) &= \int_{\mathbb{R}^D} f(\mathbf{x}-\mathbf{x}_0)\exp(-\text{j}2\pi\mathbf{u}\cdot\mathbf{x})\,\text{d}\mathbf{x} \\ &= \int_{\mathbb{R}^D} f(\mathbf{s})\exp(-\text{j}2\pi\mathbf{u}\cdot(\mathbf{s}+\mathbf{x}_0))\,\text{d}\mathbf{s} \quad (\mathbf{s}=\mathbf{x}-\mathbf{x}_0) \\ &= \left(\int_{\mathbb{R}^D} f(\mathbf{s})\exp(-\text{j}2\pi\mathbf{u}\cdot\mathbf{s})\,\text{d}\mathbf{s}\right)\exp(-\text{j}2\pi\mathbf{u}\cdot\mathbf{x}_0) \\ &= F(\mathbf{u})\exp(-\text{j}2\pi\mathbf{u}\cdot\mathbf{x}_0). \end{aligned} \tag{2.53}$$

□

Property 2.3 *Modulation*: $f(\mathbf{x})\exp(\text{j}2\pi\mathbf{u}_0\cdot\mathbf{x}) \stackrel{\text{CDFT}}{\longleftrightarrow} F(\mathbf{u}-\mathbf{u}_0)$.

Proof: Let $g(\mathbf{x}) = f(\mathbf{x})\exp(\text{j}2\pi\mathbf{u}_0\cdot\mathbf{x})$ for some $\mathbf{u}_0 \in \mathbb{R}^D$. Then

$$\begin{aligned} G(\mathbf{u}) &= \int_{\mathbb{R}^D} f(\mathbf{x})\exp(\text{j}2\pi\mathbf{u}_0\cdot\mathbf{x})\exp(-\text{j}2\pi\mathbf{u}\cdot\mathbf{x})\,\text{d}\mathbf{x} \\ &= \int_{\mathbb{R}^D} f(\mathbf{x})\exp(-\text{j}2\pi(\mathbf{u}-\mathbf{u}_0)\cdot\mathbf{x})\,\text{d}\mathbf{x} \\ &= F(\mathbf{u}-\mathbf{u}_0). \end{aligned} \tag{2.54}$$

□

Property 2.4 *Convolution*: $f(\mathbf{x}) * g(\mathbf{x}) \stackrel{\text{CDFT}}{\longleftrightarrow} F(\mathbf{u})G(\mathbf{u})$.

Proof: Let $q(\mathbf{x}) = f(\mathbf{x}) * g(\mathbf{x}) = \int_{\mathbb{R}^D} f(\mathbf{s})g(\mathbf{x}-\mathbf{s})\,\text{d}\mathbf{s}$. Then

$$\begin{aligned} Q(\mathbf{u}) &= \int_{\mathbb{R}^D}\left(\int_{\mathbb{R}^D} f(\mathbf{s})g(\mathbf{x}-\mathbf{s})\,\text{d}\mathbf{s}\right)\exp(-\text{j}2\pi\mathbf{u}\cdot\mathbf{x})\,\text{d}\mathbf{x} \\ &= \int_{\mathbb{R}^D}\int_{\mathbb{R}^D} f(\mathbf{s})g(\mathbf{t})\exp(-\text{j}2\pi\mathbf{u}\cdot(\mathbf{t}+\mathbf{s}))\,\text{d}\mathbf{s}\text{d}\mathbf{t} \quad (\mathbf{t}=\mathbf{x}-\mathbf{s}) \\ &= \int_{\mathbb{R}^D} f(\mathbf{s})\exp(-\text{j}2\pi\mathbf{u}\cdot\mathbf{s})\,\text{d}\mathbf{s}\int_{\mathbb{R}^D} g(\mathbf{t})\exp(-\text{j}2\pi\mathbf{u}\cdot\mathbf{t})\,\text{d}\mathbf{t} \\ &= F(\mathbf{u})G(\mathbf{u}). \end{aligned} \tag{2.55}$$

□

Note that using this property, commutativity and associativity of complex multiplication imply commutativity and associativity of convolution.

Property 2.5 *Multiplication*: $f(\mathbf{x})g(\mathbf{x}) \stackrel{\text{CDFT}}{\longleftrightarrow} F(\mathbf{u}) * G(\mathbf{u})$.

Proof: Let $q(\mathbf{x}) = f(\mathbf{x})g(\mathbf{x})$. Then

$$Q(\mathbf{u}) = \int_{\mathbb{R}^D} f(\mathbf{x})g(\mathbf{x})\exp(-\text{j}2\pi\mathbf{u}\cdot\mathbf{x})\,\text{d}\mathbf{x}$$

$$= \int_{\mathbb{R}^D} g(\mathbf{x}) \left(\int_{\mathbb{R}^D} F(\mathbf{w}) \exp(\mathrm{j}2\pi \mathbf{w} \cdot \mathbf{x})\, \mathrm{d}\mathbf{w} \right) \exp(-\mathrm{j}2\pi \mathbf{u} \cdot \mathbf{x})\, \mathrm{d}\mathbf{x}$$

$$= \int_{\mathbb{R}^D} F(\mathbf{w}) \left(\int_{\mathbb{R}^D} g(\mathbf{x}) \exp(-\mathrm{j}2\pi (\mathbf{u} - \mathbf{w}) \cdot \mathbf{x})\, \mathrm{d}\mathbf{x} \right) \mathrm{d}\mathbf{w}$$

$$= \int_{\mathbb{R}^D} F(\mathbf{w}) G(\mathbf{u} - \mathbf{w})\, \mathrm{d}\mathbf{w} = F(\mathbf{u}) * G(\mathbf{u}). \tag{2.56}$$

□

Property 2.6 *Linear transformation of the domain* $\mathbb{R}^D$: $f(\mathbf{Ax}) \stackrel{\mathrm{CDFT}}{\longleftrightarrow} \frac{1}{|\det(\mathbf{A})|} F(\mathbf{A}^{-T}\mathbf{u})$, where $\mathbf{A}$ is a nonsingular $D \times D$ matrix and $\mathbf{A}^{-T}$ denotes $(\mathbf{A}^T)^{-1} = (\mathbf{A}^{-1})^T$.

Proof: If we denote $g(\mathbf{x}) = f(\mathbf{Ax})$, then

$$G(\mathbf{u}) = \int_{\mathbb{R}^D} f(\mathbf{Ax}) \exp(-\mathrm{j}2\pi \mathbf{u} \cdot \mathbf{x})\, \mathrm{d}\mathbf{x}. \tag{2.57}$$

With the change of variables $\mathbf{s} = \mathbf{Ax}$, the Jacobian is $\frac{\partial(\mathbf{s})}{\partial(\mathbf{x})} = \det(\mathbf{A})$, and using standard techniques for change of variables in an integral (e.g., [(Kaplan, 1984, section 4.6)]), we obtain

$$G(\mathbf{u}) = \frac{1}{|\det(\mathbf{A})|} \int_{\mathbb{R}^D} f(\mathbf{s}) \exp(-\mathrm{j}2\pi \mathbf{u} \cdot (\mathbf{A}^{-1}\mathbf{s}))\, \mathrm{d}\mathbf{s}$$

$$= \frac{1}{|\det(\mathbf{A})|} \int_{\mathbb{R}^D} f(\mathbf{s}) \exp(-\mathrm{j}2\pi (\mathbf{A}^{-T}\mathbf{u}) \cdot \mathbf{s})\, \mathrm{d}\mathbf{s}$$

$$= \frac{1}{|\det(\mathbf{A})|} F(\mathbf{A}^{-T}\mathbf{u}) \tag{2.58}$$

where we use the identity $\mathbf{a} \cdot (\mathbf{Cb}) = (\mathbf{C}^T\mathbf{a}) \cdot \mathbf{b}$, or equivalently $\mathbf{a}^T\mathbf{Cb} = (\mathbf{C}^T\mathbf{a})^T\mathbf{b}$. □

This property can be used to determine the effect of independent scaling of x- and y-axes, of rotation of the image, or of an affine transformation of the independent variable (along with the shifting property). For any rotation matrix $\mathbf{R}$ in D dimensions, $\det(\mathbf{R}) = 1$ and $\mathbf{R}^{-1} = \mathbf{R}^T$ (so $\mathbf{R}^{-T} = \mathbf{R}$), so that in this case $f(\mathbf{Rx}) \stackrel{\mathrm{CDFT}}{\longleftrightarrow} F(\mathbf{Ru})$. Also, if $\mathbf{A} = -\mathbf{I}$, we get immediately that $f(-\mathbf{x}) \stackrel{\mathrm{CDFT}}{\longleftrightarrow} F(-\mathbf{u})$.

Property 2.7 *Differentiation*: $\nabla_{\mathbf{x}} f(\mathbf{x}) \stackrel{\mathrm{CDFT}}{\longleftrightarrow} \mathrm{j}2\pi \mathbf{u} F(\mathbf{u})$.

Proof: We define the gradient vector $\mathbf{g}(\mathbf{x}) = \nabla_{\mathbf{x}} f(\mathbf{x}) = \left[\frac{\partial f}{\partial x_1}, \dots, \frac{\partial f}{\partial x_D}\right]^T$. Taking the derivative of the synthesis equation for f to get the synthesis equation for g_i,

$$g_i(\mathbf{x}) = \frac{\partial f}{\partial x_i}(\mathbf{x}) = \int_{\mathbb{R}^D} F(\mathbf{u})(j2\pi u_i) \exp(\mathrm{j}2\pi \mathbf{u} \cdot \mathbf{x})\, \mathrm{d}\mathbf{u}. \tag{2.59}$$

Thus $G_i(\mathbf{u}) = \mathrm{j}2\pi u_i F(\mathbf{u})$ and in matrix form

$$\mathbf{G}(\mathbf{u}) = \mathrm{j}2\pi \mathbf{u} F(\mathbf{u}). \tag{2.60}$$

□

Property 2.8 *Differentiation in frequency*: $\mathbf{x}f(\mathbf{x}) \overset{\text{CDFT}}{\longleftrightarrow} \frac{\mathrm{j}}{2\pi}\nabla_{\mathbf{u}}F(\mathbf{u})$.

Proof: Let $\mathbf{g}(\mathbf{x}) = \mathbf{x}f(\mathbf{x})$. Taking the derivative of the analysis equation for f with respect to u_i,

$$\begin{aligned}\frac{\partial F}{\partial u_i}(\mathbf{u}) &= \int_{\mathbb{R}^D} f(\mathbf{x})(-\mathrm{j}2\pi x_i)\exp(-\mathrm{j}2\pi\mathbf{u}\cdot\mathbf{x})\,\mathrm{d}\mathbf{x} \\ &- \mathrm{j}2\pi\int_{\mathbb{R}^D} g_i(\mathbf{x})\exp(-\mathrm{j}2\pi\mathbf{u}\cdot\mathbf{x})\,\mathrm{d}\mathbf{x} \\ &= -\mathrm{j}2\pi G_i(\mathbf{u}).\end{aligned} \tag{2.61}$$

Thus, $\mathbf{G}(\mathbf{u}) = \frac{\mathrm{j}}{2\pi}\nabla_{\mathbf{u}}F(\mathbf{u})$. □

Property 2.9 *Complex conjugation*: $f^*(\mathbf{x}) \overset{\text{CDFT}}{\longleftrightarrow} F^*(-\mathbf{u})$

Proof: Let $g(\mathbf{x}) = f^*(\mathbf{x})$. Then

$$\begin{aligned}G(\mathbf{u}) &= \int_{\mathbb{R}^D} f^*(\mathbf{x})\exp(-\mathrm{j}2\pi\mathbf{u}\cdot\mathbf{x})\,\mathrm{d}\mathbf{x} \\ &= \int_{\mathbb{R}^D} (f(\mathbf{x})\exp(\mathrm{j}2\pi\mathbf{u}\cdot\mathbf{x}))^*\,\mathrm{d}\mathbf{x} \\ &= \left(\int_{\mathbb{R}^D} f(\mathbf{x})\exp(\mathrm{j}2\pi\mathbf{u}\cdot\mathbf{x})\,\mathrm{d}\mathbf{x}\right)^* \\ &= F^*(-\mathbf{u}).\end{aligned} \tag{2.62}$$

□

Property 2.10 *Duality*: $F(\mathbf{x}) \overset{\text{CDFT}}{\longleftrightarrow} f(-\mathbf{u})$

Proof: Let $g(\mathbf{x}) = F(\mathbf{x})$. Then

$$\begin{aligned}G(\mathbf{u}) &= \int_{\mathbb{R}^D} F(\mathbf{x})\exp(-\mathrm{j}2\pi\mathbf{u}\cdot\mathbf{x})\,\mathrm{d}\mathbf{x} \\ &= \int_{\mathbb{R}^D} F(\mathbf{x})\exp(\mathrm{j}2\pi(-\mathbf{u})\cdot\mathbf{x})\,\mathrm{d}\mathbf{x} \\ &= f(-\mathbf{u}).\end{aligned} \tag{2.63}$$

□

Property 2.11 *Separability*: $f_1(x_1)\cdots f_D(x_D) \overset{\text{CDFT}}{\longleftrightarrow} F_1(u_1)\cdots F_D(u_D)$

Proof: This follows from the separability of the complex exponentials,

$$\begin{aligned}&\int_{-\infty}^{\infty}\cdots\int_{-\infty}^{\infty} f_1(x_1)\cdots f_D(x_D)\exp(-\mathrm{j}2\pi(u_1x_1+\cdots+u_Dx_D))\,\mathrm{d}x_1\cdots\mathrm{d}x_D \\ &\quad= \left(\int_{-\infty}^{\infty} f_1(x_1)\exp(-\mathrm{j}2\pi u_1x_1)\,\mathrm{d}x_1\right)\end{aligned}$$

$$\times \cdots \times \left(\int_{-\infty}^{\infty} f_D(x_D) \exp(-j2\pi u_D x_D)\, dx_D \right)$$
$$= F_1(u_1) \cdots F_D(u_D) \tag{2.64}$$

□

Property 2.12 *Parseval relation*: $\int_{\mathbb{R}^D} f(\mathbf{x}) g^*(\mathbf{x})\, d\mathbf{x} = \int_{\mathbb{R}^D} F(\mathbf{u}) G^*(\mathbf{u})\, d\mathbf{u}$

Proof: Let $r(\mathbf{x}) = g^*(\mathbf{x})$ and $q(\mathbf{x}) = f(\mathbf{x})r(\mathbf{x})$. Then

$$Q(\mathbf{u}) = \int_{\mathbb{R}^D} F(\mathbf{w}) R(\mathbf{u} - \mathbf{w})\, d\mathbf{w}$$
$$= \int_{\mathbb{R}^D} F(\mathbf{w}) G^*(\mathbf{w} - \mathbf{u})\, d\mathbf{w} \tag{2.65}$$

Evaluating at $\mathbf{u} = 0$,

$$Q(\mathbf{0}) = \int_{\mathbb{R}^D} f(\mathbf{x}) g^*(\mathbf{x})\, d\mathbf{x} = \int_{\mathbb{R}^D} F(\mathbf{w}) G^*(\mathbf{w})\, d\mathbf{w}. \tag{2.66}$$

□

If $f = g$, we obtain

$$\int_{\mathbb{R}^D} |f(\mathbf{x})|^2\, d\mathbf{x} = \int_{\mathbb{R}^D} |F(\mathbf{w})|^2\, d\mathbf{w}. \tag{2.67}$$

2.6.2 Evaluation of Multidimensional Fourier Transforms

In general, the multiD Fourier transform is determined by direct evaluation of the defining integral (2.50) using standard methods of integral calculus. Simplifications are possible if the function $f(\mathbf{x})$ is separable or isotropic, and of course maximum use should be made of the Fourier transform properties of Table 2.1. A few examples follow, and Table 2.2 provides a number of useful two-dimensional Fourier transforms, and others are derived in the problems.

Example 2.2
Compute the 2D Fourier transform of $f(x, y) = \text{rect}(x, y)$.

Solution:

$$F(u, v) = \int_{-\infty}^{\infty} \int_{-\infty}^{\infty} \text{rect}(x, y) \exp(-j2\pi(ux + vy))\, dx\, dy$$
$$= \int_{-0.5}^{0.5} \int_{-0.5}^{0.5} \exp(-j2\pi(ux + vy))\, dx\, dy$$
$$= \int_{-0.5}^{0.5} \exp(-j2\pi ux)\, dx \int_{-0.5}^{0.5} \exp(-j2\pi vy)\, dy$$
$$= \frac{\sin \pi u}{\pi u} \frac{\sin \pi v}{\pi v}$$
$$= \text{sinc}(u)\,\text{sinc}(v),$$

Table 2.2 Two-dimensional Fourier transform of selected functions.

$f(x,y) = \int_{\mathbb{R}^2} F(u,v)\exp(\mathrm{j}2\pi(ux+vy))\,\mathrm{d}u\mathrm{d}v$	$F(u,v) = \int_{\mathbb{R}^2} f(x,y)\exp(-\mathrm{j}2\pi(ux+vy))\,\mathrm{d}x\mathrm{d}y$
$\mathrm{rect}(x,y)$	$\frac{\sin(\pi u)}{\pi u}\frac{\sin(\pi v)}{\pi v}$
$\mathrm{circ}(x,y)$	$\frac{1}{\sqrt{u^2+v^2}} J_1(2\pi\sqrt{u^2+v^2})$
$\exp(-(x^2+y^2)/2)$	$2\pi\exp(-2\pi^2(u^2+v^2))$
$\exp(-2\pi\sqrt{x^2+y^2})$	$\frac{1}{2\pi^2}\frac{1}{(1+u^2+v^2)^{3/2}}$
$\cos(\pi(x^2+y^2))$	$\sin(\pi(u^2+v^2))$
$\exp(\mathrm{j}\pi(x^2+y^2))$	$\mathrm{j}\exp(-\mathrm{j}\pi(u^2+v^2))$
$\delta(x,y)$	1

where we introduce the standard function

$$\mathrm{sinc}(t) = \frac{\sin \pi t}{\pi t}.$$

It is easy to show that $\mathrm{sinc}(0) = 1$. We note that since $\mathrm{rect}(x,y)$ is separable, the resulting 2D Fourier transform is the product of the corresponding 1D Fourier transforms. □

Example 2.3
Compute the 2D Fourier transform of $f(x,y) = c\,\mathrm{rect}(ax, by)$, where $a > 0$ and $b > 0$.

Solution:
Use Property 2.1 and Property 2.6, with

$$\mathbf{A} = \begin{bmatrix} a & 0 \\ 0 & b \end{bmatrix}.$$

Then $|\det \mathbf{A}| = ab$ and

$$\mathbf{A}^{-T} = \begin{bmatrix} \frac{1}{a} & 0 \\ 0 & \frac{1}{b} \end{bmatrix}.$$

It follows that

$$F(u,v) = \frac{c}{ab}\frac{\sin(\pi u/a)}{\pi u/a}\frac{\sin(\pi v/b)}{\pi v/b}$$
$$= \frac{c}{ab}\mathrm{sinc}(u/a)\mathrm{sinc}(v/b).$$

□

Example 2.4
Compute the 2D Fourier transform of $f(x,y) = \mathrm{circ}(x,y)$, and of $h(x,y) = \frac{1}{\pi R^2}\mathrm{circ}\left(\frac{x}{R}, \frac{y}{R}\right)$.

Solution:

$$F(u,v) = \int_{-\infty}^{\infty}\int_{-\infty}^{\infty} \text{circ}(x,y)\exp(-\text{j}2\pi(ux+vy))\,\text{d}x\,\text{d}y$$

This function is not separable but it is isotropic. We make the following changes of variables to polar coordinates:

$$x = r\cos\theta \quad y = r\sin\theta$$
$$u = w\cos\phi \quad v = w\sin\phi.$$

The region of support of circ(x,y) is the circle $0 \le r \le 1$, $-\pi \le \theta < \pi$. Thus, denoting $F'(w,\phi) = F(w\cos\phi, w\sin\phi)$, we obtain

$$\begin{aligned}
F'(w,\phi) &= \int_{-\pi}^{\pi}\int_{0}^{1} \exp(-\text{j}2\pi(rw\cos\theta\cos\phi + rw\sin\theta\sin\phi))r\,\text{d}r\,\text{d}\theta \\
&= \int_{0}^{1} r\int_{-\pi}^{\pi} \exp(-\text{j}2\pi rw\cos(\theta-\phi))\,\text{d}\theta\,\text{d}r \\
&= 2\pi\int_{0}^{1} rJ_0(2\pi rw)\,\text{d}r \\
&= \frac{1}{2\pi w^2}\int_{0}^{2\pi w} zJ_0(z)\,\text{d}z \qquad (z = 2\pi rw) \\
&= \frac{1}{w}J_1(2\pi w)
\end{aligned}$$

where $J_0(\cdot)$ and $J_1(\cdot)$ are Bessel functions of the first kind of order zero and order one respectively [Poularikas (1998)]. We have used the properties that

$$J_0(x) = \frac{1}{2\pi}\int_{-\pi}^{\pi}\exp(\text{j}x\cos(\theta-\alpha))\,\text{d}\theta \qquad \text{for any } \alpha,$$

$$xJ_1(x) = \int_0^x sJ_0(s)\,\text{d}s.$$

Thus,

$$F(u,v) = \frac{1}{\sqrt{u^2+v^2}}J_1(2\pi\sqrt{u^2+v^2}).$$

We can now conclude using properties 2.1 and 2.6 that the frequency response of the out-of-focus blur of Example 2.1 with $h(x,y) = \frac{1}{\pi R^2}\text{circ}(\frac{x}{R}, \frac{y}{R})$ is

$$\begin{aligned}
H(u,v) &= \frac{1}{\pi R^2}\cdot R^2\frac{1}{R\sqrt{u^2+v^2}}J_1(2\pi R\sqrt{u^2+v^2}) \\
&= \frac{1}{\pi R\sqrt{u^2+v^2}}J_1(2\pi R\sqrt{u^2+v^2}).
\end{aligned}$$

□

Example 2.5

Compute the impulse response of an ideal two-dimensional low-pass filter with frequency response $H(u,v) = \text{circ}(u/W, v/W)$, where W is the bandwidth of the filter.

Solution:
From Example 2.4 and Property 2.6,

$$\mathrm{circ}\left(\frac{x}{W}, \frac{y}{W}\right) \stackrel{\text{CDFT}}{\longleftrightarrow} \frac{W}{\sqrt{u^2+v^2}} J_1(2\pi W\sqrt{u^2+v^2}).$$

Noting that circ is symmetric about the origin, the duality Property 2.10 yields

$$h(x,y) = \frac{W}{\sqrt{x^2+y^2}} J_1(2\pi W\sqrt{x^2+y^2}).$$

□

Example 2.6 A reasonable model for the impulse response of a vidicon video camera sensor is a separable Gaussian in the spatial dimensions and rect function in the temporal dimension, specifically

$$\begin{aligned} h(x,y,t) &= \exp(-(x^2+y^2)/2r_0^2)\mathrm{rect}(t/T - .5) \\ &= \exp(-x^2/2r_0^2)\exp(-y^2/2r_0^2)\mathrm{rect}(t/T - 0.5). \end{aligned}$$

Using separability, and the standard results for the Fourier transform of the 1D Gaussian signal [Bracewell (2000)],

$$\exp(-\pi x^2) \stackrel{\text{CDFT}}{\longleftrightarrow} \exp(-\pi u^2)$$

$$\mathrm{rect}(t) \stackrel{\text{CDFT}}{\longleftrightarrow} \frac{\sin(\pi w)}{\pi w}$$

gives

$$\begin{aligned} H(u,v,w) &= \sqrt{2\pi}r_0\exp(-2\pi^2u^2r_0^2)\sqrt{2\pi}r_0\exp(-2\pi^2v^2r_0^2)\frac{\sin(\pi wT)}{\pi w} \\ &\quad\times\exp(-\mathrm{j}\pi Tw) \\ &= 2r_0^2\exp(-\mathrm{j}\pi Tw)\exp(-2\pi^2(u^2+v^2)r_0^2)\sin(\pi wT)/w. \end{aligned}$$

2.6.3 Two-Dimensional Fourier Transform of Polygonal Zero–One Functions

Polygonal zero–one functions are frequently encountered in the analysis of modern cameras and display devices, and their Fourier transform is required. For the rectangular region considered in Example 2.3, it is straightforward to compute the Fourier transform by direct evaluation of the integral. However, for other shapes, such as hexagons, octagons, chevrons, etc., direct computation of the Fourier transform is more involved and tedious. It is possible to convert the area integral in the direct definition of the Fourier transform to a line integral along the boundary of the region $\mathcal{A}$ using the 2D version of Gauss's divergence theorem and thereby obtain a closed form expression for the Fourier transform in the case of polygonal regions.

Let $\mathcal{A} \subset \mathbb{R}^2$ be a bounded, simply connected region in the plane and define $p_{\mathcal{A}}(\mathbf{x})$ as in Equation (2.1). Then, the Fourier transform is given by

$$P_{\mathcal{A}}(\mathbf{u}) = \iint_{\mathcal{A}} \exp(-\mathrm{j}2\pi\mathbf{u}\cdot\mathbf{x})\,\mathrm{d}\mathbf{x}. \tag{2.68}$$

Let $\partial\mathcal{A}$ be the boundary of $\mathcal{A}$, assumed to be piecewise smooth, traversed in the clockwise direction. Then let $\mathbf{Q}(\mathbf{x}) = (Q_1(\mathbf{x}), Q_2(\mathbf{x}))$ be a vector field defined on $\mathcal{A}$, assumed to be continuous with continuous first partial derivatives. The divergence theorem (see for example [(Kaplan, 1984, section 5.11)]) states that

$$\iint_{\mathcal{A}} \mathrm{div}\mathbf{Q}(\mathbf{x})\, \mathrm{d}\mathbf{x} = \oint_{\partial\mathcal{A}} \mathbf{Q}(\mathbf{x}) \cdot \mathbf{n}_{\mathbf{x}}\, \mathrm{d}S(\mathbf{x}), \tag{2.69}$$

where

$$\mathrm{div}\mathbf{Q}(\mathbf{x}) = \frac{\partial Q_1(\mathbf{x})}{\partial x} + \frac{\partial Q_2(\mathbf{x})}{\partial y}, \tag{2.70}$$

$\mathbf{n}_{\mathbf{x}}$ is a unit vector normal to $\partial\mathcal{A}$ at $\mathbf{x}$ and pointing outward, and $S(\mathbf{x})$ denotes arc length along $\partial\mathcal{A}$ at $\mathbf{x}$. This result can be applied to computing the Fourier transform in Equation (2.68) by choosing

$$\mathbf{Q}_{\mathbf{u}}(\mathbf{x}) = \frac{j\mathbf{u}}{2\pi\|\mathbf{u}\|^2} \exp(-\mathrm{j}2\pi\mathbf{u}\cdot\mathbf{x}). \tag{2.71}$$

By applying the definition of divergence,

$$\mathrm{div}\,\mathbf{Q}_{\mathbf{u}}(\mathbf{x}) = \exp(-\mathrm{j}2\pi\mathbf{u}\cdot\mathbf{x}). \tag{2.72}$$

We then find that

$$P_{\mathcal{A}}(\mathbf{u}) = \oint_{\partial\mathcal{A}} \left(\frac{\mathrm{j}\mathbf{u}}{2\pi\|\mathbf{u}\|^2} \exp(-\mathrm{j}2\pi\mathbf{u}\cdot\mathbf{x}) \right) \cdot \mathbf{n}_{\mathbf{x}}\, \mathrm{d}S(\mathbf{x}). \tag{2.73}$$

This result has been called the Abbe transform and was cited in the dissertation of Straubel in 1888 [Komrska (1982)]. As shown in Komrska (1982), the contour integral can easily be evaluated in closed form for a polygonal region as follows.

Assume that $\mathcal{A}$ is a polygon with K sides, with vertices $\mathbf{a}_1, \ldots, \mathbf{a}_K$ in the clockwise direction; for convenience, we denote $\mathbf{a}_{K+1} = \mathbf{a}_1$. We define the following quantities that are easily determined once the vertices are specified:

$$d_k = \|\mathbf{a}_{k+1} - \mathbf{a}_k\| \qquad \text{length of side } k \tag{2.74}$$

$$\mathbf{c}_k = \frac{\mathbf{a}_k + \mathbf{a}_{k+1}}{2} \qquad \text{midpoint of side } k \tag{2.75}$$

$$\mathbf{t}_k = \frac{\mathbf{a}_{k+1} - \mathbf{a}_k}{d_k} \qquad \text{unit vector parallel to side } k \tag{2.76}$$

$$\mathbf{n}_k = \mathcal{R}\mathbf{t}_k \qquad \text{unit normal to side } k \text{ pointing outward} \tag{2.77}$$

where $\mathcal{R}$ rotates counterclockwise by 90°. With these definitions, the Fourier transform expression given in Equation (2.73) can be written as a sum of the integrals over each of the polygon sides as follows.

$$\begin{aligned} P_{\mathcal{A}}(\mathbf{u}) &= \sum_{k=1}^{K} \int_{\mathbf{a}_k}^{\mathbf{a}_{k+1}} \frac{\mathrm{j}\mathbf{u}\cdot\mathbf{n}_k}{2\pi\|\mathbf{u}\|^2} \exp(-\mathrm{j}2\pi\mathbf{u}\cdot\mathbf{x})\, \mathrm{d}S(\mathbf{x}) \\ &= \frac{\mathrm{j}}{2\pi\|\mathbf{u}\|^2} \sum_{k=1}^{K} (\mathbf{u}\cdot\mathbf{n}_k) \int_{-d_k/2}^{d_k/2} \exp(-\mathrm{j}2\pi\mathbf{u}\cdot(\mathbf{c}_k + s\mathbf{t}_k))\, \mathrm{d}s \\ &= \frac{\mathrm{j}}{2\pi\|\mathbf{u}\|^2} \sum_{k=1}^{K} (\mathbf{u}\cdot\mathbf{n}_k) \exp(-\mathrm{j}2\pi\mathbf{u}\cdot\mathbf{c}_k) \int_{-d_k/2}^{d_k/2} \exp(-\mathrm{j}2\pi s\mathbf{u}\cdot\mathbf{t}_k)\, \mathrm{d}s. \end{aligned} \tag{2.78}$$

The integral can be easily evaluated to give the final result:

$$P_{\mathcal{A}}(\mathbf{u}) = \frac{j}{2\pi\|\mathbf{u}\|^2}\sum_{k=1}^{K} d_k(\mathbf{u}\cdot\mathbf{n}_k)\exp(-j\pi\mathbf{u}\cdot(\mathbf{a}_{k+1}+\mathbf{a}_k))\frac{\sin(\pi\mathbf{u}\cdot(\mathbf{a}_{k+1}-\mathbf{a}_k))}{\pi\mathbf{u}\cdot(\mathbf{a}_{k+1}-\mathbf{a}_k)}. \tag{2.79}$$

In many (but not all) cases of interest, the polygon is symmetric about the origin, i.e. $\mathbf{x}\in\mathcal{A}\Rightarrow -\mathbf{x}\in\mathcal{A}$. In this case, the number of vertices and sides is necessarily even, and the terms corresponding to the two opposite sides in Equation (2.79) can be combined to yield a real-valued Fourier transform [Lu *et al.* (2009)].

$$\begin{aligned}P_{\mathcal{A}}(\mathbf{u}) &= \frac{1}{\pi\|\mathbf{u}\|^2}\sum_{k=1}^{K/2} d_k(\mathbf{u}\cdot\mathbf{n}_k)\sin(\pi\mathbf{u}\cdot(\mathbf{a}_{k+1}+\mathbf{a}_k))\frac{\sin(\pi\mathbf{u}\cdot(\mathbf{a}_{k+1}-\mathbf{a}_k))}{\pi\mathbf{u}\cdot(\mathbf{a}_{k+1}-\mathbf{a}_k)}\\ &= \frac{1}{\pi\|\mathbf{u}\|^2}\sum_{k=1}^{K/2} d_k(\mathbf{u}\cdot\mathbf{n}_k)\sin(\pi\mathbf{u}\cdot(\mathbf{a}_{k+1}+\mathbf{a}_k))\,\mathrm{sinc}(\mathbf{u}\cdot(\mathbf{a}_{k+1}-\mathbf{a}_k)).\end{aligned} \tag{2.80}$$

This result has been extended to zero–one functions in more than two dimensions where the region of support is a polytope [Brandolini *et al.* (1997)], and applications in multiD signal processing have been described in Lu *et al.* (2009).

It is very straightforward to apply this result to determine the Fourier transform of a rect function, and this is left as an exercise. The following shows the application to a regular hexagon with unit side.

Example 2.7 Let $\mathcal{A}$ be a regular hexagon with unit side, as illustrated in Figure 2.12. Since the hexagon is symmetric about the origin, we can apply Equation (2.80). From the figure, we observe that

$$\mathbf{a}_1 = \begin{bmatrix}1\\0\end{bmatrix} \quad \mathbf{a}_2 = \begin{bmatrix}\frac{1}{2}\\ \frac{\sqrt{3}}{2}\end{bmatrix} \quad \mathbf{a}_3 = \begin{bmatrix}-\frac{1}{2}\\ \frac{\sqrt{3}}{2}\end{bmatrix} \quad \mathbf{a}_4 = \begin{bmatrix}-1\\0\end{bmatrix}.$$

By assumption, $d_k = 1$ for all k, and we can easily determine that

$$\mathbf{n}_1 = \begin{bmatrix}\frac{\sqrt{3}}{2}\\ \frac{1}{2}\end{bmatrix} \quad \mathbf{n}_2 = \begin{bmatrix}0\\1\end{bmatrix} \quad \mathbf{n}_3 = \begin{bmatrix}-\frac{\sqrt{3}}{2}\\ \frac{1}{2}\end{bmatrix}.$$

Simply inserting these into Equation (2.80) (and using symmetry of sinc) gives the result

$$\begin{aligned}P_{\mathcal{A}}(\mathbf{u}) = &\frac{1}{\pi(u^2+v^2)}\left(\frac{\sqrt{3}u+v}{2}\sin\left(\frac{\pi(3u+\sqrt{3}v)}{2}\right)\mathrm{sinc}\left(\frac{-u+\sqrt{3}v}{2}\right)\right.\\ &+ v\sin(\pi\sqrt{3}v)\mathrm{sinc}(u) + \frac{-\sqrt{3}u+v}{2}\sin\left(\frac{\pi(-3u+\sqrt{3}v)}{2}\right)\\ &\left.\times\,\mathrm{sinc}\left(\frac{u+\sqrt{3}v}{2}\right)\right).\end{aligned}$$

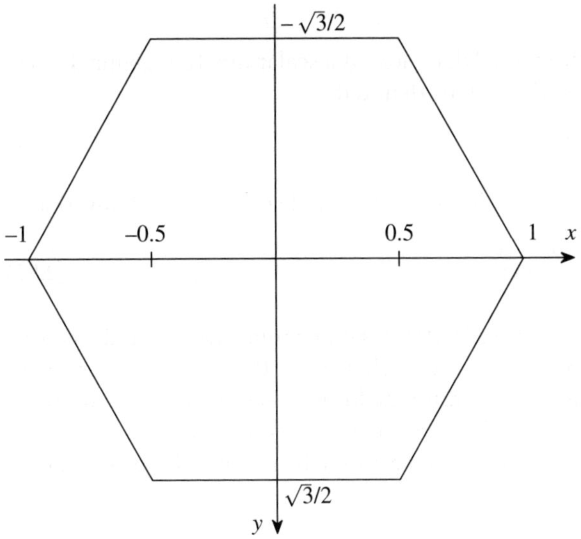

Figure 2.12 Regular hexagon with unit side.

2.6.4 Fourier Transform of a Translating Still Image

Assume that a still image $f_0(\mathbf{x})$ is moving with a uniform velocity $\mathbf{v}$ to produce the time-varying image $f(\mathbf{x}, t) = f_0(\mathbf{x} - \mathbf{v}t)$. We wish to relate the 3D Fourier transform of $f(\mathbf{x}, t)$ to the 2D Fourier transform of $f_0(\mathbf{x})$.

$$
\begin{aligned}
F(\mathbf{u}, w) &= \int_{-\infty}^{\infty} \int_{-\infty}^{\infty} \int_{-\infty}^{\infty} f_0(\mathbf{x} - \mathbf{v}t) \exp(-\mathrm{j}2\pi \mathbf{u} \cdot \mathbf{x})\, \mathrm{d}\mathbf{x} \exp(-\mathrm{j}2\pi w t)\, \mathrm{d}t \\
&= \int_{-\infty}^{\infty} \exp(-\mathrm{j}2\pi w t) \exp(-\mathrm{j}2\pi \mathbf{u} \cdot \mathbf{v}t) \int_{-\infty}^{\infty} \int_{-\infty}^{\infty} f_0(\mathbf{s}) \exp(-\mathrm{j}2\pi \mathbf{u} \cdot \mathbf{s})\, \mathrm{d}\mathbf{s}\, \mathrm{d}t \\
&= F_0(\mathbf{u}) \int_{-\infty}^{\infty} \exp(-\mathrm{j}2\pi(\mathbf{u} \cdot \mathbf{v} + w)t)\, \mathrm{d}t \\
&= F_0(\mathbf{u}) \delta(\mathbf{u} \cdot \mathbf{v} + w).
\end{aligned}
$$

Thus, the 3D Fourier transform is concentrated on the plane $\mathbf{u} \cdot \mathbf{v} + w = 0$. This leads us to conclude that the 3D Fourier transform of a typical time-varying image is not uniformly spread out in 3D frequency space, but will be largely concentrated near planes representing the dominant motions in the scene.

2.7 Further Properties of Differentiation and Related Systems

Image derivatives are frequently used in the image processing literature. Although generally applied to discrete-domain images, where derivatives are not defined, they usually presuppose an underlying continuous-domain image where derivatives *are* defined. Thus we introduce here several additional continuous-domain LSI systems based on derivatives, beyond the gradient already seen in Property 2.7.

2.7.1 Directional Derivative

Let $\mathbf{z}$ be a unit vector in $\mathbb{R}^D$. The directional derivative is a scalar function giving the rate of change of $f(\mathbf{x})$ in the direction $\mathbf{z}$. This can be denoted

$$D_{\mathbf{z}} f(\mathbf{x}) = \lim_{\alpha \to 0} \frac{f(\mathbf{x} + \alpha \mathbf{z}) - f(\mathbf{x})}{\alpha}. \tag{2.81}$$

We assume f is continuous at $\mathbf{x}$. From standard multivariable calculus, we know that

$$D_{\mathbf{z}} f(\mathbf{x}) = \nabla_{\mathbf{x}} f(\mathbf{x}) \cdot \mathbf{z} = \sum_{i=1}^{D} z_i \frac{\partial f}{\partial x_i}(\mathbf{x}). \tag{2.82}$$

The magnitude of the directional derivative $|D_{\mathbf{z}} f(\mathbf{x})|$ is maximum when the unit vector $\mathbf{z}$ is collinear with the gradient $\nabla_{\mathbf{x}} f(\mathbf{x})$, and it is zero when $\mathbf{z}$ is orthogonal to the gradient. As a result, the gradient is often used to quantify the local directionality of f. Although this result is quite evident from standard vector analysis, the following matrix proof leads to many interesting generalizations. We can express the magnitude squared of the directional derivative as

$$\begin{aligned} |D_{\mathbf{z}} f(\mathbf{x})|^2 &= \mathbf{z}^T \nabla_{\mathbf{x}} f(\mathbf{x}) \nabla_{\mathbf{x}}^T f(\mathbf{x}) \mathbf{z} \\ &= \mathbf{z}^T \mathbf{G}(\mathbf{x}) \mathbf{z} \end{aligned} \tag{2.83}$$

where $\|\mathbf{z}\| = 1$. This is maximized when $\mathbf{z}$ is the normalized eigenvector corresponding to the maximum eigenvalue of $\mathbf{G}(\mathbf{x})$. Since $\mathbf{G}(\mathbf{x})$ has rank one and thus the null space has dimension $D - 1$, it follows that $D - 1$ eigenvalues are zero. The eigenvector corresponding to the one non-zero eigenvalue is $\nabla_{\mathbf{x}} f(\mathbf{x})$, since

$$\begin{aligned} \mathbf{G}(\mathbf{x}) \nabla_{\mathbf{x}} f(\mathbf{x}) &= \nabla_{\mathbf{x}} f(\mathbf{x}) (\nabla_{\mathbf{x}}^T f(\mathbf{x}) \nabla_{\mathbf{x}} f(\mathbf{x})) \\ &= \|\nabla_{\mathbf{x}} f(\mathbf{x})\|^2 \nabla_{\mathbf{x}} f(\mathbf{x}). \end{aligned} \tag{2.84}$$

Then, the maximum value of $|D_{\mathbf{z}} f(\mathbf{x})|$ is $\|\nabla_{\mathbf{x}} f(\mathbf{x})\|$, which occurs for $\mathbf{z} = \nabla_{\mathbf{x}} f(\mathbf{x}) / \|\nabla_{\mathbf{x}} f(\mathbf{x})\|$, as stated previously. The entity $\mathbf{G}(\mathbf{x}) = \nabla_{\mathbf{x}} f(\mathbf{x}) \nabla_{\mathbf{x}}^T f(\mathbf{x})$ is usually referred to as the structure tensor, and this is a quantity that has many generalizations.

The Fourier transform of $D_{\mathbf{z}} f(\mathbf{x})$ is then given by $\mathrm{j}2\pi \mathbf{z} \cdot \mathbf{u} F(\mathbf{u})$. Thus, the directional derivative is a linear shift-invariant system with frequency response $\mathrm{j}2\pi \mathbf{z} \cdot \mathbf{u}$.

2.7.2 Laplacian

The Laplacian is a scalar system involving second order derivatives, typically denoted $\nabla_{\mathbf{x}}^2$:

$$\nabla_{\mathbf{x}}^2 f(\mathbf{x}) = \sum_{i=1}^{D} \frac{\partial^2 f}{\partial x_i^2}(\mathbf{x}). \tag{2.85}$$

If $g(\mathbf{x}) = \nabla_{\mathbf{x}}^2 f(\mathbf{x})$, the Fourier transform of the output of the Laplacian is given by

$$\begin{aligned} G(\mathbf{u}) &= \sum_{i=1}^{D} (j2\pi u_i)^2 F(\mathbf{u}) \\ &= -\left(\sum_{i=1}^{D} (2\pi u_i)^2 \right) F(\mathbf{u}) \\ &= -(2\pi)^2 \|\mathbf{u}\|^2 F(\mathbf{u}). \end{aligned} \tag{2.86}$$

Thus, the Laplacian is an LSI system, with frequency response $-(2\pi)^2\|\mathbf{u}\|^2$, which is isotropic.

2.7.3 Filtered Derivative Systems

The derivative systems presented so far have frequency responses that increase in magnitude with $\|\mathbf{u}\|$. This makes them very sensitive to high-frequency noise, and they are regularly coupled with low-pass filters. For example, the gradient and Laplacian are frequently preceded with Gaussian filters to smooth the high frequencies before applying the gradient, Laplacian or higher-order derivative operator. When using these to analyze the image, the Gaussian filter can also serve to set the scale at which the analysis takes place.

As an example, consider the Laplacian that is an isotropic scalar system. It is usually coupled with an isotropic Gaussian low-pass filter with impulse response and frequency response

$$h_G(\mathbf{x}) = \frac{1}{(2\pi r_0^2)^{D/2}} \exp\left(-\frac{\|\mathbf{x}\|^2}{2r_0^2}\right), \qquad H_G(\mathbf{u}) = \exp(-2\pi^2 r_0^2\|\mathbf{u}\|^2). \tag{2.87}$$

Thus, the filtered Laplacian has frequency response

$$H(\mathbf{u}) = -(2\pi)^2\|\mathbf{u}\|^2 \exp(-2\pi^2 r_0^2\|\mathbf{u}\|^2). \tag{2.88}$$

Using the associative property of LSI systems, the impulse response of the filtered Laplacian is given by the Laplacian of the Gaussian impulse response (so it is called a Laplacian of Gaussian or LoG filter):

$$\begin{aligned} h(\mathbf{x}) = \nabla_{\mathbf{x}}^2 h_G(\mathbf{x}) &= \frac{1}{(2\pi r_0^2)^{D/2}} \sum_{i=1}^{D} \frac{\partial^2}{\partial x_i^2} \exp\left(-\frac{\|\mathbf{x}\|^2}{2r_0^2}\right) \\ &= \frac{1}{(2\pi r_0^2)^{D/2}} \sum_{i=1}^{D} \left(\left(-\frac{x_i}{r_0^2}\right)^2 - \frac{1}{r_0^2}\right) \exp\left(-\frac{\|\mathbf{x}\|^2}{2r_0^2}\right) \\ &= \frac{1}{(2\pi r_0^2)^{D/2}} \frac{\|\mathbf{x}\|^2 - Dr_0^2}{r_0^4} \exp\left(-\frac{\|\mathbf{x}\|^2}{2r_0^2}\right). \end{aligned} \tag{2.89}$$

The filtered Laplacian is most frequently used in two dimensions, where the frequency response and the impulse response can be written

$$H(u, v) = -(2\pi)^2(u^2 + v^2)\exp(-2\pi^2 r_0^2(u^2 + v^2)), \tag{2.90}$$

$$h(x, y) = \frac{x^2 + y^2 - 2r_0^2}{2\pi r_0^6} \exp\left(-\frac{x^2 + y^2}{2r_0^2}\right). \tag{2.91}$$

In general, the absolute magnitude of h is not significant, and it can be scaled to any convenient value. The frequency response is circularly symmetric with value 0 at the origin and for large frequency, and a peak amplitude at the radial frequency $1/\sqrt{2}\pi r_0$. Figure 2.13 depicts the impulse response and magnitude frequency response of a LoG filter with $r_0 = 0.0025$ ph. The filter is scaled so that the maximum magnitude frequency

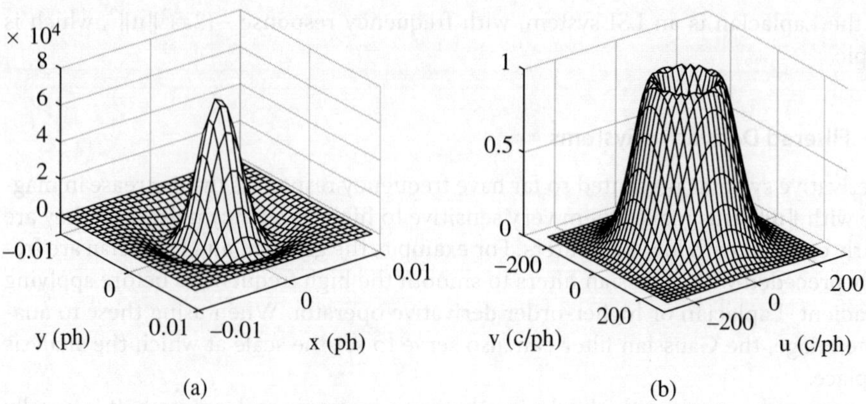

Figure 2.13 Laplacian of Gaussian filter with $r_0 = 0.0025$. (a) Negative of impulse response. (b) Magnitude of frequency response.

Figure 2.14 Laplacian of Gaussian filter with $r_0 = 0.0025$ applied to 'Barbara' image.

response is 1.0, which occurs at radial frequency 90 c/ph. Figure 2.14 shows a simulation of the filtering of the 'Barbara' image with this LoG filter. Since the mean output level of the filtered image is 0, a value of 0.5 on a scale of 0 to 1 is added to the image for the purpose of display.

Problems

1 Consider a two-dimensional sinusoidal signal $f(x, y) = A\cos(2\pi(ux + vy) + \phi)$ where x and y are in ph and u and v are in c/ph. Form the one-dimensional signal $g(z)$ by tracing $f(x, y)$ along the line $y = cx$, where c is some real constant, as a function of distance along the line, $z = \sqrt{x^2 + y^2}$.

a) Show that $g(z)$ is a sinusoidal signal $g(z) = A\cos(2\pi wz + \phi)$ and determine the spatial frequency w in c/ph, as a function of u, v and c.

b) Explain what happens when $c = 0$ and when $c \to \infty$.

c) Show that the spatial frequency w is greatest along the line $y = (v/u)x$, if $u \neq 0$. What is the value of this maximum spatial frequency? What happens if $u = 0$?

2 Show that for each of the following functions $\delta_\Delta(x, y)$,

$$\int_{-\infty}^{\infty}\int_{-\infty}^{\infty} \delta_\Delta(x, y)\, dx\, dy = 1$$

and

$$\lim_{\Delta\to 0}\int_{-\infty}^{\infty}\int_{-\infty}^{\infty} \delta_\Delta(x, y) f(x, y)\, dx\, dy = f(0, 0)$$

for any function $f(x, y)$ that is continuous at $(x, y) = (0, 0)$.

a) $\delta_\Delta(x, y) = \frac{1}{\Delta^2}\mathrm{rect}(x/\Delta, y/\Delta)$

b) $\delta_\Delta(x, y) = \frac{1}{\Delta^2}\exp(-\pi(x^2 + y^2)/\Delta^2)$

c) $\delta_\Delta(x, y) = \frac{1}{\pi\Delta^2}\mathrm{circ}(x/\Delta, y/\Delta)$.

3 Show that

$$\delta(ax, by) = \frac{1}{|ab|}\delta(x, y)$$

where $a, b \neq 0$.

4 Prove that the following systems are linear systems.

a) The shift system $\mathcal{T}_\mathbf{d}$ for any shift $\mathbf{d} \in \mathbb{R}^D$.

b) The system induced by a nonsingular transformation of the domain, $\mathcal{M}_\mathbf{A} : g = \mathcal{M}_\mathbf{A} f : g(\mathbf{x}) = f(\mathbf{A}\mathbf{x})$, where $\mathbf{A}$ is any nonsingular $D \times D$ matrix.

c) The cascade of two linear systems $\mathcal{H}_1$ and $\mathcal{H}_2$. Thus, the system induced by an affine transformation of the domain is a linear system.

d) The parallel combination of two linear systems with the same domain and range, $\mathcal{H}_1 + \mathcal{H}_2$.

e) The partial derivative systems ∂_x and ∂_y defined in Section 2.5.2.

5 Prove that the following systems are linear shift-invariant systems.

a) The shift system $\mathcal{T}_\mathbf{d}$ for any shift $\mathbf{d} \in \mathbb{R}^D$.

b) The cascade of two LSI systems $\mathcal{H}_1$ and $\mathcal{H}_2$.

c) The parallel combination of two LSI systems with the same domain and range, $\mathcal{H}_1 + \mathcal{H}_2$.
d) The partial derivative systems ∂_x and ∂_y defined in Section 2.5.2.

6 Let $f(x, y) = 0.5\,\mathrm{rect}(4(x - 0.5), 2(y - 0.25))$ and $h(x, y) = \mathrm{rect}(10x, 10y)$, where x and y are in ph.
a) Sketch the region of support of $f(x, y)$ and $h(x, y)$ in the XY-plane (i.e., the area where these two signals are nonzero).
b) Compute the two-dimensional convolution $f(x, y) * h(x, y)$ from the definition using integration in the spatial domain.
c) Suppose that $f(x, y)$ is the input to a 2D system, and the output of this system is computed as in (b). What can we say about this system?
d) Determine the continuous-space Fourier transforms $F(u, v)$, $H(u, v)$ and $G(u, v)$ of the above three signals. Make liberal use of Fourier transform properties. What are the units of u and v?
e) Continuing with question (c), what is the interpretation of $H(u, v)$?

7 Determine the response of an LSI system with impulse response $h(\mathbf{x})$ to a real sinusoidal signal $f(\mathbf{x}) = A + B\cos(2\pi\mathbf{u}\cdot\mathbf{x} + \phi)$ where $A > 0$ and $0 < B < A$.

8 A 2D continuous-space linear shift-invariant system has impulse response

$$h(x, y) = \begin{cases} \dfrac{1}{2\pi R_1 R_2}, & \left(\dfrac{x}{R_1}\right)^2 + \left(\dfrac{y}{R_2}\right)^2 \le 1 \\ 0, & \text{otherwise,} \end{cases}$$

where $R_1 = 1/1000$ ph and $R_2 = 1/500$ ph.
a) Sketch the region of support of the impulse response in the XY-plane, following the conventions used for the labelling of axes. Express $h(x, y)$ in terms of the circ function.
b) Find the frequency response $H(u, v)$ of this system, where u and v are in c/ph.
c) The image $f(x, y) = \mathrm{rect}(5(x - 0.5), 2(y - 0.5))$ is filtered with this system to produce the output $g(x, y) = f(x, y) * h(x, y)$. Determine the Fourier transform of the output, $G(u, v)$.

9 Compute the two-dimensional continuous-space Fourier transform of the following signals:
a) The separable signal $f(x, y) = h_X^{(1)}(x)h_Y^{(1)}(y)$ where

$$h_T^{(1)}(t) = \begin{cases} 1 - \dfrac{|t|}{T} & |t| \le T, \\ 0 & \text{otherwise.} \end{cases}$$

b) A Gaussian function $f(x, y) = \frac{1}{2\pi r_0^2}\mathrm{e}^{-(x^2+y^2)/2r_0^2}$. (i) Obtain the result from the entry in Table 2.2 (with $r_0 = 1$). (ii) Prove the result in Table 2.2.

c) A real zone plate, $f(x,y) = \cos(\pi(x^2+y^2)/r_0^2)$. (Hint: Find the Fourier transform of the complex zone plate $\exp(j\pi(x^2+y^2)/r_0^2)$ and use linearity. You can use $\int_{-\infty}^{\infty} e^{jy^2}\,dy = \sqrt{\pi}e^{j\pi/4}$.)

d) Diamond-shaped pulse

$$f(x,y) = \begin{cases} 1 & |x|+|y| \le 1, \\ 0 & |x|+|y| > 1. \end{cases}$$

(Hint: obtain this function from a rect function using a rotation transformation.)

e) Gabor function

$$f(x,y) = \cos(2\pi(u_0x+v_0y))\exp\left(-\frac{(x-x_0)^2+(y-y_0)^2}{2r_0^2}\right)$$

f) The 2D zero–one function $p_{\mathcal{A}}(x,y)$ where $\mathcal{A}$ is an elliptical region, with semi-minor axis X and semi-major axis $2X$, oriented at 45° as shown in Figure 2.15

10 Derive the expression for the Fourier transform of a zero–one function on a polygon symmetric about the origin, as given in Equation (2.80).

11 Use the expression in Equation (2.80) to compute the Fourier transform of the rect function.

12 Use the expression in Equation (2.80) to compute the Fourier transform of a zero-one function with a region $\mathcal{A}$ that is a regular hexagon of unit area, with vertices on the y-axis.

13 Use the expression in Equation (2.80) to compute the Fourier transform of a zero–one function with a region $\mathcal{A}$ that is a regular octagon of unit area, with two sides parallel to the x-axis.

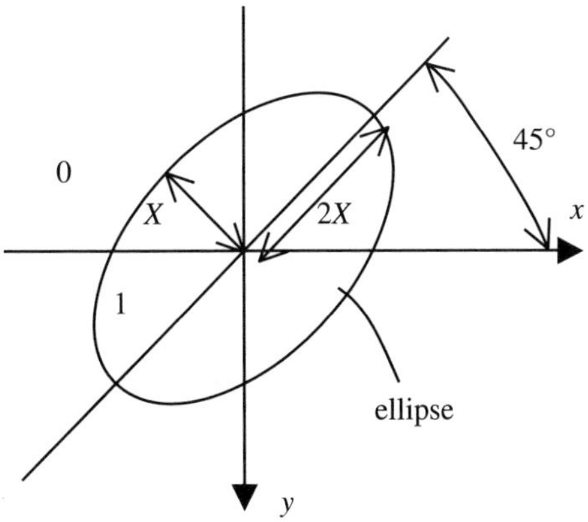

Figure 2.15 Elliptical region of support of a 2D zero–one function.

14 Consider a continuous-domain Laplacian of Gaussian (LoG) filter with impulse response

$$h(x,y) = c\frac{x^2+y^2-2r_0^2}{2\pi r_0^6}\exp\left(-\frac{x^2+y^2}{2r_0^2}\right).$$

a) Show that the magnitude frequency response has a peak at radial frequency

$$\sqrt{u^2+v^2} = \frac{1}{\sqrt{2}\pi r_0}.$$

b) What is the value of c such that the peak magnitude frequency response is 1.0, i.e.,

$$|H(u,v)| = 1 \text{ when } u^2+v^2 = \frac{1}{2\pi^2 r_0^2}.$$

c) Compute the values found in (a) and (b) when $r_0 = 0.0025$ ph.

15 Find the D-dimensional Fourier transform of the following functions:

a) A D-dimensional Gaussian $f(\mathbf{x}) = \frac{1}{(2\pi)^{D/2}}\exp(-\|\mathbf{x}\|^2/2)$ (Equation (2.87)).

b) A D-dimensional circularly symmetric exponential $f(\mathbf{x}) = \exp(-2\pi \parallel \mathbf{x} \parallel)$. Answer:

$$F(\mathbf{u}) = c_D\frac{1}{(1+\|\mathbf{u}\|^2)^{(D+1)/2}}$$

where $c_D = \Gamma((D+1)/2)/\pi^{(D+1)/2}$. $\Gamma(\cdot)$ is the Gamma function, which satifies the following properties: $\Gamma(n) = n!$ for $n = 1, 2, \ldots$, $\Gamma(x+1) = x\Gamma(x)$, $\Gamma(0.5) = \sqrt{\pi}$. Hint: The solution can be found on pages 6 and 7 in Stein and Weiss (1971).

3

Discrete-Domain Signals and Systems

3.1 Introduction

Although the original images projected on a sensor and the final images presented to a viewer's visual system are continuous in space and time, a discrete intermediate representation is required in order to carry out digital image and video processing operations. Thus images must be *sampled* in space and time for processing and eventually converted back to continuous form for presentation to the viewer. In this chapter we introduce discrete-space and discrete-space-time signals and systems under the general name of discrete-domain signals and systems.

For the sampling of one-dimensional signals, all that needs to be specified is the sampling period (or equivalently, the sampling frequency), and possibly the sampling phase. In two, three, or more dimensions, the situation is more complicated. We have to specify how the samples are arranged in space and time. The simplest arrangement is to lay out the samples on a rectangular grid, and this is the approach taken in most discussions of digital image processing. However, there are many important applications where this is not the case. For example, in standard broadcast television, the interlaced scanning method is used; scanning lines in each vertical pass of the image are midway between the scanning lines of the previous vertical pass. It follows that any scheme for sampling standard TV signals will lead to nonrectangular sampling in 3D space-time. Another important example of nonrectangular sampling is the widely used Bayer color filter array (see e.g., Gunturk *et al.* (2005)). Figure 3.1 shows the layout of red, green and blue sensor elements in such an array. We see that the red, green and blue samples are each acquired on different sampling structures with different offsets, and that the green sampling structure is not rectangular. In fact, virtually all digital cameras use some form of color filter array with nonrectangular sampling for at least some, if not all, of the components. Our approach will be to develop sampling theory from the beginning in the general multidimensional setting, which includes two and three-dimensional sampling on both rectangular and nonrectangular sampling structures. Rectangular sampling will be a special case. This is different from the usual approach of introducing rectangular sampling first, and then possibly extending the theory to nonrectangular sampling. The development will be based on the theory of *lattices*.

The application of lattices to the sampling of multidimensional signals was presented by Petersen and Middleton (1962), and many of the basic results were presented in

Multidimensional Signal and Color Image Processing Using Lattices, First Edition. Eric Dubois.

Companion website: www.wiley.com/go/Dubois/multiSP

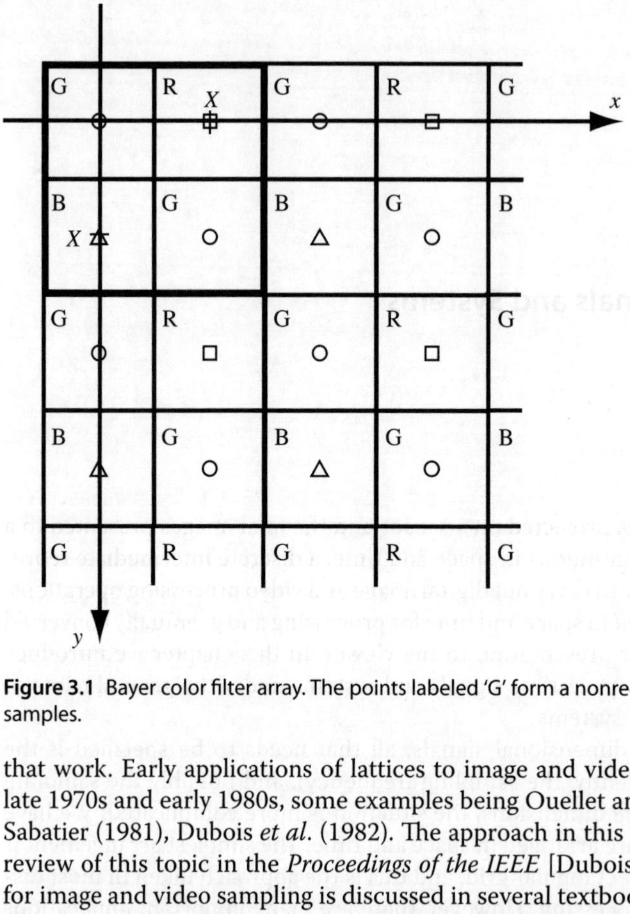

Figure 3.1 Bayer color filter array. The points labeled 'G' form a nonrectangular arrangement of samples.

that work. Early applications of lattices to image and video sampling appeared in the late 1970s and early 1980s, some examples being Ouellet and Dubois (1981), Kretz and Sabatier (1981), Dubois *et al.* (1982). The approach in this book follows from the 1985 review of this topic in the *Proceedings of the IEEE* [Dubois (1985)]. The use of lattices for image and video sampling is discussed in several textbooks, including Dudgeon and Mersereau (1984), Wang *et al.* (2002), Woods (2006). An excellent treatment is also given in Kalker (1998). Lattices have been successfully used in other aspects of information technology and communications. A review of lattice techniques in wireless communications is given in Wübben *et al.* (2011) and applications in cryptography are given in Micciancio and Goldwasser (2002), which also gives a very good, accessible overview of lattice definitions, properties and algorithms.

Since lattices are used throughout this book, a thorough treatment of lattices suitable for multidimensional signal processing is presented in a separate chapter (Chapter 13). The ideas are introduced as needed in the treatment of the various topics, and are gathered in a coherent fashion, along with detailed proofs, in Chapter 13. That chapter stands alone and can be read at any point as required.

3.2 Lattices

3.2.1 Basic Definitions

The mathematical structure most useful in describing sampling of time-varying images is the *lattice*. A lattice Λ in D dimensions is a discrete set of points that can be expressed as the set of all linear combinations with *integer* coefficients of D linearly independent vectors in $\mathbb{R}^D$ (called basis vectors),

$$\Lambda = \{n_1\mathbf{v}_1 + \cdots + n_D\mathbf{v}_D \mid n_i \in \mathbb{Z}\}, \tag{3.1}$$

where $\mathbb{Z}$ is the set of integers. For our purposes, D will generally be 1, 2 or 3 dimensions. Figure 3.2 shows an example of a lattice in two dimensions, with basis vectors $\mathbf{v}_1 = [2X\ \ 0]^T$ and $\mathbf{v}_2 = [X\ \ Y]^T$. A convenient way to represent a lattice is the *sampling matrix* $\mathbf{V} = [\mathbf{v}_1 \mid \mathbf{v}_2 \mid \cdots \mid \mathbf{v}_D]$ whose columns are the basis vectors $\mathbf{v}_i$ represented as column matrices. We denote the lattice Λ determined by the sampling matrix $\mathbf{V}$ by $\Lambda = \mathrm{LAT}(\mathbf{V})$. The sampling matrix for the lattice of Figure 3.2 with respect to the given basis vectors is

$$\mathbf{V} = \begin{bmatrix} 2X & X \\ 0 & Y \end{bmatrix}. \tag{3.2}$$

The basis or sampling matrix for any given lattice is not unique. For example, we can easily verify by inspection that the sampling matrix

$$\mathbf{V}_1 = \begin{bmatrix} X & -X \\ Y & Y \end{bmatrix}$$

also generates the lattice of Figure 3.2. It can be shown that $\mathrm{LAT}(\mathbf{V}) = \mathrm{LAT}(\mathbf{VE})$ if $\mathbf{E}$ is any unimodular ($|\det \mathbf{E}| = 1$) integer matrix (Theorem 13.1). In the above example, $\mathbf{V}_1 = \mathbf{VE}$ where

$$\mathbf{E} = \mathbf{V}^{-1}\mathbf{V}_1 = \begin{bmatrix} 0 & -1 \\ 1 & 1 \end{bmatrix}$$

and $\det \mathbf{E} = 1$. Alternatively, we can conclude that two sampling matrices $\mathbf{V}$ and $\mathbf{V}_1$ represent the same lattice if and only if $\mathbf{V}^{-1}\mathbf{V}_1$ is an integer matrix with $|\det(\mathbf{V}^{-1}\mathbf{V}_1)| = 1$.

If we represent points in $\mathbb{R}^D$ as column matrices, the points in the lattice are given by

$$\Lambda = \{\mathbf{Vn} \mid \mathbf{n} \in \mathbb{Z}^D\}. \tag{3.3}$$

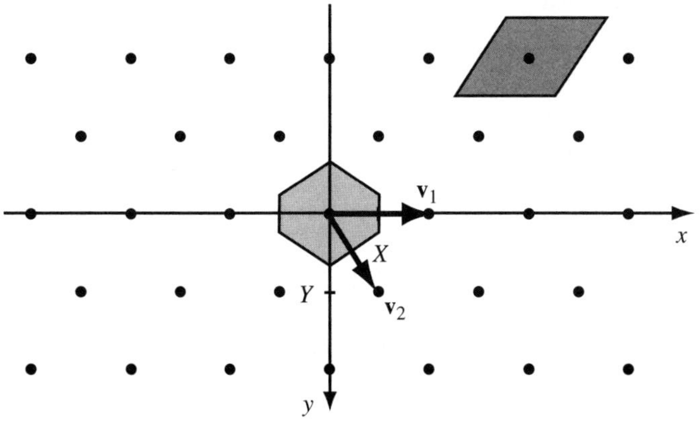

Figure 3.2 Example of a lattice in two dimensions with two possible unit cells.

For example, for the lattice defined by the sampling matrix of equation 3.2, we have

$$\Lambda = \left\{ \begin{bmatrix} 2Xn_1 + Xn_2 \\ Yn_2 \end{bmatrix} \mid n_1, n_2 \in \mathbb{Z} \right\}$$

$$= \left\{ \begin{bmatrix} (2n_1 + n_2)X \\ n_2 Y \end{bmatrix} \mid n_1, n_2 \in \mathbb{Z} \right\}.$$

A *unit cell* of a lattice Λ is a set $\mathcal{P} \subset \mathbb{R}^D$ such that copies of $\mathcal{P}$ centered on each lattice point tile the whole space without overlap: $(\mathcal{P} + \mathbf{s}_1) \cap (\mathcal{P} + \mathbf{s}_2) = \emptyset$ for $\mathbf{s}_1, \mathbf{s}_2 \in \Lambda, \mathbf{s}_1 \neq \mathbf{s}_2$, and $\cup_{\mathbf{s}\in\Lambda}(\mathcal{P} + \mathbf{s}) = \mathbb{R}^D$. The volume of a unit cell is $\mathrm{d}(\Lambda) = |\det \mathbf{V}|$, which is independent of the particular choice of sampling matrix (Theorem 13.2). We can imagine that there is a region congruent to $\mathcal{P}$ of volume $\mathrm{d}(\Lambda)$ associated with each sample in Λ, so that $\mathrm{d}(\Lambda)$ is the reciprocal of the sampling density. The quantity $\mathrm{d}(\Lambda)$ is often referred to as the *determinant* of the lattice. The unit cell of a lattice is not unique. In Figure 3.2, the shaded hexagonal region centered at the origin is a unit cell, of area $\mathrm{d}(\Lambda) = 2XY$. The shaded parallelogram in the upper right is also a possible unit cell. The hexagonal-shaped unit cell is an example of a *Voronoi* unit cell, consisting of all points closer to the origin than to any other lattice point. In the two-dimensional case, the Voronoi unit cell is a polygon whose edges are the perpendicular bisectors of lines from the origin to the nearest lattice points. Note that we are usually informal about how we treat the boundary of the unit cell, since it usually not very important, unless there is a singularity on the boundary (however, see Definition 13.5).

3.2.2 Properties of Lattices

The following properties of lattices are easily seen from the definition and are key to the theory of processing signals defined on lattices. Let Λ be any lattice.

(i) $\mathbf{0} \in \Lambda$; the origin belongs to *any* lattice.
(ii) If $\mathbf{x} \in \Lambda$ and $\mathbf{y} \in \Lambda$ then $\mathbf{x} + \mathbf{y} \in \Lambda$.
(iii) If $\mathbf{d} \in \Lambda$ then $\Lambda + \mathbf{d} = \Lambda$ where $\Lambda + \mathbf{d} = \{\mathbf{x} + \mathbf{d} \mid \mathbf{x} \in \Lambda\}$

See Section 13.3 for more details.

3.2.3 Examples of 2D and 3D Lattices

This section presents several lattices that have been used for image sampling.

2D rectangular (or orthogonal) lattice (Figure 3.3).

The rectangular lattice is defined by perpendicular basis vectors, and thus can be generated by a diagonal sampling matrix.

$$\mathbf{V} = \begin{bmatrix} X & 0 \\ 0 & Y \end{bmatrix}, \qquad \mathrm{d}(\Lambda) = XY. \tag{3.4}$$

If $X = Y$, we call it a square lattice. If $X = Y = 1$, we call it the integer lattice $\mathbb{Z}^2$.

2D hexagonal lattice (Figure 3.4). The hexagonal lattice is so named because the six nearest neighbors of any lattice point form the vertices of a hexagon. The standard form of the sampling matrix is

$$\mathbf{V} = \begin{bmatrix} X & X/2 \\ 0 & Y \end{bmatrix}, \qquad \mathrm{d}(\Lambda) = XY. \tag{3.5}$$

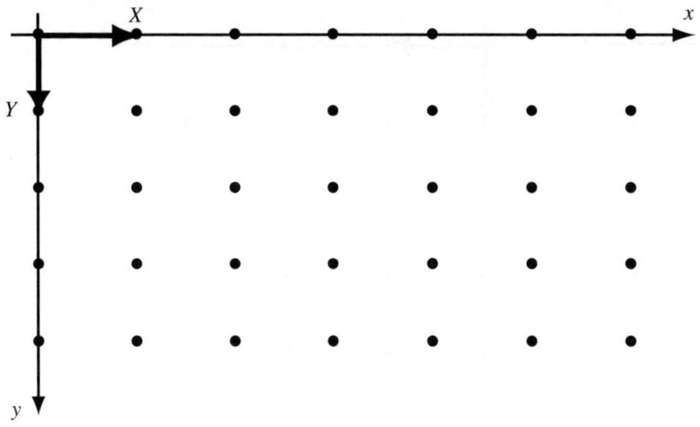

Figure 3.3 Two-dimensional rectangular lattice.

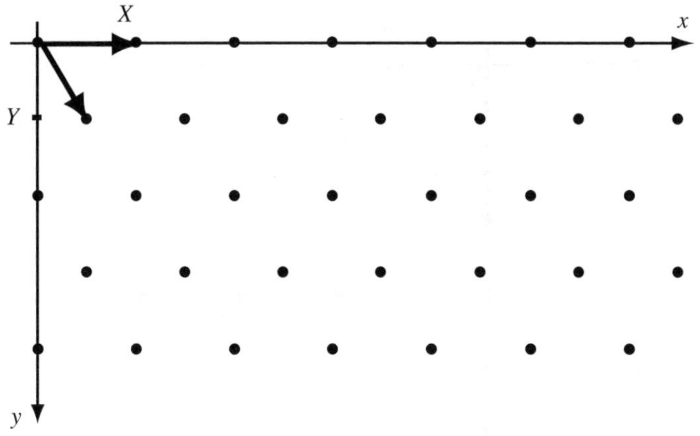

Figure 3.4 Two-dimensional hexagonal lattice.

If $Y = \frac{\sqrt{3}X}{2}$, the six nearest neighbors of any lattice point form the vertices of a *regular* hexagon.

3D rectangular lattice (Figure 3.5). The three-dimensional rectangular lattice is defined by three perpendicular basis vectors, and so is defined by a 3×3 diagonal matrix.

$$\mathbf{V} = \begin{bmatrix} X & 0 & 0 \\ 0 & Y & 0 \\ 0 & 0 & T \end{bmatrix}, \qquad \mathrm{d}(\Lambda) = XYT. \tag{3.6}$$

3D interlaced lattice (Figure 3.6). The three-dimensional interlaced lattice arises when an interlaced signal like any standard television signal is sampled with all samples aligned vertically. The sampling matrix is

$$\mathbf{V} = \begin{bmatrix} X & 0 & 0 \\ 0 & 2Y & Y \\ 0 & 0 & T/2 \end{bmatrix}, \qquad \mathrm{d}(\Lambda) = XYT. \tag{3.7}$$

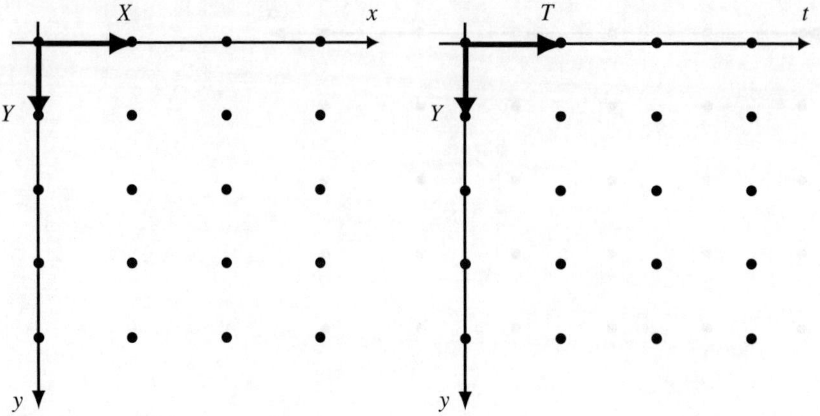

Figure 3.5 Spatial and vertical-temporal projection of a three-dimensional rectangular lattice.

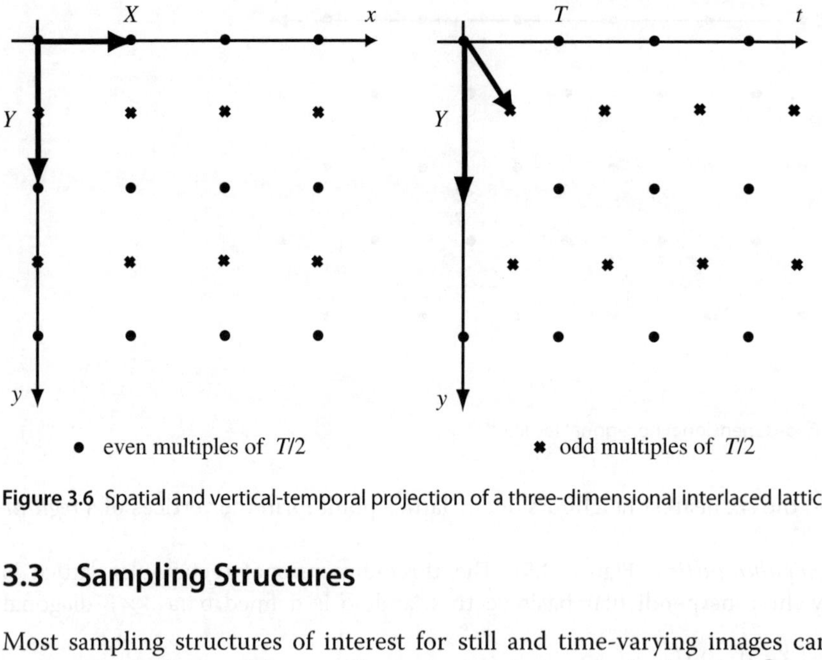

Figure 3.6 Spatial and vertical-temporal projection of a three-dimensional interlaced lattice.

3.3 Sampling Structures

Most sampling structures of interest for still and time-varying images can be constructed using lattices. The sampling structure is the set of points in $\mathbb{R}^D$ on which the image is defined. Let us consider still images first. The sampling structure determines how the sample points are laid out on the image plane. By far the most widely used structure is the rectangular lattice of Figure 3.3. However, the hexagonal lattice of Figure 3.4 is also used in a number of important applications. The sampling structure of an image is normally confined to the image window $\mathcal{W}$ of Figure 2.1. However, for convenience of analysis, we often consider the sampling structure to be of infinite extent. Outside $\mathcal{W}$, the image can be considered to be zero, periodically extended, or extrapolated in some way. This has an impact on the performance of image processing algorithms near the boundary of the image, and may or may not be important.

If the sampling phase is not important, we can assume that one point of the sampling structure lies at the origin of the coordinate system and that the sampling structure is a lattice Λ. However, if we consider the Bayer color filter array of Figure 3.1 to simultaneously sample the red, green and blue components of a color image, we see that the sampling structures for R, G, and B have no points in common. Thus, only one of them can contain the origin and be a lattice. The other two are *shifted* lattices. Assume that sample locations correspond to the centers of the cells in Figure 3.1, which are assumed to be squares of size X by X. The resulting sampling structures are indicated in Figure 3.1.

The sampling structure for the green component, Ψ_G, contains the origin and is the hexagonal lattice

$$\Psi_G = \text{LAT}\left(\begin{bmatrix} 2X & X \\ 0 & X \end{bmatrix}\right).$$

The sampling structures for the red and blue channels are *shifted* rectangular lattices:

$$\Psi_R = \begin{bmatrix} X \\ 0 \end{bmatrix} + \text{LAT}\left(\begin{bmatrix} 2X & 0 \\ 0 & 2X \end{bmatrix}\right)$$

$$\Psi_B = \begin{bmatrix} 0 \\ X \end{bmatrix} + \text{LAT}\left(\begin{bmatrix} 2X & 0 \\ 0 & 2X \end{bmatrix}\right).$$

Notice that the sampling structure for G has twice the sampling density as those for R and B $\left(\frac{1}{2X^2} \text{ versus } \frac{1}{4X^2}\right)$. The sampling density of a shifted lattice $\mathbf{d} + \Lambda$ is $1/\text{d}(\Lambda)$, i.e. the shift does not change the sampling density.

It is possible to construct a sampling structure that is the union of two or more shifted lattices, and this has found a few applications, for example in some recent digital cameras. This has been discussed in [Dubois (1985)].

Similarly, sampling structures for time-varying images can be constructed from three-dimensional lattices. The two most widely used correspond to the lattices of Figures 3.5 and 3.6. The rectangular lattice would normally be used whenever motion-picture film is digitized, and is now used with many video sources. However, some video cameras today still use *interlaced* scanning; each image *frame* consists of two *fields*. Each vertical pass of the scanning beam captures half of the lines forming the entire frame. This scanning structure is used in the three main analog TV formats (NTSC, PAL, SECAM) as well as in some HDTV formats. Typical parameters for the North American video signal in interlaced format in Figure 3.6 are $X = \frac{1}{540}$ ph, $Y = \frac{1}{480}$ ph and $T = \frac{1}{29.97}$ s. The value of the parameter Y follows directly from the fact that there are 480 active scanning lines (out of a total of 525) forming an image frame.

It is possible to consider *partially* sampled signals. For example, a motion picture film sequence can be considered to be continuous in the spatial domain and discrete in the temporal domain.

3.4 Signals Defined on Lattices

The extension of discrete-time signal theory to two and three (or more) dimensions is accomplished by considering signals defined on lattices. In fact, ordinary discrete-time signals can be thought of as signals defined on the one-dimensional lattice LAT([T]). Let Λ be a lattice in D dimensions. Then a real signal defined on the lattice Λ is denoted $f[\mathbf{x}]$, $\mathbf{x} \in \Lambda$. The signal f is understood to only be defined on the points of Λ; it is undefined elsewhere in $\mathbb{R}^D$ (it is not 0). Note that, following the convention of Oppenheim and Willsky (1997), we use *square* brackets to enclose the independent variables $\mathbf{x}$ for signals defined on a lattice, as opposed to round parentheses for continuous-domain signals.

As in Section 2.4, we can consider a signal space $\mathcal{S}$ of signals defined on Λ and denote an element of this space f. For example, the space of signals of finite energy defined on Λ is

$$l^2(\Lambda) = \left\{ f \,\middle|\, \sum_{\mathbf{x}\in\Lambda} |f[\mathbf{x}]|^2 < \infty \right\}. \tag{3.8}$$

The notation $\sum_{\mathbf{x}\in\Lambda} \cdot$ indicates a summation over all elements of the lattice. Using the definition of a lattice generated by the basis vectors $\{\mathbf{v}_1, \ldots, \mathbf{v}_D\}$, this summation can be written explicitly as

$$\sum_{\mathbf{x}\in\Lambda} |f[\mathbf{x}]|^2 = \sum_{n_1=-\infty}^{\infty} \cdots \sum_{n_D=-\infty}^{\infty} |f[n_1\mathbf{v}_1 + \cdots + n_D\mathbf{v}_D]|^2. \tag{3.9}$$

In the most common case of a rectangular lattice, a signal is often written explicitly in terms of its coordinates $f[n_1X, n_2Y]$ or $f[n_1X, n_2Y, n_3T]$, and the summation is expressed, in the 3D case, as

$$\sum_{\mathbf{x}\in\Lambda} |f[\mathbf{x}]|^2 = \sum_{n_1=-\infty}^{\infty} \sum_{n_2=-\infty}^{\infty} \sum_{n_3=-\infty}^{\infty} |f[n_1X, n_2Y, n_3T]|^2. \tag{3.10}$$

In the case of a *square* spatial lattice with equal horizontal and vertical spacing X, it is common to choose X as the unit of length. We shall call this length the pixel height, denoted px. In this case, the sampling matrix is the identity matrix $\mathbf{V} = \mathrm{diag}(1, 1)$ and $\Lambda = \mathbb{Z}^2$.

3.5 Special Multidimensional Signals on a Lattice

As in the continuous domain, many analytically defined two- and three-dimensional functions over a lattice are useful in image processing theory. A number of these are described here.

3.5.1 Unit Sample

The unit sample function is defined as

$$\delta[\mathbf{x}] = \begin{cases} 1 & \text{if } \mathbf{x} = \mathbf{0} \\ 0 & \text{if } \mathbf{x} \in \Lambda \backslash \mathbf{0}. \end{cases} \tag{3.11}$$

The notation $\mathbf{x} \in \Lambda\backslash\mathbf{0}$ signifies that $\mathbf{x}$ is any element of Λ except $\mathbf{0}$; we know that $\mathbf{0}$ is an element of *any* lattice. The unit sample has many characteristics in common with the Dirac delta function $\delta(\mathbf{x})$ of continuous space-time. However, whereas the latter is a rather complex generalized function (a distribution), the unit sample is probably the simplest of signals over a lattice. It will be key in the characterization of linear shift-invariant systems over a lattice. Note that we use the same symbol for the unit sample and the Dirac function; they are distinguished, however, by the use of square brackets versus parentheses, and the particular lattice Λ is assumed. If more than one lattice is being considered, then we denote the unit sample on Λ by δ_Λ.

3.5.2 Sinusoidal Signals

The real and complex forms of a sinusoidal signal defined on a lattice are

$$f_R[\mathbf{x}] = A\cos(2\pi\mathbf{u}\cdot\mathbf{x}+\phi), \tag{3.12}$$

$$f_C[\mathbf{x}] = B\exp(\mathrm{j}2\pi\mathbf{u}\cdot\mathbf{x}), \qquad \mathbf{x}\in\Lambda, \tag{3.13}$$

where, as before, $\mathbf{u}$ is a D-dimensional frequency vector, A is a real constant and B is a complex constant. We recall that for one-dimensional discrete-time signals, different frequencies can give the same discrete-time sinusoid. Specifically,

$$f_1[nT] = B\exp(\mathrm{j}2\pi unT)$$

$$\text{and} \quad f_2[nT] = B\exp\left(\mathrm{j}2\pi\left(u+\frac{k}{T}\right)nT\right)$$

are the same discrete-time sinusoid for any integer k. Thus all frequencies $u+\frac{k}{T}$, $k\in\mathbb{Z}$, define the same discrete-time sinusoid. There is a similar property for multidimensional sinusoids on a lattice. Notice that the one-dimensional signal is defined on the 1D lattice $\Lambda = \{nT \mid n\in\mathbb{Z}\}$. Also notice that the frequencies u and $u+r$ define the same discrete-time sinusoid if and only if $r\in\left\{k\cdot\frac{1}{T}\middle| k\in\mathbb{Z}\right\}$. We see that this is also a lattice that we call the reciprocal lattice Λ^*. This is the notion that extends in a straightforward fashion to higher-dimensional lattices.

The two multidimensional sinusoidal signals

$$f_1[\mathbf{x}] = B\exp(\mathrm{j}2\pi\mathbf{u}\cdot\mathbf{x})$$

$$\text{and} \quad f_2[\mathbf{x}] = B\exp(\mathrm{j}2\pi(\mathbf{u}+\mathbf{r})\cdot\mathbf{x})$$

will be identical if $\mathbf{r}\cdot\mathbf{x}$ is an integer for all $\mathbf{x}\in\Lambda$. The set of all such points $\mathbf{r}$ is called the reciprocal lattice:

$$\Lambda^* = \{\mathbf{r}\in\mathbb{R}^D \mid \mathbf{r}\cdot\mathbf{x}\in\mathbb{Z} \quad \text{for all} \quad \mathbf{x}\in\Lambda\}. \tag{3.14}$$

Theorem 3.1 If $\Lambda = \mathrm{LAT}(\mathbf{V})$, then

$$\{\mathbf{r} \mid \mathbf{r}\cdot\mathbf{x}\in\mathbb{Z} \quad \text{for all} \quad \mathbf{x}\in\Lambda\} = \mathrm{LAT}(\mathbf{V}^{-T}).$$

Proof: Let $\mathcal{R} = \{\mathbf{r} \mid \mathbf{r}\cdot\mathbf{x}\in\mathbb{Z} \quad \text{for all} \quad \mathbf{x}\in\Lambda\}$. Suppose that $\mathbf{r}\in\mathrm{LAT}(\mathbf{V}^{-T})$. Then $\mathbf{r} = \mathbf{V}^{-T}\mathbf{k}$ for some integer vector $\mathbf{k}\in\mathbb{Z}^D$. Thus,

$$\mathbf{r}\cdot\mathbf{x} = \mathbf{k}^T\mathbf{V}^{-1}\mathbf{V}\mathbf{n} = \mathbf{k}^T\mathbf{n}\in\mathbb{Z}$$

so that $\mathbf{r}\in\mathcal{R}$ and we can conclude that $\mathrm{LAT}(\mathbf{V}^{-T})\subset\mathcal{R}$.

Now suppose that $\mathbf{r}\in\mathcal{R}$. Let $\boldsymbol{\alpha} = \mathbf{V}^T\mathbf{r}$. Then

$$\mathbf{r}\cdot\mathbf{x} = \boldsymbol{\alpha}^T\mathbf{V}^{-1}\mathbf{V}\mathbf{n} = \boldsymbol{\alpha}^T\mathbf{n}.$$

This must be an integer for all possible $\mathbf{x} \in \Lambda$, and therefore for all possible $\mathbf{n} \in \mathbb{Z}^D$. In particular, if $\mathbf{n}$ is 1 in position k and zero elsewhere, then it follows that $\alpha_k \in \mathbb{Z}$ for any k, i.e. $\boldsymbol{\alpha} \in \mathbb{Z}^D$ and so $\mathbf{r} \in \mathrm{LAT}(\mathbf{V}^{-T})$. Thus $\mathcal{R} \subset \mathrm{LAT}(\mathbf{V}^{-T})$, and combining with the first result, $\mathcal{R} = \mathrm{LAT}(\mathbf{V}^{-T})$. □

The significance of the reciprocal lattice will become more apparent when we introduce the Fourier transform of signals defined on a lattice. Note that

$$\begin{aligned} d(\Lambda^*) &= |\det(\mathbf{V}^{-T})| = \frac{1}{|\det(\mathbf{V})|} \\ &= \frac{1}{d(\Lambda)} \end{aligned} \tag{3.15}$$

using standard properties of determinants. Thus, the density of the reciprocal lattice is the reciprocal of the density of the original lattice.

For the 2D examples presented previously, we have:

rectangular lattice

$$\begin{aligned} \mathbf{V} &= \begin{bmatrix} X & 0 \\ 0 & Y \end{bmatrix} & d(\Lambda) &= XY \\ \mathbf{V}^{-T} &= \begin{bmatrix} \frac{1}{X} & 0 \\ 0 & \frac{1}{Y} \end{bmatrix} & d(\Lambda^*) &= \frac{1}{XY} \end{aligned} \tag{3.16}$$

hexagonal lattice

$$\begin{aligned} \mathbf{V} &= \begin{bmatrix} X & \frac{X}{2} \\ 0 & Y \end{bmatrix} & d(\Lambda) &= XY \\ \mathbf{V}^{-T} &= \begin{bmatrix} \frac{1}{X} & 0 \\ -\frac{1}{2Y} & \frac{1}{Y} \end{bmatrix} & d(\Lambda^*) &= \frac{1}{XY}. \end{aligned} \tag{3.17}$$

It follows that a set of frequencies corresponding to *distinct* sinusoidal signals on a lattice Λ forms a unit cell of the reciprocal lattice Λ^*. For the hexagonal lattice we have considered, this is illustrated in Figure 3.7. For consistency with our convention in the spatial domain, the vertical frequency axis is oriented downward.

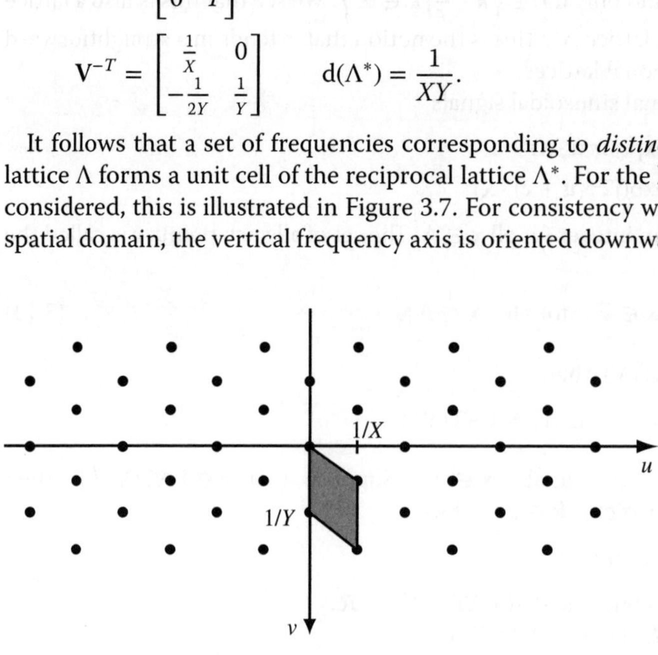

Figure 3.7 Reciprocal lattice for the hexagonal lattice with a possible unit cell (shaded).

3.6 Linear Systems Over Lattices

We now extend the familiar concept of linear systems to signals defined on a lattice. The notation of Section 2.5 for continuous space-time linear systems can be used largely unchanged. Thus, $\mathcal{S}$ denotes a vector space of signals defined on a lattice Λ and $\mathcal{H}$ is a system mapping signals in $\mathcal{S}$ to new signals in $\mathcal{S}$. The conditions for linearity are identical to the ones given in Equation (2.31) and Equation (2.32), which we repeat here for convenience. A system $\mathcal{H}$ is said to be linear if

$$\mathcal{H}(f_1 + f_2) = \mathcal{H}f_1 + \mathcal{H}f_2, \tag{3.18}$$

$$\mathcal{H}(\alpha f_1) = \alpha(\mathcal{H}f_1) \tag{3.19}$$

for all $f_1, f_2 \in \mathcal{S}$ and for all $\alpha \in \mathbb{R}$ (or for all $\alpha \in \mathbb{C}$ for complex signal spaces).

Similarly, we can define the shift operator

$$\mathcal{T}_{\mathbf{d}} : g = \mathcal{T}_{\mathbf{d}} f : g[\mathbf{x}] = f[\mathbf{x} - \mathbf{d}], \quad \mathbf{d} \in \Lambda, \tag{3.20}$$

where we impose the constraint that the shift $\mathbf{d}$ is an element of the lattice Λ. This ensures that for every $\mathbf{x} \in \Lambda$, $\mathbf{x} - \mathbf{d}$ is also in Λ and so the signal with values $f[\mathbf{x} - \mathbf{d}]$ is well defined. The shift operator is again a linear system. The definition of a linear shift-invariant system is also unchanged: $\mathcal{H}$ is shift-invariant if $\mathcal{H}(\mathcal{T}_{\mathbf{d}} f) = \mathcal{T}_{\mathbf{d}}(\mathcal{H}f)$ for any shift $\mathbf{d} \in \Lambda$.

3.6.1 Response of a Linear System

The response of a linear system to an arbitrary input is found in a similar fashion to continuous space-time systems. An arbitrary signal defined on the lattice Λ can be expressed as

$$f = \sum_{\mathbf{y} \in \Lambda} f[\mathbf{y}] \mathcal{T}_{\mathbf{y}} \delta \tag{3.21}$$

or alternatively

$$f[\mathbf{x}] = \sum_{\mathbf{y} \in \Lambda} f[\mathbf{y}] \delta[\mathbf{x} - \mathbf{y}]. \tag{3.22}$$

Applying linearity, for a linear system we have

$$\begin{aligned} \mathcal{H}f &= \sum_{\mathbf{y} \in \Lambda} f[\mathbf{y}] \mathcal{H}(\mathcal{T}_{\mathbf{y}} \delta) \\ &= \sum_{\mathbf{y} \in \Lambda} f[\mathbf{y}] h_{\mathbf{y}} \end{aligned} \tag{3.23}$$

where $h_{\mathbf{y}}$ is the response of the linear system to a unit sample at position $\mathbf{y}$.

For a linear shift-invariant system, we have $\mathcal{H}\mathcal{T}_{\mathbf{y}} = \mathcal{T}_{\mathbf{y}}\mathcal{H}$, and so

$$\mathcal{H}(\mathcal{T}_{\mathbf{y}} \delta) = \mathcal{T}_{\mathbf{y}} \mathcal{H} \delta = \mathcal{T}_{\mathbf{y}} h \tag{3.24}$$

where h is the unit sample response of the system. Written explicitly, if $g = \mathcal{H}f$, then

$$g[\mathbf{x}] = \sum_{\mathbf{y} \in \Lambda} f[\mathbf{y}] h[\mathbf{x} - \mathbf{y}]. \tag{3.25}$$

This expression defines *convolution* over a lattice, which we denote $g = f * h$. Once again, convolution is commutative, $f * h = h * f$, and so we also have

$$g[\mathbf{x}] = \sum_{\mathbf{y}\in\Lambda} h[\mathbf{y}] f[\mathbf{x} - \mathbf{y}]. \tag{3.26}$$

3.6.2 Frequency Response

Let the input to an LSI system $\mathcal{H}$ be a complex sinusoid of frequency $\mathbf{u}$,

$$f[\mathbf{x}] = \exp(\mathrm{j}2\pi\mathbf{u}\cdot\mathbf{x}), \qquad \mathbf{x}\in\Lambda.$$

Then, from Equation (3.26),

$$\begin{aligned} g[\mathbf{x}] &= \sum_{\mathbf{y}\in\Lambda} h[\mathbf{y}]\exp(\mathrm{j}2\pi\mathbf{u}\cdot(\mathbf{x}-\mathbf{y})) \\ &= \exp(\mathrm{j}2\pi\mathbf{u}\cdot\mathbf{x})\sum_{\mathbf{y}\in\Lambda} h[\mathbf{y}]\exp(-\mathrm{j}2\pi\mathbf{u}\cdot\mathbf{y}) \\ &= H(\mathbf{u})\exp(\mathrm{j}2\pi\mathbf{u}\cdot\mathbf{x}), \end{aligned} \tag{3.27}$$

where

$$H(\mathbf{u}) = \sum_{\mathbf{y}\in\Lambda} h[\mathbf{y}]\exp(-\mathrm{j}2\pi\mathbf{u}\cdot\mathbf{y}) \tag{3.28}$$

(assuming that the sum converges). Once again, the complex sinusoid is an eigenfunction of an LSI system over a lattice, with eigenvalue $H(\mathbf{u})$. We call $H(\mathbf{u})$, as a function of the frequency vector $\mathbf{u}$, the *frequency response* of the system. It is also the discrete-domain Fourier transform over the lattice Λ of the unit sample response $h[\mathbf{x}]$.

3.7 Discrete-Domain Fourier Transforms Over a Lattice

3.7.1 Definition of the Discrete-Domain Fourier Transform

Let $f[\mathbf{x}]$ be defined on the lattice Λ. We define the discrete-domain Fourier transform of $f[\mathbf{x}]$ by

$$F(\mathbf{u}) = \sum_{\mathbf{x}\in\Lambda} f[\mathbf{x}]\exp(-\mathrm{j}2\pi\mathbf{u}\cdot\mathbf{x}). \tag{3.29}$$

The property that $\exp(-\mathrm{j}2\pi\mathbf{u}\cdot\mathbf{x})$ and $\exp(-\mathrm{j}2\pi(\mathbf{u}+\mathbf{r})\cdot\mathbf{x})$ are the same signal if $\mathbf{r}\in\Lambda^*$ determines the periodicity property of the Fourier transform on Λ:

$$F(\mathbf{u}) = F(\mathbf{u}+\mathbf{r}) \quad \text{for all} \quad \mathbf{r}\in\Lambda^*. \tag{3.30}$$

We say that $F(\mathbf{u})$ is periodic with periodicity lattice Λ^*. It follows that $F(\mathbf{u})$ is completely determined by its values over one unit cell of Λ^*, which we denote $\mathcal{P}^*$ (or sometimes $\mathcal{P}_{\Lambda^*}$).

It can be shown that the inverse Fourier transform is given by

$$\begin{aligned} f[\mathbf{x}] &= \frac{1}{|\mathcal{P}^*|}\int_{\mathcal{P}^*} F(\mathbf{u})\exp(\mathrm{j}2\pi\mathbf{u}\cdot\mathbf{x})\,\mathrm{d}\mathbf{u} \\ &= \mathrm{d}(\Lambda)\int_{\mathcal{P}^*} F(\mathbf{u})\exp(\mathrm{j}2\pi\mathbf{u}\cdot\mathbf{x})\,\mathrm{d}\mathbf{u}. \end{aligned} \tag{3.31}$$

The notation $\int_{\mathcal{P}^*}$ means integration over one unit cell of the reciprocal lattice Λ^*. The reason for this form of the inverse Fourier transform will be discussed in more detail in Chapter 6. We denote that $f[\mathbf{x}]$ and $F(\mathbf{u})$ form a discrete-domain Fourier transform pair by $f[\mathbf{x}] \overset{\text{DDFT}}{\longleftrightarrow} F(\mathbf{u})$.

Consider the most common case of a 2D rectangular lattice

$$\Lambda = \text{LAT}\left(\begin{bmatrix} X & 0 \\ 0 & Y \end{bmatrix}\right), \qquad \Lambda^* = \text{LAT}\left(\begin{bmatrix} \frac{1}{X} & 0 \\ 0 & \frac{1}{Y} \end{bmatrix}\right).$$

Then

$$F(u, v) = \sum_{n_1=-\infty}^{\infty} \sum_{n_2=-\infty}^{\infty} f[n_1X, n_2Y] \exp(-\text{j}2\pi(un_1X + vn_2Y)). \tag{3.32}$$

A suitable unit cell of Λ^* is

$$\mathcal{P}^* = \left\{ (u, v) \,\middle|\, -\frac{1}{2X} \le u \le \frac{1}{2X}, -\frac{1}{2Y} \le v \le \frac{1}{2Y} \right\} \tag{3.33}$$

and so the inverse Fourier transform can be written

$$f[n_1X, n_2Y] = XY \int_{-\frac{1}{2X}}^{\frac{1}{2X}} \int_{-\frac{1}{2Y}}^{\frac{1}{2Y}} F(u, v) \exp(\text{j}2\pi(un_1X + vn_2Y))\, du dv. \tag{3.34}$$

It is common in the literature to write

$$f'[n_1, n_2] = f[n_1X, n_2Y]$$

and to use the normalized frequency variables $\omega_1 = 2\pi uX$ and $\omega_2 = 2\pi vY$. We then find

$$\begin{aligned} F'(\omega_1, \omega_2) &= F\left(\frac{\omega_1}{2\pi X}, \frac{\omega_2}{2\pi Y}\right) \\ &= \sum_{n_1=-\infty}^{\infty} \sum_{n_2=-\infty}^{\infty} f'[n_1, n_2] \exp(-\text{j}(\omega_1 n_1 + \omega_2 n_2)) \\ &- \pi \le \omega_1, \omega_2 \le \pi. \end{aligned}$$

Although this simplifies some expressions, we lose the physical dimensions of space and spatial frequency. Furthermore, if $X \neq Y$ as is often the case (e.g., television), we also introduce geometric distortion. For these reasons, we prefer the non-normalized definition of the Fourier transform in this book.

3.7.2 Properties of the Multidimensional Fourier Transform Over a Lattice Λ

The Fourier transform over a lattice has very similar properties to the continuous-domain Fourier transform, as well as to the one-dimensional discrete-time Fourier transform. In fact, these are all just different versions of properties of a general Fourier transform in an algebraic setting. Table 3.1 lists some of the most commonly used properties.

Although for the most part, the proofs are very similar to the proofs for the continuous-domain case, we provide them here for completeness. The discrete-domain signals involved are all assumed to be absolutely summable, so that the Fourier transforms involved are all well-defined.

Table 3.1 Properties of the multidimensional Fourier transform over a lattice Λ.

	$f[\mathbf{x}] = d(\Lambda)\int_{\mathcal{P}^*} F(\mathbf{u})\exp(j2\pi\mathbf{u}\cdot\mathbf{x})\,d\mathbf{u}$	$F(\mathbf{u}) = \sum_{\mathbf{x}\in\Lambda} f[\mathbf{x}]\exp(-j2\pi\mathbf{u}\cdot\mathbf{x})$
(3.1)	$Af[\mathbf{x}] + Bg[\mathbf{x}]$	$AF(\mathbf{u}) + BG(\mathbf{u})$
(3.2)	$f[\mathbf{x}-\mathbf{x}_0]$	$F(\mathbf{u})\exp(-j2\pi\mathbf{u}\cdot\mathbf{x}_0)$
(3.3)	$f[\mathbf{x}]\exp(j2\pi\mathbf{u}_0\cdot\mathbf{x})$	$F(\mathbf{u}-\mathbf{u}_0)$
(3.4)	$f[\mathbf{x}] * g[\mathbf{x}]$	$F(\mathbf{u})G(\mathbf{u})$
(3.5)	$f[\mathbf{x}]g[\mathbf{x}]$	$d(\Lambda)\int_{\mathcal{P}^*} F(\mathbf{r})G(\mathbf{u}-\mathbf{r})\,d\mathbf{r}$
(3.6)	$f[\mathbf{A}\mathbf{x}]$	$F(\mathbf{A}^{-T}\mathbf{u})$
(3.7)	$\mathbf{x}f[\mathbf{x}]$	$\frac{j}{2\pi}\nabla_{\mathbf{u}}F(\mathbf{u})$
(3.8)	$f^*[\mathbf{x}]$	$F^*(-\mathbf{u})$
(3.9)	$\tilde{F}[\mathbf{x}]$	$d(\Gamma)\tilde{f}(-\mathbf{u})$
(3.10)	$\sum_{\mathbf{x}\in\Lambda} f[\mathbf{x}]g^*[\mathbf{x}] = d(\Lambda)\int_{\mathcal{P}^*} F(\mathbf{u})G^*(\mathbf{u})\,d\mathbf{u}$	

Property 3.1 *Linearity*: $Af[\mathbf{x}] + Bg[\mathbf{x}] \overset{\text{DDFT}}{\longleftrightarrow} AF(\mathbf{u}) + BG(\mathbf{u})$.

Proof: Let $q[\mathbf{x}] = Af[\mathbf{x}] + Bg[\mathbf{x}]$. Then

$$\begin{aligned}
Q(\mathbf{u}) &= \sum_{\mathbf{x}\in\Lambda}(Af[\mathbf{x}] + Bg[\mathbf{x}])\exp(-j2\pi\mathbf{u}\cdot\mathbf{x}) \\
&= A\sum_{\mathbf{x}\in\Lambda} f[\mathbf{x}]\exp(-j2\pi\mathbf{u}\cdot\mathbf{x}) + B\sum_{\mathbf{x}\in\Lambda} g[\mathbf{x}]\exp(-j2\pi\mathbf{u}\cdot\mathbf{x}) \\
&= AF(\mathbf{u}) + BG(\mathbf{u}).
\end{aligned} \tag{3.35}$$

□

Property 3.2 *Shift*: $f[\mathbf{x}-\mathbf{x}_0] \overset{\text{DDFT}}{\longleftrightarrow} F(\mathbf{u})\exp(-j2\pi\mathbf{u}\cdot\mathbf{x}_0)$.

Proof: Let $g[\mathbf{x}] = f[\mathbf{x}-\mathbf{x}_0]$ for some $\mathbf{x}_0 \in \Lambda$. Then

$$\begin{aligned}
G(\mathbf{u}) &= \sum_{\mathbf{x}\in\Lambda} f[\mathbf{x}-\mathbf{x}_0]\exp(-j2\pi\mathbf{u}\cdot\mathbf{x}) \\
&= \sum_{\mathbf{s}\in\Lambda} f[\mathbf{s}]\exp(-j2\pi\mathbf{u}\cdot(\mathbf{s}+\mathbf{x}_0)) \quad (\mathbf{s} = \mathbf{x}-\mathbf{x}_0) \\
&= \left(\sum_{\mathbf{s}\in\Lambda} f[\mathbf{s}]\exp(-j2\pi\mathbf{u}\cdot\mathbf{s})\right)\exp(-j2\pi\mathbf{u}\cdot\mathbf{x}_0) \\
&= F(\mathbf{u})\exp(-j2\pi\mathbf{u}\cdot\mathbf{x}_0).
\end{aligned} \tag{3.36}$$

□

Property 3.3 *Modulation*: $f[\mathbf{x}]\exp(j2\pi\mathbf{u}_0\cdot\mathbf{x}) \overset{\text{DDFT}}{\longleftrightarrow} F(\mathbf{u}-\mathbf{u}_0)$.

Proof: Let $g[\mathbf{x}] = f[\mathbf{x}]\exp(j2\pi\mathbf{u}_0\cdot\mathbf{x})$ for some $\mathbf{u}_0 \in \mathbb{R}^D$. Then

$$G(\mathbf{u}) = \sum_{\mathbf{x}\in\Lambda} f[\mathbf{x}]\exp(j2\pi\mathbf{u}_0\cdot\mathbf{x})\exp(-j2\pi\mathbf{u}\cdot\mathbf{x})$$

$$= \sum_{\mathbf{x}\in\Lambda} f[\mathbf{x}] \exp(-\mathrm{j}2\pi(\mathbf{u}-\mathbf{u}_0)\cdot\mathbf{x})$$
$$= F(\mathbf{u}-\mathbf{u}_0). \tag{3.37}$$

□

Property 3.4 *Convolution*: $f[\mathbf{x}] * g[\mathbf{x}] \xleftrightarrow{\text{DDFT}} F(\mathbf{u})G(\mathbf{u})$.

Proof: Let $q[\mathbf{x}] = f[\mathbf{x}] * g[\mathbf{x}] = \sum_{\mathbf{s}\in\Lambda} f[\mathbf{s}]g[\mathbf{x}-\mathbf{s}]$. Then

$$\begin{aligned}
Q(\mathbf{u}) &= \sum_{\mathbf{x}\in\Lambda}\left(\sum_{\mathbf{s}\in\Lambda} f[\mathbf{s}]g[\mathbf{x}-\mathbf{s}]\right)\exp(-\mathrm{j}2\pi\mathbf{u}\cdot\mathbf{x}) \\
&= \sum_{\mathbf{s}\in\Lambda}\sum_{\mathbf{t}\in\Lambda} f[\mathbf{s}]g[\mathbf{t}]\exp(-\mathrm{j}2\pi\mathbf{u}\cdot(\mathbf{t}+\mathbf{s}) \quad (\mathbf{t}=\mathbf{x}-\mathbf{s}) \\
&= \sum_{\mathbf{s}\in\Lambda} f[\mathbf{s}]\exp(-\mathrm{j}2\pi\mathbf{u}\cdot\mathbf{s})\sum_{\mathbf{t}\in\Lambda} g[\mathbf{t}]\exp(-\mathrm{j}2\pi\mathbf{u}\cdot\mathbf{t}) \\
&= F(\mathbf{u})G(\mathbf{u}).
\end{aligned} \tag{3.38}$$

□

Note that using this property, commutativity and associativity of complex multiplication imply commutativity and associativity of convolution.

Property 3.5 *Multiplication*: $f[\mathbf{x}]g[\mathbf{x}] \xleftrightarrow{\text{DDFT}} F(\mathbf{u}) * G(\mathbf{u})$.

Proof: Let $q[\mathbf{x}] = f[\mathbf{x}]g[\mathbf{x}]$. Then

$$\begin{aligned}
Q(\mathbf{u}) &= \sum_{\mathbf{x}\in\Lambda} f[\mathbf{x}]g[\mathbf{x}]\exp(-\mathrm{j}2\pi\mathbf{u}\cdot\mathbf{x}) \\
&= \sum_{\mathbf{x}\in\Lambda} g[\mathbf{x}]\left(d(\Lambda)\int_{\mathcal{P}^*} F(\mathbf{w})\exp(\mathrm{j}2\pi\mathbf{w}\cdot\mathbf{x})\,d\mathbf{w}\right)\exp(-\mathrm{j}2\pi\mathbf{u}\cdot\mathbf{x}) \\
&= d(\Lambda)\int_{\mathcal{P}^*} F(\mathbf{w})\left(\sum_{\mathbf{x}\in\Lambda} g[\mathbf{x}]\exp(-\mathrm{j}2\pi(\mathbf{u}-\mathbf{w})\cdot\mathbf{x})\right)d\mathbf{w} \\
&= d(\Lambda)\int_{\mathcal{P}^*} F(\mathbf{w})G(\mathbf{u}-\mathbf{w})\,d\mathbf{w} = F(\mathbf{u}) * G(\mathbf{u}).
\end{aligned} \tag{3.39}$$

Note that the convolution in the frequency domain is of continuous-domain periodic functions, with periodicity given by the reciprocal lattice. Continuous-domain periodic signals will be studied in Chapter 5. □

Property 3.6 *Automorphism of the domain*: if $\mathbf{A}\Lambda = \Lambda$, then $f[\mathbf{A}\mathbf{x}] \xleftrightarrow{\text{DDFT}} F(\mathbf{A}^{-T}\mathbf{u})$.

Proof: The matrix $\mathbf{A}$ is constrained to satisfy $\{\mathbf{A}\mathbf{x} \mid \mathbf{x}\in\Lambda\} = \Lambda$, which implies that $\Lambda = \mathrm{LAT}(\mathbf{A}\mathbf{V})$. Thus $\mathbf{A}\mathbf{V} = \mathbf{V}\mathbf{E}$ where $\mathbf{E}$ is a unimodular integer matrix. It follows that $|\det(\mathbf{A})| = 1$ and that $\mathbf{A}$ must be of the form $\mathbf{A} = \mathbf{V}\mathbf{E}\mathbf{V}^{-1}$ for unimodular $\mathbf{E}$. Thus, the set of admissible matrices $\mathbf{A}$ is in one-to-one correspondence with the set of integer unimodular matrices. Now, let $g[\mathbf{x}] = f[\mathbf{A}\mathbf{x}]$ for an admissible $\mathbf{A}$. Then

$$
\begin{aligned}
G(\mathbf{u}) &= \sum_{\mathbf{x}\in\Lambda} f[\mathbf{A}\mathbf{x}] \exp(-\mathrm{j}2\pi\mathbf{u}\cdot\mathbf{x}) \\
&= \sum_{\mathbf{s}\in\Lambda} f[\mathbf{s}] \exp(-\mathrm{j}2\pi\mathbf{u}\cdot(\mathbf{A}^{-1}\mathbf{s})) \\
&= \sum_{\mathbf{s}\in\Lambda} f[\mathbf{s}] \exp(-\mathrm{j}2\pi(\mathbf{A}^{-T}\mathbf{u})\cdot\mathbf{s}) \\
&= F(\mathbf{A}^{-T}\mathbf{u}).
\end{aligned} \tag{3.40}
$$

□

Property 3.7 *Differentiation in frequency*: $\mathbf{x}f[\mathbf{x}] \overset{\text{DDFT}}{\longleftrightarrow} \frac{j}{2\pi}\nabla_{\mathbf{u}}F(\mathbf{u})$.

Proof: Let $\mathbf{g}[\mathbf{x}] = \mathbf{x}f[\mathbf{x}]$. Taking the derivative of the analysis equation for f with respect to u_i,

$$
\begin{aligned}
\frac{\partial F}{\partial u_i}(\mathbf{u}) &= \sum_{\mathbf{x}\in\Lambda} f[\mathbf{x}](-j2\pi x_i) \exp(-\mathrm{j}2\pi\mathbf{u}\cdot\mathbf{x}) \\
&\quad - j2\pi \sum_{\mathbf{x}\in\Lambda} g_i(\mathbf{x}) \exp(-\mathrm{j}2\pi\mathbf{u}\cdot\mathbf{x}) \\
&= -j2\pi G_i(\mathbf{u}).
\end{aligned} \tag{3.41}
$$

Thus, $\mathbf{G}(\mathbf{u}) = \frac{j}{2\pi}\nabla_{\mathbf{u}}F(\mathbf{u})$. □

Property 3.8 *Complex conjugation*: $f^*[\mathbf{x}] \overset{\text{DDFT}}{\longleftrightarrow} F^*(-\mathbf{u})$

Proof: Let $g[\mathbf{x}] = f^*[\mathbf{x}]$. Then

$$
\begin{aligned}
G(\mathbf{u}) &= \sum_{\mathbf{x}\in\Lambda} f^*[\mathbf{x}] \exp(-\mathrm{j}2\pi\mathbf{u}\cdot\mathbf{x}) \\
&= \sum_{\mathbf{x}\in\Lambda} (f[\mathbf{x}] \exp(\mathrm{j}2\pi\mathbf{u}\cdot\mathbf{x}))^* \\
&= \left(\sum_{\mathbf{x}\in\Lambda} f[\mathbf{x}] \exp(\mathrm{j}2\pi\mathbf{u}\cdot\mathbf{x})\right)^* \\
&= F^*(-\mathbf{u}).
\end{aligned} \tag{3.42}
$$

□

Property 3.9 *Duality*: if $\tilde{f}(\mathbf{x}) \overset{\text{CDFS}}{\longleftrightarrow} \tilde{F}[\mathbf{u}]$ then $\tilde{F}[\mathbf{x}] \overset{\text{DDFT}}{\longleftrightarrow} d(\Gamma)\tilde{f}(-\mathbf{u})$.

Proof: Since the Fourier transform for periodic continuous-domain signals has not been introduced yet, the meaning and proof of this property is deferred to Chapter 5. □

Property 3.10 *Parseval relation*: $\sum_{\mathbf{x}\in\Lambda} f[\mathbf{x}]g^*[\mathbf{x}] = d(\Lambda)\int_{\mathcal{P}^*} F(\mathbf{u})G^*(\mathbf{u})\,d\mathbf{u}$

Proof: Let $r[\mathbf{x}] = g^*[\mathbf{x}]$ and $q[\mathbf{x}] = f[\mathbf{x}]r[\mathbf{x}]$. Then

$$\begin{aligned} Q(\mathbf{u}) &= \int_{\mathcal{P}^*} F(\mathbf{w})R(\mathbf{u}-\mathbf{w})\, d\mathbf{w} \\ &= \int_{\mathcal{P}^*} F(\mathbf{w})G^*(\mathbf{w}-\mathbf{u})\, d\mathbf{w} \end{aligned} \tag{3.43}$$

Evaluating at $\mathbf{u} = \mathbf{0}$,

$$Q(\mathbf{0}) = \sum_{\mathbf{x}\in\Lambda} f[\mathbf{x}]g^*[\mathbf{x}] = d(\Lambda)\int_{\mathcal{P}^*} F(\mathbf{w})G^*(\mathbf{w})\, d\mathbf{w}. \tag{3.44}$$

□

If $f = g$, we obtain

$$\sum_{\mathbf{x}\in\Lambda} |f[\mathbf{x}]|^2 = d(\Lambda)\int_{\mathcal{P}^*} |F(\mathbf{w})|^2\, d\mathbf{w}. \tag{3.45}$$

3.7.3 Evaluation of Forward and Inverse Discrete-Domain Fourier Transforms

Many of the discrete-domain signals of interest are of finite extent, and the Fourier transform is obtained by direct evaluation of the defining sum (Equation 3.29). Examples are given in the next section on finite impulse response filters. In a few cases, the sum can be evaluated analytically, and we give a few examples here.

Example 3.1
Compute the 2D Fourier transform of the discrete rect function p_{MN} on a rectangular lattice with horizontal and vertical spacing X and Y:

$$p_{MN}[mX, nY] = \begin{cases} 1 & -M \le m \le M \text{ and } -N \le n \le N \\ 0 & \text{otherwise} \end{cases}.$$

Solution:

$$\begin{aligned} P_{MN}(u, v) &= \sum_{m=-M}^{M}\sum_{n=-N}^{N} \exp(-j2\pi(umX + vnY)) \\ &= \sum_{m=-M}^{M} \exp(-j2\pi umX) \sum_{n=-N}^{N} \exp(-j2\pi vnY), \end{aligned}$$

which is separable. Both terms have the same form. Let

$$S = \sum_{k=-L}^{L} \exp(-j2\pi wkZ),$$

which is seen to be a geometric series of the form

$$a + ar + ar^2 + \cdots + ar^{n-1} = \frac{a(1-r^n)}{1-r} = \frac{a - r\ell}{1-r}.$$

Here, $a = \exp(\mathrm{j}2\pi wLZ)$, $r = \exp(-\mathrm{j}2\pi wZ)$ and $\ell = ar^{n-1} = \exp(-\mathrm{j}2\pi wLZ)$. Thus

$$\begin{aligned}S &= \frac{\exp(\mathrm{j}2\pi wLZ) - \exp(-\mathrm{j}2\pi wZ)\exp(-\mathrm{j}2\pi wLZ)}{1-\exp(-\mathrm{j}2\pi wZ)}\\ &= \frac{\exp(-\mathrm{j}\pi wZ)(\exp(\mathrm{j}\pi w(2L+1)Z) - \exp(-\mathrm{j}\pi w(2L+1)Z)}{\exp(-\mathrm{j}\pi wZ)(\exp(\mathrm{j}\pi wZ) - \exp(-\mathrm{j}\pi wZ))}\\ &= \frac{2j\sin(\pi w(2L+1)Z)}{2j\sin(\pi wZ)}.\end{aligned}$$

Substituting in the two terms

$$P_{MN}(u,v) = \frac{\sin(\pi u(2M+1)X)\sin(\pi v(2N+1)Y)}{\sin(\pi uX)\sin(\pi vY)}.$$

□

Example 3.2

Let $\Lambda = \mathrm{LAT}\left(\left[\begin{smallmatrix}2&1\\0&1\end{smallmatrix}\right]\right)$ and $f[x,y] = ca^{|x+y|}b^{|x-y|}$ for $(x,y)\in\Lambda$, with $|a|<1$, $|b|<1$. Find c so that $F(0,0)=1$, and find and plot $F(u,v)$ for this value of c, and $a=b=0.7$, for $-0.75 \le u,v \le 0.75$. The spatial unit is the px and so spatial frequencies are in units of c/px.

Solution:

We note that Λ is a hexagonal lattice, and the points of Λ can be explicitly enumerated as

$$\Lambda = \left\{\begin{bmatrix}2&1\\0&1\end{bmatrix}\begin{bmatrix}n_1\\n_2\end{bmatrix} = \begin{bmatrix}2n_1+n_2\\n_2\end{bmatrix} \mid n_1,n_2\in\mathbb{Z}\right\}. \tag{3.46}$$

Then the Fourier transform is given by

$$\begin{aligned}F(u,v) &= \sum_{(x,y)\in\Lambda} f[x,y]\exp(-\mathrm{j}2\pi(ux+vy))\\ &= \sum_{n_1=-\infty}^{\infty}\sum_{n_2=-\infty}^{\infty} f[2n_1+n_2,n_2]\exp(-\mathrm{j}2\pi(u(2n_1+n_2)+vn_2))\\ &= c\sum_{n_1=-\infty}^{\infty}\sum_{n_2=-\infty}^{\infty} a^{2|n_1+n_2|}b^{2|n_1|}\exp(-\mathrm{j}2\pi(u(2n_1+n_2)+vn_2))\\ &= c\sum_{n_1=-\infty}^{\infty} b^{2|n_1|}\exp(-\mathrm{j}2\pi u2n_1)\sum_{n_2=-\infty}^{\infty} a^{2|n_1+n_2|}\exp(-\mathrm{j}2\pi(u+v)n_2).\end{aligned} \tag{3.47}$$

For each fixed n_1 in the first sum, we let $m_2 = n_1+n_2$ in the second sum, where m_2 also runs over all of $\mathbb{Z}$. Then

$$\begin{aligned}F(u,v) &= c\sum_{n_1=-\infty}^{\infty} b^{2|n_1|}\exp(-\mathrm{j}2\pi u2n_1)\sum_{m_2=-\infty}^{\infty} a^{2|m_2|}\exp(-\mathrm{j}2\pi(u+v)(m_2-n_1))\\ &= c\sum_{n_1=-\infty}^{\infty} b^{2|n_1|}\exp(-\mathrm{j}2\pi(u-v)n_1)\sum_{m_2=-\infty}^{\infty} a^{2|m_2|}\exp(-\mathrm{j}2\pi(u+v)m_2).\end{aligned} \tag{3.48}$$

Both sums have the same form, $S = \sum_{n=-\infty}^{\infty} \alpha^{|n|} \exp(-j2\pi wn)$, where $|\alpha| < 1$. We evaluate this by breaking it up into three terms, two of which are geometric series.

$$\begin{aligned} S &= 1 + \sum_{n=1}^{\infty} (\alpha e^{-j2\pi w})^n + \sum_{n=-1}^{-\infty} (\alpha e^{j2\pi w})^{-n} \\ &= 1 + \sum_{n=1}^{\infty} (\alpha e^{-j2\pi w})^n + \sum_{m=1}^{\infty} (\alpha e^{j2\pi w})^m. \end{aligned} \tag{3.49}$$

These geometric series converge since $|\alpha e^{\pm j2\pi w}| = |\alpha| < 1$. Then,

$$\begin{aligned} S &= 1 + \frac{\alpha \exp(-j2\pi w)}{1 - \alpha \exp(-j2\pi w)} + \frac{\alpha \exp(j2\pi w)}{1 - \alpha \exp(j2\pi w)} \\ &= \frac{1 - \alpha^2}{1 + \alpha^2 - 2\alpha \cos(2\pi w)}. \end{aligned} \tag{3.50}$$

Inserting this expression in the formula for $F(u, v)$, we find

$$F(u, v) = c \frac{1 - a^4}{1 + a^4 - 2a^2 \cos(2\pi(u + v))} \frac{1 - b^4}{1 + b^4 - 2b^2 \cos(2\pi(u - v))}. \tag{3.51}$$

Evaluating $F(0, 0)$, we find

$$F(0, 0) = c \frac{(1 + a^2)(1 + b^2)}{(1 - a^2)(1 - b^2)} = 1, \tag{3.52}$$

so that we choose

$$c = \frac{(1 - a^2)(1 - b^2)}{(1 + a^2)(1 + b^2)}. \tag{3.53}$$

This gives the final result

$$F(u, v) = \frac{(1 - a^2)^2(1 - b^2)^2}{(1 + a^4 - 2a^2 \cos(2\pi(u + v)))(1 + b^4 - 2b^2 \cos(2\pi(u - v)))}. \tag{3.54}$$

Figure 3.8 shows the perspective view and the contour plot of $F(u, v)$ over the requested region. We note that $\Lambda^* = \text{LAT}\left(\begin{bmatrix} 0.5 & 0 \\ -0.5 & 1 \end{bmatrix}\right)$ and $F(u, v)$ is Λ^*-periodic, where one period is a unit cell of Λ^*. The figure shows the Voronoi unit cell of Λ^*, a diamond shaped region. □

3.8 Finite Impulse Response (FIR) Filters

A multidimensional filter over a lattice Λ is said to be a finite impulse filter if its unit sample response $h[\mathbf{x}]$ has a finite number of nonzero values, i.e. the set

$$\mathcal{A} = \{\mathbf{x} \in \Lambda \mid h[\mathbf{x}] \neq 0\} \tag{3.55}$$

has a finite number of elements. In this case

$$H(\mathbf{u}) = \sum_{\mathbf{x} \in \mathcal{A}} h[\mathbf{x}] \exp(-j2\pi \mathbf{u} \cdot \mathbf{x}). \tag{3.56}$$

The set $\mathcal{A}$ is called the region of support of the FIR filter. If $\mathcal{A}$ has an infinite number of elements, the filter is called infinite impulse response (IIR). In image processing, FIR

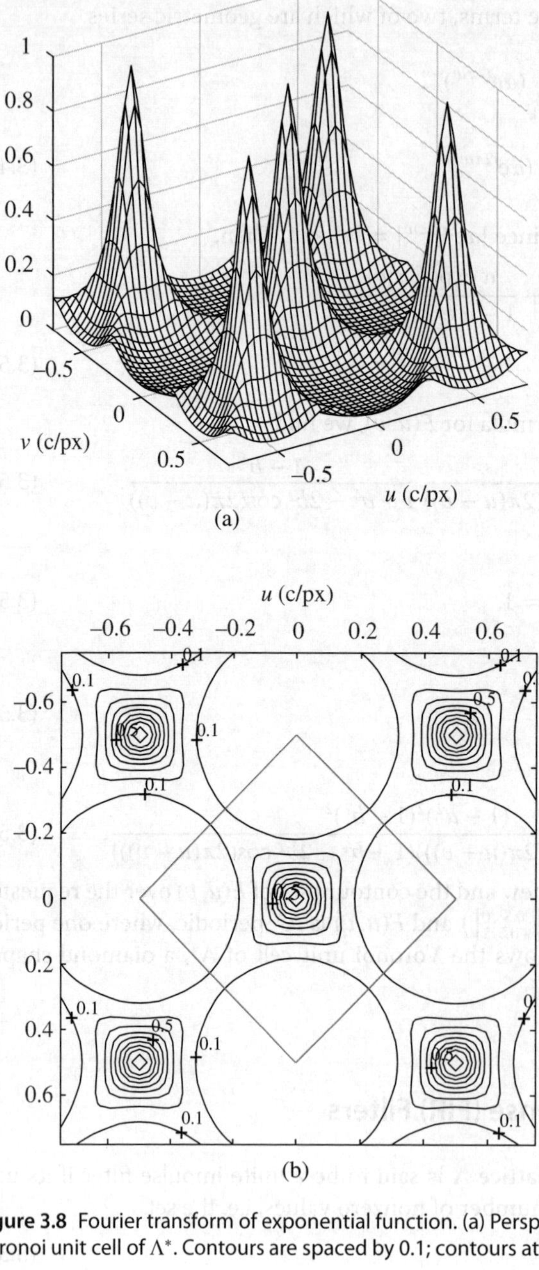

Figure 3.8 Fourier transform of exponential function. (a) Perspective plot. (b) Contour plot, showing a Voronoi unit cell of Λ^*. Contours are spaced by 0.1; contours at levels 0.1 and 0.5 are labelled.

filters are used almost exclusively, mainly because they do not have stability problems and they can be designed to have linear phase or, more often, zero phase.

Example 3.3 Let $\Lambda = \text{LAT}\left(\left[\begin{smallmatrix} X & 0 \\ 0 & X \end{smallmatrix}\right]\right)$ and

$$h[n_1X, n_2X] = \begin{cases} \frac{1}{2} & \text{if } n_1 = n_2 = 0; \\ \frac{1}{8} & \text{if } (n_1, n_2) = (\pm 1, 0) \text{ or } (0, \pm 1); \\ 0 & \text{otherwise.} \end{cases}$$

The unit sample response is illustrated in Figure 3.9.

If the input to the filter is $f[\mathbf{x}]$, the output $g[\mathbf{x}]$ can be computed by direct evaluation of the convolution

$$g[\mathbf{x}] = \sum_{\mathbf{s}\in\Lambda} h[\mathbf{s}]f[\mathbf{x} - \mathbf{s}]$$

where there are only five nonzero terms in the summation. Explicitly

$$g[x, y] = \frac{1}{2}f[x, y] + \frac{1}{8}f[x - X, y] + \frac{1}{8}f[x + X, y] + \frac{1}{8}f[x, y - X] + \frac{1}{8}f[x, y + X], \qquad (x, y) \in \Lambda$$

or alternatively,

$$g[n_1X, n_2X] = \frac{1}{2}f[n_1X, n_2X] + \frac{1}{8}f[(n_1 - 1)X, n_2X] + \frac{1}{8}f[(n_1 + 1)X, n_2X] + \frac{1}{8}f[n_1X, (n_2 - 1)X] + \frac{1}{8}f[n_1X, (n_2 + 1)X],$$
$$(n_1, n_2) \in \mathbb{Z}^2.$$

The frequency response of this filter is

$$H(\mathbf{u}) = \sum_{\mathbf{x}\in\Lambda} h[\mathbf{x}] \exp(-\mathrm{j}2\pi\mathbf{u} \cdot \mathbf{x})$$

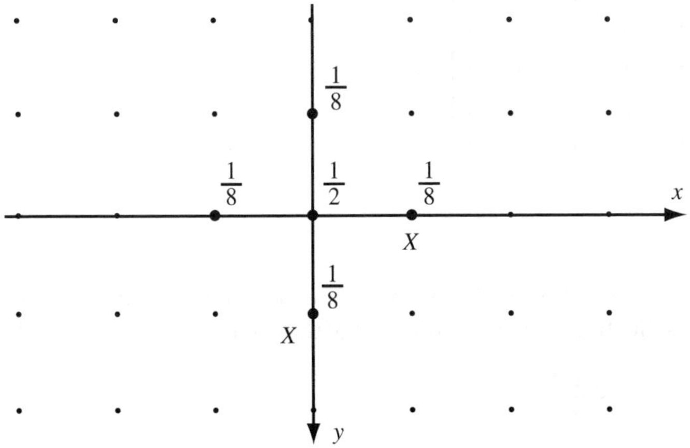

Figure 3.9 Unit sample response of a simple two-dimensional FIR filter. Small dots have a value of 0.0.

or alternatively

$$H(u,v) = \sum_{n_1=-\infty}^{\infty} \sum_{n_2=-\infty}^{\infty} h[n_1X, n_2X] \exp(-j2\pi(un_1X + vn_2X))$$

where there are only five nonzero terms in the double summation. Thus,

$$\begin{aligned} H(u,v) &= \frac{1}{2} + \frac{1}{8}\exp(-j2\pi uX) + \frac{1}{8}\exp(j2\pi uX) + \frac{1}{8}\exp(-j2\pi vX) \\ &\quad + \frac{1}{8}\exp(j2\pi vX) \\ &= \frac{1}{2} + \frac{1}{4}\cos(2\pi uX) + \frac{1}{4}\cos(2\pi vX). \end{aligned}$$

Note that $H(0,0) = 1$. A constant input is unchanged by this filter; we say that the *DC gain* of the filter is one. Also note that $H(\frac{1}{2X}, \frac{1}{2X}) = 0$. The input $f[n_1X, n_2X] = \cos(\pi(n_1 + n_2)) = (-1)^{n_1+n_2}$ yields an identically zero output. The frequency response of this filter is real and positive, so $|H(u,v)| = H(u,v)$ and $\angle H(u,v) = 0$. We call this a zero-phase filter.

Any real FIR filter satisfying the symmetry condition $h[\mathbf{x}] = h[-\mathbf{x}]$ will have a purely real frequency response, since combining terms in the expression for the frequency response gives

$$\begin{aligned} &h[\mathbf{x}]\exp(-j2\pi\mathbf{u}\cdot\mathbf{x}) + h[-\mathbf{x}]\exp(-j2\pi\mathbf{u}\cdot(-\mathbf{x})) \\ &\quad = h[\mathbf{x}]\exp(-j2\pi\mathbf{u}\cdot\mathbf{x}) + h[\mathbf{x}]\exp(j2\pi\mathbf{u}\cdot\mathbf{x}) \\ &\quad = 2h[\mathbf{x}]\cos(2\pi\mathbf{u}\cdot\mathbf{x}) \end{aligned}$$

which is real, so the overall frequency response is real.

Example 3.4 Another example of a 2D FIR filter on the integer sampling lattice $\mathbb{Z}^2$ has unit sample response

$$h[n_1,n_2] = \begin{array}{c|ccccc} & -2 & -1 & 0 & 1 & 2 \;\; n_1 \\ -2 & 1 & 1 & 1 & 1 & 1 \\ -1 & 1 & 2 & 2 & 2 & 1 \\ 0 & 1 & 2 & 3 & 2 & 1 \\ 1 & 1 & 2 & 2 & 2 & 1 \\ 2 & 1 & 1 & 1 & 1 & 1 \\ n_2 & & & & & \end{array} /35.$$

□

The frequency response of this filter is shown in perspective and contour views in Figure 3.10. The response is illustrated over one unit cell of the reciprocal lattice, namely

$$\mathcal{P}^* = \left\{ (u,v) \,\middle|\, -\frac{1}{2} \le u,v \le \frac{1}{2} \right\},$$

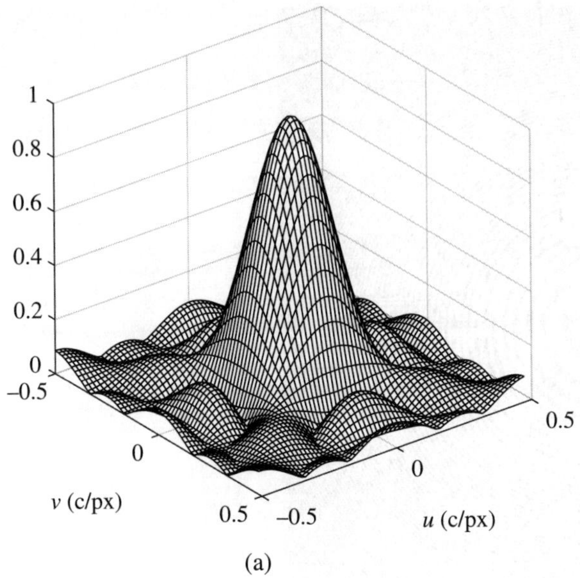

(a)

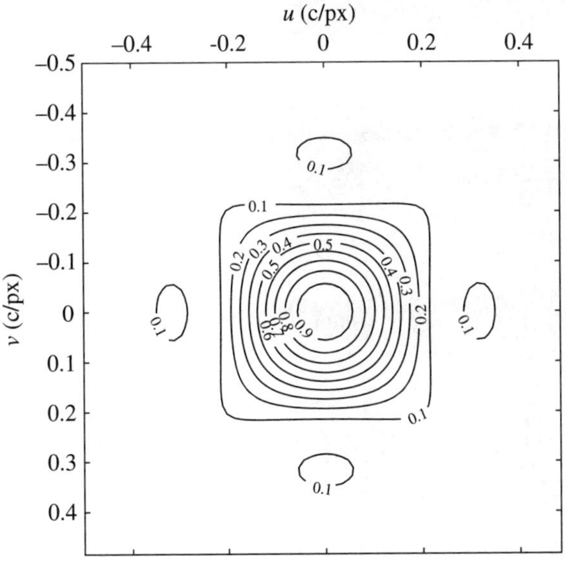

(b)

Figure 3.10 Frequency response of the FIR filter of Example 3.4. (a) Perspective view. (b) Contour plot.

where spatial frequency units are c/px. Figure 3.11 shows the result of filtering a zoneplate test pattern with this filter, while the result of filtering the Barbara image is shown in Figure 3.12. The low-pass characteristic of the filter can be clearly seen from the result on the zoneplate. The result on the Barbara image demonstrates the blurring effect associated with a low-pass filter.

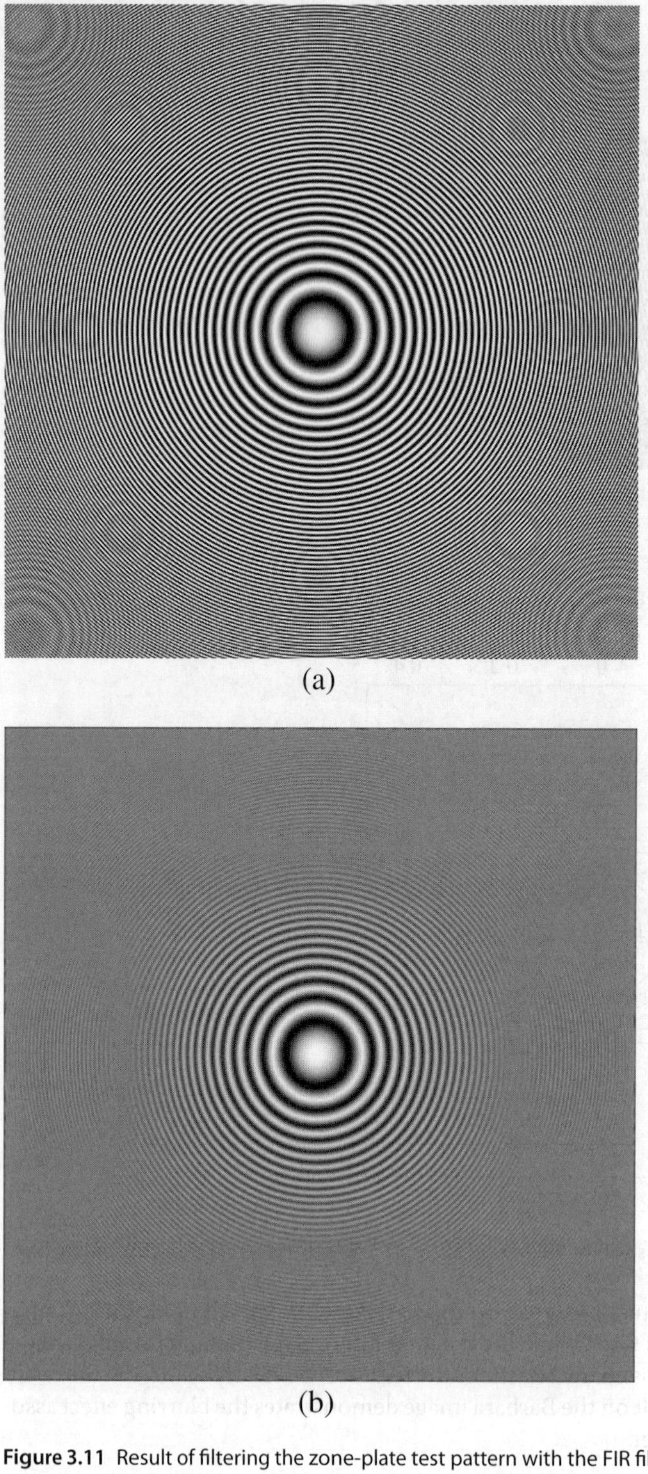

Figure 3.11 Result of filtering the zone-plate test pattern with the FIR filter of Example 3.4. (a) Original zone plate. (b) Filtered zone plate. Note: view the digital file at high magnification to remove artifacts in the zone plate.

(a)

(b)

Figure 3.12 Result of filtering Barbara image with the FIR filter of Example 3.4. (a) Original image. (b) Filtered image.

3.8.1 Separable Filters

We consider here filters defined on a 2D rectangular lattice $\Lambda = \mathrm{LAT}\left(\left[\begin{smallmatrix} X & 0 \\ 0 & Y \end{smallmatrix}\right]\right)$. A filter is separable if its unit sample response satisfies $h[n_1X, n_2Y] = h_1[n_1X] \cdot h_2[n_2Y]$ for two 1D filters h_1 and h_2. Separability can significantly reduce the implementation complexity of a filter, as we will see. The frequency response of a separable filter is given by

$$\begin{aligned} H(u,v) &= \sum_{n_1=-\infty}^{\infty} \sum_{n_2=-\infty}^{\infty} h[n_1X, n_2Y] \exp(-j2\pi(un_1X + vn_2Y)) \\ &= \sum_{n_1=-\infty}^{\infty} h_1[n_1X] \exp(-j2\pi un_1X) \sum_{n_2=-\infty}^{\infty} h_2[n_2Y] \exp(-j2\pi vn_2Y) \\ &= H_1(u)H_2(v) \end{aligned} \tag{3.57}$$

which is also a separable function in the frequency domain.

Now consider the implementation of the 2D convolution for a separable FIR system. Assume that the region of support of h_1 is $[P_1X, Q_1X]$ and the region of support of h_2 is $[P_2Y, Q_2Y]$, so that the region of support of the separable filter is

$$\begin{aligned} \mathcal{A} &= [P_1X, Q_1X] \times [P_2Y, Q_2Y] \\ &= \{(n_1X, n_2Y) \mid P_1 \le n_1 \le Q_1, P_2 \le n_2 \le Q_2\}. \end{aligned} \tag{3.58}$$

The filter output, by direct implementation of the convolution, is

$$g[n_1X, n_2Y] = \sum_{k_1=P_1}^{Q_1} \sum_{k_2=P_2}^{Q_2} h[k_1X, k_2Y] f[(n_1 - k_1)X, (n_2 - k_2)Y]. \tag{3.59}$$

This double sum contains $(Q_1 - P_1 + 1)(Q_2 - P_2 + 1)$ terms so that direct implementation requires $(Q_1 - P_1 + 1)(Q_2 - P_2 + 1)$ multiplications per output point and $(Q_1 - P_1 + 1)(Q_2 - P_2 + 1) - 1$ additions.

Now, using separability, we can express the filter output as

$$\begin{aligned} g[n_1X, n_2Y] &= \sum_{k_1=P_1}^{Q_1} \sum_{k_2=P_2}^{Q_2} h_1[k_1X] h_2[k_2Y] f[(n_1 - k_1)X, (n_2 - k_2)Y] \\ &= \sum_{k_1=P_1}^{Q_1} h_1[k_1X] \sum_{k_2=P_2}^{Q_2} h_2[k_2Y] f[(n_1 - k_1)X, (n_2 - k_2)Y]. \end{aligned} \tag{3.60}$$

Let us define

$$r[jX, n_2Y] = \sum_{k_2=P_2}^{Q_2} h_2[k_2Y] f[jX, (n_2 - k_2)Y]. \tag{3.61}$$

This intermediate image r is obtained by filtering each *column* of the input image f with the 1D filter h_2, and requires $Q_2 - P_2 + 1$ multiplications per output point. Then, the overall filter output is

$$g[n_1X, n_2Y] = \sum_{k_1=P_1}^{Q_1} h_1[k_1X] r[(n_1 - k_1)X, n_2Y] \tag{3.62}$$

which is obtained by filtering each *row* of the intermediate image r with the 1D filter h_1, requiring $Q_1 - P_1 + 1$ multiplications per output point. Thus overall, separable filtering requires $(Q_1 - P_1 + 1) + (Q_2 + P_2 + 1)$ multiplications per output point. This can be significantly smaller than the $(Q_1 - P_1 + 1)(Q_2 - P_2 + 1)$ required for the nonseparable implementation, especially as the filter size increases. For example, if $P_1 = P_2 = -5$ and $Q_1 = Q_2 = 5$, we have an 11×11 filter impulse response. The nonseparable implementation requires 121 multiplications per output point, while the separable implementation requires 22 multiplications, a factor of 5.5 less.

Problems

1 Prove the properties of lattices given in Section 3.2.2.

2 Prove that convolution on a lattice is commutative, $f * h = h * f$.

3 A linear shift-invariant filter defined on the hexagonal lattice

$$\Lambda = \text{LAT}\left(\begin{bmatrix} 2X & X \\ 0 & 1.5X \end{bmatrix}\right)$$

has unit-sample response given by

$$h[\mathbf{x}] = \begin{cases} \frac{1}{4} & \mathbf{x} = (0,0) \\ \frac{1}{8} & \mathbf{x} = (X, 1.5X)\text{or } (-X, -1.5X)\text{or } (0, 3X)\text{or } (0, -3X) \\ \frac{1}{16} & \mathbf{x} = (2X, 0)\text{or } (-2X, 0)\text{or } (X, -1.5X)\text{or } (-X, 1.5X) \\ 0 & \text{otherwise} \end{cases}.$$

Determine the frequency response $H(u, v)$ of this filter. Express it in real form. What is the DC gain of this filter?

4 For each of the following two-dimensional lattices Λ given by their sampling matrix, sketch the lattice to scale in the space domain, determine and sketch the reciprocal lattice and a Voronoi unit cell of the reciprocal lattice.

(a) $V_\Lambda = \begin{bmatrix} 2X & 0 \\ 0 & 2X \end{bmatrix}$

(b) $V_\Lambda = \begin{bmatrix} 3X & X \\ 0 & X \end{bmatrix}$

(c) $V_\Lambda = \begin{bmatrix} X & X \\ X & -X \end{bmatrix}$.

5 A two-dimensional FIR filter defined on the rectangular lattice $\Lambda = \text{LAT}(\text{diag}(X, X))$ has unit-sample response shown in Figure 3.13.

(a) Compute the frequency response $H(u, v)$. Express it in real form.

(b) What is the output $g[x, y]$ of this filter if the input is

$$f[x, y] = \delta[x - X, y + X] - \delta[x + X, y - X]?$$

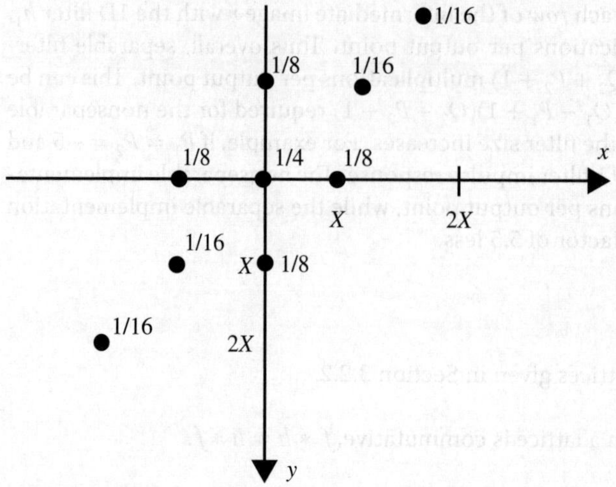

Figure 3.13 Unit-sample response $h[x, y]$. Nonzero values are shown; all others are zero.

Carefully sketch the output signal $g[x, y]$ to scale in the same manner as in Figure 3.13.

6 Consider an ideal discrete-space circularly symmetric low-pass filter defined on the rectangular lattice with horizontal and vertical sample spacing X and Y. The passband is $C_W = \{(u, v) \mid u^2 + v^2 \leq W^2\}$ and the unit cell of the reciprocal lattice is $\mathcal{P}^* = \{(u, v) \mid -1/2X \leq u < 1/2X, -1/2Y \leq v < 1/2Y\}$. Assume that $W < \min(1/2X, 1/2Y)$.

$$H(u, v) = \begin{cases} 1 & (u, v) \in C_W \\ 0 & (u, v) \in \mathcal{P}^* \backslash C_W \end{cases}$$

where of course $H(u + k/X, v + l/Y) = H(u, v)$ for all integers k, l. Show that the unit sample response of this filter is given by

$$h[mX, nY] = \frac{WXY}{\sqrt{X^2m^2 + Y^2n^2}} J_1(2\pi W \sqrt{X^2m^2 + Y^2n^2})$$

where $J_1(s)$ is the Bessel function of the first kind and first order. You may use the following identities:

$$J_0(s) = \frac{1}{2\pi} \int_0^{2\pi} \exp[js\cos(\theta + \phi)]\, d\theta, \qquad \text{for any } \phi$$

$$\int sJ_0(s)\, ds = sJ_1(s)$$

Simplify the expression in the case $X = Y$.

4

Discrete-Domain Periodic Signals

4.1 Introduction

This chapter is concerned with discrete-domain signals, defined on a lattice Λ, that are periodic. This means that there is another lattice Γ that is a sublattice of Λ that defines the periodicity. We can develop a complete theory for signal spaces of periodic signals, and linear shift-invariant systems acting on elements of these signal spaces. Once again, the Fourier transform is a key tool for the analysis of such signals and systems. The resulting Fourier transform in this case is the multidimensional extension of the well-known discrete Fourier transform (DFT), and also has available the same family of fast computation algorithms, namely the fast Fourier transform (FFT).

4.2 Periodic Signals

Let Λ be a lattice in D dimensions and Γ a D-dimensional sublattice of Λ, a subset of Λ that is also a lattice, denoted $\Gamma \subset \Lambda$. Section 13.5 contains details on the definition and properties of sublattices that we use here. In particular, Theorem 13.4 shows that $\mathrm{LAT}(\mathbf{V}_\Gamma) \subset \mathrm{LAT}(\mathbf{V}_\Lambda)$ if and only if $\mathbf{V}_\Lambda^{-1}\mathbf{V}_\Gamma$ is an integer matrix. It follows that $\mathrm{d}(\Gamma)/\mathrm{d}(\Lambda)$ must be an integer.

A signal defined on Λ is said to be Γ-periodic if

$$\tilde{f}[\mathbf{x}+\mathbf{t}] = \tilde{f}[\mathbf{x}] \quad \forall \mathbf{t} \in \Gamma, \quad \forall \mathbf{x} \in \Lambda. \tag{4.1}$$

Following the convention used by Oppenheim *et al.* (1999), we denote periodic signals with the tilde symbol ($\tilde{\cdot}$) when we wish to emphasize the periodic nature of the signal. Figure 4.1 shows an example of a periodic signal that is representative of a wallpaper pattern. The periodicity is given by the lattice Γ shown on the figure. The underlying lattice Λ is dense on this scale and is not shown, but Γ must be a sublattice.

The integer $K = \mathrm{d}(\Gamma)/\mathrm{d}(\Lambda)$ is called the index of Γ in Λ. The set $\mathbf{b} + \Gamma$ for any $\mathbf{b} \in \Lambda$ is called a coset of Γ in Λ (Definition 13.7). Two points $\mathbf{x}, \mathbf{y} \in \Lambda$ belong to the same coset if and only if $\mathbf{y} - \mathbf{x} \in \Gamma$. There are K distinct cosets that partition Λ (Theorem 13.6).

Multidimensional Signal and Color Image Processing Using Lattices, First Edition. Eric Dubois.

Companion website: www.wiley.com/go/Dubois/multiSP

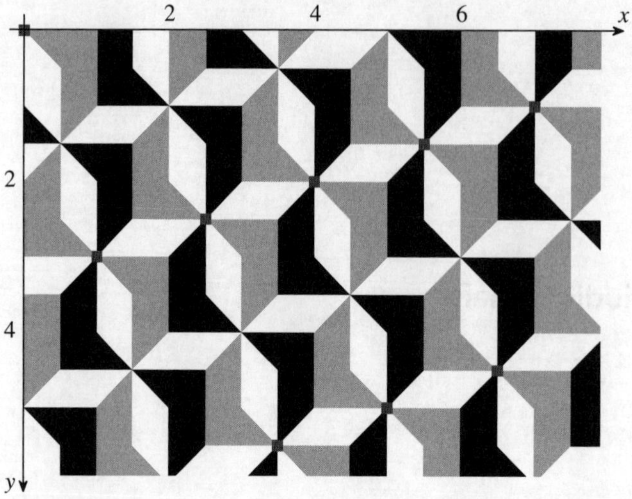

Figure 4.1 Portion of a periodic image representative of a wallpaper pattern. The periodicity is given by the lattice Γ shown, which is a sublattice of the underlying lattice Λ (not shown). (*See color plate section for the color representation of this figure.*)

Let $\mathbf{b}_0, \ldots, \mathbf{b}_{K-1}$ be K elements, one arbitrarily selected from each coset, called coset representatives. Then

$$\Lambda = \bigcup_{k=0}^{K-1} (\mathbf{b}_k + \Gamma). \tag{4.2}$$

A periodic signal is constant on each coset and is specified by its K values at the coset representatives, $\tilde{f}[\mathbf{b}_k], k = 0, \ldots, K-1$. The domain of the periodic signal is essentially the set of cosets, which has a group structure. This is the quotient group Λ/Γ, which is a finite group with K elements. Thus, the space of Γ-periodic signals on Λ is denoted here as $S_{\Lambda/\Gamma}$. As a vector space, the signal space $S_{\Lambda/\Gamma}$ is isomorphic to $\mathbb{C}^K$, the vector space of complex K-tuples. Since $S_{\Lambda/\Gamma}$ is finite dimensional, the only condition we require for all operations to be well defined is that $|\tilde{f}[\mathbf{b}_k]| < \infty, k = 0, \ldots, K-1$. We illustrate with an example.

Example 4.1 Let

$$\Lambda = \mathrm{LAT}\left(\begin{bmatrix} 2 & 1 \\ 0 & 1 \end{bmatrix}\right) \qquad \Gamma = \mathrm{LAT}\left(\begin{bmatrix} 10 & 0 \\ 0 & 6 \end{bmatrix}\right).$$

We can verify that

$$\mathbf{V}_\Lambda^{-1}\mathbf{V}_\Gamma = \begin{bmatrix} \frac{1}{2} & -\frac{1}{2} \\ 0 & 1 \end{bmatrix} \begin{bmatrix} 10 & 0 \\ 0 & 6 \end{bmatrix} = \begin{bmatrix} 5 & -3 \\ 0 & 6 \end{bmatrix},$$

is an integer matrix, with $|\det(\mathbf{V}_\Lambda^{-1}\mathbf{V}_\Gamma)| = 30$. In this example, $\mathrm{d}(\Lambda) = 2$, $\mathrm{d}(\Gamma) = 60$ and so $K = 30$, and we need to identify 30 coset representatives. Figure 4.2 shows the two

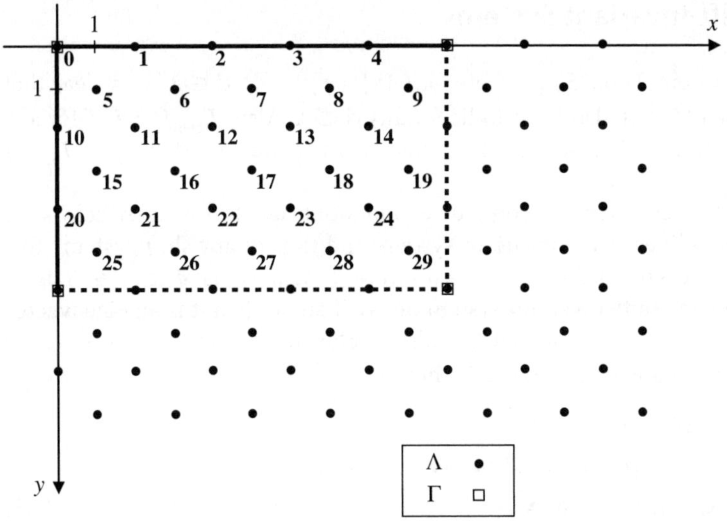

Figure 4.2 Portion of lattices Λ and Γ showing a rectangular unit cell of Γ and the corresponding coset representatives, where the numbers shown indicate the indexes of $\mathbf{b}_i$. The solid edges are included in the unit cell, while the dashed edges are not.

lattices and a unit cell of Γ. The 30 coset representatives within the unit cell are identified arbitrarily from 0 to 29, in row-by-row order.

There are two basic types of signals that we use frequently in this work. The first is the periodic unit sample, defined as

$$\tilde{\delta}_{\Lambda/\Gamma}[\mathbf{x}] = \begin{cases} 1 & \mathbf{x} \in \Gamma \\ 0 & \mathbf{x} \in \Lambda\backslash\Gamma. \end{cases} \tag{4.3}$$

We can also express this as a sum of shifted versions of the unit sample on Λ,

$$\tilde{\delta}_{\Lambda/\Gamma} = \sum_{\mathbf{s}\in\Gamma} \mathcal{T}_{\mathbf{s}}\delta_{\Lambda}. \tag{4.4}$$

The second type is the class of sinusoidal signals, with the usual form $f[\mathbf{x}] = \exp(\mathrm{j}2\pi\mathbf{u}\cdot\mathbf{x})$ for some $\mathbf{u} \in \mathcal{P}_{\Lambda^*}$. However, for f to be Γ-periodic, we must have

$$\exp(\mathrm{j}2\pi\mathbf{u}\cdot(\mathbf{x}+\mathbf{t})) = \exp(\mathrm{j}2\pi\mathbf{u}\cdot\mathbf{x}), \quad \mathbf{t}\in\Gamma, \mathbf{x}\in\Lambda. \tag{4.5}$$

This holds if $\exp(\mathrm{j}2\pi\mathbf{u}\cdot\mathbf{t}) = 1$ for all $\mathbf{t}\in\Gamma$, i.e. $\mathbf{u}\cdot\mathbf{t}\in\mathbb{Z}$ for all $\mathbf{t}\in\Gamma$, which means $\mathbf{u}\in\Gamma^*$. Thus there are only a finite number of admissible $\mathbf{u}$ vectors in $\Gamma^*\cap\mathcal{P}_{\Lambda^*}$, of course not unique since $\mathcal{P}_{\Lambda^*}$ is not unique. In fact, $\Lambda^*\subset\Gamma^*$, so there are K cosets of Λ^* in Γ^*. Let $\mathbf{d}_0,\ldots,\mathbf{d}_{K-1}$ be any selected coset representatives. Then

$$\Gamma^* = \bigcup_{k=0}^{K-1}(\mathbf{d}_k + \Lambda^*). \tag{4.6}$$

In summary, there are K distinct sinusoidal signals in $\mathcal{S}_{\Lambda/\Gamma}$, namely $\exp(\mathrm{j}2\pi\mathbf{d}_i\cdot\mathbf{x})$, $i = 0,\ldots,K-1$. We emphasize that the choice of coset representatives within each coset is arbitrary to this definition, and is based on convenience.

4.3 Linear Shift-Invariant Systems

Consider the shift system $\mathcal{T}_{\mathbf{d}}$ on $\mathcal{S}_{\Lambda/\Gamma}$, where $(\mathcal{T}_{\mathbf{d}}\tilde{f})[\mathbf{x}] = \tilde{f}[\mathbf{x} - \mathbf{d}]$, $\mathbf{d} \in \Lambda$. It is clear that if $\tilde{f}$ is Γ-periodic, then $\mathcal{T}_{\mathbf{d}}\tilde{f}$ is also Γ-periodic for any $\mathbf{d} \in \Lambda$. Also, $\mathcal{T}_{\mathbf{d}+\mathbf{c}}\tilde{f} = \mathcal{T}_{\mathbf{d}}\tilde{f}$ for any $\mathbf{c} \in \Gamma$. Thus, there are only K distinct shift systems, which we can identify as $\mathcal{T}_{\mathbf{b}_k}$ for $k = 0, \dots, K-1$.

We recall from Chapter 3 that the complex exponential signals are eigenvectors (or eigenfunctions) of any linear shift-invariant system, and thus of any shift system. This is easily seen to be necessary and sufficient: a discrete-domain signal $\phi[\mathbf{x}]$, $\mathbf{x} \in \Lambda$, is an eigenvector of any linear shift-invariant system on $\mathcal{S}_\Lambda$ if and only if it is an eigenvector of any shift system on $\mathcal{S}_\Lambda$. We show that this also applies to LSI systems on $\mathcal{S}_{\Lambda/\Gamma}$. Let $\tilde{\phi}_{\mathbf{u}}[\mathbf{x}] = \exp(\mathrm{j}2\pi\mathbf{u}\cdot\mathbf{x})$ for some fixed $\mathbf{u} \in \Gamma^*$. Then

$$\begin{aligned}(\mathcal{T}_{\mathbf{d}}\tilde{\phi}_{\mathbf{u}})[\mathbf{x}] &= \exp(\mathrm{j}2\pi\mathbf{u}\cdot(\mathbf{x}-\mathbf{d}))\\ &= \exp(-\mathrm{j}2\pi\mathbf{u}\cdot\mathbf{d})\exp(\mathrm{j}2\pi\mathbf{u}\cdot\mathbf{x})\\ &= \exp(-\mathrm{j}2\pi\mathbf{u}\cdot\mathbf{d})\tilde{\phi}_{\mathbf{u}}[\mathbf{x}],\end{aligned} \tag{4.7}$$

so that $\tilde{\phi}_{\mathbf{u}}$ is an eigenvector of the shift system $\mathcal{T}_{\mathbf{d}}$ with eigenvalue $\exp(-\mathrm{j}2\pi\mathbf{u}\cdot\mathbf{d})$. There are K such distinct eigenvectors, $\tilde{\phi}_{\mathbf{d}_i}$ for $i = 0, 1, \dots, K-1$.

Let $\mathcal{H}$ be a linear, shift-invariant system on $\mathcal{S}_{\Lambda/\Gamma}$. The shifted versions of the unit sample, $\mathcal{T}_{\mathbf{b}_k}\tilde{\delta}_{\Lambda/\Gamma}$ for $k = 0, \dots, K-1$ form an elementary basis for $\mathcal{S}_{\Lambda/\Gamma}$. The unique representation of a signal $\tilde{f}$ in this basis is

$$\tilde{f} = \sum_{k=0}^{K-1} \tilde{f}[\mathbf{b}_k]\mathcal{T}_{\mathbf{b}_k}\tilde{\delta}_{\Lambda/\Gamma}. \tag{4.8}$$

Using the LSI property, the output of the system $\mathcal{H}$ to the input $\tilde{f}$ is

$$\begin{aligned}\tilde{g} = \mathcal{H}\tilde{f} &= \sum_{k=0}^{K-1} \tilde{f}[\mathbf{b}_k]\mathcal{H}(\mathcal{T}_{\mathbf{b}_k}\tilde{\delta}_{\Lambda/\Gamma})\\ &= \sum_{k=0}^{K-1} \tilde{f}[\mathbf{b}_k]\mathcal{T}_{\mathbf{b}_k}(\mathcal{H}\tilde{\delta}_{\Lambda/\Gamma})\\ &= \sum_{k=0}^{K-1} \tilde{f}[\mathbf{b}_k]\mathcal{T}_{\mathbf{b}_k}\tilde{h},\end{aligned} \tag{4.9}$$

where $\tilde{h}$ is the unit-sample response of $\mathcal{H}$, which is necessarily Γ-periodic. Evaluating this at any $\mathbf{x} \in \Lambda$, we find

$$\tilde{g}[\mathbf{x}] = \sum_{k=0}^{K-1} \tilde{f}[\mathbf{b}_k]\tilde{h}[\mathbf{x} - \mathbf{b}_k] \tag{4.10}$$

which is a periodic convolution. Of course, this result is Γ-periodic, and only needs to be evaluated at $\mathbf{b}_i$ for $i = 0, \dots, K-1$:

$$\tilde{g}[\mathbf{b}_i] = \sum_{k=0}^{K-1} \tilde{f}[\mathbf{b}_k]\tilde{h}[\mathbf{b}_i - \mathbf{b}_k], \quad i = 0, \dots, K-1. \tag{4.11}$$

In this equation, we must use the periodicity of $\tilde{h}$ since, in general, $\mathbf{b}_i - \mathbf{b}_k$ will not be one of the coset representatives. To limit the expression to the coset representatives, for any $\mathbf{b} \in \Lambda$, we define

$$\langle \mathbf{b} \rangle = \mathbf{b}_\ell \quad \text{such that} \quad \mathbf{b} \in \mathbf{b}_\ell + \Gamma; \tag{4.12}$$

there is one and only one $\mathbf{b}_\ell$ that satisfies this. With this notation, the LSI system output can be written in a form involving only coset representatives

$$\tilde{g}[\mathbf{b}_i] = \sum_{k=0}^{K-1} \tilde{f}[\mathbf{b}_k]\tilde{h}[\langle \mathbf{b}_i - \mathbf{b}_k \rangle], \quad i = 0, \dots, K-1. \tag{4.13}$$

Now, for any i, as k runs from 0 to $K-1$, $\langle \mathbf{b}_i - \mathbf{b}_k \rangle$ runs over all the coset representatives in some order. Thus we can equivalently write

$$\tilde{g}[\mathbf{b}_i] = \sum_{k=0}^{K-1} \tilde{h}[\mathbf{b}_k]\tilde{f}[\langle \mathbf{b}_i - \mathbf{b}_k \rangle], \quad i = 0, \dots, K-1, \tag{4.14}$$

showing that periodic convolution is commutative. In terms of signals

$$\tilde{g} = \sum_{k=0}^{K-1} \tilde{h}[\mathbf{b}_k]\mathcal{T}_{\mathbf{b}_k}\tilde{f} \tag{4.15}$$

and thus the LSI system can be expressed as a linear combination of shift systems

$$\mathcal{H} = \sum_{k=0}^{K-1} \tilde{h}[\mathbf{b}_k]\mathcal{T}_{\mathbf{b}_k}. \tag{4.16}$$

Now, if we apply the complex exponential $\tilde{\phi}_{\mathbf{d}_i}$ as input to the LSI system, the output is

$$\begin{aligned}
\mathcal{H}\tilde{\phi}_{\mathbf{d}_i} &= \sum_{k=0}^{K-1} \tilde{h}[\mathbf{b}_k]\mathcal{T}_{\mathbf{b}_k}\tilde{\phi}_{\mathbf{d}_i} \\
&= \left(\sum_{k=0}^{K-1} \tilde{h}[\mathbf{b}_k] \exp(-\mathrm{j}2\pi \mathbf{d}_i \cdot \mathbf{b}_k) \right) \tilde{\phi}_{\mathbf{d}_i} \\
&= \widetilde{H}[\mathbf{d}_i]\tilde{\phi}_{\mathbf{d}_i}
\end{aligned} \tag{4.17}$$

Thus, $\tilde{\phi}_{\mathbf{d}_i}$ is indeed an eigenfunction of $\mathcal{H}$ with eigenvalue

$$\widetilde{H}[\mathbf{d}_i] = \sum_{k=0}^{K-1} \tilde{h}[\mathbf{b}_k] \exp(-\mathrm{j}2\pi \mathbf{d}_i \cdot \mathbf{b}_k), \quad \text{for } i = 0, \dots, K-1. \tag{4.18}$$

This is the discrete-domain periodic Fourier transform of the periodic unit sample response. Note that $\widetilde{H}$ is also a discrete-domain periodic function in the frequency domain, defined on the lattice Γ^*, with periodicity given by the sublattice Λ^*.

4.4 Discrete-Domain Periodic Fourier Transform

If $\tilde{f}$ is any element of $\mathcal{S}_{\Lambda/\Gamma}$, we can similarly define the discrete-domain periodic Fourier transform

$$\widetilde{F}[\mathbf{d}_i] = \sum_{k=0}^{K-1} \tilde{f}[\mathbf{b}_k] \exp(-\mathrm{j}2\pi \mathbf{d}_i \cdot \mathbf{b}_k), \quad \text{for } i = 0, \dots, K-1. \tag{4.19}$$

We will also use the term discrete-domain Fourier series, since the term Fourier series is generally used when analyzing periodic signals, where the inverse Fourier transform is a series and not an integral, as we will see in both the discrete domain and the continuous domain. We will denote a Fourier transform pair in this case by

$$\tilde{f}[\mathbf{x}] \xleftrightarrow{\text{DDFS}} \tilde{F}[\mathbf{u}]. \tag{4.20}$$

From its definition in Equation (4.19), the discrete-domain periodic Fourier transform is easily seen to be a linear transformation from $\mathbb{C}^K$ to $\mathbb{C}^K$. If we denote $\tilde{\mathbf{f}} = [\tilde{f}[\mathbf{b}_0], \dots, \tilde{f}[\mathbf{b}_{K-1}]]^T$ and $\tilde{\mathbf{F}} = [\tilde{F}[\mathbf{d}_0], \dots, \tilde{F}[\mathbf{d}_{K-1}]]^T$, then we have $\tilde{\mathbf{F}} = \mathbf{E}\tilde{\mathbf{f}}$, where $E_{ik} = \exp(-\mathrm{j}2\pi\mathbf{d}_i \cdot \mathbf{b}_k)$. It is convenient to represent the sets of coset representative with $D \times K$ matrices, $\mathbf{B} = [\mathbf{b}_0 \mid \dots \mid \mathbf{b}_{K-1}]$ and $\mathbf{D} = [\mathbf{d}_0 \mid \dots \mid \mathbf{d}_{K-1}]$, so that $E_{ik} = \exp(-\mathrm{j}2\pi[\mathbf{D}^T\mathbf{B}]_{ik})$.

Equation (4.19) is written in terms of explicit sets of coset representatives for Γ in Λ and Λ^* in Γ^* respectively. Since the coset representatives are arbitrary, we can write an expression for the discrete-domain periodic Fourier transform that does not explicitly depend on the coset representatives. Specifically, we write

$$\tilde{F}[\mathbf{u}] = \sum_{\mathbf{x}\in\mathcal{B}} \tilde{f}[\mathbf{x}] \exp(-\mathrm{j}2\pi\mathbf{u}\cdot\mathbf{x}), \quad \mathbf{u} \in \Gamma^*, \tag{4.21}$$

where $\mathcal{B}$ is an arbitrary set of coset representatives. The Fourier transform $\tilde{F}[\mathbf{u}]$ is itself periodic, with $\tilde{F}[\mathbf{u}+\mathbf{r}] = \tilde{F}[\mathbf{u}]$ for all $\mathbf{r} \in \Lambda^*$. Thus, it is sufficient to evaluate it at a set of coset representatives of Λ^* in Γ^*, as in Equation (4.19).

If $\tilde{f}$ is applied to an LSI system with Γ-periodic unit-sample response $\tilde{h}$, the Fourier transform of the output is

$$\begin{aligned}
\tilde{G}[\mathbf{d}_i] &= \sum_{k=0}^{K-1}\left(\sum_{\ell=0}^{K-1} \tilde{f}[\mathbf{b}_\ell]\tilde{h}[\langle \mathbf{b}_k - \mathbf{b}_\ell\rangle]\right)\exp(-\mathrm{j}2\pi\mathbf{d}_i\cdot\mathbf{b}_k)\\
&= \sum_{\ell=0}^{K-1}\tilde{f}[\mathbf{b}_\ell]\sum_{k=0}^{K-1}\tilde{h}[\langle \mathbf{b}_k - \mathbf{b}_\ell\rangle]\exp(-\mathrm{j}2\pi\mathbf{d}_i\cdot\mathbf{b}_k)\\
&= \sum_{\ell=0}^{K-1}\tilde{f}[\mathbf{b}_\ell]\sum_{m=0}^{K-1}\tilde{h}[\mathbf{b}_m]\exp(-\mathrm{j}2\pi\mathbf{d}_i\cdot(\mathbf{b}_\ell+\mathbf{b}_m))\\
&= \sum_{\ell=0}^{K-1}\tilde{f}[\mathbf{b}_\ell]\exp(-\mathrm{j}2\pi\mathbf{d}_i\cdot\mathbf{b}_\ell)\sum_{m=0}^{K-1}\tilde{h}[\mathbf{b}_m]\exp(-\mathrm{j}2\pi\mathbf{d}_i\cdot\mathbf{b}_m)\\
&= \tilde{F}[\mathbf{d}_i]\tilde{H}[\mathbf{d}_i].
\end{aligned} \tag{4.22}$$

Thus, the Fourier transform for this signal space maintains its characterizing property of transforming convolution in the signal domain to multiplication in the frequency domain. This is just one of the standard properties of the discrete-domain periodic Fourier transform that will be covered shortly.

In the previous chapters, Fourier transform inversion has been stated but not proved. In the case of discrete-domain periodic Fourier transforms, establishing the inverse transform is relatively straightforward, as we will now show. Other inverse transforms can then be derived using limiting arguments. We first show that the $\phi_{\mathbf{d}_i}$ form an orthogonal basis of $\mathcal{S}_{\Lambda/\Gamma}$ and thus the discrete Fourier transform inversion follows in a straightforward fashion. As an inner product, we adopt the standard inner product on $\mathbb{C}^K$,

$$\langle \tilde{f}_1 \mid \tilde{f}_2 \rangle = \sum_{k=0}^{K-1} \tilde{f}_1[\mathbf{b}_k] \tilde{f}_2^*[\mathbf{b}_k]. \tag{4.23}$$

Theorem 4.1 The signals $\tilde{\phi}_{\mathbf{d}_i}(\mathbf{x}) = \exp(\mathrm{j}2\pi \mathbf{d}_i \cdot \mathbf{x})$ for $i = 0, 1, \ldots, K-1$ form an orthogonal basis for $\mathcal{S}_{\Lambda/\Gamma}$.

Proof: We want to show that $\langle \tilde{\phi}_{\mathbf{d}_i} \mid \tilde{\phi}_{\mathbf{d}_\ell} \rangle = 0$ if $i \neq \ell$.

The proof rests on the Smith normal form. As shown in Section 13.8, there exist bases for Λ and Γ that are collinear: $\mathbf{V}_\Gamma = \mathbf{V}_\Lambda \mathbf{D}$ where $\mathbf{D}$ is a positive, diagonal integer matrix, such that $d_{i+1,i+1}$ divides d_{ii} for $i = 1, 2, \ldots, D-1$. It follows that $K = \det(\mathbf{D}) = \prod_{i=1}^{D} d_{ii}$. This allows us to explicitly write a set of coset representatives as

$$\mathbf{b}_{n_1,n_2,\ldots,n_D} = n_1 \mathbf{v}_{\Lambda 1} + \cdots + n_D \mathbf{v}_{\Lambda D}, \quad 0 \leq n_m \leq d_{mm} - 1. \tag{4.24}$$

With this representation,

$$\begin{aligned} \tilde{\phi}_{\mathbf{d}_i}[\mathbf{b}_{n_1,n_2,\ldots,n_D}] &= \exp(\mathrm{j}2\pi \mathbf{d}_i \cdot (n_1 \mathbf{v}_{\Lambda 1} + \cdots + n_D \mathbf{v}_{\Lambda D})) \\ &= (\exp(\mathrm{j}2\pi \mathbf{d}_i \cdot \mathbf{v}_{\Lambda 1}))^{n_1} \cdots (\exp(\mathrm{j}2\pi \mathbf{d}_i \cdot \mathbf{v}_{\Lambda D}))^{n_D}. \end{aligned} \tag{4.25}$$

Then

$$\begin{aligned} &\langle \tilde{\phi}_{\mathbf{d}_i} \mid \tilde{\phi}_{\mathbf{d}_\ell} \rangle \\ &= \sum_{n_1=0}^{d_{11}-1} \cdots \sum_{n_D=0}^{d_{DD}-1} (\exp(\mathrm{j}2\pi(\mathbf{d}_i - \mathbf{d}_\ell) \cdot \mathbf{v}_{\Lambda 1}))^{n_1} \cdots (\exp(\mathrm{j}2\pi(\mathbf{d}_i - \mathbf{d}_\ell) \cdot \mathbf{v}_{\Lambda D}))^{n_D} \\ &= \left(\sum_{n_1=0}^{d_{11}-1} (\exp(\mathrm{j}2\pi(\mathbf{d}_i - \mathbf{d}_\ell) \cdot \mathbf{v}_{\Lambda 1}))^{n_1} \right) \cdots \left(\sum_{n_D=0}^{d_{DD}-1} (\exp(\mathrm{j}2\pi(\mathbf{d}_i - \mathbf{d}_\ell) \cdot \mathbf{v}_{\Lambda D}))^{n_D} \right). \end{aligned} \tag{4.26}$$

Let $\exp(\mathrm{j}2\pi(\mathbf{d}_i - \mathbf{d}_\ell) \cdot \mathbf{v}_{\Lambda m}) = r_m$ for $m = 1, \ldots, D$. Since $\mathbf{d}_i \neq \mathbf{d}_\ell$ and $\mathbf{v}_{\Lambda m} \notin \Gamma$, $r_m \neq 1$. However, since $d_{mm} \mathbf{v}_{\Lambda m} = \mathbf{v}_{\Gamma m}$, thus $r_m^{d_{mm}} = 1$. Thus

$$\begin{aligned} \langle \tilde{\phi}_{\mathbf{d}_i} \mid \tilde{\phi}_{\mathbf{d}_\ell} \rangle &= \left(\sum_{n_1=0}^{d_{11}-1} r_1^{n_1} \right) \cdots \left(\sum_{n_D=0}^{d_{DD}-1} r_D^{n_D} \right) \\ &= \frac{1 - r_1^{d_{11}}}{1 - r_1} \cdots \frac{1 - r_D^{d_{DD}}}{1 - r_D} \\ &= 0, \end{aligned} \tag{4.27}$$

as required. It is also easy to see that $\langle \tilde{\phi}_{\mathbf{d}_i} \mid \tilde{\phi}_{\mathbf{d}_i} \rangle = \|\tilde{\phi}_{\mathbf{d}_i}\|^2 = K$.

Since the K vectors $\tilde{\phi}_{\mathbf{d}_i}$ are mutually orthogonal, they necessarily form a basis for the K-dimensional vector space $\mathcal{S}_{\Lambda/\Gamma}$. □

Thus, any $\tilde{f} \in \mathcal{S}_{\Lambda/\Gamma}$ can be expressed

$$\tilde{f} = \sum_{i=0}^{K-1} \alpha_i \tilde{\phi}_{\mathbf{d}_i} \tag{4.28}$$

for some unique $\alpha_i \in \mathbb{C}$. To determine α_i, we note that

$$\begin{aligned}\langle \tilde{f} \mid \tilde{\phi}_{\mathbf{d}_k} \rangle &= \left\langle \sum_{i=0}^{K-1} \alpha_i \tilde{\phi}_{\mathbf{d}_i} \mid \tilde{\phi}_{\mathbf{d}_k} \right\rangle \\ &= \sum_{i=0}^{K-1} \alpha_i \langle \tilde{\phi}_{\mathbf{d}_i} \mid \tilde{\phi}_{\mathbf{d}_k} \rangle \\ &= \alpha_k K. \end{aligned} \tag{4.29}$$

Thus

$$\begin{aligned}\alpha_k &= \frac{1}{K} \langle \tilde{f} \mid \tilde{\phi}_{\mathbf{d}_k} \rangle \\ &= \frac{1}{K} \sum_{\ell=0}^{K-1} \tilde{f}[\mathbf{b}_\ell] \exp(-\mathrm{j}2\pi \mathbf{d}_k \cdot \mathbf{b}_\ell) \\ &= \frac{1}{K} \tilde{F}[\mathbf{d}_k]. \end{aligned} \tag{4.30}$$

Substituting in the representation of $\tilde{f}$,

$$\tilde{f}[\mathbf{b}_k] = \frac{1}{K} \sum_{i=0}^{K-1} \tilde{F}[\mathbf{d}_i] \exp(\mathrm{j}2\pi \mathbf{d}_i \cdot \mathbf{b}_k). \tag{4.31}$$

This is the inverse multidimensional discrete Fourier transform, also called the discrete-domain Fourier series. Again, since the set of coset representatives of Λ^* in Γ^* is arbitrary, we can write more generally

$$\tilde{f}[\mathbf{x}] = \frac{1}{K} \sum_{\mathbf{u} \in \mathcal{D}} \tilde{F}[\mathbf{u}] \exp(\mathrm{j}2\pi \mathbf{u} \cdot \mathbf{x}), \quad \mathbf{x} \in \Lambda, \tag{4.32}$$

which will automatically have the correct periodicity. The inverse Fourier transform provides a *synthesis* of the periodic signal as a linear combination of discrete-domain periodic sinusoidal signals.

The orthogonality conditions can be written in terms of $\mathbf{E}$ as $\mathbf{E}\mathbf{E}^H = K\mathbf{I}_K$, so that $\mathbf{E}^{-1} = \frac{1}{K}\mathbf{E}^H$. In this way, the inverse Fourier transform can be expressed as $\tilde{\mathbf{f}} = \frac{1}{K}\mathbf{E}^H\tilde{\mathbf{F}}$.

Before going further, a few examples can show how to concretely apply these representations and relate them to standard DFT expansions.

Example 4.2 One-dimensional DFT: $D = 1$, $\Lambda = \mathbb{Z}$, $\Gamma = K\mathbb{Z}$. A natural choice for the discrete-domain coset representatives is $\mathbf{b}_k = k$ for $k = 0, \dots, K-1$. For the reciprocal lattices, $\Lambda^* = \mathbb{Z}$ and $\Gamma^* = \frac{1}{K}\mathbb{Z}$. Thus, a natural choice for the frequency domain coset representatives is $\mathbf{d}_i = \frac{i}{K}$ for $i = 0, \dots, K-1$. Then

$$\tilde{\phi}_{\mathbf{d}_i}[\mathbf{b}_k] = \exp\left(\mathrm{j}2\pi \frac{ik}{K}\right), \qquad i = 0, \dots, K-1; k = 0, \dots, K-1. \tag{4.33}$$

The Fourier transform of $\tilde{f}$ can be written

$$\tilde{F}\left[\frac{i}{K}\right] = \sum_{k=0}^{K-1} \tilde{f}[k] \exp\left(-\mathrm{j}2\pi \frac{ik}{K}\right), \qquad i = 0, \dots, K-1, \tag{4.34}$$

the usual one-dimensional DFT. Similarly, for the inverse DFT,

$$\tilde{f}[k] = \frac{1}{K}\sum_{i=0}^{K-1} \tilde{F}\left[\frac{i}{K}\right] \exp\left(j2\pi\frac{ik}{K}\right), \qquad k = 0, \dots, K-1. \tag{4.35}$$

Example 4.3 D-dimensional rectangular DFT: $\Lambda = \mathbb{Z}^D$, $\Gamma = \text{LAT}(\text{diag}(K_1, K_2, \dots, K_D))$, where $K = K_1K_2 \cdots K_D$. In this case, it is convenient to index the coset representatives with D indexes along each of the orthogonal directions of $\mathbb{Z}^D$:

$$\mathbf{b}_{k_1,k_2,\dots,k_D} = [k_1\ k_2\ \cdots\ k_D]^T, \qquad 0 \le k_\ell \le K_\ell - 1. \tag{4.36}$$

The reciprocal lattices are $\Lambda^* = \mathbb{Z}^D$ and $\Gamma^* = \text{LAT}(\text{diag}(\frac{1}{K_1}, \frac{1}{K_2}, \dots, \frac{1}{K_D}))$. In this case, the coset representatives can be chosen as

$$\mathbf{d}_{i_1,i_2,\dots,i_D} = \left[\frac{i_1}{K_1}\ \frac{i_2}{K_2}\ \cdots\ \frac{i_D}{K_D}\right]^T, \qquad 0 \le i_\ell \le K_\ell - 1. \tag{4.37}$$

Then, applying the results above, the D-dimensional rectangular DFT can be written

$$\tilde{F}\left[\frac{i_1}{K_1}, \dots, \frac{i_D}{K_D}\right] = \sum_{k_1=0}^{K_1-1} \cdots \sum_{k_D=0}^{K_D-1} \tilde{f}[k_1, k_2, \dots, k_D] \exp\left(-j2\pi\left(\frac{i_1k_1}{K_1} + \cdots + \frac{i_Dk_D}{K_D}\right)\right), \tag{4.38}$$

with inverse DFT given by

$$\tilde{f}[k_1, k_2, \dots, k_D] = \frac{1}{K}\sum_{i_1=0}^{K_1-1} \cdots \sum_{i_D=0}^{K_D-1} \tilde{F}\left[\frac{i_1}{K_1}, \dots, \frac{i_D}{K_D}\right] \exp\left(j2\pi\left(\frac{i_1k_1}{K_1} + \cdots + \frac{i_Dk_D}{K_D}\right)\right). \tag{4.39}$$

4.5 Properties of the Discrete-Domain Periodic Fourier Transform

The discrete-domain periodic Fourier transform has a set of properties very similar to those already seen for the continuous-domain and discrete-domain aperiodic cases. These properties are also the general multidimensional extension of the standard DFT properties. Table 4.1 lists the main properties that we shall prove here. In the table and in the proofs, $\mathcal{B}$ is a set of coset representatives for Γ in Λ and $\mathcal{D}$ is a set of coset representatives for Λ^* in Γ^*.

We provide the proofs of these properties for completeness, although in many cases, the proofs are very simple modifications of the ones already presented. Since all signals consist of K distinct values, and all sums are finite with K terms, there are no conditions imposed on the signals for these properties to hold, other than that they be finite valued.

Table 4.1 Properties of the multidimensional discrete-domain periodic Fourier transform over a lattice Λ.

	$\tilde{f}[\mathbf{x}] = \frac{1}{K}\sum_{\mathbf{u}\in\mathcal{D}}\tilde{F}[\mathbf{u}]\exp(\mathrm{j}2\pi\mathbf{u}\cdot\mathbf{x})$	$\tilde{F}[\mathbf{u}] = \sum_{\mathbf{x}\in\mathcal{B}}\tilde{f}[\mathbf{x}]\exp(-\mathrm{j}2\pi\mathbf{u}\cdot\mathbf{x})$
(4.1)	$A\tilde{f}[\mathbf{x}] + B\tilde{g}[\mathbf{x}]$	$A\tilde{F}[\mathbf{u}] + B\tilde{G}[\mathbf{u}]$
(4.2)	$\tilde{f}[\mathbf{x}-\mathbf{x}_0]$	$\tilde{F}[\mathbf{u}]\exp(-\mathrm{j}2\pi\mathbf{u}\cdot\mathbf{x}_0)$
(4.3)	$\tilde{f}[\mathbf{x}]\exp(\mathrm{j}2\pi\mathbf{u}_0\cdot\mathbf{x})$	$\tilde{F}[\mathbf{u}-\mathbf{u}_0]$
(4.4)	$\tilde{f}[\mathbf{x}] * \tilde{g}[\mathbf{x}]$	$\tilde{F}[\mathbf{u}]\tilde{G}[\mathbf{u}]$
(4.5)	$\tilde{f}[\mathbf{x}]\tilde{g}[\mathbf{x}]$	$\frac{1}{K}\sum_{\mathbf{r}\in\mathcal{D}}\tilde{F}[\mathbf{r}]\tilde{G}[\mathbf{u}-\mathbf{r}]$
(4.6)	$\tilde{f}[\mathbf{A}\mathbf{x}]$	$\tilde{F}[\mathbf{A}^{-T}\mathbf{u}]$
(4.7)	$\tilde{f}^*[\mathbf{x}]$	$\tilde{F}^*[-\mathbf{u}]$
(4.8)	$\sum_{\mathbf{x}\in\mathcal{B}}\tilde{f}[\mathbf{x}]\tilde{g}^*[\mathbf{x}] = \frac{1}{K}\sum_{\mathbf{u}\in\mathcal{D}}\tilde{F}[\mathbf{u}]\tilde{G}^*[\mathbf{u}]$	

Property 4.1 *Linearity*: $A\tilde{f}[\mathbf{x}] + B\tilde{g}[\mathbf{x}] \xleftrightarrow{\text{DDFS}} A\tilde{F}[\mathbf{u}] + B\tilde{G}[\mathbf{u}]$.

Proof: Let $\tilde{q}[\mathbf{x}] = A\tilde{f}[\mathbf{x}] + B\tilde{g}[\mathbf{x}]$. Then

$$\begin{aligned}\tilde{Q}[\mathbf{u}] &= \sum_{\mathbf{x}\in\mathcal{B}}(A\tilde{f}[\mathbf{x}] + B\tilde{g}[\mathbf{x}])\exp(-\mathrm{j}2\pi\mathbf{u}\cdot\mathbf{x})\\ &= A\sum_{\mathbf{x}\in\mathcal{B}}\tilde{f}[\mathbf{x}]\exp(-\mathrm{j}2\pi\mathbf{u}\cdot\mathbf{x}) + B\sum_{\mathbf{x}\in\mathcal{B}}\tilde{g}[\mathbf{x}]\exp(-\mathrm{j}2\pi\mathbf{u}\cdot\mathbf{x})\\ &= A\tilde{F}[\mathbf{u}] + B\tilde{G}[\mathbf{u}].\end{aligned} \tag{4.40}$$

□

Property 4.2 *Shift*: $\tilde{f}[\mathbf{x}-\mathbf{x}_0] \xleftrightarrow{\text{DDFS}} \tilde{F}[\mathbf{u}]\exp(-\mathrm{j}2\pi\mathbf{u}\cdot\mathbf{x}_0)$.

Proof: Let $\tilde{g}[\mathbf{x}] = \tilde{f}[\mathbf{x}-\mathbf{x}_0]$ for some $\mathbf{x}_0 \in \Lambda$. Then

$$\begin{aligned}\tilde{G}[\mathbf{u}] &= \sum_{\mathbf{x}\in\mathcal{B}}\tilde{f}[\mathbf{x}-\mathbf{x}_0]\exp(-\mathrm{j}2\pi\mathbf{u}\cdot\mathbf{x})\\ &= \sum_{\mathbf{s}\in\mathcal{B}-\mathbf{x}_0}\tilde{f}[\mathbf{s}]\exp(-\mathrm{j}2\pi\mathbf{u}\cdot(\mathbf{s}+\mathbf{x}_0)) \quad (\mathbf{s} = \mathbf{x}-\mathbf{x}_0)\\ &= \left(\sum_{\mathbf{s}\in\mathcal{B}}f[\mathbf{s}]\exp(-\mathrm{j}2\pi\mathbf{u}\cdot\mathbf{s})\right)\exp(-\mathrm{j}2\pi\mathbf{u}\cdot\mathbf{x}_0)\\ &= \tilde{F}[\mathbf{u}]\exp(-\mathrm{j}2\pi\mathbf{u}\cdot\mathbf{x}_0)\end{aligned} \tag{4.41}$$

since $\mathcal{B}-\mathbf{x}_0$ is also a set of coset representatives. □

Property 4.3 *Modulation*: $\tilde{f}[\mathbf{x}]\exp(\mathrm{j}2\pi\mathbf{u}_0\cdot\mathbf{x}) \xleftrightarrow{\text{DDFS}} \tilde{F}[\mathbf{u}-\mathbf{u}_0]$.

Proof: Let $\tilde{g}[\mathbf{x}] = \tilde{f}[\mathbf{x}]\exp(\mathrm{j}2\pi\mathbf{u}_0\cdot\mathbf{x})$ for some $\mathbf{u}_0 \in \Gamma^*$. Then

$$\begin{aligned}\tilde{G}[\mathbf{u}] &= \sum_{\mathbf{x}\in\mathcal{B}} \tilde{f}[\mathbf{x}]\exp(\mathrm{j}2\pi\mathbf{u}_0\cdot\mathbf{x})\exp(-\mathrm{j}2\pi\mathbf{u}\cdot\mathbf{x})\\ &= \sum_{\mathbf{x}\in\mathcal{B}} \tilde{f}[\mathbf{x}]\exp(-\mathrm{j}2\pi(\mathbf{u}-\mathbf{u}_0)\cdot\mathbf{x})\\ &= \tilde{F}[\mathbf{u}-\mathbf{u}_0].\end{aligned} \tag{4.42}$$

□

Property 4.4 *Convolution*: $\tilde{f}[\mathbf{x}] * \tilde{g}[\mathbf{x}] \xleftrightarrow{\text{DDFS}} \tilde{F}[\mathbf{u}]\tilde{G}[\mathbf{u}]$.

Proof: Let $\tilde{q}[\mathbf{x}] = \tilde{f}[\mathbf{x}] * \tilde{g}[\mathbf{x}] = \sum_{\mathbf{s}\in\mathcal{B}}\tilde{f}[\mathbf{s}]\tilde{g}[\mathbf{x}-\mathbf{s}]$. Then

$$\begin{aligned}\tilde{Q}[\mathbf{u}] &= \sum_{\mathbf{x}\in\mathcal{B}}\left(\sum_{\mathbf{s}\in\mathcal{B}}\tilde{f}[\mathbf{s}]\tilde{g}[\mathbf{x}-\mathbf{s}]\right)\exp(-\mathrm{j}2\pi\mathbf{u}\cdot\mathbf{x})\\ &= \sum_{\mathbf{s}\in\mathcal{B}}\sum_{\mathbf{t}\in\mathcal{B}-\mathbf{s}}\tilde{f}[\mathbf{s}]\tilde{g}[\mathbf{t}]\exp(-\mathrm{j}2\pi\mathbf{u}\cdot(\mathbf{t}+\mathbf{s}))\quad (\mathbf{t}=\mathbf{x}-\mathbf{s})\\ &= \sum_{\mathbf{s}\in\mathcal{B}}\tilde{f}[\mathbf{s}]\exp(-\mathrm{j}2\pi\mathbf{u}\cdot\mathbf{s})\sum_{\mathbf{t}\in\mathcal{B}}\tilde{g}[\mathbf{t}]\exp(-\mathrm{j}2\pi\mathbf{u}\cdot\mathbf{t})\\ &= \tilde{F}[\mathbf{u}]\tilde{G}[\mathbf{u}],\end{aligned} \tag{4.43}$$

since $\mathcal{B}-\mathbf{s}$ is also a set of coset representatives. □

Note that using this property, commutativity and associativity of complex multiplication implies commutativity and associativity of convolution.

Property 4.5 *Multiplication*: $\tilde{f}[\mathbf{x}]\tilde{g}[\mathbf{x}] \xleftrightarrow{\text{DDFS}} \frac{1}{K}\tilde{F}[\mathbf{u}] * \tilde{G}[\mathbf{u}]$.

Proof: Let $\tilde{q}[\mathbf{x}] = \tilde{f}[\mathbf{x}]\tilde{g}[\mathbf{x}]$. Then

$$\begin{aligned}\tilde{Q}[\mathbf{u}] &= \sum_{\mathbf{x}\in\mathcal{B}}\tilde{f}[\mathbf{x}]\tilde{g}[\mathbf{x}]\exp(-\mathrm{j}2\pi\mathbf{u}\cdot\mathbf{x})\\ &= \sum_{\mathbf{x}\in\mathcal{B}}\tilde{g}[\mathbf{x}]\left(\frac{1}{K}\sum_{\mathbf{w}\in\mathcal{D}}\tilde{F}[\mathbf{w}]\exp(\mathrm{j}2\pi\mathbf{w}\cdot\mathbf{x})\right)\exp(-\mathrm{j}2\pi\mathbf{u}\cdot\mathbf{x})\\ &= \frac{1}{K}\sum_{\mathbf{w}\in\mathcal{D}}\tilde{F}[\mathbf{w}]\left(\sum_{\mathbf{x}\in\mathcal{B}}\tilde{g}[\mathbf{x}]\exp(-\mathrm{j}2\pi(\mathbf{u}-\mathbf{w})\cdot\mathbf{x})\right)\\ &= \frac{1}{K}\sum_{\mathbf{w}\in\mathcal{D}}\tilde{F}[\mathbf{w}]\tilde{G}[\mathbf{u}-\mathbf{w}] = \frac{1}{K}\tilde{F}[\mathbf{u}] * \tilde{G}[\mathbf{u}].\end{aligned} \tag{4.44}$$

□

Property 4.6 *Automorphism of the domain*: If $\mathbf{A}\Lambda = \Lambda$ and $\mathbf{A}\Gamma = \Gamma$, then $\tilde{f}[\mathbf{A}\mathbf{x}] \xleftrightarrow{\text{DDFS}} \tilde{F}[\mathbf{A}^{-T}\mathbf{u}]$.

Proof: As for the case of automorphism of a lattice, we require $\mathbf{A}$ to be of the form $\mathbf{A} = \mathbf{V}_\Lambda \mathbf{E}_1 \mathbf{V}_\Lambda^{-1}$, where $\mathbf{V}_\Lambda$ is a sampling matrix for Λ and $\mathbf{E}_1$ is a unimodular integer matrix. For the transformation to also preserve periodicity, we also need $\mathbf{A}\Gamma = \Gamma$, i.e. $\mathbf{A} = \mathbf{V}_\Gamma \mathbf{E}_2 \mathbf{V}_\Gamma^{-1}$, where $\mathbf{V}_\Gamma$ is a sampling matrix for Γ and $\mathbf{E}_2$ is a second unimodular integer matrix. Now, let $\tilde{g}[\mathbf{x}] = \tilde{f}[\mathbf{A}\mathbf{x}]$ for an admissible $\mathbf{A}$. Then

$$\begin{aligned}
\tilde{G}[\mathbf{u}] &= \sum_{\mathbf{x}\in\mathcal{B}} \tilde{f}[\mathbf{A}\mathbf{x}] \exp(-\mathrm{j}2\pi\mathbf{u}\cdot\mathbf{x}) \\
&= \sum_{\mathbf{s}\in\mathbf{A}\mathcal{B}} \tilde{f}[\mathbf{s}] \exp(-\mathrm{j}2\pi\mathbf{u}\cdot(\mathbf{A}^{-1}\mathbf{s})) \\
&= \sum_{\mathbf{s}\in\mathbf{A}\mathcal{B}} \tilde{f}[\mathbf{s}] \exp(-\mathrm{j}2\pi(\mathbf{A}^{-T}\mathbf{u})\cdot\mathbf{s}) \\
&= \sum_{\mathbf{s}\in\mathcal{B}} \tilde{f}[\mathbf{s}] \exp(-\mathrm{j}2\pi(\mathbf{A}^{-T}\mathbf{u})\cdot\mathbf{s}) \\
&= F[\mathbf{A}^{-T}\mathbf{u}]
\end{aligned} \tag{4.45}$$

since $\mathbf{A}\mathcal{B}$ is also a set of coset representatives and $\mathbf{A}^{-T}\mathbf{u} \in \Gamma^*$. □

Property 4.7 *Complex conjugation:* $\tilde{f}^*[\mathbf{x}] \xleftrightarrow{\text{DDFS}} \tilde{F}^*[-\mathbf{u}]$.

Proof: Let $\tilde{g}[\mathbf{x}] = \tilde{f}^*[\mathbf{x}]$. Then

$$\begin{aligned}
\tilde{G}[\mathbf{u}] &= \sum_{\mathbf{x}\in\mathcal{B}} \tilde{f}^*[\mathbf{x}] \exp(-\mathrm{j}2\pi\mathbf{u}\cdot\mathbf{x}) \\
&= \sum_{\mathbf{x}\in\mathcal{B}} (\tilde{f}[\mathbf{x}] \exp(\mathrm{j}2\pi\mathbf{u}\cdot\mathbf{x}))^* \\
&= \left(\sum_{\mathbf{x}\in\mathcal{B}} \tilde{f}[\mathbf{x}] \exp(\mathrm{j}2\pi\mathbf{u}\cdot\mathbf{x})\right)^* \\
&= \tilde{F}^*[-\mathbf{u}].
\end{aligned} \tag{4.46}$$

□

Property 4.8 *Parseval relation:* $\sum_{\mathbf{x}\in\mathcal{B}} \tilde{f}[\mathbf{x}]\tilde{g}^*[\mathbf{x}] = \frac{1}{K}\sum_{\mathbf{w}\in\mathcal{D}} \tilde{F}[\mathbf{u}]\tilde{G}^*[\mathbf{u}]$.

Proof: Let $\tilde{r}[\mathbf{x}] = \tilde{g}^*[\mathbf{x}]$ and $\tilde{q}[\mathbf{x}] = \tilde{f}[\mathbf{x}]\tilde{r}[\mathbf{x}]$. Then

$$\begin{aligned}
\tilde{Q}[\mathbf{u}] &= \frac{1}{K}\sum_{\mathbf{w}\in\mathcal{D}} \tilde{F}[\mathbf{w}]\tilde{R}[\mathbf{u}-\mathbf{w}] \\
&= \frac{1}{K}\sum_{\mathbf{w}\in\mathcal{D}} \tilde{F}[\mathbf{w}]\tilde{G}^*[\mathbf{w}-\mathbf{u}].
\end{aligned} \tag{4.47}$$

Evaluating at $\mathbf{u} = 0$,

$$\tilde{Q}[\mathbf{0}] = \sum_{\mathbf{x}\in\mathcal{B}} \tilde{f}[\mathbf{x}]\tilde{g}^*[\mathbf{x}] = \frac{1}{K}\sum_{\mathbf{w}\in\mathcal{D}} \tilde{F}[\mathbf{w}]\tilde{G}^*[\mathbf{w}]. \tag{4.48}$$

□

If $\tilde{f} = \tilde{g}$, we obtain

$$\sum_{\mathbf{x}\in\mathcal{B}} |\tilde{f}[\mathbf{x}]|^2 = \frac{1}{K}\sum_{\mathbf{w}\in\mathcal{D}} |\tilde{F}[\mathbf{w}]|^2. \tag{4.49}$$

4.6 Computation of the Discrete-Domain Periodic Fourier Transform

4.6.1 Direct Computation

Since the data in the signal domain and in the frequency domain contain only K distinct values, determination of the discrete-domain periodic Fourier transform is generally just a matter of carrying out the computation of the finite sum that defines the transform. Nevertheless some simple examples can be formulated to illustrate the ideas. However, for the general case, some methods are required to organize the computations, especially regarding selection of coset representatives in the signal domain and in the frequency domain.

Example 4.4 Consider the green channel in a Bayer color filter array, as illustrated in Figure 3.1. For simplicity, assume that $X = 1$, so that the lattice of all points on the sensor is $\Lambda = \mathbb{Z}^2$. Values of g are observed on the lattice $\Lambda_G = \text{LAT}\left(\left[\begin{smallmatrix} 2 & 1 \\ 0 & 1 \end{smallmatrix}\right]\right)$. Assume that the value of g at the points labeled r and b is 0, i.e. $g(\mathbf{x}) = 0$ for $\mathbf{x} \in \Lambda \backslash \Lambda_G$. If there is an underlying signal g_o defined on all of Λ, we have

$$g[\mathbf{x}] = \begin{cases} g_o[\mathbf{x}] & \mathbf{x} \in \Lambda_G \\ 0 & \mathbf{x} \in \Lambda \backslash \Lambda_G \end{cases}. \tag{4.50}$$

Alternatively, we can write $g[\mathbf{x}] = g_o[\mathbf{x}]m[\mathbf{x}]$, where

$$m[\mathbf{x}] = \begin{cases} 1 & \mathbf{x} \in \Lambda_G \\ 0 & \mathbf{x} \in \Lambda \backslash \Lambda_G \end{cases}. \tag{4.51}$$

$m[\mathbf{x}]$ is a very simple periodic signal defined on Λ with periodicity given by Λ_G, i.e. $m[\mathbf{x} + \mathbf{s}] = m[\mathbf{x}]$ for all $\mathbf{s} \in \Lambda_G$. In this case, $K = \text{d}(\Lambda_G)/\text{d}(\Lambda) = 2$, so there are two samples in a period. As a set of coset representatives, we can take $\mathcal{B} = \{[0, 0], [1, 0]\}$. The reciprocal lattices are $\Lambda^* = \mathbb{Z}^2$ and $\Lambda_G^* = \text{LAT}\left(\left[\begin{smallmatrix} 0.5 & 0 \\ -0.5 & 1 \end{smallmatrix}\right]\right)$. As coset representatives in the frequency domain, we can choose $\mathcal{D} = \{[0, 0], [0.5, 0.5]\}$. The Fourier transform and inverse Fourier transform of m are given by

$$M[u, v] = m[0, 0] \exp(-\text{j}2\pi(u \cdot 0 + v \cdot 0)) + m[1, 0] \exp(-\text{j}2\pi(u \cdot 1 + v \cdot 0)) = 1 \tag{4.52}$$

for all $(u, v) \in \mathcal{D}$, and

$$\begin{aligned} m[n_1, n_2] &= \frac{1}{2}(1 \exp(\text{j}2\pi(0 \cdot n_1 + 0 \cdot n_2)) + 1 \exp(\text{j}2\pi(0.5 \cdot n_1 + 0.5 \cdot n_2))) \\ &= \frac{1}{2}(1 + \exp(\text{j}\pi(n_1 + n_2))) \\ &= \frac{1}{2}(1 + (-1)^{n_1+n_2}). \end{aligned} \tag{4.53}$$

Although we could write the synthesis equation by inspection in this simple case, this would not be possible for more complex periodic signals. If the discrete domain signals g and g_o (these are not periodic) have Fourier transforms $G(\mathbf{u})$ and $G_o(\mathbf{u})$, then using the discrete-domain modulation Property 3.3, we find that $G(\mathbf{u}) = 0.5(G_o(\mathbf{u}) + G_o(\mathbf{u} - [0.5, 0.5]))$.

4.6.2 Selection of Coset Representatives

In order to implement the Fourier transform of Equation (4.21) and the inverse Fourier transform of Equation (4.32), we need to identify sets of coset representatives $\mathcal{B}$ and $\mathcal{D}$. In the case of rectangular sampling, the possible choices for the set of coset representatives in both the signal domain and the frequency domain are quite straightforward, as demonstrated in Example 4.3. However, when one or both of the signal domain lattice and the periodicity lattice are not rectangular, another approach is needed. Several approaches are described next.

The most straightforward approach is manual, based on unit cells of Γ in the signal domain, and of Λ^* in the frequency domain. Since the approaches are the same in either domain, we illustrate them in the signal domain. Let $\mathcal{P}_\Gamma$ be a convenient unit cell of Γ. We generally like it to be a connected region with as simple a shape as possible. By the definition of unit cell, no two points of Λ in $\mathcal{P}_\Gamma$ can belong to the same coset of Γ in Λ. Since $\mathrm{d}(\Gamma)/\mathrm{d}(\Lambda) = K$, $\mathcal{P}_\Gamma$ must contain K distinct points of Λ. Thus, we can denote these as $\mathbf{b}_0, \dots, \mathbf{b}_{K-1}$ in any convenient order. We illustrate by returning to Example 4.1.

Example 4.5 Let

$$\Lambda = \mathrm{LAT}\left(\begin{bmatrix} 2 & 1 \\ 0 & 1 \end{bmatrix}\right) \qquad \Gamma = \mathrm{LAT}\left(\begin{bmatrix} 10 & 0 \\ 0 & 6 \end{bmatrix}\right).$$

Again, $\mathrm{d}(\Lambda) = 2$, $\mathrm{d}(\Gamma) = 60$ and so $K = 30$, and we need to identify 30 coset representatives. As we have already seen, Figure 4.2 shows the two lattices and the fundamental parallelepiped unit cell of Γ. The 30 coset representatives within the unit cell are identified arbitrarily from 0 to 29, in row-by-row order.

Note that we could also identify the coset representatives with two indexes as follows

$$\mathbf{b}_{i\ell} = \begin{bmatrix} 2i + (\ell \bmod 2) \\ \ell \end{bmatrix}, \qquad 0 \le i \le 4, 0 \le \ell \le 5. \tag{4.54}$$

With this set of coset representatives, the group structure of Λ/Γ is not at all evident.

Another manual approach uses a different unit cell of Γ, which is aligned with the basis vectors of Λ (Figure 4.3), specifically

$$\mathcal{P}_\Gamma^{(2)} = \left\{ \begin{bmatrix} x + y \\ y \end{bmatrix} \mid 0 \le x < 10, 0 \le y < 6 \right\}. \tag{4.55}$$

The 30 coset representatives are enumerated in Figure 4.3 row by row. In this case, we can choose the coset representatives to be integer multiples of the basis vectors of Λ:

$$\mathbf{b}_{i\ell}^{(2)} = \begin{bmatrix} 2i + \ell \\ \ell \end{bmatrix} = \mathbf{V}_\Lambda \begin{bmatrix} i \\ \ell \end{bmatrix} \qquad 0 \le i \le 4, 0 \le \ell \le 5. \tag{4.56}$$

A more systematic approach is to use the Smith normal form [Newman (1972)] for the integer matrix $\mathbf{M} = \mathbf{V}_\Lambda^{-1}\mathbf{V}_\Gamma$. More details on the Smith form can be found in Section 13.8. From, for example, theorem II.9 of Newman (1972) or theorem 2.4.12 of Cohen (1993), for any square, non-singular integer matrix $\mathbf{M}$, we can find unimodular matrices $\mathbf{E}_1$ and $\mathbf{E}_2$ such that $\mathbf{E}_1\mathbf{M}\mathbf{E}_2 = \mathbf{D}$ where $\mathbf{D} = \mathrm{diag}(d_{11}, \dots, d_{DD})$ is a positive integer diagonal matrix such that $d_{i+1,i+1}$ divides d_{ii} for $i = 1, \dots, D-1$. Efficient algorithms to determine this decomposition are given in Cohen (1993). Note that this decomposition is

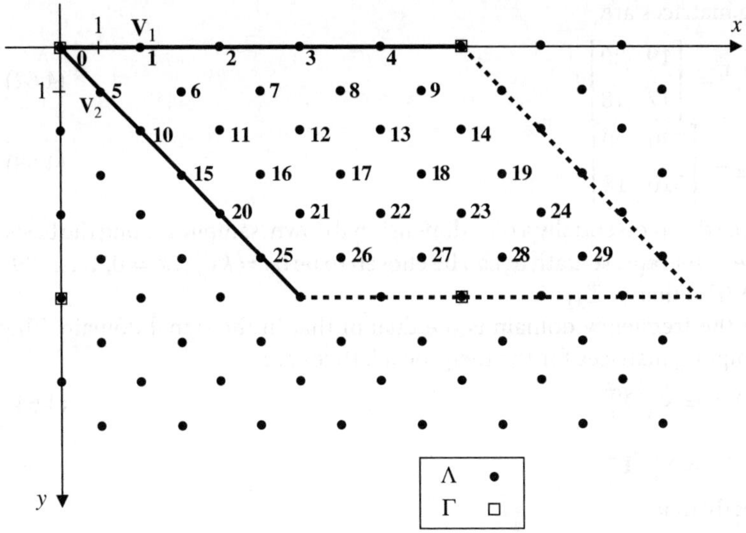

Figure 4.3 Portion of lattices Λ and Γ showing a parallelogram unit cell of Γ and the corresponding coset representatives. The solid edges are included in the unit cell, while the dashed edges are not.

not unique. This means that given lattices Λ and Γ, where $\Gamma \subset \Lambda$, defined by sampling matrices $\mathbf{V}_\Lambda$ and $\mathbf{V}_\Gamma$, we can find new bases for Λ and Γ such that the corresponding basis vectors are collinear. Using the Smith normal form algorithm, we determine for the matrix $\mathbf{M} = \mathbf{V}_\Lambda^{-1}\mathbf{V}_\Gamma$ the matrices $\mathbf{E}_1$, $\mathbf{E}_2$ and $\mathbf{D}$ as described above. From this, we can define the new bases for Λ and Γ as

$$\begin{aligned} \mathbf{V}_\Lambda^{(S)} &= \mathbf{V}_\Lambda \mathbf{E}_1^{-1} \\ \mathbf{V}_\Gamma^{(S)} &= \mathbf{V}_\Gamma \mathbf{E}_2 \end{aligned} \tag{4.57}$$

so that

$$\mathbf{V}_\Gamma^{(S)} = \mathbf{V}_\Lambda^{(S)}\mathbf{D}. \tag{4.58}$$

With these bases, the choice of coset representatives is analogous to the case of rectangular lattices. We can index the coset representatives with D indexes, and define them as

$$\mathbf{b}_{k_1,k_2,\ldots,k_D} = \sum_{\ell=1}^{D} k_\ell \mathbf{v}_{\Lambda,\ell}^{(S)}, \quad 0 \le k_\ell \le d_{\ell\ell} - 1. \tag{4.59}$$

If any of the $d_{\ell\ell} = 1$, the corresponding terms and subscripts are not needed, as in the example to follow. These bases are not necessarily 'nice' or intuitive, but can be very useful in establishing certain results, or for computational purposes.

Continuing with the previous example, using the Smith normal form algorithm, we find that

$$\mathbf{M} = \mathbf{V}_\Lambda^{-1}\mathbf{V}_\Gamma = \begin{bmatrix} 5 & 3 \\ 0 & 6 \end{bmatrix} \tag{4.60}$$

$$= \mathbf{E}_1^{-1}\mathbf{D}\mathbf{E}_2^{-1} = \begin{bmatrix} 18 & 19 \\ 17 & 18 \end{bmatrix} \begin{bmatrix} 30 & 0 \\ 0 & 1 \end{bmatrix} \begin{bmatrix} 3 & -2 \\ -85 & 57 \end{bmatrix}. \tag{4.61}$$

The new sampling matrices are

$$\mathbf{V}_{\Lambda}^{(S)} = \mathbf{V}_{\Lambda}\mathbf{E}_1^{-1} = \begin{bmatrix} 19 & 20 \\ 17 & 18 \end{bmatrix} \tag{4.62}$$

$$\mathbf{V}_{\Gamma}^{(S)} = \mathbf{V}_{\Gamma}\mathbf{E}_2 = \begin{bmatrix} 570 & 20 \\ 510 & 18 \end{bmatrix}. \tag{4.63}$$

Since $d_{22} = 1$, we see this is essentially a one-dimensional down-sampling along the basis vector $\mathbf{v}_{\Lambda,1}^{(S)}$, and the coset representatives can be chosen to be $\mathbf{b}_k = k\mathbf{v}_{\Lambda,1}^{(S)}, k = 0, 1, \ldots, 29$. The structure of Λ/Γ is that of $\mathbb{Z}_{30}$.

The situation in the frequency domain is the dual of that in the signal domain. The corresponding sampling matrices for the reciprocal lattices are

$$\mathbf{V}_{\Lambda^*}^{(S)} = (\mathbf{V}_{\Lambda}^{(S)})^{-T} = \mathbf{V}_{\Lambda}^{-T}\mathbf{E}_1^{T} \tag{4.64}$$

$$\mathbf{V}_{\Gamma^*}^{(S)} = (\mathbf{V}_{\Gamma}^{(S)})^{-T} = \mathbf{V}_{\Gamma}^{-T}\mathbf{E}_2^{-T} \tag{4.65}$$

and it follows directly that

$$\mathbf{V}_{\Lambda^*}^{(S)} = \mathbf{V}_{\Gamma^*}^{(S)}\mathbf{D}. \tag{4.66}$$

Thus, we can choose the coset representatives in the frequency domain to be

$$\mathbf{d}_{i_1,i_2,\ldots,i_D} = \sum_{\ell=1}^{D} i_\ell \mathbf{v}_{\Gamma^*,\ell}^{(S)}, \quad 0 \le i_\ell \le d_{\ell\ell} - 1 \tag{4.67}$$

and the structure of Γ^*/Λ^* is identical to that of Λ/Γ. With these choices for coset representatives in the signal domain and the frequency domain, the multidimensional DFT has the same form as in Example 4.3. Specifically, let us define

$$\tilde{f}_S[k_1, k_2, \ldots, k_D] = \tilde{f}[\mathbf{b}_{k_1,k_2,\ldots,k_D}] \tag{4.68}$$

$$\tilde{F}_S[i_1, i_2, \ldots, i_D] = \tilde{F}[\mathbf{d}_{i_1,i_2,\ldots,i_D}]. \tag{4.69}$$

The term $\mathbf{d}_{i_1,i_2,\ldots,i_D} \cdot \mathbf{b}_{k_1,k_2,\ldots,k_D}$ in the Fourier transform relations can be expressed as

$$\begin{aligned}
\mathbf{d}_{i_1,i_2,\ldots,i_D} \cdot \mathbf{b}_{k_1,k_2,\ldots,k_D} &= \left(\sum_{\ell=1}^{D} i_\ell \mathbf{v}_{\Gamma^*,\ell}^{(S)}\right)^T \left(\sum_{m=1}^{D} k_m \mathbf{v}_{\Lambda,m}^{(S)}\right) \\
&= \begin{bmatrix} i_1 & i_2 & \ldots & i_D \end{bmatrix} (\mathbf{V}_{\Gamma^*}^{(S)})^T \mathbf{V}_{\Lambda}^{(S)} \begin{bmatrix} k_1 \\ k_2 \\ \vdots \\ k_D \end{bmatrix} \\
&= \begin{bmatrix} i_1 & i_2 & \ldots & i_D \end{bmatrix} \mathbf{D}^{-1} (\mathbf{V}_{\Lambda^*}^{(S)})^T \mathbf{V}_{\Lambda}^{(S)} \begin{bmatrix} k_1 \\ k_2 \\ \vdots \\ k_D \end{bmatrix} \\
&= \sum_{\ell=1}^{D} \frac{i_\ell k_\ell}{d_{\ell\ell}}.
\end{aligned} \tag{4.70}$$

Then the multidimensional DFT is given by

$$\tilde{F}_S[i_1, i_2, \ldots, i_D] = \sum_{k_1=0}^{d_{11}-1} \cdots \sum_{k_D=0}^{d_{DD}-1} \tilde{f}_S[k_1, k_2, \ldots, k_D] \exp\left(-\mathrm{j}2\pi\left(\frac{i_1 k_1}{d_{11}} + \ldots \frac{i_D k_D}{d_{DD}}\right)\right). \tag{4.71}$$

This is identical in form to the rectangular case in Equation (4.38).

For the example considered previously, the multidimensional DFT for this configuration is equivalent to a one-dimensional 30-point DFT, and can be evaluated using any efficient 1D FFT algorithm for this data size (e.g., Winograd, Cooley-Tukey, etc.). FFT algorithms have been extensively studied in the one-dimensional case, and in the multidimensional rectangular case (see for example Rao *et al.* (2010) for a recent example). With the above Smith normal form, the general discrete domain periodic Fourier transform is converted to an equivalent rectangular form, Equation (4.71), on which standard FFT algorithms can be applied. Thus, we do not pursue FFT algorithms further in this book, but we do assume they will be used.

We now present a detailed example involving a non-rectangular lattice Λ.

Example 4.6 Let Λ be a hexagonal lattice with sampling matrix $\mathbf{V}_\Lambda = \left[\begin{smallmatrix} 2 & 1 \\ 0 & 1 \end{smallmatrix}\right]$, where we denote the unit of distance as 1 px. Let the periodicity lattice Γ be rectangular, with sampling matrix $\mathbf{V}_\Gamma = \left[\begin{smallmatrix} 4 & 0 \\ 0 & 2 \end{smallmatrix}\right]$. We have that $\mathrm{d}(\Lambda) = 2$ and $\mathrm{d}(\Gamma) = 8$, so that $K = \mathrm{d}(\Gamma)/\mathrm{d}(\Lambda) = 4$. The lattice and sublattice are shown in Figure 4.4(a). We must choose four coset representatives for cosets of Γ in Λ; we arbitrarily choose (according to our preference and convenience) $\mathcal{B} = \{(0,0), (2,0), (1,1), (3,1)\}$, as indicated in Figure 4.4(a). The corresponding matrix $\mathbf{B}$ is

$$\mathbf{B} = \begin{bmatrix} 0 & 2 & 1 & 3 \\ 0 & 0 & 1 & 1 \end{bmatrix}.$$

In the frequency domain, the reciprocal lattices Λ^* and Γ^* are given by sampling matrices

$$\mathbf{V}_{\Lambda^*} = \mathbf{V}_\Lambda^{-T} = \begin{bmatrix} 0.5 & 0.0 \\ -0.5 & 1.0 \end{bmatrix}, \qquad \mathrm{d}(\Lambda^*) = 0.5$$

$$\mathbf{V}_{\Gamma^*} = \mathbf{V}_\Gamma^{-T} = \begin{bmatrix} 0.25 & 0.0 \\ 0.0 & 0.5 \end{bmatrix}, \qquad \mathrm{d}(\Gamma^*) = 0.125,$$

with $\mathrm{d}(\Lambda^*)/\mathrm{d}(\Gamma^*) = 4 = K$ as required. The reciprocal lattices are illustrated in Figure 4.4(b). As four representatives for cosets of Λ^* in Γ^*, we choose (arbitrarily) $\mathcal{D} = \{(0,0), (0.25,0), (-0.25,0), (0.5,0)\}$, as shown in Figure 4.4(b), with corresponding $\mathbf{D}$ matrix

$$\mathbf{D} = \begin{bmatrix} 0.0 & 0.25 & -0.25 & 0.5 \\ 0.0 & 0.0 & 0.0 & 0.0 \end{bmatrix}.$$

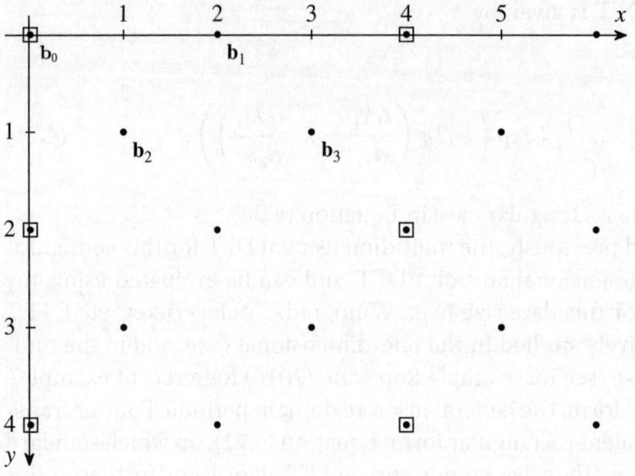

(a)

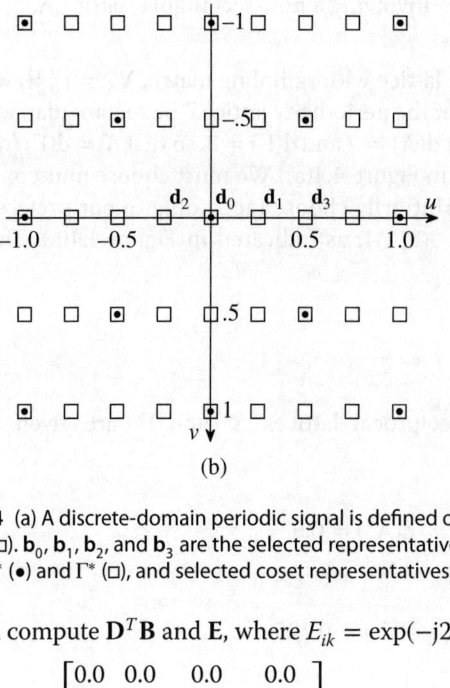

(b)

Figure 4.4 (a) A discrete-domain periodic signal is defined on lattice Λ (•) and has periodicity given by lattice Γ (□). $\mathbf{b}_0$, $\mathbf{b}_1$, $\mathbf{b}_2$, and $\mathbf{b}_3$ are the selected representatives for the cosets of Γ in Λ. (b) Reciprocal lattices Λ^* (•) and Γ^* (□), and selected coset representatives of Λ^* in Γ^*, $\mathbf{d}_0$, $\mathbf{d}_1$, $\mathbf{d}_2$, and $\mathbf{d}_3$.

We then compute $\mathbf{D}^T\mathbf{B}$ and $\mathbf{E}$, where $E_{ik} = \exp(-j2\pi[\mathbf{D}^T\mathbf{B}]_{ik})$,

$$\mathbf{D}^T\mathbf{B} = \begin{bmatrix} 0.0 & 0.0 & 0.0 & 0.0 \\ 0.0 & 0.5 & 0.25 & 0.25 \\ 0.0 & -0.5 & -0.25 & -0.75 \\ 0.0 & 1.0 & 0.5 & 1.5 \end{bmatrix}$$

$$\mathbf{E} = \begin{bmatrix} 1 & 1 & 1 & 1 \\ 1 & -1 & -j & j \\ 1 & -1 & j & -j \\ 1 & 1 & -1 & -1 \end{bmatrix}$$

and verify that $\mathbf{E}\mathbf{E}^H = 4\mathbf{I}_4$.

The discrete-domain periodic Fourier transform is given explicitly by $\tilde{\mathbf{F}} = \mathbf{E}\tilde{\mathbf{f}}$, i.e.

$$\begin{aligned}
\tilde{F}[0,0] &= \tilde{f}[0,0] + \tilde{f}[2,0] + \tilde{f}[1,1] + \tilde{f}[3,1] \\
\tilde{F}[0.25,0] &= \tilde{f}[0,0] - \tilde{f}[2,0] - j\tilde{f}[1,1] + j\tilde{f}[3,1] \\
\tilde{F}[-0.25,0] &= \tilde{f}[0,0] - \tilde{f}[2,0] + j\tilde{f}[1,1] - j\tilde{f}[3,1] \\
\tilde{F}[0.5,0] &= \tilde{f}[0,0] + \tilde{f}[2,0] - \tilde{f}[1,1] - \tilde{f}[3,1].
\end{aligned}$$

The inverse transform is given by $\tilde{\mathbf{f}} = \frac{1}{4}\mathbf{E}^H\tilde{\mathbf{F}}$, i.e.

$$\begin{aligned}
\tilde{f}[0,0] &= \frac{1}{4}(\tilde{F}[0,0] + \tilde{F}[0.25,0] + \tilde{F}[-0.25,0] + \tilde{F}[0.5,0]) \\
\tilde{f}[2,0] &= \frac{1}{4}(\tilde{F}[0,0] - \tilde{F}[0.25,0] - \tilde{F}[-0.25,0] + \tilde{F}[0.5,0]) \\
\tilde{f}[1,1] &= \frac{1}{4}(\tilde{F}[0,0] + j\tilde{F}[0.25,0] - j\tilde{F}[-0.25,0] - \tilde{F}[0.5,0]) \\
\tilde{f}[3,1] &= \frac{1}{4}(\tilde{F}[0,0] - j\tilde{F}[0.25,0] + j\tilde{F}[-0.25,0] - \tilde{F}[0.5,0]).
\end{aligned}$$

The corresponding synthesis equation is

$$\begin{aligned}
\tilde{f}[x,y] &= \frac{1}{4}(\tilde{F}[0,0] + \tilde{F}[0.25,0]\exp(\mathrm{j}2\pi(0.25x)) \\
&\quad + \tilde{F}[-0.25,0]\exp(\mathrm{j}2\pi(-0.25x)) + \tilde{F}[0.5,0]\exp(\mathrm{j}2\pi(0.5x))) \\
&= \frac{1}{4}(\tilde{F}[0,0] + (j)^x\tilde{F}[0.25,0] + (-j)^x\tilde{F}[-0.25,0] + (-1)^x\tilde{F}[0.5,0]) \\
&\quad (x,y) \in \Lambda.
\end{aligned}$$

Although this appears to only depend on x, the values of x belonging to Λ are different for even and odd values of y.

4.7 Vector Space Representation of Images Based on the Discrete-Domain Periodic Fourier Transform

4.7.1 Vector Space Representation of Signals with Finite Extent

The set of discrete-domain periodic signals on a lattice Λ with periodicity lattice Γ has been seen to form a finite-dimensional vector space of dimension $K = \mathrm{d}(\Gamma)/\mathrm{d}(\Lambda)$. The elementary basis for this vector space is composed of the shifted unit samples $\mathcal{T}_{\mathbf{b}_k}\tilde{\delta}_{\Lambda/\Gamma}, k = 0, K-1$ and the expansion of an arbitrary signal with this basis is given by Equation (4.8), $\tilde{f} = \sum_{k=0}^{K-1}\tilde{f}[\mathbf{b}_k]\mathcal{T}_{\mathbf{b}_k}\tilde{\delta}_{\Lambda/\Gamma}$. The orthogonal sinusoidal basis for this vector space is given by $\tilde{\phi}_{\mathbf{d}_i}[\mathbf{x}] = \exp(\mathrm{j}2\pi\mathbf{d}_i \cdot \mathbf{x}), i = 0, K-1$ (Theorem 4.1), leading to the expansion

$$\tilde{f} = \frac{1}{K}\sum_{i=1}^{K-1}\tilde{F}[\mathbf{d}_i]\tilde{\phi}_{\mathbf{d}_i}, \tag{4.72}$$

with $\tilde{F}[\mathbf{d}_i]$ given by Equation (4.19). This is a change of basis operation where the signal is represented by the Fourier coefficients $\tilde{F}[\mathbf{d}_i]$ rather than the signal values $\tilde{f}[\mathbf{b}_k]$. This change of basis can be implemented as a matrix multiplication. If we define the $K \times 1$ column matrices $\tilde{\mathbf{f}} = [\tilde{f}[\mathbf{b}_0], \dots, \tilde{f}[\mathbf{b}_{K-1}]^T]$ and $\tilde{\mathbf{F}} = [\tilde{F}[\mathbf{d}_0], \dots, \tilde{F}[\mathbf{d}_{K-1}]]^T$, then $\tilde{\mathbf{F}} = \mathbf{A}\tilde{\mathbf{f}}$, where $A_{ik} = \exp(-\mathrm{j}2\pi \mathbf{d}_{i-1} \cdot \mathbf{b}_{k-1})$, $i, k = 1, K$. Of course, we can return to the original representation with the inverse transformation $\tilde{\mathbf{f}} = \mathbf{A}^{-1}\widetilde{\mathbf{F}}$.

This expansion can also be used to represent finite-extent signals. Assume that $f[\mathbf{x}]$ is a signal defined on a lattice Λ having a finite region of support $\mathcal{A}$, so that $f[\mathbf{x}] = 0$ for $\mathbf{x} \in \Lambda \backslash \mathcal{A}$. Let Γ be a sublattice of Λ such that there exists a unit cell $\mathcal{P}_\Gamma$ of Γ that contains $\mathcal{A}$. Then, we can consider f within $\mathcal{A}$ to be one period of the periodic signal

$$\tilde{f}[\mathbf{x}] = \sum_{\mathbf{t}\in\Gamma} f[\mathbf{x} - \mathbf{t}]. \tag{4.73}$$

(This operation is known as periodization and will be studied in detail in Chapter 6.) We note that these shifted versions of f do not overlap, so $\tilde{f}[\mathbf{x}] = f[\mathbf{x}]$ for $\mathbf{x} \in \mathcal{P}_\Gamma$. Thus we have the representation

$$f = \frac{1}{K} \sum_{i=0}^{K-1} F[\mathbf{d}_i] \phi_{\mathbf{d}_i}, \tag{4.74}$$

where

$$\phi_{\mathbf{d}_i}[\mathbf{x}] = \begin{cases} \tilde{\phi}_{\mathbf{d}_i}[\mathbf{x}] & \mathbf{x} \in \mathcal{P}_\Gamma, \\ 0 & \mathbf{x} \in \Lambda \backslash \mathcal{P}_\Gamma, \end{cases} \tag{4.75}$$

and $F[\mathbf{d}_i] = \widetilde{F}[\mathbf{d}_i]$ for $i = 0, K-1$. Thus, the discrete-domain periodic Fourier transform can be used in a straightforward way to represent finite extent signals, with the expansion coefficients $F[\mathbf{d}_i]$ efficiently computed using the FFT algorithm.

4.7.2 Block-Based Vector-Space Representation

The representation described above can be applied to an entire image, which can of course be very large (megapixels). In many cases, this tends to *not* be very effective. Such a representation does not capture local properties of the image, which can vary substantially over the full extent of the image. A better approach is often to partition the image into disjoint blocks that are relatively small (8×8 and 16×16 are quite common sizes) and apply the DFT representation independently to each block.

Let f be a signal defined on lattice Λ, with region of support $\mathcal{L}$. Let Γ be a sublattice of Λ which defines the block size, so that a unit cell of Γ would be a prototype block. In most applications in still images, $\Lambda = \mathbb{Z}^2$ and $\Gamma = (N\mathbb{Z})^2$, where N is the block size in each dimension, typically 8 or 16 and $\mathcal{P}_\Gamma$ is a rectangular $N \times N$ block. Let us denote $\mathcal{A} = \mathcal{P}_\Gamma$ and

$$\mathcal{T}_{\mathbf{z}}\mathcal{A} = \{\mathbf{x} - \mathbf{z} \mid \mathbf{x} \in \mathcal{A}\} \quad \text{where } \mathbf{z} \in \Gamma. \tag{4.76}$$

These shifted versions of $\mathcal{A}$ partition Λ and a finite number M of them cover $\mathcal{L}$. The signal within each block can be represented using the DFT basis to get an overall representation of the image using a local basis (i.e. all basis vectors have support in a relatively small local area of the image).

We number the blocks that cover $\mathcal{L}$ from 0 to $M-1$ in some arbitrary fixed order and denote the corresponding points of Γ in each block as $\mathbf{z}^{(m)}$. Then, we define the K orthogonal vectors with region of support $\mathcal{T}_{\mathbf{z}^{(m)}}\mathcal{A}$ as

$$\phi_i^{(m)} = \mathcal{T}_{\mathbf{z}^{(m)}} \phi_{\mathbf{d}_i}, \qquad i = 1, K. \tag{4.77}$$

The set of all such vectors $\{\phi_i^{(m)}, i = 0, K-1; m = 0, M-1\}$ forms an orthogonal basis for the space of images with support on $\mathcal{L}$. The fact that $\langle \phi_i^{(m)} \mid \phi_j^{(n)} \rangle = 0$ if $m \neq n$ is evident since $\phi_i^{(m)}$ and $\phi_j^{(n)}$ are nonzero on disjoint blocks. If $m = n$, then $\langle \phi_i^{(m)} \mid \phi_j^{(m)} \rangle = \langle \phi_{\mathbf{d}_i} \mid \phi_{\mathbf{d}_j} \rangle = K\delta_{ij}$. The resulting representation of the image is

$$f = \frac{1}{K} \sum_{m=0}^{M-1} \sum_{i=0}^{K-1} F^{(m)}[\mathbf{d}_i] \phi_i^{(m)}, \tag{4.78}$$

where

$$\begin{aligned} F^{(m)}[\mathbf{d}_i] &= \langle f \mid \phi_i^{(m)} \rangle \\ &= \sum_{k=0}^{K-1} f[\mathbf{z}^{(m)} + \mathbf{b}_k] \exp(-j2\pi \mathbf{d}_i \cdot \mathbf{b}_k). \end{aligned} \tag{4.79}$$

This approach can be used with any orthogonal basis for blocks of size K. In applications, the DFT is often not the best choice. In particular, the discrete cosine transform (DCT) is better able to capture more of the block's information in fewer coefficients and is used in numerous compression standards. We will return to the DCT in Chapter 12.

Example 4.7 The usual case for application of the block based local basis is when $\Lambda = \mathbb{Z}^2$ and $\Gamma = (N\mathbb{Z})^2$, so that $K = N^2$. Suppose that the image size is $L_1 \times L_2$ and that both L_1 and L_2 are integer multiples of N. Then the number of blocks is $M = (L_1/N)(L_2/N) = M_1 M_2$. In this separable case, it is more convenient to label the blocks with horizontal and vertical indexes (m_1, m_2), with $\mathbf{z}^{(m_1,m_2)} = (m_1 N, m_2 N)$. The block-based partition of the image is illustrated in Figure 4.5.

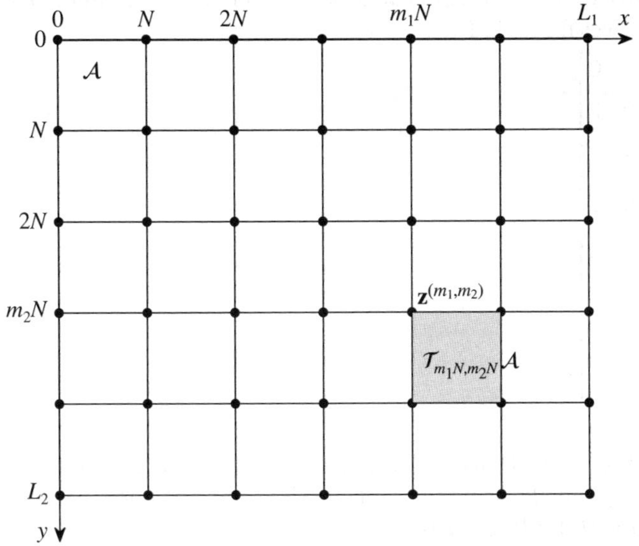

Figure 4.5 Block based partition of an image in separable rectangular case. The image raster $\mathcal{L}$ is partitioned into $M_1 M_2$ copies of the basic block $\mathcal{A}$. One typical block $\mathcal{T}_{m_1N,m_2N}\mathcal{A}$ located at $\mathbf{z}^{(m_1,m_2)}$ is indicated.

The block-based image representation using this double indexing is then

$$f = \frac{1}{N^2} \sum_{m_1=0}^{M_1-1} \sum_{m_2=0}^{M_2-1} \sum_{i_1=0}^{N-1} \sum_{i_2=0}^{N-1} F^{(m_1,m_2)}\left[\frac{i_1}{N}, \frac{i_2}{N}\right] \phi_{i_1,i_2}^{(m_1,m_2)}, \tag{4.80}$$

where

$$F^{(m_1,m_2)}\left[\frac{i_1}{N}, \frac{i_2}{N}\right] = \sum_{k_1=0}^{N-1} \sum_{k_2=0}^{N-1} f[m_1N + k_1, m_2N + k_2] \exp(-\mathrm{j}2\pi(i_1k_1 + i_2k_2)/N). \tag{4.81}$$

The basis vectors are

$$\phi_{i_1,i_2}^{(m_1,m_2)} = \mathcal{T}_{(m_1N, m_2N)} \phi_{i_1,i_2} \tag{4.82}$$

with

$$\phi_{i_1,i_2}[k_1,k_2] = \begin{cases} \exp(\mathrm{j}2\pi(i_1k_1 + i_2k_2)/N) & \text{if } 0 \le k_1, k_2 \le N-1 \\ 0 & \text{otherwise.} \end{cases} \tag{4.83}$$

Problems

1 For each of the following pairs of lattice Λ and sublattice Γ given by their sampling matrices $\mathbf{V}_\Lambda$ and $\mathbf{V}_\Gamma$ respectively: (i) verify that Γ is indeed a sublattice of Λ; (ii) compute the index K of Γ in Λ; (iii) enumerate a set of K coset representatives for Γ in Λ; (iv) find sampling matrices for the reciprocal lattices Λ^* and Γ^*; (v) enumerate a set of K coset representatives for Λ^* in Γ^*.

a) $\mathbf{V}_\Lambda = \begin{bmatrix} 1 & 0 \\ 0 & 1 \end{bmatrix}, \mathbf{V}_\Gamma = \begin{bmatrix} 4 & 0 \\ 0 & 2 \end{bmatrix}.$

b) $\mathbf{V}_\Lambda = \begin{bmatrix} 1 & 0 \\ 0 & 1 \end{bmatrix}, \mathbf{V}_\Gamma = \begin{bmatrix} 4 & 2 \\ 0 & 1 \end{bmatrix}.$

c) $\mathbf{V}_\Lambda = \begin{bmatrix} 2 & 1 \\ 0 & 1 \end{bmatrix}, \mathbf{V}_\Gamma = \begin{bmatrix} 4 & 0 \\ 0 & 2 \end{bmatrix}.$

d) $\mathbf{V}_\Lambda = \begin{bmatrix} 4 & 2 \\ 0 & 1 \end{bmatrix}, \mathbf{V}_\Gamma = \begin{bmatrix} 4 & 2 \\ 0 & 3 \end{bmatrix}.$

2 Find the analysis and synthesis equations for the discrete-domain Fourier series representation of the following signals, using the signal lattices and periodicity lattices of corresponding Problems 1(a)–(d) respectively.

a) $\tilde{f}[\mathbf{x}] = \begin{cases} 1 & \mathbf{x} \in \Gamma \\ 0 & \mathbf{x} \in \Lambda \backslash \Gamma \end{cases}$. Note that here, $\tilde{f} = \tilde{\delta}_{\Lambda/\Gamma}$.

b) $\tilde{f}[\mathbf{x}] = \begin{cases} 1 & \mathbf{x} \in \Gamma \\ 0 & \mathbf{x} \in \Lambda \backslash \Gamma \end{cases}$. Note that here again, $\tilde{f} = \tilde{\delta}_{\Lambda/\Gamma}$.

c) $\tilde{f}[\mathbf{x}] = \begin{cases} 1 & \mathbf{x} \in \Gamma \text{or } \mathbf{x} \in [2,0]^T + \Gamma \\ 0 & \text{otherwise} \end{cases}$.

d) $\tilde{f}[\mathbf{x}] = \begin{cases} 1 & \mathbf{x} \in \Gamma \\ 2 & \mathbf{x} \in [2,1]^T + \Gamma. \\ 3 & \mathbf{x} \in [0,2]^T + \Gamma \end{cases}$

3 Let Λ be a lattice and Γ a sublattice. Let $\mathcal{B}$ be any set of coset representatives for Γ in Λ.

a) Let $\mathbf{x}_0$ be any element of Λ. Show that $\mathcal{B} - \mathbf{x}_0 = \{\mathbf{b} - \mathbf{x}_0 \mid \mathbf{b} \in \mathcal{B}\}$ is also a set of coset representatives for Γ in Λ. This is used in the proof of Property 4.2.

b) Let $\mathbf{A}$ be a trasformation of $\mathbb{R}^D$ such that $\mathbf{A}\Lambda = \Lambda$ and $\mathbf{A}\Gamma = \Gamma$. Show that $\mathbf{A}\mathcal{B} = \{\mathbf{A}\mathbf{b} \mid \mathbf{b} \in \mathcal{B}\}$ is also a set of coset representatives for Γ in Λ. This is used in the proof of Property 4.6.

5

Continuous-Domain Periodic Signals

5.1 Introduction

Chapters 2, 3 and 4 covered the cases of continuous-domain aperiodic, discrete-domain aperiodic, and discrete-domain periodic signals and systems respectively. This chapter concludes the series with the case of continuous-domain periodic signals. The results are very similar in flavor to the previous three cases, so a fairly brief rundown will be presented. This topic constitutes the multidimensional extension of one-dimensional Fourier series representations of periodic signals. In the multidimensional case, periodicity is specified by a lattice Γ. A periodic signal $\tilde{f}$ satisfies

$$\tilde{f}(\mathbf{x}+\mathbf{t}) = \tilde{f}(\mathbf{x}) \quad \text{for all } \mathbf{t} \in \Gamma, \quad \text{and for all } \mathbf{x} \in \mathbb{R}^D. \tag{5.1}$$

Our only constraint here is that Γ is a D-dimensional lattice in $\mathbb{R}^D$. Again we use the $\tilde{}$ to emphasize that a signal is periodic, and round parentheses to denote that it is a continuous-domain signal.

It is possible to have signals with periodicity specified by a lattice of dimension less than D. For example, a two-dimensional signal could be periodic in the horizontal direction only (or any direction). This would be one example among many types of hybrid signals whose domain could be partly continuous, partly discrete (e.g., a scanned analog TV signal is discrete vertically and temporally and continuous horizontally), as well as periodic in some dimensions (or subspaces of $\mathbb{R}^D$) and not others. This interesting topic is beyond the scope of our treatment, but the extension is generally quite straightforward. Many such cases are discussed in a general setting in Cariolaro (2011).

5.2 Continuous-Domain Periodic Signals

Let Γ be a D-dimensional lattice in $\mathbb{R}^D$. A Γ-periodic signal satisfies Equation (5.1). To completely specify the periodic signal, we need its values in one unit cell of Γ, denoted $\mathcal{P}_\Gamma$. We can denote this essential domain of the signal as $\mathbb{R}^D/\Gamma$. It represents the set of cosets of Γ in $\mathbb{R}^D$, with addition modulo Γ. Of course, we can describe any number of specific Γ-periodic signals, but, as in the previous chapters, two are of particular note: the complex sinusoidal and Dirac delta Γ-periodic signals.

Multidimensional Signal and Color Image Processing Using Lattices, First Edition. Eric Dubois.

Companion website: www.wiley.com/go/Dubois/multiSP

A complex sinusoidal signal with frequency vector $\mathbf{u} \in \mathbb{R}^D$ is given by

$$\tilde{\phi}_{\mathbf{u}}(\mathbf{x}) = \exp(\mathrm{j}2\pi\mathbf{u}\cdot\mathbf{x}) \tag{5.2}$$

where $\mathbf{u}$ and $\mathbf{x}$ both belong to $\mathbb{R}^D$. However, for the signal to be Γ-periodic, we require

$$\exp(\mathrm{j}2\pi\mathbf{u}\cdot(\mathbf{x}+\mathbf{t})) = \exp(\mathrm{j}2\pi\mathbf{u}\cdot\mathbf{x}) \quad \text{for all } \mathbf{t}\in\Gamma, \tag{5.3}$$

for all $\mathbf{x}$. This in turn implies that $\exp(\mathrm{j}2\pi\mathbf{u}\cdot\mathbf{t}) = 1$ for all $\mathbf{t}\in\Gamma$, which can occur if $\mathbf{u}\cdot\mathbf{t}\in\mathbb{Z}$ for all $\mathbf{t}\in\Gamma$, placing a restriction on possible values of $\mathbf{u}$. As we have seen, the set of all $\mathbf{u}$ such that $\mathbf{u}\cdot\mathbf{t}\in\mathbb{Z}$ for all $\mathbf{t}\in\Gamma$ forms a lattice, namely the reciprocal lattice Γ^* (see Theorem 13.3 for a proof). Thus, the set of Γ-periodic complex sinusoids is $\{\exp(\mathrm{j}2\pi\mathbf{u}\cdot\mathbf{x}) \mid \mathbf{u}\in\Gamma^*\}$. As we shall see, any suitably regular Γ-periodic signal can be synthesized as a superposition of such Γ-periodic complex sinusoids.

The second category of useful signals are the Γ-periodic Dirac deltas, denoted

$$\tilde{\delta}(\mathbf{x}) = \sum_{\mathbf{t}\in\Gamma}\delta(\mathbf{x}-\mathbf{t}) \tag{5.4}$$

where $\delta(\mathbf{x})$ is the conventional Dirac delta in $\mathbb{R}^D$. This is seen to be an array of Dirac deltas, situated on the points of Γ. If we need to emphasize the periodicity lattice, we denote the Γ-periodic Dirac delta as $\tilde{\delta}_\Gamma$. Similarly to previous cases, any linear shift-invariant system for Γ-periodic signals will be characterized by its response to this signal.

5.3 Linear Shift-Invariant Systems

Let $\mathcal{S}_{\mathbb{R}^D/\Gamma}$ denote a vector space of D-dimensional Γ-periodic signals. It could be as simple as the space of all bounded signals, although we would generally like to impose some regularity on these signals. A system $\mathcal{H}$ is any operator that produces a well-defined output element of $\mathcal{S}_{\mathbb{R}^D/\Gamma}$ for any input from this space, symbolically, $\mathcal{H} : \mathcal{S}_{\mathbb{R}^D/\Gamma} \to \mathcal{S}_{\mathbb{R}^D/\Gamma}$. The system is linear if $\mathcal{H}(\alpha_1\tilde{f}_1 + \alpha_2\tilde{f}_2) = \alpha_1\mathcal{H}(\tilde{f}_1) + \alpha_2\mathcal{H}(\tilde{f}_2)$ for all $\alpha_1, \alpha_2 \in \mathbb{C}$ and for all $\tilde{f}_1, \tilde{f}_2 \in \mathcal{S}_{\mathbb{R}^D/\Gamma}$. The usual shift system $\mathcal{T}_{\mathbf{d}}$ for any $\mathbf{d}\in\mathbb{R}^D$ is such a linear system. As before, we define the shift system by

$$(\mathcal{T}_{\mathbf{d}}\tilde{f})(\mathbf{x}) = \tilde{f}(\mathbf{x}-\mathbf{d}). \tag{5.5}$$

We note that

$$(\mathcal{T}_{\mathbf{d}}\tilde{f})(\mathbf{x}+\mathbf{t}) = \tilde{f}(\mathbf{x}+\mathbf{t}-\mathbf{d}) = \tilde{f}(\mathbf{x}-\mathbf{d}) = (\mathcal{T}_{\mathbf{d}}\tilde{f})(\mathbf{x}), \qquad \mathbf{t}\in\Gamma, \tag{5.6}$$

so that the output of the shift system is also Γ-periodic. A linear system on $\mathcal{S}_{\mathbb{R}^D/\Gamma}$ is said to be shift-invariant if

$$\mathcal{H}(\mathcal{T}_{\mathbf{d}}\tilde{f}) = \mathcal{T}_{\mathbf{d}}(\mathcal{H}\tilde{f}) \quad \text{for all } \mathbf{d}\in\mathbb{R}^D. \tag{5.7}$$

However, it is sufficient to specify this condition for just $\mathbf{d}\in\mathcal{P}_\Gamma$.

Following the same line of reasoning as for the continuous-domain aperiodic case, let $\tilde{h} = \mathcal{H}\tilde{\delta}$ be the response of $\mathcal{H}$ to the Γ-periodic Dirac delta. If we represent an arbitrary periodic signal $\tilde{f}$ as

$$\tilde{f} = \int_{\mathcal{P}_\Gamma}(\tilde{f}(\mathbf{s})\,d\mathbf{s})\mathcal{T}_{\mathbf{s}}\tilde{\delta}, \tag{5.8}$$

and apply linearity, we find

$$\mathcal{H}\tilde{f} = \int_{\mathcal{P}_\Gamma} (\tilde{f}(\mathbf{s})\, d\mathbf{s})\mathcal{H}\mathcal{T}_{\mathbf{s}}\tilde{\delta}. \tag{5.9}$$

Assuming that $\mathcal{H}$ is shift invariant, so that $\mathcal{H}\mathcal{T}_{\mathbf{s}}\tilde{\delta} = \mathcal{T}_{\mathbf{s}}\mathcal{H}\tilde{\delta} = \mathcal{T}_{\mathbf{s}}\tilde{h}$, then

$$\mathcal{H}\tilde{f} = \int_{\mathcal{P}_\Gamma} (\tilde{f}(\mathbf{s})\, d\mathbf{s})\mathcal{T}_{\mathbf{s}}\tilde{h}. \tag{5.10}$$

Evaluating at $\mathbf{x}$, we find

$$(\mathcal{H}\tilde{f})(\mathbf{x}) = \int_{\mathcal{P}_\Gamma} \tilde{f}(\mathbf{s})\tilde{h}(\mathbf{x}-\mathbf{s})\, d\mathbf{s}. \tag{5.11}$$

This is the Γ-periodic, D-dimensional convolution. Note that for every $\mathbf{x} \in \mathbb{R}^D$, there is a unique $\langle \mathbf{x} \rangle_\Gamma$ in $\mathcal{P}_\Gamma$ such that $\mathbf{x} - \langle \mathbf{x} \rangle_\Gamma \in \Gamma$, and of course $\tilde{f}(\langle \mathbf{x} \rangle_\Gamma) = \tilde{f}(\mathbf{x})$. Thus we can rewrite the convolution integral Equation (5.11) to use only signal values in the basic period $\mathcal{P}_\Gamma$.

$$(\mathcal{H}\tilde{f})(\mathbf{x}) = \int_{\mathcal{P}_\Gamma} \tilde{f}(\mathbf{s})\tilde{h}(\langle \mathbf{x}-\mathbf{s}\rangle_\Gamma)\, d\mathbf{s}. \tag{5.12}$$

The periodic convolution can be shown to commutative, just like all the other forms of convolution we have seen.

$$\begin{aligned}
\int_{\mathcal{P}_\Gamma} \tilde{f}(\mathbf{s})\tilde{h}(\mathbf{x}-\mathbf{s})\, d\mathbf{s} &= (-1)^D \int_{\mathbf{x}-\mathcal{P}_\Gamma} \tilde{f}(\mathbf{x}-\mathbf{t})\tilde{h}(\mathbf{t})\, d\mathbf{t} \qquad (\mathbf{t} = \mathbf{x}-\mathbf{s}) \\
&= \int_{\mathcal{P}_\Gamma - \mathbf{x}} \tilde{f}(\mathbf{x}-\mathbf{t})\tilde{h}(\mathbf{t})\, d\mathbf{t} \\
&= \int_{\mathcal{P}_\Gamma} \tilde{f}(\mathbf{x}-\mathbf{t})\tilde{h}(\mathbf{t})\, d\mathbf{t}
\end{aligned} \tag{5.13}$$

since $\mathcal{P}_\Gamma - \mathbf{x}$ is also a unit cell of Γ.

Now we show that the complex exponentials are again the eigenfunctions of linear shift-invariant systems and derive the frequency response. Let $\tilde{f}(\mathbf{x}) = \tilde{\phi}_{\mathbf{u}}(\mathbf{x}) = \exp(\mathrm{j}2\pi\mathbf{u}\cdot\mathbf{x})$ for some $\mathbf{u} \in \Gamma^*$. Then

$$\begin{aligned}
(\mathcal{H}\tilde{\phi}_{\mathbf{u}})(\mathbf{x}) &= \int_{\mathcal{P}_\Gamma} \tilde{h}(\mathbf{s})\tilde{\phi}_{\mathbf{u}}(\mathbf{x}-\mathbf{s})\, d\mathbf{s} \\
&= \int_{\mathcal{P}_\Gamma} \tilde{h}(\mathbf{s}) \exp(\mathrm{j}2\pi\mathbf{u}\cdot(\mathbf{x}-\mathbf{s}))\, d\mathbf{s} \\
&= \left(\int_{\mathcal{P}_\Gamma} \tilde{h}(\mathbf{s}) \exp(-\mathrm{j}2\pi\mathbf{u}\cdot\mathbf{s})\, d\mathbf{s} \right) \tilde{\phi}_{\mathbf{u}}(\mathbf{x}) \\
&= \tilde{H}[\mathbf{u}]\tilde{\phi}_{\mathbf{u}}(\mathbf{x})
\end{aligned} \tag{5.14}$$

i.e. the complex sinusoid is an eigenfunction of the LSI system, with corresponding eigenvalue

$$\tilde{H}[\mathbf{u}] = \int_{\mathcal{P}_\Gamma} \tilde{h}(\mathbf{s}) \exp(-\mathrm{j}2\pi\mathbf{u}\cdot\mathbf{s})\, d\mathbf{s} \quad \mathbf{u} \in \Gamma^*. \tag{5.15}$$

Note that the frequency variable is discrete, so that $\tilde{H}[\mathbf{u}]$ is a discrete-domain aperiodic function of frequency. This has defined the continuous-domain periodic Fourier transform. Although $\tilde{H}[\mathbf{u}]$ is not periodic, it inherits the tilde from $\tilde{h}$; this is consistent with our previous usage.

5.4 Continuous-Domain Periodic Fourier Transform

As in all previous cases, the eigenvalues of the LSI system are given by a transformation of the impulse response, which defines the Fourier transform for this domain. Thus for any Γ-periodic signal $\tilde{f} \in S_{\mathbb{R}^D/\Gamma}$, we define

$$\tilde{F}[\mathbf{u}] = \int_{\mathcal{P}_\Gamma} \tilde{f}(\mathbf{x}) \exp(-\mathrm{j}2\pi\mathbf{u}\cdot\mathbf{x})\, \mathrm{d}\mathbf{x} \quad \mathbf{u} \in \Gamma^*. \tag{5.16}$$

We give, without proof at this time, the corresponding inverse Fourier transform

$$\tilde{f}(\mathbf{x}) = \frac{1}{\mathrm{d}(\Gamma)} \sum_{\mathbf{u}\in\Gamma^*} \tilde{F}[\mathbf{u}] \exp(\mathrm{j}2\pi\mathbf{u}\cdot\mathbf{x}), \tag{5.17}$$

where $\mathrm{d}(\Gamma)$ is the determinant of the lattice Γ. This automatically has the correct periodicity, due to the periodicity of each term $\exp(\mathrm{j}2\pi\mathbf{u}\cdot\mathbf{x})$.

5.5 Properties of the Continuous-Domain Periodic Fourier Transform

We provide here the properties of the continuous-domain periodic Fourier transform, all very similar to the three cases already presented. These properties are the multidimensional extension of properties of the ordinary one-dimensional continuous-domain Fourier series. For this reason, we refer to a transform pair as a continuous-domain Fourier series pair, denoted $\tilde{f}(\mathbf{x}) \xleftrightarrow{\text{CDFS}} \tilde{F}[\mathbf{u}]$. Table 5.1 lists the main properties that we shall prove here. In the table and in the proofs, $\mathcal{P}_\Gamma$ is an arbitrary unit cell of the D-dimensional lattice Γ.

We provide proofs of these properties, which are for the most part, direct adaptations of proofs already presented for the other cases. We assume that the signal $\tilde{f}$ is bounded, and since the integrals have finite support, convergence should not be an issue; we suppose that the signals are sufficiently regular for this to be the case.

Property 5.1 *Linearity*: $A\tilde{f}(\mathbf{x}) + B\tilde{g}(\mathbf{x}) \xleftrightarrow{\text{CDFS}} A\tilde{F}[\mathbf{u}] + B\tilde{G}[\mathbf{u}]$.

Proof: Let $\tilde{q}(\mathbf{x}) = A\tilde{f}(\mathbf{x}) + B\tilde{g}(\mathbf{x})$. Then

$$\begin{aligned}\tilde{Q}[\mathbf{u}] &= \int_{\mathcal{P}_\Gamma} (A\tilde{f}(\mathbf{x}) + B\tilde{g}(\mathbf{x})) \exp(-\mathrm{j}2\pi\mathbf{u}\cdot\mathbf{x})\, \mathrm{d}\mathbf{x} \\ &= A\int_{\mathcal{P}_\Gamma} \tilde{f}[\mathbf{x}] \exp(-\mathrm{j}2\pi\mathbf{u}\cdot\mathbf{x})\, \mathrm{d}\mathbf{x} + B\int_{\mathcal{P}_\Gamma} \tilde{g}(\mathbf{x}) \exp(-\mathrm{j}2\pi\mathbf{u}\cdot\mathbf{x})\, \mathrm{d}\mathbf{x} \\ &= A\tilde{F}[\mathbf{u}] + B\tilde{G}[\mathbf{u}]. \end{aligned} \tag{5.18}$$

□

Table 5.1 Properties of the multidimensional continuous-domain periodic Fourier transform with periodicity lattice Γ.

	$\tilde{f}(\mathbf{x}) = \frac{1}{d(\Gamma)}\sum_{\mathbf{u}\in\Gamma^*}\tilde{F}[\mathbf{u}]\exp(j2\pi\mathbf{u}\cdot\mathbf{x})$	$\tilde{F}[\mathbf{u}] = \int_{\mathcal{P}_\Gamma}\tilde{f}(\mathbf{x})\exp(-j2\pi\mathbf{u}\cdot\mathbf{x})$
(5.1)	$A\tilde{f}(\mathbf{x}) + B\tilde{g}(\mathbf{x})$	$A\tilde{F}[\mathbf{u}] + B\tilde{G}[\mathbf{u}]$
(5.2)	$\tilde{f}(\mathbf{x}-\mathbf{x}_0)$	$\tilde{F}[\mathbf{u}]\exp(-j2\pi\mathbf{u}\cdot\mathbf{x}_0)$
(5.3)	$\tilde{f}(\mathbf{x})\exp(j2\pi\mathbf{u}_0\cdot\mathbf{x})$	$\tilde{F}[\mathbf{u}-\mathbf{u}_0]$
(5.4)	$\int_{\mathcal{P}_\Gamma}\tilde{f}(\mathbf{s})\tilde{g}(\mathbf{x}-\mathbf{s})\,d\mathbf{s}$	$\tilde{F}[\mathbf{u}]\tilde{G}[\mathbf{u}]$
(5.5)	$\tilde{f}(\mathbf{x})\tilde{g}(\mathbf{x})$	$\frac{1}{d(\Gamma)}\sum_{\mathbf{w}\in\Gamma^*}\tilde{F}[\mathbf{w}]\tilde{G}[\mathbf{u}-\mathbf{w}]$
(5.6)	$\tilde{f}(\mathbf{A}\mathbf{x})$	$\tilde{F}[\mathbf{A}^{-T}\mathbf{u}]$
(5.7)	$\nabla_{\mathbf{x}}\tilde{f}(\mathbf{x})$	$j2\pi\mathbf{u}\tilde{F}[\mathbf{u}]$
(5.8)	$\tilde{f}^*(\mathbf{x})$	$\tilde{F}^*[-\mathbf{u}]$
(5.9)	$F(\mathbf{x})$	$\frac{1}{d(\Gamma^*)}f[-\mathbf{u}]$
(5.10)	$\int_{\mathcal{P}_\Gamma}\tilde{f}(\mathbf{x})\tilde{g}^*(\mathbf{x})\,d\mathbf{x} = \frac{1}{d(\Gamma)}\sum_{\mathbf{u}\in\Gamma^*}\tilde{F}[\mathbf{u}]\tilde{G}^*[\mathbf{u}]$	

Property 5.2 *Shift*: $\tilde{f}(\mathbf{x}-\mathbf{x}_0) \xleftrightarrow{\text{CDFS}} \tilde{F}[\mathbf{u}]\exp(-j2\pi\mathbf{u}\cdot\mathbf{x}_0)$.

Proof: Let $\tilde{g}(\mathbf{x}) = \tilde{f}(\mathbf{x}-\mathbf{x}_0)$ for some $\mathbf{x}_0 \in \mathbb{R}^D$ (although it is sufficient to consider $\mathbf{x}_0 \in \mathcal{P}_\Gamma$). Then

$$\begin{aligned}\tilde{G}[\mathbf{u}] &= \int_{\mathcal{P}_\Gamma}\tilde{f}(\mathbf{x}-\mathbf{x}_0)\exp(-j2\pi\mathbf{u}\cdot\mathbf{x})\,d\mathbf{x}\\ &= \int_{\mathcal{P}_\Gamma-\mathbf{x}_0}\tilde{f}(\mathbf{s})\exp(-j2\pi\mathbf{u}\cdot(\mathbf{s}+\mathbf{x}_0))\,d\mathbf{s} \quad (\mathbf{s}=\mathbf{x}-\mathbf{x}_0)\\ &= \left(\int_{\mathcal{P}_\Gamma}f(\mathbf{s})\exp(-j2\pi\mathbf{u}\cdot\mathbf{s})\,d\mathbf{s}\right)\exp(-j2\pi\mathbf{u}\cdot\mathbf{x}_0)\\ &= \tilde{F}[\mathbf{u}]\exp(-j2\pi\mathbf{u}\cdot\mathbf{x}_0).\end{aligned} \tag{5.19}$$

since $\mathcal{P}_\Gamma - \mathbf{x}_0$ is also a unit cell of Γ. □

Property 5.3 *Modulation*: $\tilde{f}(\mathbf{x})\exp(j2\pi\mathbf{u}_0\cdot\mathbf{x}) \xleftrightarrow{\text{CDFS}} \tilde{F}[\mathbf{u}-\mathbf{u}_0]$.

Proof: Let $\tilde{g}(\mathbf{x}) = \tilde{f}(\mathbf{x})\exp(j2\pi\mathbf{u}_0\cdot\mathbf{x})$ for some $\mathbf{u}_0 \in \Gamma^*$. Then

$$\begin{aligned}\tilde{G}[\mathbf{u}] &= \int_{\mathcal{P}_\Gamma}\tilde{f}(\mathbf{x})\exp(j2\pi\mathbf{u}_0\cdot\mathbf{x})\exp(-j2\pi\mathbf{u}\cdot\mathbf{x})\,d\mathbf{x}\\ &= \int_{\mathcal{P}_\Gamma}\tilde{f}(\mathbf{x})\exp(-j2\pi(\mathbf{u}-\mathbf{u}_0)\cdot\mathbf{x})\,d\mathbf{x}\\ &= \tilde{F}[\mathbf{u}-\mathbf{u}_0].\end{aligned} \tag{5.20}$$

□

Property 5.4 *Convolution:* $\int_{\mathcal{P}_\Gamma} \tilde{f}(\mathbf{s})\tilde{g}(\mathbf{x}-\mathbf{s})\,d\mathbf{s} \stackrel{\text{CDFS}}{\longleftrightarrow} \tilde{F}[\mathbf{u}]\tilde{G}[\mathbf{u}]$.

Proof: Let $\tilde{q}[\mathbf{x}] = \int_{\mathcal{P}_\Gamma} \tilde{f}(\mathbf{s})\tilde{g}(\mathbf{x}-\mathbf{s})\,d\mathbf{s}$. Then

$$\begin{aligned}
\tilde{Q}[\mathbf{u}] &= \int_{\mathcal{P}_\Gamma}\left(\int_{\mathcal{P}_\Gamma} \tilde{f}(\mathbf{s})\tilde{g}(\mathbf{x}-\mathbf{s})\,d\mathbf{s}\right)\exp(-j2\pi\mathbf{u}\cdot\mathbf{x})\,d\mathbf{x}\\
&= \int_{\mathcal{P}_\Gamma}\int_{\mathcal{P}_\Gamma-\mathbf{s}} \tilde{f}(\mathbf{s})\tilde{g}(\mathbf{t})\exp(-j2\pi\mathbf{u}\cdot(\mathbf{t}+\mathbf{s}))\,d\mathbf{t}ds \quad (\mathbf{t}=\mathbf{x}-\mathbf{s})\\
&= \int_{\mathcal{P}_\Gamma} \tilde{f}(\mathbf{s})\exp(-j2\pi\mathbf{u}\cdot\mathbf{s})\,d\mathbf{s}\int_{\mathcal{P}_\Gamma}\tilde{g}(\mathbf{t})\exp(-j2\pi\mathbf{u}\cdot\mathbf{t})\,d\mathbf{t}\\
&= \tilde{F}[\mathbf{u}]\tilde{G}[\mathbf{u}],
\end{aligned} \tag{5.21}$$

since $\mathcal{P}_\Gamma - \mathbf{s}$ is also a unit cell of Γ. □

Note that, using this property, commutativity and associativity of complex multiplication imply commutativity and associativity of convolution.

Property 5.5 *Multiplication:* $\tilde{f}(\mathbf{x})\tilde{g}(\mathbf{x}) \stackrel{\text{CDFS}}{\longleftrightarrow} \frac{1}{d(\Gamma)}\sum_{\mathbf{w}\in\Gamma^*}\tilde{F}[\mathbf{w}]\tilde{G}[\mathbf{u}-\mathbf{w}]$.

Proof: Let $\tilde{q}(\mathbf{x}) = \tilde{f}(\mathbf{x})\tilde{g}(\mathbf{x})$. Then

$$\begin{aligned}
\tilde{Q}[\mathbf{u}] &= \int_{\mathcal{P}_\Gamma} \tilde{f}(\mathbf{x})\tilde{g}(\mathbf{x})\exp(-j2\pi\mathbf{u}\cdot\mathbf{x})\,d\mathbf{x}\\
&= \int_{\mathcal{P}_\Gamma} \tilde{g}(\mathbf{x})\left(\frac{1}{d(\Gamma)}\sum_{\mathbf{w}\in\Gamma^*}\tilde{F}[\mathbf{w}]\exp(j2\pi\mathbf{w}\cdot\mathbf{x})\right)\exp(-j2\pi\mathbf{u}\cdot\mathbf{x})\,d\mathbf{x}\\
&= \frac{1}{d(\Gamma)}\sum_{\mathbf{w}\in\Gamma^*}\tilde{F}[\mathbf{w}]\left(\int_{\mathcal{P}_\Gamma}\tilde{g}(\mathbf{x})\exp(-j2\pi(\mathbf{u}-\mathbf{w})\cdot\mathbf{x})\,d\mathbf{x}\right)\\
&= \frac{1}{d(\Gamma)}\sum_{\mathbf{w}\in\Gamma^*}\tilde{F}[\mathbf{w}]\tilde{G}[\mathbf{u}-\mathbf{w}].
\end{aligned} \tag{5.22}$$

□

Property 5.6 *Automorphism of the domain:* If $\mathbf{A}\Gamma = \Gamma$, then $\tilde{f}(\mathbf{A}\mathbf{x}) \stackrel{\text{CDFS}}{\longleftrightarrow} \tilde{F}[\mathbf{A}^{-T}\mathbf{u}]$.

Proof: For $\mathbf{A}$ to be an admissible automorphism of the domain, we need $\tilde{f}(\mathbf{A}\mathbf{x})$ to be Γ-periodic, i.e. $\tilde{f}(\mathbf{A}(\mathbf{x}+\mathbf{s})) = \tilde{f}(\mathbf{A}\mathbf{x})$ for all $\mathbf{s}\in\Gamma$ and thus $\mathbf{A}\mathbf{s}\in\Gamma$ for all $\mathbf{s}\in\Gamma$. Since $\mathbf{A}$ is an automorphism, it must be a point symmetry of Γ (see Chapter 12), and $|\det(\mathbf{A})| = 1$.

Let $\tilde{g}(\mathbf{x}) = \tilde{f}(\mathbf{A}\mathbf{x})$ for an admissible $\mathbf{A}$. Then

$$\tilde{G}[\mathbf{u}] = \int_{\mathcal{P}_\Gamma}\tilde{f}(\mathbf{A}\mathbf{x})\exp(-j2\pi\mathbf{u}\cdot\mathbf{x})\,d\mathbf{x}. \tag{5.23}$$

Since $\mathbf{A}$ is a point symmetry of Γ, $\{\mathbf{A}\mathbf{x} \mid \mathbf{x}\in\mathcal{P}_\Gamma\}$ is a unit cell of Γ; call it $\mathcal{P}'_\Gamma$. We make the change of variables $\mathbf{s} = \mathbf{A}\mathbf{x}$ in the integral; note that the Jacobian $\frac{\partial\mathbf{s}}{\partial\mathbf{x}} = \det(\mathbf{A})$ and thus $d\mathbf{s} = |\det(\mathbf{A})|d\mathbf{x} = d\mathbf{x}$. Thus

$$\begin{aligned}
\tilde{G}[\mathbf{u}] &= \int_{\mathcal{P}'_\Gamma}\tilde{f}(\mathbf{s})\exp(-j2\pi\mathbf{u}\cdot(\mathbf{A}^{-1}\mathbf{s}))d\mathbf{s}\\
&= \int_{\mathcal{P}'_\Gamma}\tilde{f}(\mathbf{s})\exp(-j2\pi(\mathbf{A}^{-T}\mathbf{u})\cdot\mathbf{s})\,d\mathbf{s}.
\end{aligned} \tag{5.24}$$

Note that $\mathbf{A}^{-T}\mathbf{u} \in \Gamma^*$, the argument of the integral is Γ-periodic, so the integral can be over any unit cell. Thus

$$\begin{aligned}\tilde{G}[\mathbf{u}] &= \int_{\mathcal{P}_\Gamma} \tilde{f}(\mathbf{x}) \exp(-j2\pi(\mathbf{A}^{-T}\mathbf{u}) \cdot \mathbf{x})\, d\mathbf{x} \\ &= \tilde{F}[\mathbf{A}^{-T}\mathbf{u}]. \qquad (5.25)\end{aligned}$$

□

Property 5.7 *Differentiation*: $\nabla_{\mathbf{x}}\tilde{f}(\mathbf{x}) \overset{\text{CDFS}}{\longleftrightarrow} j2\pi\mathbf{u}\tilde{F}[\mathbf{u}]$.

Proof: We define the gradient vector $\tilde{\mathbf{g}}(\mathbf{x}) = \nabla_{\mathbf{x}}\tilde{f}(\mathbf{x}) = \left[\frac{\partial\tilde{f}}{\partial x_1}, \dots, \frac{\partial\tilde{f}}{\partial x_D}\right]^T$. Taking the derivative of the synthesis equation for $\tilde{f}$ to get the synthesis equation for $\tilde{g}_i$,

$$g_i(\mathbf{x}) = \frac{\partial\tilde{f}}{\partial x_i}(\mathbf{x}) = \frac{1}{d(\Gamma)} \sum_{\mathbf{u}\in\Gamma^*} \tilde{F}[\mathbf{u}](j2\pi u_i) \exp(j2\pi\mathbf{u} \cdot \mathbf{x}). \qquad (5.26)$$

Thus $\tilde{G}_i[\mathbf{u}] = j2\pi u_i\tilde{F}(\mathbf{u})$ and in matrix form

$$\tilde{\mathbf{G}}(\mathbf{u}) = j2\pi\mathbf{u}\tilde{F}[\mathbf{u}]. \qquad (5.27)$$

□

Property 5.8 *Complex conjugation*: $\tilde{f}^*(\mathbf{x}) \overset{\text{CDFS}}{\longleftrightarrow} \tilde{F}^*[-\mathbf{u}]$

Proof: Let $\tilde{g}(\mathbf{x}) = \tilde{f}^*(\mathbf{x})$. Then

$$\begin{aligned}\tilde{G}[\mathbf{u}] &= \int_{\mathcal{P}_\Gamma} \tilde{f}^*(\mathbf{x}) \exp(-j2\pi\mathbf{u} \cdot \mathbf{x})\, d\mathbf{x} \\ &= \int_{\mathcal{P}_\Gamma} (\tilde{f}(\mathbf{x}) \exp(j2\pi\mathbf{u} \cdot \mathbf{x}))^*\, d\mathbf{x} \\ &= \left(\int_{\mathcal{P}_\Gamma} \tilde{f}(\mathbf{x}) \exp(j2\pi\mathbf{u} \cdot \mathbf{x})\, d\mathbf{x}\right)^* \\ &= \tilde{F}^*[-\mathbf{u}]. \qquad (5.28)\end{aligned}$$

□

Property 5.9 *Duality*: $F(\mathbf{x}) \overset{\text{CDFS}}{\longleftrightarrow} \frac{1}{d(\Gamma^*)}f[-\mathbf{u}]$ where $f[\mathbf{x}]$ is a discrete-domain aperiodic signal on lattice Γ^* with Fourier transform $F(\mathbf{u})$. (Note that $F(\mathbf{u})$ is periodic, but we don't use the $\tilde{}$ since it is the Fourier transform of an aperiodic discrete-domain signal.)

Proof: Let $\tilde{g}(\mathbf{x}) = F(\mathbf{x})$ for $\mathbf{x} \in \mathbb{R}^D$, which is Γ-periodic. Then

$$\begin{aligned}\tilde{G}[\mathbf{u}] &= \int_{\mathcal{P}_\Gamma} F(\mathbf{x}) \exp(-j2\pi\mathbf{u} \cdot \mathbf{x})\, d\mathbf{x} \\ &= \frac{1}{d(\Gamma^*)}f[-\mathbf{u}]. \qquad (5.29)\end{aligned}$$

□

Property 3.9. *Duality*: If $\tilde{f}(\mathbf{x}) \overset{\text{CDFS}}{\longleftrightarrow} \tilde{F}[\mathbf{u}]$, where $\tilde{f}$ is a continuous-domain, Γ-periodic signal, then $\tilde{F}[\mathbf{x}] \overset{\text{DDFT}}{\longleftrightarrow} d(\Gamma)\tilde{f}(-\mathbf{u})$. (This is Property 3.9 that we deferred from Chapter 3.)

Proof: Let $g[\mathbf{x}] = \tilde{F}[\mathbf{x}]$, for $\mathbf{x} \in \Gamma^*$. Then

$$\begin{aligned} G(\mathbf{u}) &= \sum_{\mathbf{x}\in\Gamma^*} \tilde{F}[\mathbf{x}] \exp(-j2\pi\mathbf{u}\cdot\mathbf{x}) \\ &= \sum_{\mathbf{x}\in\Gamma^*} \tilde{F}[\mathbf{x}] \exp(j2\pi(-\mathbf{u})\cdot\mathbf{x}) \\ &= d(\Gamma)\tilde{f}(-\mathbf{u}) \\ &= \frac{1}{d(\Gamma^*)}\tilde{f}(-\mathbf{u}). \end{aligned} \tag{5.30}$$

Note the similar form of these two duality properties. □

Property 5.10 *Parseval relation:* $\int_{\mathcal{P}_\Gamma} \tilde{f}(\mathbf{x})\tilde{g}^*(\mathbf{x})\,d\mathbf{x} = \frac{1}{d(\Gamma)}\sum_{\mathbf{u}\in\Gamma^*} \tilde{F}[\mathbf{u}]\tilde{G}^*[\mathbf{u}]$

Proof: Let $\tilde{r}(\mathbf{x}) = \tilde{g}^*(\mathbf{x})$ and $\tilde{q}(\mathbf{x}) = \tilde{f}(\mathbf{x})\tilde{r}(\mathbf{x})$. Then

$$\begin{aligned} \tilde{Q}[\mathbf{u}] &= \frac{1}{d(\Gamma)} \sum_{\mathbf{w}\in\Gamma^*} \tilde{F}[\mathbf{w}]\tilde{R}[\mathbf{u}-\mathbf{w}] \\ &= \frac{1}{d(\Gamma)} \sum_{\mathbf{w}\in\Gamma^*} \tilde{F}[\mathbf{w}]\tilde{G}^*[\mathbf{w}-\mathbf{u}]. \end{aligned} \tag{5.31}$$

Evaluating at $\mathbf{u} = 0$,

$$\tilde{Q}[\mathbf{0}] = \int_{\mathcal{P}_\Gamma} \tilde{f}(\mathbf{x})\tilde{g}^*(\mathbf{x})\,d\mathbf{x} = \frac{1}{d(\Gamma)} \sum_{\mathbf{w}\in\Gamma^*} \tilde{F}[\mathbf{w}]\tilde{G}^*[\mathbf{w}]. \tag{5.32}$$

□

If $\tilde{f} = \tilde{g}$, we obtain

$$\int_{\mathcal{P}_\Gamma} |\tilde{f}(\mathbf{x})|^2\,d\mathbf{x} = \frac{1}{d(\Gamma^*)} \sum_{\mathbf{w}\in\Gamma^*} |\tilde{F}[\mathbf{w}]|^2. \tag{5.33}$$

5.6 Evaluation of the Continuous-Domain Periodic Fourier Transform

Evaluation of continuous-domain periodic Fourier transforms is generally a matter of evaluating the defining integral of Equation (5.16) and judiciously applying the Fourier transform properties. We provide here a few examples to illustrate some of the issues.

Example 5.1
Consider the periodic signal $\tilde{f}$ consisting of an $A \times A$ square patch replicated on a square lattice with spacing Z, where $Z > A$, as shown in Figure 5.1. We wish to find the continuous-domain Fourier series representation of $\tilde{f}$.

Solution:
The periodicity lattice Γ is generated by the sampling matrix $\mathbf{V}_\Gamma = \begin{bmatrix} Z & 0 \\ 0 & Z \end{bmatrix}$. Because $A < Z$, we can express $\tilde{f}$ as

$$\tilde{f}(x, y) = \sum_{k_1=-\infty}^{\infty} \sum_{k_2=-\infty}^{\infty} \operatorname{rect}\left(\frac{x - k_1 Z}{A}, \frac{y - k_2 Z}{A}\right),$$

where at any point (x, y), only one term in this infinite sum is nonzero.

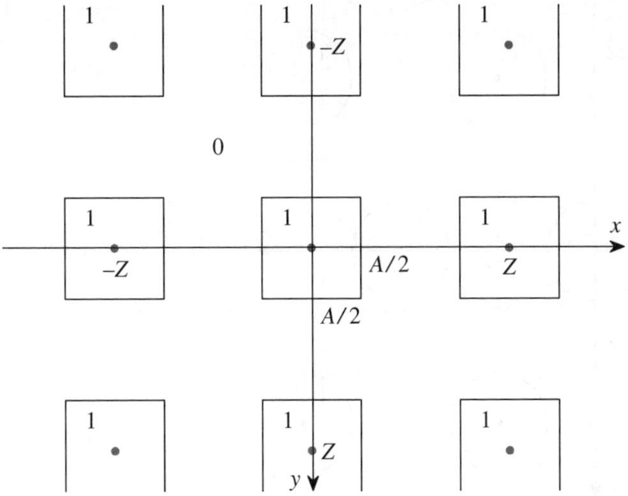

Figure 5.1 Continuous-domain periodic signal consisting of a square $A \times A$ patch repeated at the points of a square lattice Γ with spacing Z.

Taking the standard Voronoi unit cell of Γ, $\mathcal{P}_\Gamma = \left\{(x, y) \mid -\frac{Z}{2} \leq x < \frac{Z}{2}, -\frac{Z}{2} \leq y < \frac{Z}{2}\right\}$, the continuous-domain periodic Fourier transform is given by

$$\begin{aligned}\tilde{F}[u, v] &= \int_{-\frac{Z}{2}}^{\frac{Z}{2}} \int_{-\frac{Z}{2}}^{\frac{Z}{2}} \tilde{f}(x, y) \exp(-j2\pi(ux + vy))\, dxdy \\ &= \int_{-\frac{A}{2}}^{\frac{A}{2}} \int_{-\frac{A}{2}}^{\frac{A}{2}} \exp(-j2\pi(ux + vy))\, dxdy \\ &= \frac{\sin(\pi uA)\sin(\pi vA)}{\pi^2 uv} \qquad (u, v) \in \Gamma^*,\end{aligned}$$

where $\Gamma^* = \text{LAT}\left(\begin{bmatrix} 1/Z & 0 \\ 0 & 1/Z \end{bmatrix}\right)$. Note that this is the same integral that was evaluated in Example 2.3.

The continuous-domain periodic Fourier transform of $\tilde{f}$ is the same as the aperiodic Fourier transform of $\text{rect}\left(\frac{x}{A}, \frac{y}{A}\right)$, evaluated at the discrete points of Γ^*. The continuous-domain Fourier series representation of $\tilde{f}$ is then given by

$$\tilde{f}(x, y) = Z^2 \sum_{k_1=-\infty}^{\infty} \sum_{k_2=-\infty}^{\infty} \frac{\sin\left(\frac{\pi k_1 A}{Z}\right)\sin\left(\frac{\pi k_2 A}{Z}\right)}{\pi^2 k_1 k_2} \exp\left(-j2\pi\left(\frac{k_1 x}{Z} + \frac{k_2 y}{Z}\right)\right).$$

□

Example 5.2

Consider the mask shown in Figure 5.2, which consists of a circular target or aperture of radius A, replicated on a regular hexagonal lattice Γ, given by the sampling matrix $\mathbf{V}_\Gamma = \begin{bmatrix} 2Z & Z \\ 0 & \sqrt{3}Z \end{bmatrix}$, where $A < Z$. Find the continuous-domain periodic Fourier transform of this signal.

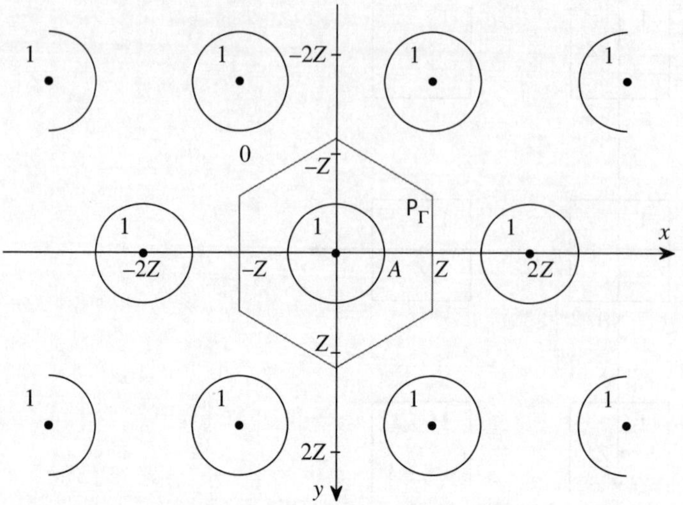

Figure 5.2 Continuous-domain periodic signal consisting of a circle of radius *A* repeated at the points of a regular hexagonal lattice Γ. The Voronoi unit cell of Γ is shown.

Solution:
The continuous-domain periodic Fourier transform is defined on the reciprocal lattice Γ^*, generated by

$$\mathbf{V}_{\Gamma^*} = \mathbf{V}_{\Gamma}^{-T} = \begin{bmatrix} \frac{1}{2Z} & 0 \\ -\frac{1}{2\sqrt{3}Z} & \frac{1}{\sqrt{3}Z} \end{bmatrix}.$$

Since $A < Z$, the circular target is entirely contained within a Voronoi unit cell of Γ, as shown in Figure 5.2. Thus we can write

$$\tilde{f}(\mathbf{x}) = \sum_{\mathbf{t}\in\Gamma} \mathrm{circ}\left(\frac{\mathbf{x}-\mathbf{t}}{A}\right),$$

where at each point $\mathbf{x} \in \mathbb{R}^2$, only one term of the infinite sum is nonzero.

The Fourier transform is given by

$$\begin{aligned}\tilde{F}[u,v] &= \iint_{\mathcal{P}_\Gamma} \tilde{f}(x,y)\exp(-\mathrm{j}2\pi(ux+vy))\,\mathrm{d}x\mathrm{d}y \\ &= \iint_{x^2+y^2\le A^2} \exp(-\mathrm{j}2\pi(ux+vy))\,\mathrm{d}x\mathrm{d}y \\ &= \frac{A}{\sqrt{u^2+v^2}} J_1(2\pi A\sqrt{u^2+v^2}), \qquad (u,v)\in\Gamma^*,\end{aligned}$$

where we refer to Example 2.4 for the evaluation of this integral. □

The two above examples illustrate a general property: the Fourier transform of a continuous-domain Γ-periodic signal $\tilde{f}$ is equal to the continuous-domain Fourier transform of a signal equal to $\tilde{f}$ on one unit cell of Γ and equal to zero elsewhere, evaluated at the points of Γ^*. This will be taken up again in Chapter 6.

Example 5.3

Figure 5.3(a) shows an image of a periodic function which could represent a simple test pattern. Find the Fourier series representation of this signal.

Solution:

Figure 5.3(b) shows this signal in the form of a line drawing, indicating signal values in each region. Inspecting the figure, we can identify that the periodicity is given by periodicity lattice Γ with sampling matrix $\mathbf{V}_\Gamma = \left[\begin{smallmatrix} 4 & 2 \\ 0 & 1.5 \end{smallmatrix}\right]$; the points of this lattice are shown in Figure 5.3(b). The Fourier series coefficients can be obtained directly by integrating $\tilde{f}(x, y)\exp(-\mathrm{j}2\pi(ux + vy))$ over a unit cell of Γ. The Voronoi unit cell is shown in Figure 5.3(b). While doable, this would be rather tedious. Instead, we will use Fourier transform properties to simplify the process. Let us express $\tilde{f} = \tilde{f}_1 + \tilde{f}_2$, where $\tilde{f}_1$ represents the pattern of repeating diamonds and $\tilde{f}_2$ represents the pattern of repeating squares. Let

$$f_\mathrm{D}(x, y) = \begin{cases} 1 & |x| + |y| \leq 1 \\ 0 & \text{otherwise} \end{cases}$$

which can be obtained by rotating a square of side $\sqrt{2}$ by 45°. Thus

$$f_\mathrm{D}(x, y) = \mathcal{M}_{\mathbf{A}_{\pi/4}}\mathrm{rect}\left(\frac{x}{\sqrt{2}}, \frac{y}{\sqrt{2}}\right)$$

and

$$\tilde{f}_1(x, y) = \sum_{(s,t)\in\Gamma} f_\mathrm{D}(x - s, y - t).$$

To get the correct value in the square-shaped regions, we must add 0.5 in the background regions and subtract 0.5 in the diamond-shaped regions. If we define $f_\mathrm{S}(s, y) = \mathrm{rect}(2x, 2y)$, and

$$\tilde{f}_S(x, y) = \sum_{(s,t)\in\Gamma} f_\mathrm{S}(x - s, y - t),$$

then $\tilde{f}_2(x, y) = -0.5\tilde{f}_S(x, y) + 0.5\tilde{f}_S(x - 2, y)$.

Since $f_\mathrm{D}(x, y)$ is completely contained in a unit cell of Γ,

$$\begin{aligned} \tilde{F}_1[u, v] &= \iint_{\mathcal{P}_\Gamma} f_\mathrm{D}(x, y)\exp(-\mathrm{j}2\pi(ux + vy))\, \mathrm{d}x\mathrm{d}y \\ &= \int_{-\infty}^{\infty}\int_{-\infty}^{\infty} f_\mathrm{D}(x, y)\exp(-\mathrm{j}2\pi(ux + vy))\, \mathrm{d}x\mathrm{d}y, \qquad (u, v) \in \Gamma^*. \end{aligned}$$

This is the continuous-domain Fourier transform of f_D evaluated at the points of Γ^*. Now $F_\mathrm{D}(u, v)$ can be evaluated using the scaling and rotation properties of the continuous-domain Fourier transform applied to the rect function (see Example 2.3 and Problem 2.8(d)). This gives

$$\begin{aligned} F_\mathrm{D}(u, v) &= 2\mathrm{sinc}(u + v)\,\mathrm{sinc}(u - v), &&\text{and thus} \\ \tilde{F}_1[u, v] &= 2\mathrm{sinc}(u + v)\,\mathrm{sinc}(u - v), &&(u, v) \in \Gamma^*. \end{aligned}$$

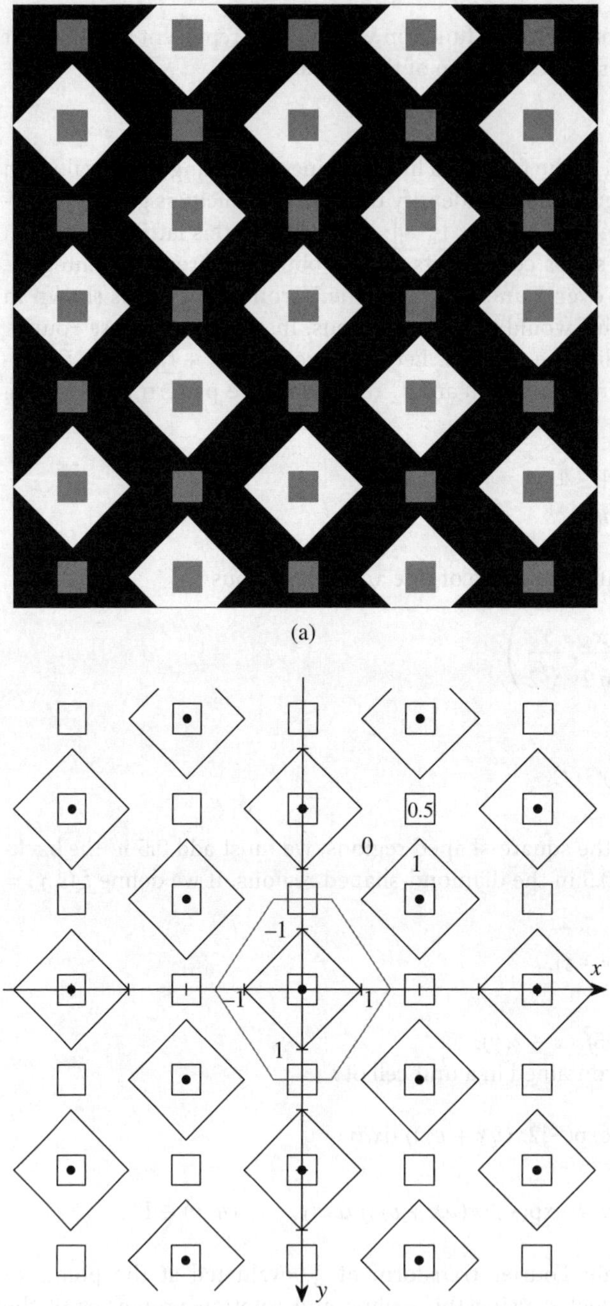

Figure 5.3 (a) Portion of a continuous-domain periodic signal consisting of repeating diamonds and squares shown as an image. (b) Test pattern shown as a line drawing with the periodicity lattice Γ (•), signal values (0 in background, 0.5 in squares and 1 in diamonds outside the squares), and the Voronoi cell of Γ.

Similarly, since $f_S(x,y)$ has a region of support completely contained within $\mathcal{P}_\Gamma$,

$$\tilde{F}_S[u,v] = \frac{1}{4}\mathrm{sinc}(u/2)\mathrm{sinc}(v/2), \qquad (u,v) \in \Gamma^*.$$

Using Properties (5.1) and (5.2),

$$\begin{aligned}\tilde{F}[u,v] &= \tilde{F}_1[u,v] - 0.5\tilde{F}_S[u,v] + 0.5\tilde{F}_S[u,v]\exp(-\mathrm{j}2\pi(2u)) \\ &= 2\mathrm{sinc}(u+v)\mathrm{sinc}(u-v) + \frac{1}{8}\mathrm{sinc}(u/2)\mathrm{sinc}(v/2)(\exp(-\mathrm{j}4\pi u) - 1), \\ &\quad (u,v) \in \Gamma^*.\end{aligned}$$

To evaluate this concretely at the points of Γ^*,

$$\mathbf{V}_{\Gamma^*} = \mathbf{V}_\Gamma^{-T} = \begin{bmatrix} \frac{1}{4} & 0 \\ -\frac{1}{3} & \frac{2}{3} \end{bmatrix}, \quad \text{so that}$$

$$\Gamma^* = \left\{ \begin{bmatrix} \frac{k_1}{4} \\ -\frac{k_1}{3} + \frac{2k_2}{3} \end{bmatrix} \mid k_1, k_2 \in \mathbb{Z} \right\}.$$

Thus,

$$\begin{aligned}\tilde{F}\left[\frac{k_1}{4}, -\frac{k_1}{3} + \frac{2k_2}{3}\right] &= 2\,\mathrm{sinc}\left(\frac{-k_1 + 8k_2}{12}, \frac{7k_1 - 8k_2}{12}\right) \\ &\quad + \frac{1}{8}\mathrm{sinc}\left(\frac{k_1}{8}\right)\mathrm{sinc}\left(\frac{-k_1 + 2k_2}{6}\right)((-1)^{k_1} - 1), \quad k_1, k_2 \in \mathbb{Z}.\end{aligned}$$

We see that the second term will vanish when k_1 is even. □

Problems

1 Find the continuous-domain Fourier series representation of the following periodic signals.

a) A circ function with rectangular periodicity

$$\tilde{f}(x,y) = \sum_{k_1=-\infty}^{\infty} \sum_{k_2=-\infty}^{\infty} \mathrm{circ}\left(\frac{x - k_1 Z}{A}, \frac{y - k_2 Z}{A}\right)$$

where $A, Z > 0$ and $A < Z$.

b) A hexagonal function with rectangular periodicity

$$\tilde{f}(x,y) = \sum_{k_1=-\infty}^{\infty} \sum_{k_2=-\infty}^{\infty} \mathrm{hex}\left(\frac{x - k_1 Z}{A}, \frac{y - k_2 Z}{A}\right)$$

where $A, Z > 0$, $A < Z/2$ and $\mathrm{hex}(x,y)$ is the the zero-one function with hexagonal region of support of unit side as defined in Example 2.7.

6

Sampling, Reconstruction and Sampling Theorems for Multidimensional Signals

6.1 Introduction

Physical real-world images are continuous in space and time, while we must use sampled signals for digital image processing. Thus we must be able to convert continuous-domain signals to discrete-domain signals, a process known as sampling. Likewise, discrete-domain signals may need to be converted to the continuous domain for viewing, a process known as reconstruction. This chapter presents the theory of ideal sampling and reconstruction, which is strongly based on Fourier analysis. Some of the issues involved in practical sampling and reconstruction are then discussed.

Sampling of an aperiodic continuous-domain signal produces an aperiodic discrete-domain signal. For periodic continuous-domain signals, if the periodicity lattice is a sub-lattice of the sampling lattice, sampling produces a periodic discrete-domain signal with the same periodicity lattice. We can also sample the Fourier transform of an aperiodic continuous-domain signal, yielding a periodic continuous-domain signal. Finally, we can sample the Fourier transform of an aperiodic discrete-domain signal (with conditions on the lattices), which yields a periodic discrete-domain signal. Thus, all the categories of signals we have studied can be related though sampling in the signal domain or the frequency domain. This will prove very useful in evaluating Fourier transforms and carrying out processing in the frequency domain. These cases and their relationships are all covered in this chapter.

6.2 Ideal Sampling and Reconstruction of Continuous-Domain Signals

Let $f_c(\mathbf{x})$ be a continuous-domain multidimensional signal, which could be for example a still image $f_c(x, y)$ or time-varying video $f_c(x, y, t)$. We wish to convert f_c to a discrete-domain signal for digital processing and subsequent conversion back to continuous form. The ideal sampling operation can be simply expressed

$$f[\mathbf{x}] = f_c(\mathbf{x}), \qquad \mathbf{x} \in \Lambda \tag{6.1}$$

Multidimensional Signal and Color Image Processing Using Lattices, First Edition. Eric Dubois.

Companion website: www.wiley.com/go/Dubois/multiSP

where Λ is the sampling lattice. Assume that $f_c(\mathbf{x})$ has Fourier transform

$$F_c(\mathbf{u}) = \int_{\mathbb{R}^D} f_c(\mathbf{x}) \exp(-j2\pi\mathbf{u}\cdot\mathbf{x})\, d\mathbf{x}, \qquad \mathbf{u} \in \mathbb{R}^D \tag{6.2}$$

with corresponding inverse Fourier transform

$$f_c(\mathbf{x}) = \int_{\mathbb{R}^D} F_c(\mathbf{u}) \exp(j2\pi\mathbf{u}\cdot\mathbf{x})\, d\mathbf{u}, \qquad \mathbf{x} \in \mathbb{R}^D. \tag{6.3}$$

Let $\mathcal{P}^*$ be a unit cell of the reciprocal lattice Λ^*. We know that $\mathcal{P}^*$ tiles $\mathbb{R}^D$ when centered on the points of Λ^*, so that we can write

$$f[\mathbf{x}] = \sum_{\mathbf{r}\in\Lambda^*} \int_{\mathcal{P}^*} F_c(\mathbf{u}-\mathbf{r}) \exp(j2\pi(\mathbf{u}-\mathbf{r})\cdot\mathbf{x})\, d\mathbf{u}, \quad \mathbf{x} \in \Lambda. \tag{6.4}$$

Since $\mathbf{r}\cdot\mathbf{x} \in \mathbb{Z}$ for all $\mathbf{r} \in \Lambda^*$, $\mathbf{x} \in \Lambda$, this becomes

$$\begin{aligned} f[\mathbf{x}] &= \sum_{\mathbf{r}\in\Lambda^*} \int_{\mathcal{P}^*} F_c(\mathbf{u}-\mathbf{r}) \exp(j2\pi\mathbf{u}\cdot\mathbf{x})\, d\mathbf{u} \\ &= \int_{\mathcal{P}^*} \left(\sum_{\mathbf{r}\in\Lambda^*} F_c(\mathbf{u}-\mathbf{r}) \right) \exp(j2\pi\mathbf{u}\cdot\mathbf{x})\, d\mathbf{u}, \quad \mathbf{x} \in \Lambda. \end{aligned} \tag{6.5}$$

We know that

$$f[\mathbf{x}] = d(\Lambda) \int_{\mathcal{P}^*} F(\mathbf{u}) \exp(j2\pi\mathbf{u}\cdot\mathbf{x})\, d\mathbf{u}, \quad \mathbf{x} \in \Lambda \tag{6.6}$$

and that the Fourier transform is unique. It follows that

$$F(\mathbf{u}) = \frac{1}{d(\Lambda)} \sum_{\mathbf{r}\in\Lambda^*} F_c(\mathbf{u}-\mathbf{r}). \tag{6.7}$$

We assume in the above that the sums and integrals converge and that exchanging the order of the sum and integrals is permitted; these conditions generally hold in the cases of interest to us. It usually suffices that $F_c(\mathbf{u})$ be continuous and absolutely integrable. A more rigorous but accessible discussion can be found in standard books on Fourier analysis such as Kammler (2000).

Figure 6.1 illustrates the sampling on a hexagonal lattice of a 2D continuous-space signal, whose Fourier transform is limited to a circular region. If the signal $f_{c1}(\mathbf{x})$ whose Fourier transform $F_{c1}(\mathbf{u})$ is limited to the circular region in Figure 6.1(a) is sampled on this lattice, the replicas on the points of the reciprocal lattice will overlap, as shown in Figure 6.1(b). If the signal $f_{c2}(\mathbf{x})$ whose Fourier transform $F_{c2}(\mathbf{u})$ is limited to the smaller circular region in Figure 6.1(c) is sampled on the same lattice, the replicas on the points of the reciprocal lattice will not overlap, as shown in Figure 6.1(d).

It will be possible to reconstruct $f_c(\mathbf{x})$ exactly from the samples if the shifted copies $F_c(\mathbf{u}-\mathbf{r})$ in the sum do not overlap. Assume that $F_c(\mathbf{u}) = 0$ if $\mathbf{u} \notin \mathcal{V}$, where $\mathcal{V}$ is some bounded set in the frequency domain. If the lattice Λ is chosen so that $\mathcal{V} \subset \mathcal{P}^*$ for *some* unit cell $\mathcal{P}^*$ of Λ^*, then $F_c(\mathbf{u}) = 0$ for $\mathbf{u} \notin \mathcal{P}^*$. It follows that $F_c(\mathbf{u})$ and $F_c(\mathbf{u}-\mathbf{r})$ do not overlap for $\mathbf{r} \in \Lambda^*\backslash\mathbf{0}$. This implies that

$$F(\mathbf{u}) = \frac{1}{d(\Lambda)} F_c(\mathbf{u}) \quad \text{for } \mathbf{u} \in \mathcal{P}^*. \tag{6.8}$$

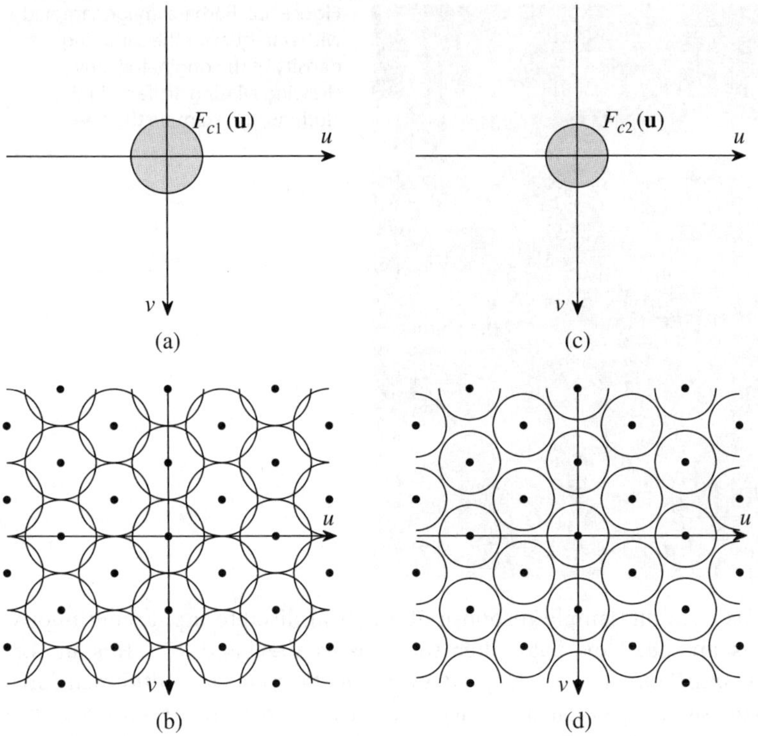

Figure 6.1 Frequency domain representation of sampling of a 2D continuous-space signal on a hexagonal lattice. (a) Region of support of Fourier transform $F_{c1}(\mathbf{u})$ of wider-band signal. (b) Region of support of signal of (a) sampled on a hexagonal lattice showing spectral overlap leading to aliasing. (c) Region of support of Fourier transform $F_{c2}(\mathbf{u})$ of narrower-band signal. (d) Region of support of signal of (c) sampled on the same hexagonal lattice showing no spectral overlap and so no aliasing.

Thus

$$
\begin{aligned}
f_c(\mathbf{x}) &= \int_{\mathbb{R}^D} F_c(\mathbf{u}) \exp(\mathrm{j}2\pi \mathbf{u} \cdot \mathbf{x})\, d\mathbf{u} \\
&= \int_{\mathcal{P}^*} F_c(\mathbf{u}) \exp(\mathrm{j}2\pi \mathbf{u} \cdot \mathbf{x})\, d\mathbf{u} \\
&= d(\Lambda) \int_{\mathcal{P}^*} F(\mathbf{u}) \exp(\mathrm{j}2\pi \mathbf{u} \cdot \mathbf{x})\, d\mathbf{u} \\
&= d(\Lambda) \int_{\mathcal{P}^*} \left(\sum_{\mathbf{s}\in\Lambda} f[\mathbf{s}] \exp(-\mathrm{j}2\pi \mathbf{u} \cdot \mathbf{s}) \right) \exp(\mathrm{j}2\pi \mathbf{u} \cdot \mathbf{x})\, d\mathbf{u} \\
&= d(\Lambda) \sum_{\mathbf{s}\in\Lambda} f[\mathbf{s}] \int_{\mathcal{P}^*} \exp(\mathrm{j}2\pi \mathbf{u} \cdot (\mathbf{x} - \mathbf{s}))\, d\mathbf{u} \\
&= \sum_{\mathbf{s}\in\Lambda} f[\mathbf{s}] t(\mathbf{x} - \mathbf{s}), \qquad \mathbf{x} \in \mathbb{R}^D
\end{aligned}
\tag{6.9}
$$

where

$$
t(\mathbf{x}) = d(\Lambda) \int_{\mathcal{P}^*} \exp(\mathrm{j}2\pi \mathbf{u} \cdot \mathbf{x})\, d\mathbf{u}. \tag{6.10}
$$

Figure 6.2 Barbara image sampled with one-quarter the sampling density of the original, clearly showing aliasing artifacts in the clothing, for example the knee.

We identify $t(\mathbf{x})$ as the unit-sample response of a hybrid discrete-input, continuous-output filter. It is an ideal low-pass filter that passes the band $\mathcal{P}^*$. This important result is the *multidimensional Nyquist sampling theorem*: a multidimensional continuous-domain signal $f_c(\mathbf{x})$ can be exactly reconstructed from its samples on a lattice Λ if the Fourier transform $F_c(\mathbf{u})$ is zero everywhere outside some unit cell of the reciprocal lattice Λ^*.

On the other hand, if $f_c(\mathbf{x})$ is not bandlimited to a unit cell of Λ^*, the replicas $F_c(\mathbf{u} + \mathbf{r})$ will overlap and perfect reconstruction of $f_c(\mathbf{x})$ cannot be achieved from its samples on Λ. High frequencies in the input are mapped into lower frequencies in the sampled signal, a phenomenon called *aliasing*.

A good illustration of aliasing is shown in Figure 6.2, where the Barbara image of Figure 3.12(a) is sampled with half the sampling spacing in each dimension, so one quarter of the sampling density. We see that high-frequency patterns in the original as seen for example on the clothes have spatial frequencies beyond the unit cell of the reciprocal lattice and are mapped into different lower frequencies, with different orientations.

6.3 Practical Sampling

The situation discussed in the previous section does not correspond to sampling and reconstruction in real applications for several reasons:

- We cannot measure the intensity of an image at a single point in space-time. We must collect light over some spatial neighborhood of the desired point, and for some period of time.
- We don't want to measure the intensity at a single point anyway since that would inevitably result in aliasing.
- We cannot reconstruct the continuous space-time image using an ideal low-pass filter with a physical system.

In practice, light is collected over some spatiotemporal neighborhood of the point to be sampled using a *sampling aperture*:

$$f[\mathbf{x}] = \int_{\mathbb{R}^D} f_c(\mathbf{x} + \mathbf{s}) a(\mathbf{x}, \mathbf{s})\, d\mathbf{s}, \quad \mathbf{x} \in \Lambda. \tag{6.11}$$

Here, $a(\mathbf{x}, \mathbf{s})$ is a space-time variant aperture centered at point $\mathbf{x}$ and generally limited to a relatively small neighborhood of $\mathbf{x}$. If we assume that the aperture is space-time invariant, $a(\mathbf{x}, \mathbf{s}) = a(\mathbf{s})$, then

$$f[\mathbf{x}] = \int_{\mathbb{R}^D} f_c(\mathbf{x} + \mathbf{s}) a(\mathbf{s})\, d\mathbf{s}, \quad \mathbf{x} \in \Lambda. \tag{6.12}$$

Let $h_a(\mathbf{s}) = a(-\mathbf{s})$, and let $\mathbf{t} = -\mathbf{s}$. Then

$$f[\mathbf{x}] = \int_{\mathbb{R}^D} h_a(\mathbf{t}) f_c(\mathbf{x} - \mathbf{t})\, d\mathbf{t}, \quad \mathbf{x} \in \Lambda. \tag{6.13}$$

This can be modeled as a two-step process. First, $f_c(\mathbf{x})$ is filtered with a continuous space-time filter with impulse response $h_a(\mathbf{x})$ to give

$$f_p(\mathbf{x}) = \int_{\mathbb{R}^D} h_a(\mathbf{t}) f_c(\mathbf{x} - \mathbf{t})\, d\mathbf{t}, \quad \mathbf{x} \in \mathbb{R}^D. \tag{6.14}$$

Then, $f_p(\mathbf{x})$ undergoes *ideal* sampling on the lattice Λ,

$$f[\mathbf{x}] = f_p(\mathbf{x}), \quad \mathbf{x} \in \Lambda. \tag{6.15}$$

In this case, we have

$$\begin{aligned} F(\mathbf{u}) &= \frac{1}{d(\Lambda)} \sum_{\mathbf{r} \in \Lambda^*} F_p(\mathbf{u} - \mathbf{r}) \\ &= \frac{1}{d(\Lambda)} \sum_{\mathbf{r} \in \Lambda^*} H_a(\mathbf{u} - \mathbf{r}) F_c(\mathbf{u} - \mathbf{r}). \end{aligned} \tag{6.16}$$

The ideal role of $H_a(\mathbf{u})$ is to bandlimit $F_c(\mathbf{u})$ to $\mathcal{P}^*$, while preserving the frequency content within this band as much as possible.

In typical image acquisition systems, the spatial aperture is either a rect function or a Gaussian function. Of course, neither of these is an ideal low-pass filter, so typically $f_p(\mathbf{x})$ will not be bandlimited to a unit cell of Λ^* and some aliasing will occur, while desired signal components within Λ^* will be attenuated. The temporal aperture is typically a rect function, corresponding to integration over some period T seconds. Consider for example the sampling of the green component of a color image with the Bayer array of Figure 3.1. Then, the sampling aperture is

$$a(x, y) = \frac{1}{X^2} \operatorname{rect}(x/X, y/X) = h_a(x, y) \tag{6.17}$$

since the aperture is symmetric about the origin. From Example 2.3,

$$H_a(u, v) = \frac{\sin(\pi u X) \sin(\pi v X)}{\pi^2 X^2 u v}. \tag{6.18}$$

Recall that a sampling matrix for the green-channel sampling lattice is $\mathbf{V} = \begin{bmatrix} 2X & X \\ 0 & X \end{bmatrix}$, and a sampling matrix for the reciprocal lattice is

$$\mathbf{V}^{-T} = \begin{bmatrix} \frac{1}{2X} & 0 \\ -\frac{1}{2X} & \frac{1}{X} \end{bmatrix}.$$

Figure 6.3 shows contours of $|H_a(u, v)|$ along with the reciprocal lattice Λ^* and a possible unit cell, where we take $X = 1$ px. It is clear that if the image on the sensor has a lot of detail or high-frequency components, the sampling aperture will let much of that through and aliasing will be present. Compare with an ideal low-pass filter whose passband is the unit cell. Aliasing can be reduced by limiting the high-frequency components on the sensor using the camera lens. Note that the aliasing artifacts described in this example can be reduced by jointly reconstructing the three RGB components on the Bayer array, as will be discussed in later chapters.

The case of a video camera with a Gaussian spatial aperture and a rect temporal aperture was considered in Example 2.6.

6.4 Practical Reconstruction

We have seen that ideal reconstruction is achieved with an ideal low-pass filter whose passband is the unit cell. Practical display devices do not have such a characteristic. A suitable model for the reconstruction process is

$$f_r(\mathbf{x}) = \sum_{\mathbf{s}\in\Lambda} f[\mathbf{s}]\mathrm{d}(\mathbf{x} - \mathbf{s}) \tag{6.19}$$

where $\mathrm{d}(\mathbf{x})$ is called the display aperture. Thus the sample at position $\mathbf{s}$ contributes the display aperture centered at position $\mathbf{s}$ and weighted by the value $f[\mathbf{s}]$ to the reconstructed output. Combining practical reconstruction with practical sampling, we can relate the Fourier transform of the displayed image to that of the original image by

$$F_r(\mathbf{u}) = \frac{1}{d(\Lambda)} D(\mathbf{u}) \sum_{\mathbf{r}\in\Lambda^*} H_a(\mathbf{u} - \mathbf{r}) F_c(\mathbf{u} - \mathbf{r}), \tag{6.20}$$

where $D(\mathbf{u})$ is the continuous-domain Fourier transform of the display aperture.

In a cathode ray tube (CRT) display, we can model $\mathrm{d}(\mathbf{x})$ by a Gaussian in the spatial domain and an exponential in the temporal domain. Liquid crystal displays may be modeled with a display aperture that is a rect function in both spatial and temporal dimensions. It is clear that neither of these is an ideal low-pass filter, and so a significant portion of the spectral replicas remain in the reconstructed signal. However, if the image is then viewed by a human observer at a normal viewing distance, further filtering takes place in the human visual system. The reconstruction of color images by a typical display device is discussed further in Chapter 8.

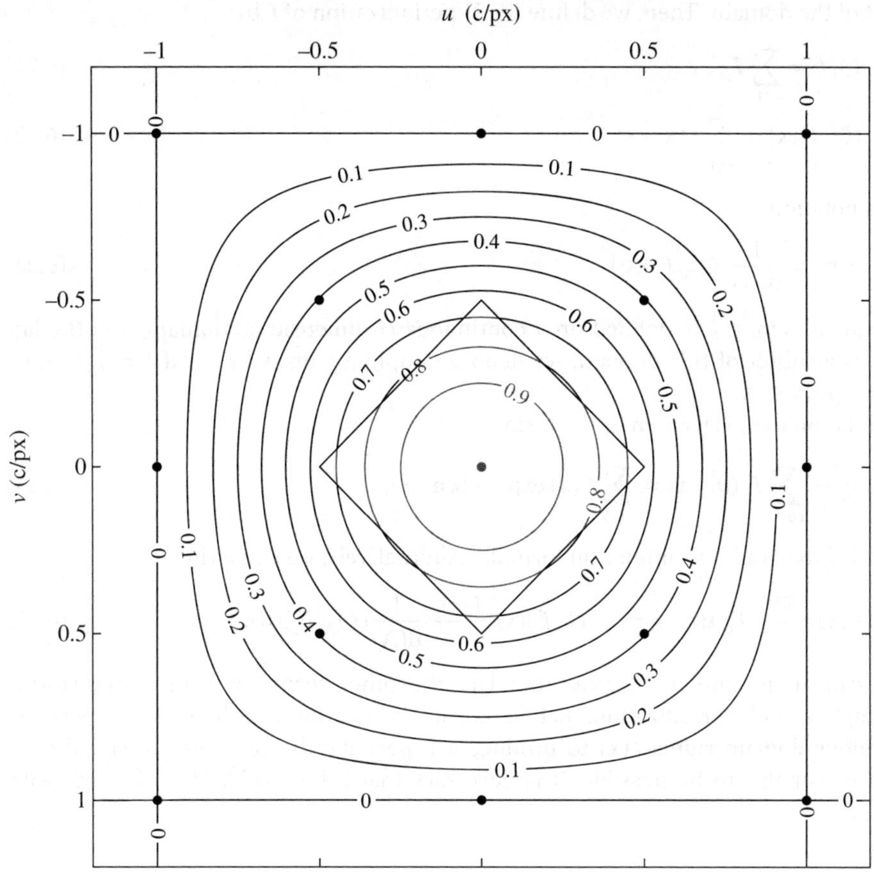

Figure 6.3 Frequency response of the rectangular aperture for the green channel of a Bayer color filter array. Also shown are the points of the reciprocal lattice Λ^* (•) and the diamond-shaped Voronoi unit cell of Λ^*.

6.5 Sampling and Periodization of Multidimensional Signals and Transforms

We have seen that if a continuous-domain signal f_c is sampled on a lattice Λ, the discrete-domain Fourier transform of the sampled signal satisfies Equation (6.7), which we repeat here

$$F(\mathbf{u}) = \frac{1}{d(\Lambda)} \sum_{\mathbf{r}\in\Lambda^*} F_c(\mathbf{u} - \mathbf{r}).$$

The right-hand side of this equation is called the Λ^*-*periodization* of F_c, and we assume that F_c is such that this converges (e.g., continuous and absolutely integrable). In general, let f be defined on a D-dimensional domain; it could be $\mathbb{R}^D$ or a D-dimensional lattice, in the signal domain or the frequency domain. Let Γ be a D-dimensional lattice that is a subset of the domain. Then, we define the Γ-periodization of f by

$$\circlearrowleft_{\Gamma} f = \sum_{\mathbf{r}\in\Gamma} \mathcal{T}_{\mathbf{r}} f; \tag{6.21}$$

$$(\circlearrowleft_{\Gamma} f)(\mathbf{x}) = \sum_{\mathbf{r}\in\Gamma} f(\mathbf{x}-\mathbf{r}). \tag{6.22}$$

In this notation,

$$F(\mathbf{u}) = \frac{1}{\mathrm{d}(\Lambda)}(\circlearrowleft_{\Lambda^*} F_{\mathrm{c}})(\mathbf{u}). \tag{6.23}$$

In a similar vein, if f is defined on a continuous D-dimensional domain, and the lattice Λ is a subset of this domain, we denote sampling f on Λ by $\downarrow_{\Lambda}$: if $f = \downarrow_{\Lambda} f_{\mathrm{c}}$, then $f[\mathbf{x}] = f_{\mathrm{c}}(\mathbf{x}), \mathbf{x} \in \Lambda$.

If Equation (6.7) is written in the form

$$\frac{1}{\mathrm{d}(\Lambda)} \sum_{\mathbf{r}\in\Lambda^*} F_{\mathrm{c}}(\mathbf{u}-\mathbf{r}) = \sum_{\mathbf{x}\in\Lambda} f_{\mathrm{c}}(\mathbf{x}) \exp(-\mathrm{j}2\pi\mathbf{u}\cdot\mathbf{x}), \tag{6.24}$$

it is called the Poisson summation formula. Alternatively, we can write

$$f_{\mathrm{c}}(\mathbf{x}) \xleftrightarrow{\text{CDFT}} F_{\mathrm{c}}(\mathbf{u}) \quad \Longrightarrow \quad (\downarrow_{\Lambda} f_{\mathrm{c}})(\mathbf{x}) \xleftrightarrow{\text{DDFT}} \frac{1}{\mathrm{d}(\Lambda)}(\circlearrowleft_{\Lambda^*} F_{\mathrm{c}})(\mathbf{u}). \tag{6.25}$$

We can apply similar analyses to relate the other forms of Fourier transforms by sampling and periodization. Let us consider first the sampling of a Γ-periodic continuous-domain signal $\tilde{f}(\mathbf{x})$ to produce a Γ-periodic discrete-domain signal on a lattice Λ. For this to be possible, it is necessary that Γ be a sublattice of Λ, say with index K.

Thus,

$$\tilde{f}[\mathbf{x}] = \tilde{f}_c(\mathbf{x}), \quad \mathbf{x} \in \Lambda, \tag{6.26}$$

or in the notation introduced above, $\tilde{f} = \downarrow_{\Lambda} \tilde{f}_c$. Using the Fourier series synthesis equation for $\tilde{f}_c$

$$\tilde{f}[\mathbf{x}] = \tilde{f}_c(\mathbf{x}) = \frac{1}{\mathrm{d}(\Gamma)} \sum_{\mathbf{u}\in\Gamma^*} \tilde{F}[\mathbf{u}] \exp(\mathrm{j}2\pi\mathbf{u}\cdot\mathbf{x}), \qquad \mathbf{x} \in \Lambda. \tag{6.27}$$

Let $\mathcal{D} = \{\mathbf{d}_0, \ldots, \mathbf{d}_{K-1}\}$ be a set of coset representatives of Λ^* in Γ^*. We can partition the sum over Γ^* into K summations over each of the cosets of Λ^* in Γ^*.

$$\begin{aligned}\tilde{f}[\mathbf{x}] &= \frac{1}{\mathrm{d}(\Gamma)} \sum_{k=0}^{K-1} \sum_{\mathbf{r}\in\Lambda^*} \tilde{F}_c[\mathbf{d}_k - \mathbf{r}] \exp(\mathrm{j}2\pi(\mathbf{d}_k - \mathbf{r})\cdot\mathbf{x}) \\ &= \frac{1}{\mathrm{d}(\Gamma)} \sum_{k=0}^{K-1} \left(\sum_{\mathbf{r}\in\Lambda^*} \tilde{F}_c[\mathbf{d}_k - \mathbf{r}] \right) \exp(\mathrm{j}2\pi\mathbf{d}_k\cdot\mathbf{x}),\end{aligned} \tag{6.28}$$

since $\mathbf{r}\cdot\mathbf{x} \in \mathbb{Z}$. However, from Equation (4.31),

$$\tilde{f}[\mathbf{x}] = \frac{1}{K} \sum_{k=0}^{K-1} \tilde{F}[\mathbf{d}_k] \exp(\mathrm{j}2\pi\mathbf{d}_k\cdot\mathbf{x}). \tag{6.29}$$

From this we conclude that

$$\tilde{F}[\mathbf{d}_k] = \frac{K}{d(\Gamma)} \sum_{\mathbf{r}\in\Lambda^*} \tilde{F}_c[\mathbf{d}_k - \mathbf{r}] \tag{6.30}$$

or simply

$$\tilde{F}[\mathbf{u}] = \frac{K}{d(\Gamma)} \sum_{\mathbf{r}\in\Lambda^*} \tilde{F}_c[\mathbf{u} - \mathbf{r}], \qquad \mathbf{u} \in \Gamma^* \tag{6.31}$$

which is the periodization of the periodic continuous-domain Fourier transform. Alternatively, we can write

$$\tilde{f}_c \overset{\text{CDFS}}{\longleftrightarrow} \tilde{F}_c \quad \Longrightarrow \quad \downarrow_\Lambda \tilde{f}_c \overset{\text{DDFS}}{\longleftrightarrow} \frac{K}{d(\Gamma)} \circlearrowright_{\Lambda^*} \tilde{F}_c. \tag{6.32}$$

The corresponding Poisson summation formula is

$$\frac{K}{d(\Gamma)} \sum_{\mathbf{r}\in\Lambda^*} \tilde{F}_c[\mathbf{u} - \mathbf{r}] = \sum_{\mathbf{x}\in\mathcal{B}} \tilde{f}_c(\mathbf{x}) \exp(-j2\pi\mathbf{u}\cdot\mathbf{x}). \tag{6.33}$$

These examples illustrate that sampling in the signal domain corresponds to periodization in the frequency domain. We now show that periodization in the signal domain corresponds to sampling in the frequency domain. Specifically, consider the case of an aperiodic discrete-domain signal f defined on a D-dimensional lattice Λ. Let Γ be a D-dimensional sublattice of Λ of index K and let $\tilde{g} = \circlearrowright_\Gamma f$ be the Γ-periodization of f. We will assume that f is absolutely summable on Λ and thus has a convergent discrete-domain Fourier transform. Then

$$\begin{aligned} |\tilde{g}[\mathbf{x}]| &= \left| \sum_{\mathbf{t}\in\Gamma} f[\mathbf{x} - \mathbf{t}] \right| \\ &\leq \sum_{\mathbf{t}\in\Gamma} |f[\mathbf{x} - \mathbf{t}]| \\ &\leq \sum_{\mathbf{t}\in\Lambda} |f[\mathbf{x} - \mathbf{t}]| \\ &< \infty. \end{aligned} \tag{6.34}$$

Thus, the Γ-periodization of f is bounded, and has K distinct values on the cosets of Γ in Λ. We compute the discrete-domain periodic Fourier transform of $\tilde{g}$ (Equation (4.19),

$$\begin{aligned} \tilde{G}[\mathbf{d}_i] &= \sum_{k=0}^{K-1} \tilde{g}[\mathbf{b}_k] \exp(-j2\pi\mathbf{d}_i \cdot \mathbf{b}_k) \\ &= \sum_{k=0}^{K-1} \left(\sum_{\mathbf{t}\in\Gamma} f[\mathbf{b}_k - \mathbf{t}] \right) \exp(-j2\pi\mathbf{d}_i \cdot \mathbf{b}_k) \\ &= \sum_{\mathbf{t}\in\Gamma} \sum_{k=0}^{K-1} f[\mathbf{b}_k - \mathbf{t}] \exp(-j2\pi\mathbf{d}_i \cdot \mathbf{b}_k). \end{aligned} \tag{6.35}$$

As usual, $\mathcal{B} = \{\mathbf{b}_0, \dots, \mathbf{b}_{K-1}\}$ is a set of coset representatives for Γ in Λ, and $\mathcal{D} = \{\mathbf{d}_0, \dots, \mathbf{d}_{K-1}\}$ is a set of coset representatives for Λ^* in Γ^*. Since $\mathbf{t} \in \Gamma$ and

$\mathbf{d}_i \in \Gamma^*$, it follows that $\mathbf{t} \cdot \mathbf{d}_i \in \mathbb{Z}$ and thus

$$\begin{aligned}\tilde{G}[\mathbf{d}_i] &= \sum_{\mathbf{t}\in\Gamma}\sum_{k=0}^{K-1} f[\mathbf{b}_k - \mathbf{t}]\exp(-\mathrm{j}2\pi\mathbf{d}_i\cdot(\mathbf{b}_k-\mathbf{t}))\\ &= \sum_{\mathbf{s}\in\Lambda} f[\mathbf{s}]\exp(-\mathrm{j}2\pi\mathbf{d}_i\cdot\mathbf{s})\\ &= F(\mathbf{d}_i),\end{aligned} \tag{6.36}$$

i.e. we simply sample $F(\mathbf{u})$ at the $\mathbf{d}_i$ to get $G(\mathbf{d}_i)$. Thus we have

$$f \overset{\text{DDFT}}{\longleftrightarrow} F \quad \Longrightarrow \quad \circlearrowright_{\Gamma} f \overset{\text{DDFS}}{\longleftrightarrow} \downarrow_{\Gamma^*} F. \tag{6.37}$$

The corresponding Poisson summation formula is

$$\sum_{\mathbf{t}\in\Gamma} f[\mathbf{b}_k - \mathbf{t}] = \frac{1}{K}\sum_{i=0}^{K-1} F(\mathbf{d}_i)\exp(\mathrm{j}2\pi\mathbf{d}_i\cdot\mathbf{b}_k). \tag{6.38}$$

The last case to consider is the Γ-periodization of a continuous-domain aperiodic signal f_c. Let $\tilde{g}_c = \circlearrowright_{\Gamma} f_c$. Applying the same strategy as for the previous case

$$\begin{aligned}\tilde{G}_c[\mathbf{u}] &= \int_{\mathcal{P}_\Gamma}\left(\sum_{\mathbf{t}\in\Gamma} f_c(\mathbf{x}-\mathbf{t})\right)\exp(-\mathrm{j}2\pi\mathbf{u}\cdot\mathbf{x})\,\mathrm{d}\mathbf{x}, \quad \mathbf{u}\in\Gamma^*\\ &= \sum_{\mathbf{t}\in\Gamma}\int_{\mathcal{P}_\Gamma} f_c(\mathbf{x}-\mathbf{t})\exp(-\mathrm{j}2\pi\mathbf{u}\cdot(\mathbf{x}-\mathbf{t}))\,\mathrm{d}\mathbf{x} \quad (\text{since } \mathbf{u}\cdot\mathbf{t}\in\mathbb{Z})\\ &= \int_{\mathbb{R}^D} f_c(\mathbf{s})\exp(\mathrm{j}2\pi\mathbf{u}\cdot\mathbf{s})\,\mathrm{d}\mathbf{s}\\ &= F_c(\mathbf{u}), \qquad \mathbf{u}\in\Gamma^*.\end{aligned} \tag{6.39}$$

Thus, our final case is

$$f_c \overset{\text{CDFT}}{\longleftrightarrow} F_c \quad \Longrightarrow \quad \circlearrowright_{\Gamma} f_c \overset{\text{CDFS}}{\longleftrightarrow} \downarrow_{\Gamma^*} F_c. \tag{6.40}$$

The Poisson summation formula for this final case is

$$\sum_{\mathbf{t}\in\Gamma} f_c(\mathbf{x}-\mathbf{t}) = \frac{1}{d(\Gamma)} F_c(\mathbf{u})\exp(\mathrm{j}2\pi\mathbf{u}\cdot\mathbf{x}). \tag{6.41}$$

The four cases of sampling/periodization that we have just examined are summarized in Figure 6.4. A similar figure for the one-dimensional case, called the Fourier–Poisson cube, is given in figure 1.22 of Kammler (2000).

6.6 Inverse Fourier Transforms

In the preceding chapters, inverse Fourier transforms have been provided without proof for all cases except the discrete-domain periodic signals. In this case, the signal space is of finite dimension and the Fourier basis is orthogonal, so inversion is straightforward. Here, we apply the relationships between transforms in different domains to deduce the form of the inverse Fourier transform in the other cases.

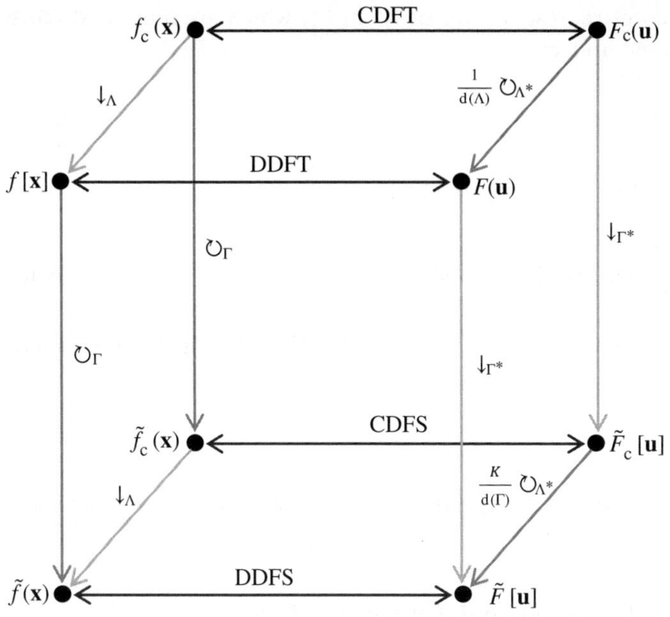

Figure 6.4 Fourier–Poisson cube illustrating Fourier transform relationships for sampling and periodization of multidimensional signals. (*See color plate section for the color representation of this figure.*)

6.6.1 Inverse Discrete-Domain Aperiodic Fourier Transform

We begin with the discrete-domain aperiodic case. Let $f[\mathbf{x}], \mathbf{x} \in \Lambda$ be absolutely summable so that its discrete-domain Fourier transform $F(\mathbf{u})$ converges. Let Γ_L be a sublattice of Λ that is expanded by a factor of L along each of the D vectors of a basis of Λ, i.e. $\mathbf{V}_{\Gamma_L} = L\mathbf{V}_\Lambda$. Since f is absolutely summable, we know that the left-hand side of the Poisson summation formula Eq (6.38) converges for all $\mathbf{x}$. In addition, absolute summability implies that for any $\epsilon > 0$, we can find a value of R such that $\sum_{\mathbf{x}:|\mathbf{x}|>R} |f[\mathbf{x}]| < \epsilon$. For each $\mathbf{x} \in \Lambda$, fix ϵ, choose R as above, and choose L sufficiently large such that $|\mathbf{x} - \mathbf{t}| > R$ for all $\mathbf{t} \in \Gamma_L \backslash \mathbf{0}$ and $\mathbf{x}$ lies in a Voronoi unit cell of Γ_L. Then

$$\sum_{\mathbf{t}\in\Gamma_L\backslash\mathbf{0}} |f[\mathbf{x} - \mathbf{t}]| < \epsilon, \tag{6.42}$$

or in other words, for the left-hand side of Equation (6.38)

$$\lim_{L\to\infty} \sum_{\mathbf{t}\in\Gamma_L} f[\mathbf{x} - \mathbf{t}] = f[\mathbf{x}]. \tag{6.43}$$

The limit of the right-hand side of Equation (6.38) gives the expression for the discrete-domain aperiodic inverse Fourier transform, i.e.

$$f[\mathbf{x}] = \lim_{L\to\infty} \frac{1}{L^D} \sum_{i=0}^{L^D-1} F(\mathbf{d}_i) \exp(\mathrm{j}2\pi \mathbf{d}_i \cdot \mathbf{x}), \tag{6.44}$$

where the $\mathbf{d}_i$ are the L^D coset representatives of Λ^* in Γ_L^*, which are more and more densely filling the unit cell Λ^*. In fact,

$$\mathcal{P}_{\Lambda^*} = \bigcup_{i=0}^{L^D-1} (\mathbf{d}_i + \mathcal{P}_{\Gamma_L^*}), \tag{6.45}$$

where the volume of $\mathcal{P}_{\Gamma_L^*}$ is $\mathrm{d}(\Gamma_L^*) = \mathrm{d}(\Lambda^*)/L^D$. Thus,

$$f[\mathbf{x}] = \mathrm{d}(\Lambda) \lim_{L\to\infty} \sum_{i=0}^{L^D-1} F(\mathbf{d}_i) \exp(\mathrm{j}2\pi \mathbf{d}_i \cdot \mathbf{x}) \mathrm{d}(\Gamma_L^*), \tag{6.46}$$

where $\mathrm{d}(\Gamma_L^*)$ is the incremental volume in $\mathbb{R}^D$ associated with $F(\mathbf{d}_i)$. The limit is seen to be the Riemann integal over $\mathcal{P}_{\Lambda^*}$, so that

$$f[\mathbf{x}] = \mathrm{d}(\Lambda) \int_{\mathcal{P}_{\Lambda^*}} F(\mathbf{u}) \exp(\mathrm{j}2\pi \mathbf{u} \cdot \mathbf{x})\, \mathrm{d}\mathbf{u}, \quad \mathbf{x} \in \Lambda, \tag{6.47}$$

which is the inverse discrete-domain Fourier transform, as was stated without proof in Equation (3.31).

6.6.2 Inverse Continuous-Domain Periodic Fourier Transform

A similar approach can be used to deduce the inverse Fourier transform for continuous-domain periodic signals, although this case is much more complicated. Continuous domain functions can exhibit very complicated behavior. Let $\tilde{f}(\mathbf{x})$ be Γ-periodic with Fourier transform $\tilde{F}[\mathbf{u}]$, $\mathbf{u} \in \Gamma^*$, where we assume that $\tilde{F}$ is absolutely summable over Γ^*. This would hold for example if $\tilde{f}(\mathbf{x})$ is continuous, with continuous partial derivatives $\frac{\partial \tilde{f}}{\partial x_i}$; any physical signal could be approximated as closely as we wish by such a function.

Let us define

$$g[\mathbf{s}] = \mathrm{d}(\Gamma^*)\tilde{F}[-s], \qquad \mathbf{s} \in \Gamma^*, \tag{6.48}$$

which is also absolutely summable. From Section 6.6.1, we know that

$$g[\mathbf{s}] = \mathrm{d}(\Gamma^*) \int_{\mathcal{P}_\Gamma} G(\mathbf{w}) \exp(\mathrm{j}2\pi \mathbf{w} \cdot \mathbf{s})\, \mathrm{d}\mathbf{w}, \qquad \mathbf{s} \in \Gamma^*, \tag{6.49}$$

where

$$G(\mathbf{w}) = \sum_{\mathbf{s}\in\Gamma^*} g[\mathbf{s}] \exp(-\mathrm{j}2\pi \mathbf{w} \cdot \mathbf{s}), \qquad \mathbf{w} \in \mathbb{R}^D. \tag{6.50}$$

Replacing $g[\mathbf{s}]$ with its definition,

$$\tilde{F}[\mathbf{s}] = \frac{1}{\mathrm{d}(\Gamma^*)} g[-\mathbf{s}] = \int_{\mathcal{P}_\Gamma} G(\mathbf{w}) \exp(-\mathrm{j}2\pi \mathbf{w} \cdot \mathbf{s})\, \mathrm{d}\mathbf{w}, \qquad \mathbf{s} \in \Gamma^*, \tag{6.51}$$

where

$$G(\mathbf{w}) = \mathrm{d}(\Gamma^*) \sum_{\mathbf{s}\in\Gamma^*} \tilde{F}[\mathbf{s}] \exp(\mathrm{j}2\pi \mathbf{w} \cdot \mathbf{s}), \qquad \mathbf{w} \in \mathbb{R}^D. \tag{6.52}$$

Putting this in the usual form, by substituting **u** for **s** and **x** for **w**,

$$\tilde{F}[\mathbf{u}] = \int_{\mathcal{P}_\Gamma} G(\mathbf{x}) \exp(-\mathrm{j}2\pi\mathbf{u}\cdot\mathbf{x})\, \mathrm{d}\mathbf{x}, \qquad \mathbf{u} \in \Gamma^* \tag{6.53}$$

$$G(\mathbf{x}) = \frac{1}{\mathrm{d}(\Gamma)} \sum_{\mathbf{u}\in\Gamma^*} \tilde{F}[\mathbf{u}] \exp(\mathrm{j}2\pi\mathbf{u}\cdot\mathbf{x}). \tag{6.54}$$

Since $\tilde{f}$ and G have the same Fourier transform, $\tilde{f} - G = 0$. Thus we can conclude that

$$\tilde{f}(\mathbf{x}) = \frac{1}{\mathrm{d}(\Gamma)} \sum_{\mathbf{u}\in\Gamma^*} \tilde{F}[\mathbf{u}] \exp(\mathrm{j}2\pi\mathbf{u}\cdot\mathbf{x}), \tag{6.55}$$

which is the inverse continuous-domain periodic Fourier transform as stated without proof in Equation (5.17).

6.6.3 Inverse Continuous-Domain Fourier Transform

Inversion of the continuous-domain aperiodic Fourier transform is largely beyond the scope of this book, due to the potential complicated behavior of continuous-domain signals. Regularity conditions including smoothness and decay at infinity must be imposed on the signals. We refer the reader to standard texts on Fourier analysis for the proof of validity of the continuous-domain inverse Fourier transform (Equation (2.51)) for various classes of signals, in particular see Stein and Weiss (1971).

6.7 Signals and Transforms with Finite Support

We have seen in Section 6.2 that if the Fourier transform $F_c(\mathbf{u})$ of a continuous-domain aperiodic signal $f_c(\mathbf{x})$ has a bounded region of support $\mathcal{V}$, we can choose a sampling lattice Λ sufficiently dense that $\mathcal{V}$ is contained in some unit cell of Λ^*. Then, the shifted copies of $F_c(\mathbf{u})$ in the periodization $\circlearrowright_{\Lambda^*} F_c$ do not overlap and we can fully reconstruct f from its samples on Λ. Similar observations can be made for other types of periodization as follows.

6.7.1 Continuous-Domain Signals with Finite Support

Suppose that $f_c(\mathbf{x})$ has a finite region of support $\mathcal{A}$, so that $f_c(\mathbf{x}) = 0$ if $\mathbf{x} \in \mathbb{R}^D \backslash \mathcal{A}$. If the periodization lattice Γ is sufficiently spaced out such that there is a unit cell $\mathcal{P}_\Gamma$ of Γ with $\mathcal{A} \subset \mathcal{P}_\Gamma$, then the shifted copies of f_c in $\circlearrowright_\Gamma f_c$ do not overlap. It follows that

$$\tilde{f}_c(\mathbf{x}) = (\circlearrowright_\Gamma f_c)(\mathbf{x}) = \sum_{\mathbf{t}\in\Gamma} f_c(\mathbf{x}-\mathbf{t}) = f_c(\mathbf{x}), \qquad \text{for } \mathbf{x} \in \mathcal{A}, \tag{6.56}$$

and thus

$$\tilde{F}_c[\mathbf{u}] = F_c(\mathbf{u}), \qquad \mathbf{u} \in \Gamma^*. \tag{6.57}$$

This was used in Examples 5.1 and 5.2.

In this case, we can evaluate $F_c(\mathbf{u})$ for all $\mathbf{u} \in \mathbb{R}^D$ from its samples on Γ^*.

$$\begin{aligned}
F_c(\mathbf{u}) &= \int_{\mathbb{R}^D} f_c(\mathbf{x}) \exp(-j2\pi\mathbf{u} \cdot \mathbf{x})\, d\mathbf{x} \\
&= \int_{\mathcal{P}_\Gamma} f_c(\mathbf{x}) \exp(-j2\pi\mathbf{u} \cdot \mathbf{x})\, d\mathbf{x} \\
&= \int_{\mathcal{P}_\Gamma} \tilde{f}_c(\mathbf{x}) \exp(-j2\pi\mathbf{u} \cdot \mathbf{x})\, d\mathbf{x} \\
&= \int_{\mathcal{P}_\Gamma} \frac{1}{d(\Gamma)} \sum_{\mathbf{w}\in\Gamma^*} \tilde{F}_c(\mathbf{w}) \exp(j2\pi\mathbf{w} \cdot \mathbf{x}) \exp(-j2\pi\mathbf{u} \cdot \mathbf{x})\, d\mathbf{x} \\
&= \frac{1}{d(\Gamma)} \sum_{\mathbf{w}\in\Gamma^*} \tilde{F}_c(\mathbf{w}) \int_{\mathcal{P}_\Gamma} \exp(-j2\pi(\mathbf{u}-\mathbf{w}) \cdot \mathbf{x})\, d\mathbf{x} \\
&= \sum_{\mathbf{w}\in\Gamma^*} \tilde{F}_c(\mathbf{w}) T(\mathbf{u}-\mathbf{w}),
\end{aligned} \tag{6.58}$$

where

$$T(\mathbf{u}) = \frac{1}{d(\Gamma)} \int_{\mathcal{P}_\Gamma} \exp(-j2\pi\mathbf{u} \cdot \mathbf{x})\, d\mathbf{x}. \tag{6.59}$$

This is an exact formula to interpolate $F_c(\mathbf{u})$ from its samples, and can be seen as a type of hybrid convolution in the frequency domain between the discrete-domain function $\tilde{F}_c$ and the continuous-domain function T. Here, $T(\mathbf{u})$ is seen to be the Fourier transform of the indicator function for $\mathcal{P}_\Gamma$, scaled by $1/d(\Gamma)$. In the two-dimensional case, if $\mathcal{P}_\Gamma$ is a polygon, this can be evaluated using the methods of Section 2.6.3.

6.7.2 Discrete-Domain Aperiodic Signals with Finite Support

Suppose that $f[\mathbf{x}]$, defined on Λ has a finite region of support $\mathcal{A}$ such that $f[\mathbf{x}] = 0$ if $\mathbf{x} \in \Lambda\backslash\mathcal{A}$. The unit sample response of the finite impulse response filters introduced in Section 3.8 are an example of such a case. If the periodization sublattice Γ is sufficiently spaced out such that there is a unit cell of $\mathcal{P}_\Gamma$ containing all the points of $\mathcal{A}$, $\mathcal{A} \subset \mathcal{P}_\Gamma$, then the shifted versions of f in $\circlearrowright_\Gamma f$ will not overlap. Thus,

$$\tilde{f}[\mathbf{x}] = (\circlearrowright_\Gamma f)[\mathbf{x}] = \sum_{\mathbf{t}\in\Gamma} f[\mathbf{x}-\mathbf{t}] = f[\mathbf{x}| \qquad \text{for } \mathbf{x} \in \mathcal{A}, \tag{6.60}$$

and so

$$\tilde{F}[\mathbf{u}] = F(\mathbf{u}), \qquad \mathbf{u} \in \Gamma^*. \tag{6.61}$$

Thus, with $K = d(\Gamma)/d(\Lambda)$, we obtain K values of $F(\mathbf{u})$ using a discrete-domain Fourier series (a.k.a. multidimensional DFT). We can also reconstruct $F(\mathbf{u})$ for all $\mathbf{u} \in \mathcal{P}_{\Lambda^*}$ from these K samples as follows.

$$\begin{aligned}
F(\mathbf{u}) &= \sum_{\mathbf{x}\in\Lambda} f[\mathbf{x}] \exp(-j2\pi\mathbf{u} \cdot \mathbf{x}) \\
&= \sum_{\mathbf{x}\in\mathcal{P}_\Gamma\cap\Lambda} f[\mathbf{x}] \exp(-j2\pi\mathbf{u} \cdot \mathbf{x}) \\
&= \sum_{\mathbf{x}\in\mathcal{B}} \tilde{f}[\mathbf{x}] \exp(-j2\pi\mathbf{u} \cdot \mathbf{x})
\end{aligned}$$

$$= \sum_{\mathbf{x}\in\mathcal{B}} \frac{1}{K} \sum_{\mathbf{w}\in\mathcal{D}} \tilde{F}[\mathbf{w}] \exp(\mathrm{j}2\pi\mathbf{w}\cdot\mathbf{x}) \exp(-\mathrm{j}2\pi\mathbf{u}\cdot\mathbf{x})$$
$$= \sum_{\mathbf{w}\in\mathcal{D}} \tilde{F}[\mathbf{w}]\tilde{T}[\mathbf{u}-\mathbf{w}], \qquad \mathbf{u}\in\Lambda^*, \tag{6.62}$$
$$\text{where } \tilde{T}[\mathbf{u}] = \frac{1}{K}\sum_{\mathbf{x}\in\mathcal{B}} \exp(-\mathrm{j}2\pi\mathbf{u}\cdot\mathbf{x}). \tag{6.63}$$

In the about equations, $\mathcal{B}$ is a set of coset representatives for Γ in Λ and $\mathcal{D}$ is a set of coset representatives for Λ^* in Γ^*.

6.7.3 Band-Limited Continuous-Domain Γ-Periodic Signals

In this last case, suppose that $\tilde{F}_c[\mathbf{u}] = 0$ if $\mathbf{u} \in \Gamma^*\backslash\mathcal{V}$ for some bounded region $\mathcal{V} \subset \mathbb{R}^D$. If $\tilde{f}$ is sampled on a lattice Λ such that $\mathcal{V}$ is a subset of some unit cell $\mathcal{P}_{\Lambda^*}$ of Λ^*, then $\tilde{f}_c(\mathbf{x})$ can be exactly reconstructed from its samples on Λ. In this case

$$\tilde{F}[\mathbf{u}] = (\mho_{\Lambda^*}\tilde{F}_c)[\mathbf{u}] = \sum_{\mathbf{w}\in\Lambda^*} \tilde{F}_c[\mathbf{u}-\mathbf{w}] = \tilde{F}_c[\mathbf{u}], \qquad \mathbf{u}\in\Gamma^*\cap\mathcal{P}_{\Lambda^*}, \tag{6.64}$$

or equivalently, for $\mathbf{u}\in\mathcal{D}$ where $\mathcal{D}$ is a set of coset representatives for Λ^* in Γ^*. The reconstruction formula in this case is

$$\tilde{f}_c(\mathbf{x}) = \frac{1}{d(\Gamma)} \sum_{\mathbf{u}\in\Gamma^*} \tilde{F}_c[\mathbf{u}]\ \exp(\mathrm{j}2\pi\mathbf{u}\cdot\mathbf{x})$$
$$= \frac{1}{d(\Gamma)} \sum_{\mathbf{u}\in\mathcal{D}} \tilde{F}_c[\mathbf{u}]\ \exp(\mathrm{j}2\pi\mathbf{u}\cdot\mathbf{x})$$
$$= \frac{1}{d(\Gamma)} \sum_{\mathbf{u}\in\mathcal{D}} \tilde{F}[\mathbf{u}]\ \exp(\mathrm{j}2\pi\mathbf{u}\cdot\mathbf{x})$$
$$= \frac{1}{d(\Gamma)} \sum_{\mathbf{u}\in\mathcal{D}} \left(\sum_{\mathbf{s}\in\mathcal{B}} \tilde{f}[\mathbf{s}] \exp(-\mathrm{j}2\pi\mathbf{u}\cdot\mathbf{s})\right) \exp(\mathrm{j}2\pi\mathbf{u}\cdot\mathbf{x})$$
$$= \frac{1}{d(\Gamma)} \sum_{\mathbf{s}\in\mathcal{B}} \tilde{f}[\mathbf{s}] \sum_{\mathbf{u}\in\mathcal{D}} \exp(\mathrm{j}2\pi\mathbf{u}\cdot(\mathbf{x}-\mathbf{s}))$$
$$= \sum_{\mathbf{s}\in\mathcal{B}} \tilde{f}[\mathbf{s}]\tilde{t}(\mathbf{x}-\mathbf{s}), \tag{6.65}$$
$$\text{where} \qquad \tilde{t}(\mathbf{x}) = \frac{1}{d(\Gamma)} \sum_{\mathbf{u}\in\mathcal{D}} \exp(\mathrm{j}2\pi\mathbf{u}\cdot\mathbf{x}). \tag{6.66}$$

Here we see that the reconstructed continuous domain periodic signal is a hybrid convolution between the discrete-domain period signal $\tilde{f}$ and the continuous-domain periodic signal $\tilde{t}$, where $\tilde{t}$ is the impulse response of a continuous-domain periodic ideal low-pass filter.

Problems

1 A two-dimensional continuous-domain signal $f_c(x, y)$ has Fourier transform

$$F_c(u, v) = \begin{cases} c e^{-\alpha(|u|+|v|)} & u^2 + v^2 < W^2 \\ 0 & u^2 + v^2 \geq W^2 \end{cases}$$

for some real number W. The signal is sampled on a hexagonal lattice Λ with sampling matrix

$$\mathbf{V} = \begin{bmatrix} X & X/2 \\ 0 & \sqrt{3}X/2 \end{bmatrix}$$

to give the sampled signal $f[x, y]$, $(x, y) \in \Lambda$, with Fourier transform $F(u, v)$.

(a) What is the expression for $F(u, v)$ in terms of $F_c(u, v)$?

(b) Find the largest possible value of X such that there is no aliasing? Sketch the region of support of the Fourier transform of the sampled signal in this case (including all replicas), and also indicate a unit cell of the reciprocal lattice Λ^*.

2 The face-centered cubic lattice is the most efficient lattice for the packing of spheres in three dimensions. A sampling matrix for this lattice is given by

$$\mathbf{V} = K \begin{bmatrix} 2 & 1 & 1 \\ 0 & 1 & 0 \\ 0 & 0 & 1 \end{bmatrix}$$

where K is some real constant. Suppose that a bandlimited three-dimensional signal $f(\mathbf{x})$ satisfying $F(\mathbf{u}) = 0$ for $|\mathbf{u}| > W$ is sampled on a lattice whose reciprocal lattice is face-centered cubic. Find the least dense lattice such that there is no aliasing. Compare the resulting sampling density with the best orthogonal sampling for which there is no aliasing.

3 Figure 6.5 illustrates the sensor in a hypothetical digital camera using a sensor element that is hexagonal in shape. There are $M = 740$ sensor elements in each horizontal row and there are $N = 480$ rows of sensor elements, for a total of 480×740

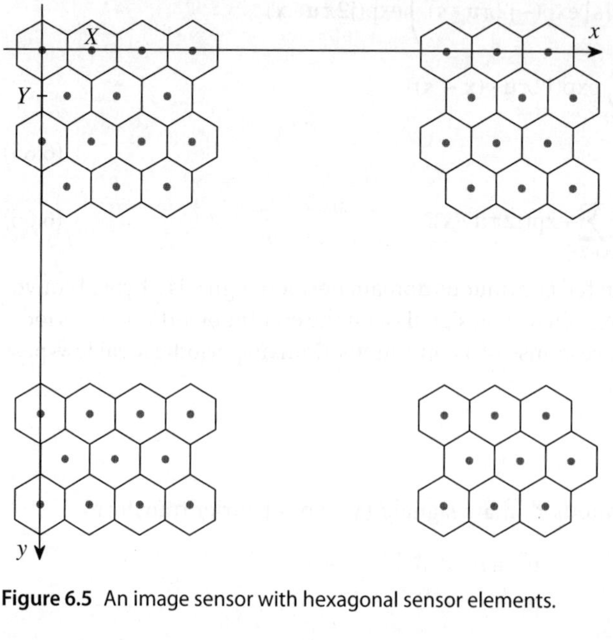

Figure 6.5 An image sensor with hexagonal sensor elements.

sensor elements. The centers of the sensor elements lie on a hexagonal lattice Λ, and each sensor element is a unit cell of this lattice. The output of each sensor element is the integral of light irradiance over the sensor element for some arbitrary exposure time, and it is associated with the lattice point at the center of the sensor element. Assume that the picture width is MX and the picture height (ph) is NY. We use the picture height as the unit of length. The sensor element is a regular hexagon with $Y = \sqrt{3}X/2$. (Note that Figure 6.5 is just a sketch and is not drawn to scale.)

(a) Give a sampling matrix for the lattice shown in Figure 6.5 in units of ph.

(b) What is the area of a sensor element, with correct units? What is the sampling density, with correct units?

(c) What is the aspect ratio of the sensor? Is it approximately 4/3 or approximately 16/9?

(d) The sampling process carried out by this sensor can be modeled by a linear shift-invariant (LSI) continuous-space filter followed by ideal sampling on Λ. Give an expression for the impulse response $h_a(x, y)$ of this LSI filter with the correct gain. Assume that if the image irradiance is a constant value over a sensor element (in arbitrary normalized units), the sampled value is that same value, i.e. the DC gain of $h_a(x, y)$ is 1.0.

(e) Give an expression for the frequency response $H_a(u, v)$ corresponding to the camera aperture impulse response $h_a(x, y)$.

(f) Assume that the continuous-space input light irradiance $f_c(x, y)$ has a Fourier transform $F_c(u, v)$. Give an expression for the Fourier transform of the sampled image $f[x, y]$, $(x, y) \in \Lambda$ in terms of in terms of $F_c(u, v)$ and $H_a(u, v)$; you should explicitly evaluate the reciprocal lattice Λ^*.

7

Light and Color Representation in Imaging Systems

7.1 Introduction

Visual images are created by patterns of light falling on the retina of the human eye. The goal of an imaging system is to use an image sensor to capture the light from a scene, to convert the sensor response to electronic form, and to store and/or transmit this data. At the end of the chain, a display device converts the electronic data back to an optical image that should appear as similar as possible to the original scene when viewed by a human viewer, i.e. patterns and colors should be accurately reproduced. Alternatively, we may wish to accurately reproduce on the display the intent of a content producer, which may contain a mix of natural and synthetic imagery.

In the previous chapters, image signals have simply been represented as a scalar value between 0 and 1, where this scalar value can represent a level of gray ranging from black to white. In this chapter, the exact nature of the image signal value is revealed for both grayscale and color images. In the case of color images, it is shown to be a three-dimensional vector quantity. We consider in this chapter the representation of light and color in large fixed patches. The study of color signals, i.e. images and video, is addressed in the next chapter. This chapter and the next are based on the book by the author [Dubois (2010)], which can be consulted for more details and references to the literature. We follow the notation used in that book.

7.2 Light

Visible light is electromagnetic radiation with wavelengths roughly in the band from 350 to 780 nm. This is the range of wavelengths to which the human eye is sensitive and has a response (although response below 400 nm and above 700 nm is very small). A broader range of wavelengths extending into infrared and ultraviolet regions of the electromagnetic spectrum may be considered when dealing with various types of imaging sensors and displays. The study of light as radiant energy, independently of its effect on the human visual system, is called *radiometry*. This radiant energy is denoted Q_e, measured in Joules (J). A good overview of the relevant aspects of radiometry can be found in Germer *et al.* (2014), particularly chapter 2.

Multidimensional Signal and Color Image Processing Using Lattices, First Edition. Eric Dubois.

Companion website: www.wiley.com/go/Dubois/multiSP

In imaging, we are generally more interested in the power of light passing through a given surface (say a sensor element) at a given place and time than in total energy. There are a number of standard units that measure various forms of radiant power. *Radiant flux* (also called radiant power), denoted P_e, is the radiant energy emitted, transferred or received by a given surface per unit of time. We can write $P_e = \frac{dQ_e}{dt}$, measured in Watts (W). The derivative notation is simply a symbolic representation of the incremental radiant energy transferred in an incremental interval of time.

The spectral distribution of light as a function of optical wavelength is described by the *power density spectrum* $P_e(\lambda)$, where $P_e(\lambda)\,d\lambda$ is the incremental power contained in the range of wavelengths $[\lambda, \lambda + d\lambda)$. Of course power is non-negative, so $P_e(\lambda) \geq 0$. Total power is given by $P_e = \int_0^\infty P_e(\lambda)\,d\lambda$, but we often write $P_e = \int_{\lambda_{min}}^{\lambda_{max}} P_e(\lambda)\,d\lambda$ to limit our consideration to wavelengths in a given range of interest, e.g., 350 nm to 780 nm.

Radiant flux refers to an entire surface, say a whole image sensor or CRT display. We are usually interested in more localized measures. For example *irradiance* E_e measures the radiant flux per unit area falling on a surface at a particular location, $E_e = \frac{dP_e}{dA}$, measured in W m^{-2}. This would be relevant in measuring the relative intensity of light falling at a particular point on an image sensor. A closely related concept is *radiant exitance*, denoted $M_e = \frac{dP_e}{dA}$ in W m^{-2}, measuring the radiant flux per unit area emitted by a surface at a particular location. This would be relevant in measuring the relative intensity of light emitted by a CRT or LCD display at a particular point on the screen.

Two final quantities of interest are *radiant intensity* and *radiance*. Radiant intensity, denoted I_e, is the radiant flux emitted by a point source of light in a given direction, per unit of solid angle, $I_e = \frac{dP_e}{d\omega}$, measured in W sr^{-1}, where *sr* stands for steradian, the unit of solid angle. Radiance is the radiant flux emitted by a surface in a given direction per unit of area and solid angle. Radiance is particularly important in describing the light coming from surfaces in the environment as seen from the point of view of an observer, and is thus a key concept in computer vision. These concepts are discussed in great detail in Germer *et al.* (2014). All of these quantities can be represented as power density spectra as a function of λ, for example $E_e(\lambda)$ for irradiance.

Figure 7.1 illustrates some typical power density spectra of light sources: an LED white light bulb, a helium–neon laser, the light from red, green, and blue phosphors of a CRT display, and D65, a standardized white illuminant discussed in Section 7.5.3. As we shall see, the relative power as a function of wavelength determines the perceived color. Note that the power density spectrum of light from a red helium–neon laser is concentrated at a single wavelength λ_0, about 633 nm. This can be approximated by a Dirac delta function $P_e\delta(\lambda - \lambda_0)$.

Figure 7.2(a) shows the transmission as a function of wavelength, $t(\lambda)$, of a plastic filter described as "dark lemon". This would also be the power density spectrum of light exiting the filter if white light with a flat spectrum were shone through it. If the irradiance were $E_e(\lambda)$ without the filter, it would be $E_e(\lambda)t(\lambda)$ with the filter in the path. This concept is used extensively in color filter arrays used in both sensors and displays. In a similar vein, Figure 7.2(b) shows the reflectance as a function of wavelength, $r(\lambda)$, of a yellow patch on a Macbeth Color Checker chart. If the incident light has power density spectrum $E_e(\lambda)$, the reflected light has power density spectrum $E_e(\lambda)r(\lambda)$. However, this is a big simplification since in general the reflectance depends on both the angle of incidence of the illumination and the viewing angle relative to the orientation of the surface in the

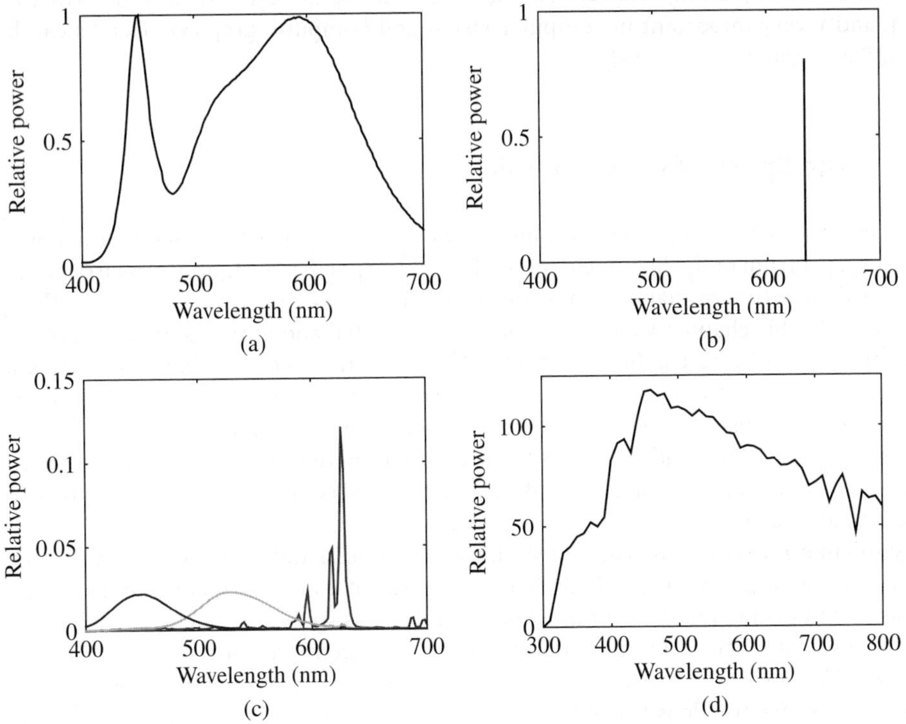

Figure 7.1 Illustration of power density spectra in arbitrary units for several light sources of interest. (a) LED T8 white lamp. (b) Helium–neon laser at 633 nm. (c) Red, green and blue phosphors of an EIZO CRT display. (d) Illuminant D65 white light.

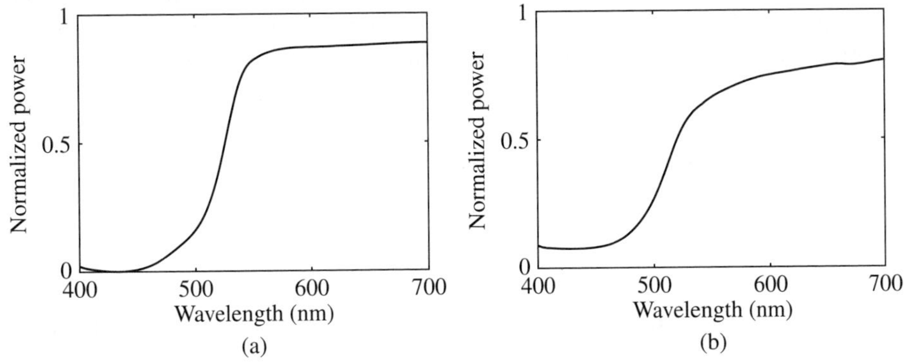

Figure 7.2 (a) Spectral transmission of dark lemon plastic filter. (b) Spectral reflectance of yellow square on a Macbeth Color Checker chart.

scene. This information is captured in the bidirectional reflectance distribution (BDRF), a quantity very important in computer vision and computer graphics. Details can be found in Germer *et al.* (2014).

7.3 The Space of Light Stimuli

The color sensation perceived by a human viewer is largely determined by the power density spectrum of the light incident on the retina, specified by the spectral irradiance. Let us denote the irradiance at position $\mathbf{x}$ and time t, as a function of wavelength, by $f(\mathbf{x}, t, \lambda)$. In this chapter, we do not consider the spatial and temporal variation of this signal; this is addressed in the next chapter. Thus we denote the light stimulus by $f(\lambda)$ and it is assumed constant over some extended spatial area and over time. The human visual system is only sensitive to electromagnetic radiation in a certain range $\mathcal{V} = [\lambda_{\min}, \lambda_{\max}]$, where $\lambda_{\min} \approx 350$ nm and $\lambda_{\max} \approx 780$ nm, although there is no consensus in the literature on what the precise range should be. This is not very critical, since the response of the human visual system is very low both below 400 nm and above 700 nm.

We define a space of physical light stimuli that is representative of the spectra encountered in practice. We wish to include both spectra that are continuous almost everywhere, with a finite number of finite discontinuities, as well as monochromatic light. A monochromatic light is for all practical purposes concentrated at a single wavelength, say λ_0. We model this by the Dirac delta $\alpha\delta(\lambda - \lambda_0)$, which has total power α. Thus a typical light spectral density can be written

$$f(\lambda) = f_c(\lambda) + \sum_{k=1}^{K} \alpha_k \delta(\lambda - \lambda_k), \quad f_c(\lambda) \geq 0, \lambda_k \in \mathcal{V}, \alpha_k > 0. \tag{7.1}$$

If there are no monochromatic components, we set $K = 0$.

Denote the integrated spectral density $F_c(\lambda) = \int_{\lambda_{\min}}^{\lambda} f_c(\mu)\, d\mu$. We assume that F_c is non-decreasing (since $f_c(\lambda) \geq 0$) and absolutely continuous, i.e. continuous everywhere and differentiable almost everywhere on $\mathcal{V}$. Given these assumptions, we define the set of physical light stimuli by

$$\mathcal{P} = \left\{ f_c + \sum_{k=1}^{K} \alpha_k \delta_{\lambda_k} \mid F_c \text{ is non-decreasing and absolutely continuous on } \mathcal{V}, \right.$$
$$\left. \alpha_k > 0, \lambda_k \in \mathcal{V}, K \in \mathbb{N} \right\}. \tag{7.2}$$

We assume that the light rays consist of incoherent radiation. This means that if we superimpose two lights with spectral densities $f_1(\lambda)$ and $f_2(\lambda)$, the spectral density of the resulting light is $f_1(\lambda) + f_2(\lambda)$, which also belongs to $\mathcal{P}$. Also, for any positive real α and $f(\lambda) \in \mathcal{P}$, $\alpha f(\lambda)$ also belongs to $\mathcal{P}$. Thus $\mathcal{P}$ is closed under addition and multiplication by a non-negative real constant.

It is convenient to embed $\mathcal{P}$ in a vector space where operations of both addition and subtraction, as well as multiplication by negative constants, are permitted. The difference between two elements of $\mathcal{P}$ would belong to such a space. Thus, similar to the definition of $\mathcal{P}$, we define

$$\mathcal{A} = \left\{ f_c + \sum_{k=1}^{K} \alpha_k \delta_{\lambda_k} \mid F_c \text{ is absolutely continuous on } \mathcal{V}, \right. \\ \left. \alpha_k \in \mathbb{R}, \lambda_k \in \mathcal{V}, K \in \mathbb{N} \right\}. \tag{7.3}$$

It is straightforward to verify that $\mathcal{A}$ satisfies all the properties of a real vector space. Also, every element of $\mathcal{A}$ can be written as the difference of two elements of $\mathcal{P}$ (this representation is not unique). For example, if $f^+(\lambda) = \max(f(\lambda), 0)$ and $f^-(\lambda) = -\min(f(\lambda), 0)$, then $f(\lambda) = f^+(\lambda) - f^-(\lambda)$, where $f^+(\lambda), f^-(\lambda) \in \mathcal{P}$. See Chapter 2 of Dubois (2010) for more details.

7.4 The Color Vector Space

It is a *psychophysical* observation that different spectral densities can give rise to identical color sensations in a human viewer. Figure 7.3 shows an example of the spectral densities of two very different lights that appear identical to a human viewer. There are physiological explanations for this, but the theory that follows is based strictly on psychophysical observations. If two spectral densities $C_1(\lambda)$ and $C_2(\lambda)$ produce an identical color sensation in a human viewer under the same viewing conditions, they are said to

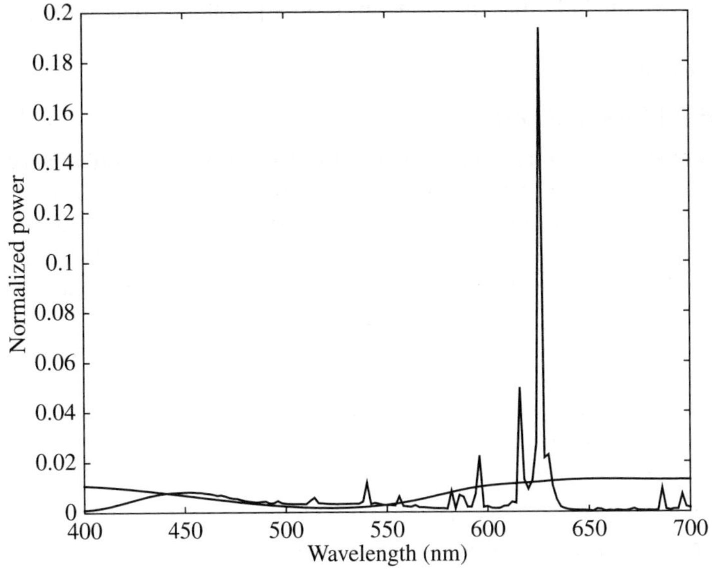

Figure 7.3 Spectral densities of two lights that appear as the same medium pink color to a normal trichomatic human viewer. The smooth curve is obtained by shining equal-energy white light through a pink plastic filter, while the peaky curve is obtained by displaying an identical looking color on a CRT monitor.

form a *metameric pair*. We denote this

$$C_1(\lambda) \triangleq C_2(\lambda). \tag{7.4}$$

The relation in Equation (7.4) applies to a *specific* observer, and may not hold *exactly* for other observers, although differences may be small for "normal" observers. Empirical psychophysical observations about such metameric equivalence allows us to establish a strong mathematical structure for color measurement. We will present these observations as needed and develop the consequences. These properties are sometimes called Grassmann's laws due to the presentation of the initial form of these properties in Grassmann's landmark paper [Grassmann (1854)].

7.4.1 Properties of Metamerism

We present several axioms of color matching, named G1–G5 after Grassmann, although the numbering is our own. These axioms represent an idealization of the properties of human color vision as described by Grassmann and others that followed. These axioms hold well in normal viewing conditions but will break down in unusual or extreme conditions, for example, very high intensity. Nevertheless, to develop the theory, we assume that they hold unconditionally.

Metamerism, as expressed by Equation (7.4), is a *relation* on the set $\mathcal{P}$. The first empirical observation is transitivity:

Transitivity (G1). If $C_1(\lambda) \triangleq C_2(\lambda)$ and $C_2(\lambda) \triangleq C_3(\lambda)$, then $C_1(\lambda) \triangleq C_3(\lambda)$.

If this is coupled with the obvious facts that $C(\lambda) \triangleq C(\lambda)$ and that $C_1(\lambda) \triangleq C_2(\lambda)$ implies $C_2(\lambda) \triangleq C_1(\lambda)$, we conclude that metamerism is an equivalence relation on $\mathcal{P}$ (see Appendix A).

For any $C(\lambda) \in \mathcal{P}$, we denote

$$[\mathbf{C}] = \{C_1(\lambda) \in \mathcal{P} \mid C_1(\lambda) \triangleq C(\lambda)\}, \tag{7.5}$$

the set of all spectral densities metamerically equivalent to $C(\lambda)$; this is an equivalence class. Metameric classes are either identical or disjoint, and form a partition of $\mathcal{P}$ (Appendix A). We refer to $[\mathbf{C}]$ as a *color*. Sometimes we write $[C(\lambda)]$ to indicate the metameric equivalence class containing the particular spectral density $C(\lambda)$. Any arbitrary element $C(\lambda) \in [\mathbf{C}]$ can be used as a representative. Equality of colors is denoted with an ordinary = sign, $[\mathbf{C}_1] = [\mathbf{C}_2]$, meaning that the equivalence classes are identical.

The following two empirical facts show that we can scale and add colors.

Scaling (G2). If $C_1(\lambda) \triangleq C_2(\lambda)$ then $\alpha C_1(\lambda) \triangleq \alpha C_2(\lambda)$ for any real, nonnegative α.

Property G2 means that if $[\mathbf{C}] = \{C_1(\lambda) \in \mathcal{P} \mid C_1(\lambda) \triangleq C(\lambda)\}$ then $[\alpha\mathbf{C}] = \{\alpha C_1(\lambda) \in \mathcal{P} \mid C_1(\lambda) \triangleq C(\lambda)\}$. We denote this color $\alpha[\mathbf{C}]$ without ambiguity.

Addition (G3). $C(\lambda) + C_1(\lambda) \triangleq C(\lambda) + C_2(\lambda)$ for arbitrary $C(\lambda)$ if and only if $C_1(\lambda) \triangleq C_2(\lambda)$.

Corollary to (G3). If $C_1(\lambda) \triangleq C_2(\lambda)$ then $C_1(\lambda) + C_3(\lambda) \triangleq C_2(\lambda) + C_4(\lambda)$ if and only if $C_3(\lambda) \triangleq C_4(\lambda)$.

Proof: Apply (G3) twice and use transitivity. By (G3), $C_1(\lambda) + C_3(\lambda) \triangleq C_2(\lambda) + C_3(\lambda)$, and again by (G3), $C_2(\lambda) + C_3(\lambda) \triangleq C_2(\lambda) + C_4(\lambda)$ if and only if $C_3(\lambda) \triangleq C_4(\lambda)$. Transitivity (G1) establishes the result. □

Property (G3) allows us to define addition of colors. Let $[\mathbf{C}_1]$ and $[\mathbf{C}_2]$ be two colors. Define the sum of $[\mathbf{C}_1]$ and $[\mathbf{C}_2]$ to be

$$[\mathbf{C}_1] + [\mathbf{C}_2] = [C_1(\lambda) + C_2(\lambda)] \tag{7.6}$$

where $C_1(\lambda)$ is any element of the class $[\mathbf{C}_1]$ and $C_2(\lambda)$ is any element of the class $[\mathbf{C}_2]$. This assignment is well-defined if it is independent of the particular choices $C_1(\lambda)$ and $C_2(\lambda)$ within $[\mathbf{C}_1]$ and $[\mathbf{C}_2]$ respectively. Indeed, let $C_{11}(\lambda)$ and $C_{12}(\lambda)$ be two arbitrary elements of $[\mathbf{C}_1]$ and $C_{21}(\lambda)$ and $C_{22}(\lambda)$ be two arbitrary elements of $[\mathbf{C}_2]$. Then

$$\begin{aligned} C_{11}(\lambda) + C_{21}(\lambda) &\triangleq C_{11}(\lambda) + C_{22}(\lambda) \\ &\triangleq C_{12}(\lambda) + C_{22}(\lambda) \end{aligned}$$

by applying (G3) twice. Thus addition of colors as defined by Equation (7.6) is well defined.

The next empirical observation allows us to establish the dimension of the eventual color vector space that will be defined. This observation applies to a *trichromatic* observer, the usual case for the human visual system.

Dimension (G4). Four colors are always linearly dependent. For $C(\lambda), C_1(\lambda), C_2(\lambda), C_3(\lambda)$ arbitrary elements of $\mathcal{P}$, there always exist $\alpha \geq 0$ and non-negative $\alpha_1, \alpha_2, \alpha_3$, not all zero, such that

$$\alpha C(\lambda) + \sum_{i \in A} \alpha_i C_i(\lambda) \triangleq \sum_{i \in \overline{A}} \alpha_i C_i(\lambda)$$

where $A \subset \{1, 2, 3\}$ and $\overline{A} = \{1, 2, 3\} \setminus A$. This is *not* the case for three colors.

In other words, either $C(\lambda)$ matches a weighted combination of the three $C_i(\lambda)$, or a weighted combination of $C(\lambda)$ and one of the $C_i(\lambda)$ matches a weighted combination of the other two, or a weighted combination of $C(\lambda)$ and two of the $C_i(\lambda)$ matches the third.

Property (G4) directly states that there exist three colors that are linearly independent, i.e. none of these three can be metamerically matched with a linear combination of the other two. The colors red, green, and blue are a familiar example of such a linearly independent set. Let us denote such a linearly independent set as $[\mathbf{P}_1]$, $[\mathbf{P}_2]$, and $[\mathbf{P}_3]$. We refer to a set of three linearly independent colors as *primaries*. From (G4), we can represent an arbitrary color using these primaries.

Theorem 7.1 Given a set of primaries $[\mathbf{P}_i], i = 1, 2, 3$, an arbitrary color $[\mathbf{C}]$ can be expressed

$$[\mathbf{C}] = C_1[\mathbf{P}_1] + C_2[\mathbf{P}_2] + C_3[\mathbf{P}_3], \tag{7.7}$$

where the C_i are *unique* real numbers. In this equation, if any of the C_i are negative, the corresponding term is added to [**C**] to form the physical match. For example, if $C_1 < 0$, the physical match is $[\mathbf{C}] + (-C_1)[\mathbf{P}_1] = C_2[\mathbf{P}_2] + C_3[\mathbf{P}_3]$, and so on.

Proof: Identifying $[\mathbf{P}_i] = [C_i(\lambda)]$ in (G4), since the $[\mathbf{P}_i]$ are linearly independent, α cannot be 0. Thus we can take $\alpha = 1$, and (G4) can be written in the equivalent form

$$[C(\lambda)] + \sum_{i \in A} \alpha_i[\mathbf{P}_i] = \sum_{i \in \bar{A}} \alpha_i[\mathbf{P}_i], \tag{7.8}$$

where $\alpha_i \geq 0$. If $i \in A$, take $C_i = -\alpha_i$ and if $i \in \bar{A}$, take $C_i = \alpha_i$.

The C_i are unique since the $[\mathbf{P}_i]$ are linearly independent. If C_i and C'_i, $i = 1, 2, 3$, are two different representations, then $\sum_{i=1}^{3}(C_i - C'_i)[\mathbf{P}_i] = 0$ and thus $C_i - C'_i = 0$, $i = 1, 2, 3$. □

This means that *given a choice of primaries*, every color can be associated with a unique point in $\mathbb{R}^3$.

The last axiom of Grassmann that we introduce relates to continuity. As stated by Grassmann, if $C_2(\lambda)$ varies continuously in a mixture $C_1(\lambda) + C_2(\lambda)$, then the perceived color $[C_1(\lambda) + C_2(\lambda)]$ varies continuously. Here, we must state what we mean by *varies continuously*. We don't need to deal with a mixture; we can state that if a spectral density $C(\lambda)$ varies continuously in $\mathcal{P}$, then the color [**C**] varies continuously. Let $C_t(\lambda)$ for $t \in (0, 1)$ be a mapping from (0, 1) to $\mathcal{P}$ (we could use any open interval in $\mathbb{R}$). We say that the mapping is continuous if for any $t_0 \in (0, 1)$, $\lim_{t \to t_0} C_t(\lambda) = C_{t_0}(\lambda)$, where convergence is in the sense of distributions since $\mathcal{P}$ contains Dirac deltas. By convergence in the sense of distributions, we mean that for any test function $\phi(\lambda)$ (an infinitely differentiable function with bounded support),

$$\lim_{t \to t_0} \int_{\mathcal{V}} \phi(\lambda) C_t(\lambda) \, d\lambda = \int_{\mathcal{V}} \phi(\lambda) C_{t_0}(\lambda) \, d\lambda. \tag{7.9}$$

For the perceived color to vary continuously, we mean that $\lim_{t \to t_0}(C_{t,1}, C_{t,2}, C_{t,3}) = (C_{t_0,1}, C_{t_0,2}, C_{t_0,3})$, where this is the usual convergence in $\mathbb{R}^3$.

Continuity (G5). Let $C_t(\lambda)$ be a continuous mapping from (0, 1) to $\mathcal{P}$ and $C(\lambda) \in \mathcal{P}$ arbitrary. Then $\lim_{t \to t_0}[C(\lambda) + C_t(\lambda)] = [C(\lambda)] + [C_{t_0}(\lambda)]$.

7.4.2 Algebraic Condition for Metameric Equivalence

As we have described it, metameric equivalence can be verified only by carrying out a physical experiment with the viewer being modeled: can two different spectral densities be distinguished visually under fixed test conditions? However, we now introduce a numerical technique to verify metameric equivalence between elements of $\mathcal{P}$.

Theorem 7.2 For a given metameric equivalence satisfying (G1)–(G5), there exist three continuous functions $\bar{p}_i(\lambda)$, $\lambda \in \mathcal{V}$, $i = 1, 2, 3$, such that two spectral densities $C_1(\lambda)$ and $C_2(\lambda)$ in $\mathcal{P}$ are metamerically equivalent if and only if

$$\int_{\mathcal{V}} C_1(\lambda)\bar{p}_i(\lambda) \, d\lambda = \int_{\mathcal{V}} C_2(\lambda)\bar{p}_i(\lambda) \, d\lambda, \qquad i = 1, 2, 3. \tag{7.10}$$

Proof: Choose three linearly independent spectral densities $P_1(\lambda)$, $P_2(\lambda)$, $P_3(\lambda)$ in the sense of axiom (G4), which define three linearly independent colors $[\mathbf{P}_i] = [P_i(\lambda)]$, $i = 1, 2, 3$. We can match each monochromatic light in the visible spectrum with a linear combination of the three primaries $[\mathbf{P}_i]$ as

$$[\delta(\lambda - \mu)] = \overline{p}_1(\mu)[\mathbf{P}_1] + \overline{p}_2(\mu)[\mathbf{P}_2] + \overline{p}_3(\mu)[\mathbf{P}_3] \tag{7.11}$$

for each $\mu \in \mathcal{V}$. This would be established experimentally with the given observer. The continuity axiom (G5) ensures that $\overline{p}_i(\mu)$ are continuous functions of μ.

Let us first consider elements of $\mathcal{P}$ formed as linear combinations of monochromatic lights,

$$f_d(\lambda) = \sum_{k=1}^{K} \alpha_k \delta(\lambda - \lambda_k). \tag{7.12}$$

We note that

$$\int_{\mathcal{V}} f_d(\lambda)\overline{p}_i(\lambda)\, \mathrm{d}\lambda = \sum_{k=1}^{K} \alpha_k \overline{p}_i(\lambda_k). \tag{7.13}$$

Then

$$\begin{aligned}
[f_d(\lambda)] &= \sum_{k=1}^{K} \alpha_k [\delta(\lambda - \lambda_k)] \\
&= \sum_{k=1}^{K} \alpha_k \left(\sum_{i=1}^{3} \overline{p}_i(\lambda_k)[\mathbf{P}_i] \right) \\
&= \sum_{i=1}^{3} \left(\sum_{k=1}^{K} \alpha_k \overline{p}_i(\lambda_k) \right) [\mathbf{P}_i] \\
&= \sum_{i=1}^{3} \left(\int_{\mathcal{V}} f_d(\lambda)\overline{p}_i(\lambda)\, \mathrm{d}\lambda \right) [\mathbf{P}_i].
\end{aligned} \tag{7.14}$$

Thus by the uniqueness condition of Theorem 7.1, the result holds for spectral densities that are linear combinations of monochromatic lights.

Now let $f_c(\lambda)$ be a continuous spectral density. This can be expressed in terms of incrementally small monochromatic lights as

$$f_c(\lambda) = \int_{\mathcal{V}} f_c(\mu)\delta(\lambda - \mu)\, \mathrm{d}\mu. \tag{7.15}$$

By the linearity axioms of color matching

$$\begin{aligned}
[f_c(\lambda)] &= \int_{\mathcal{V}} f_c(\mu)\, \mathrm{d}\mu[\delta(\lambda - \mu)] \\
&= \int_{\mathcal{V}} f_c(\mu)\, \mathrm{d}\mu \left(\sum_{i=1}^{3} \overline{p}_i(\mu)[\mathbf{P}_i] \right) \\
&= \sum_{i=1}^{3} \left(\int_{\mathcal{V}} f_c(\mu)\overline{p}_i(\mu)\, \mathrm{d}\mu \right) [\mathbf{P}_i].
\end{aligned} \tag{7.16}$$

Thus, the theorem holds for continuous spectral densities as well, and thus for arbitrary elements of $\mathcal{P}$. □

Note that in this proof we have implicitly assumed that the continuous spectral density $f_c(\lambda)$ can be approximated by the sum of a finite number of narrowband lights, and that this approximation converges to the integral of Equation (7.15) in the sense of distributions. Then by the continuity axiom (G5), the corresponding color representation also converges. We can expect that the equivalence defined by Equation (7.10) does not depend on the specific choice of primaries $[\mathbf{P}_i]$, and we will explicitly confirm this in a later section. Our development has also established the following key result.

Theorem 7.3 For any color $[\mathbf{C}]$, we have the representation

$$[\mathbf{C}] = [C(\lambda)] = C_1[\mathbf{P}_1] + C_2[\mathbf{P}_2] + C_3[\mathbf{P}_3], \text{ where} \tag{7.17}$$

$$C_i = \int_{\mathcal{V}} C(\lambda)\overline{p}_i(\lambda)\,\mathrm{d}\lambda. \tag{7.18}$$

The C_i obtained are independent of the particular representative $C(\lambda)$ taken from $[\mathbf{C}]$.

Proof: For an arbitrary $C(\lambda) \in \mathcal{P}$, adding Equation (7.14) and Equation (7.16), we have

$$[C(\lambda)] = \sum_{i=1}^{3}\left(\int_{\mathcal{V}} C(\lambda)\overline{p}_i(\lambda)\,\mathrm{d}\lambda\right)[\mathbf{P}_i], \tag{7.19}$$

and since the representation of Equation (7.17) is unique by Theorem 7.1, the result follows. □

We have shown that the metameric equivalence of human vision can be specified by three continuous functions $\overline{p}_i(\lambda) \in C^0(\mathcal{V})$, where $C^0(\mathcal{V})$ is the vector space of continuous functions on $\mathcal{V}$. Thus, we can equivalently *define* metameric equivalence by specifying the $\overline{p}_i(\lambda)$ and saying that $C_1(\lambda) \triangleq C_2(\lambda)$ if Equation (7.10) holds. Following the terminology of colorimetry, we refer to the $\overline{p}_i(\lambda)$ as color-matching functions.

Theorem 7.4 The functions $\overline{p}_i(\lambda)$ are linearly independent in $C^0(\mathcal{V})$.

Proof: Let $P_j(\lambda)$ be an arbitrary representative from the equivalence class $[\mathbf{P}_j]$. From Theorem 7.3, it follows that

$$\int_{\mathcal{V}} P_j(\lambda)\overline{p}_i(\lambda)\,\mathrm{d}\lambda = \delta_{ij}. \tag{7.20}$$

Suppose that $a_1\overline{p}_1(\lambda) + a_2\overline{p}_2(\lambda) + a_3\overline{p}_3(\lambda) = 0$. Then

$$\begin{aligned} a_j &= \int_{\mathcal{V}} P_j(\lambda)(a_1\overline{p}_1(\lambda) + a_2\overline{p}_2(\lambda) + a_3\overline{p}_3(\lambda))\,\mathrm{d}\lambda \\ &= 0, \qquad j = 1, 2, 3 \end{aligned} \tag{7.21}$$

so the $\overline{p}_i(\lambda)$ are linearly independent. □

7.4.3 Extension of Metameric Equivalence to $\mathcal{A}$

Given the characterization of Theorem 7.2, we can extend metameric equivalence from $\mathcal{P}$ to all of $\mathcal{A}$ in a straightforward and consistent fashion as follows.

Definition 7.1 Extended metameric equivalence, denoted $\boxed{\equiv}$: for two elements $C_1(\lambda), C_2(\lambda) \in \mathcal{A}$, we say that $C_1(\lambda) \boxed{\equiv} C_2(\lambda)$ if and only if

$$\int_{\mathcal{V}} C_1(\lambda)\overline{p}_i(\lambda)\, \mathrm{d}\lambda = \int_{\mathcal{V}} C_2(\lambda)\overline{p}_i(\lambda)\, \mathrm{d}\lambda, \qquad i = 1, 2, 3, \tag{7.22}$$

where $\overline{p}_i(\lambda)$ are continuous, linearly independent functions defined by metameric equivalence $\triangleq$.

It is straightforward to verify that this defines an equivalence on $\mathcal{A}$. Also, if $C_1(\lambda)$, $C_2(\lambda) \in \mathcal{P}$, then from Theorem 7.2, $C_1(\lambda) \boxed{\equiv} C_2(\lambda) \Rightarrow C_1(\lambda) \triangleq C_2(\lambda)$. Thus $\boxed{\equiv}$ is a consistent extension of $\triangleq$.

As stated in Section 7.3, every element of $\mathcal{A}$ can be written (not uniquely) as the difference of two elements of $\mathcal{P}$. Suppose that $C_1(\lambda), C_2(\lambda) \in \mathcal{A}$, with $C_1(\lambda) = C_{1a}(\lambda) - C_{1b}(\lambda)$ and $C_2(\lambda) = C_{2a}(\lambda) - C_{2b}(\lambda)$, with $C_{1a}(\lambda), C_{1b}(\lambda), C_{2a}(\lambda), C_{2b}(\lambda) \in \mathcal{P}$. Then, it is straightforward to see from the definitions that $C_1(\lambda) \boxed{\equiv} C_2(\lambda)$ if and only if $C_{1a}(\lambda) + C_{2b}(\lambda) \triangleq C_{2a}(\lambda) + C_{1b}(\lambda)$. This alternate approach to define $\boxed{\equiv}$ is developed in detail in Dubois (2010).

Once again, it is straightforward to see from the definition that extensions of axioms (G1), (G2), and (G3) hold for the relation $\boxed{\equiv}$ on elements of $\mathcal{A}$. We denote these as (G1′), (G2′), (G3′), and state them here as a theorem to emphasize that the properties are a provable consequence of the definition of $\boxed{\equiv}$ and not axioms.

Theorem 7.5 Let $C_1(\lambda), C_2(\lambda), C_3(\lambda) \in \mathcal{A}$. Then the properties of transitivity, scaling and addition apply to the relation $\boxed{\equiv}$, specifically:

Transitivity (G1′). If $C_1(\lambda) \boxed{\equiv} C_2(\lambda)$ and $C_2(\lambda) \boxed{\equiv} C_3(\lambda)$, then $C_1(\lambda) \boxed{\equiv} C_3(\lambda)$.

Scaling (G2′). If $C_1(\lambda) \boxed{\equiv} C_2(\lambda)$ then $\alpha C_1(\lambda) \boxed{\equiv} \alpha C_2(\lambda)$ for any real α.

Addition (G3′). $C(\lambda) + C_1(\lambda) \boxed{\equiv} C(\lambda) + C_2(\lambda)$ for arbitrary $C(\lambda) \in \mathcal{A}$ if and only if $C_1(\lambda) \boxed{\equiv} C_2(\lambda)$.

Proof: Transitivity follows from transitivity of equality (=) for real numbers; scaling and addition follow from the linearity of integration. □

7.4.4 Definition of the Color Vector Space

We have now shown that all functions of wavelength in the vector space $\mathcal{A}$ can be partitioned into equivalence classes by the equivalence relation $\boxed{\equiv}$. We can denote the equivalence classes by $[\mathbf{C}]_{\boxed{=}}$ or $[C(\lambda)]_{\boxed{=}}$, where

$$[C(\lambda)]_{\boxed{=}} = \{C_1(\lambda) \in \mathcal{A} \mid C_1(\lambda) \boxed{=} C(\lambda)\}. \tag{7.23}$$

We can in the same spirit denote the equivalence classes on $\mathcal{P}$ defined by Equation (7.5) as $[\mathbf{C}]_{\triangleq}$ if we wish to distinguish the two. Note that even if $C(\lambda) \in \mathcal{P}$, $[C(\lambda)]_{\boxed{=}}$ is not the same as $[C(\lambda)]_{\triangleq}$, since it will contain elements of $\mathcal{A}$ that do not belong to $\mathcal{P}$. More on this soon.

We now define the color vector space.

Theorem 7.6 The set of equivalence classes of the equivalence relation $\boxed{=}$ on the vector space $\mathcal{A}$ forms a real vector space

$$\mathcal{C} = \{[C(\lambda)]_{\boxed{=}} \mid C(\lambda) \in \mathcal{A}\} \tag{7.24}$$

with operations of addition and multiplication by a real scalar given by

$$[C_1(\lambda)]_{\boxed{=}} + [C_2(\lambda)]_{\boxed{=}} = [C_1(\lambda) + C_2(\lambda)]_{\boxed{=}} \tag{7.25}$$

$$\alpha[C(\lambda)]_{\boxed{=}} = [\alpha C(\lambda)]_{\boxed{=}}. \tag{7.26}$$

Proof: The operations of addition and scaling are well defined by properties (G3′) and (G2′), respectively. The neutral or zero element of the vector space is $[\mathbf{0}]_{\boxed{=}}$, which we refer to as *black*, since $C(\lambda) = 0$ denotes the absence of any visible light. The negative of any element is $-[C(\lambda)]_{\boxed{=}} = [-C(\lambda)]_{\boxed{=}}$. The other properties of a vector space (see Appendix C) follow from the properties of real numbers. □

Again, we emphasize that this vector space is associated with a specific viewer, and that each viewer has their own color space. Later, we will adopt a single color space to represent all viewers with normal trichromatic vision. For such a viewer, the dimension of the color space, as expected, is 3.

Theorem 7.7 The dimension of $\mathcal{C}$ for a viewer satisfying (G4) is 3.

Proof: From (G4), we can find three elements of $\mathcal{P}$, say $P_1(\lambda)$, $P_2(\lambda)$, $P_3(\lambda)$, such that none of them can match a linear combination of the other two. It follows that the $[P_i(\lambda)]$ are linearly independent in $\mathcal{A}$, since otherwise there would exist α_1, α_2, α_3, not all zero, such that

$$\alpha_1[P_1(\lambda)]_{\boxed{=}} + \alpha_2[P_2(\lambda)]_{\boxed{=}} + \alpha_3[P_3(\lambda)]_{\boxed{=}} = [\mathbf{0}]_{\boxed{=}}. \tag{7.27}$$

Arranging so that all coefficients are positive,

$$\sum_{i\in I}(-\alpha_i)P_i(\lambda) \boxed{=} \sum_{i\in \bar{I}} \alpha_i P_i(\lambda), \tag{7.28}$$

where $I = \{i \mid \alpha_i < 0\}$ and $\bar{I} = \{i \mid \alpha_i \geq 0\}$. Since the scaled $P_i(\lambda)$ are in $\mathcal{P}$,

$$\sum_{i\in I}(-\alpha_i)P_i(\lambda) \triangleq \sum_{i\in \bar{I}} \alpha_i P_i(\lambda) \tag{7.29}$$

which contradicts the assumption. Thus, $\dim(\mathcal{C}) \geq 3$.

We can further show that the $[P_i(\lambda)]_{\boxed{\equiv}}$ span $\mathcal{C}$. First, let $C(\lambda) \in \mathcal{P}$. From (G4), we can find $\alpha_1, \alpha_2, \alpha_3 \in \mathbb{R}$ such that $C(\lambda) \boxed{\equiv} \sum_{i=1}^{3} \alpha_i P_i(\lambda)$ and $-C(\lambda) \boxed{\equiv} \sum_{i=1}^{3}(-\alpha_i)P_i(\lambda)$. Now, any general element of $C(\lambda) \in \mathcal{A}$ can be written $C(\lambda) = C_1(\lambda) - C_2(\lambda)$ where $C_1(\lambda), C_2(\lambda) \in \mathcal{P}$. Since both $C_1(\lambda)$ and $C_2(\lambda)$ match a linear combination of the $P_i(\lambda)$, so does $C_1(\lambda) - C_2(\lambda)$ from (G3′) and thus the $[P_i(\lambda)]_{\boxed{\equiv}}$, $i = 1, 2, 3$, span $\mathcal{C}$. □

Since $[C(\lambda)]_{\triangleq}$ and $[C(\lambda)]_{\boxed{\equiv}}$ denote the same color if $C(\lambda) \in \mathcal{P}$, and we will generally have no need to restrict $C(\lambda)$ to $\mathcal{P}$ in the equivalence class, we will, going forward, write $[\mathbf{C}]$ instead of $[\mathbf{C}]_{\boxed{\equiv}}$. We use $[\mathbf{C}]_{\triangleq}$ to denote the equivalence class of Equation (7.5) if the distinction is needed.

Now [0] is a subspace of the vector space $\mathcal{A}$, since if $C(\lambda) \in [0]$, i.e. $C(\lambda) \boxed{\equiv} 0$, then $\alpha C(\lambda) \boxed{\equiv} 0, \forall \alpha \in \mathbb{R}$ by (G2′), and if $C_1(\lambda), C_2(\lambda) \in [0]$ then $C_1(\lambda) + C_2(\lambda) \in [0]$ (apply (G3′) twice). We denote $[0] = \mathcal{K}$ and refer to it as black space. It is easy to see that $C_1(\lambda) \boxed{\equiv} C_2(\lambda)$ if and only if $C_1(\lambda) - C_2(\lambda) \in \mathcal{K}$. Thus, we can equivalently express the equivalence classes representing a color as $[C(\lambda)] = C(\lambda) + \mathcal{K}$, which is a coset of the subspace $\mathcal{K}$ in $\mathcal{A}$. More detail of this representation and how to identify elements of $\mathcal{K}$ can be found in Dubois (2010).

We define a *physically realizable color* $[\mathbf{C}]$ to be an element of the color vector space such that the corresponding equivalence class $\{C_1(\lambda) \mid C_1(\lambda) \boxed{\equiv} C(\lambda)\}$ contains at least one element of $\mathcal{P}$, i.e. $[C] \cap \mathcal{P} \neq \emptyset$. The set of physical colors forms a *subset* of $\mathcal{C}$ (not a subspace) denoted $\mathcal{C}_R$, which we will see has the form of a convex cone.

7.4.5 Bases for the Vector Space $\mathcal{C}$

Three linearly independent colors $[\mathbf{P}_1]$, $[\mathbf{P}_2]$ and $[\mathbf{P}_3]$ form a basis for $\mathcal{C}$ and are referred to as *primaries*. By convention, the intensities of primaries are chosen so that the sum of one unit of each matches a selected *reference white*, e.g., equal energy white. Thus, if $[\mathbf{P}'_1]$, $[\mathbf{P}'_2]$ and $[\mathbf{P}'_3]$ are linearly independent elements of $\mathcal{C}$ and $[\mathbf{W}]$ is the reference white, we would get the unique representation of $[\mathbf{W}]$

$$[\mathbf{W}] = W_1[\mathbf{P}'_1] + W_2[\mathbf{P}'_2] + W_3[\mathbf{P}'_3]. \tag{7.30}$$

We then choose as primaries $[\mathbf{P}_i] = W_i[\mathbf{P}'_i]$, $i = 1, 2, 3$. An example of a set of primaries would be the equivalence classes containing the spectral densities of the light emitted by the red, green and blue phosphors used in a CRT television display. There is no requirement for the $[\mathbf{P}_i]$ to be physical colors, and in fact they are not for several widely used color coordinates systems.

The coefficients in the expansion of a color with respect to a set of primaries are called *tristimulus values*. For any $[\mathbf{C}] \in \mathcal{C}$, we have

$$[\mathbf{C}] = C_1[\mathbf{P}_1] + C_2[\mathbf{P}_2] + C_3[\mathbf{P}_3] \tag{7.31}$$

where the C_i are tristimulus values. We generally use uppercase letters with an appropriate subscript related to the primaries to denote tristimulus values. Any color can be represented by a triple of tristimulus values with respect to a set of primaries and thus could be plotted on a three-dimensional set of coordinate axes. It should be emphasized that in such a diagram, usual Cartesian concepts such as distance and orthogonality are not meaningful.

If we are given $C(\lambda) \in \mathcal{A}$, then we can still compute the tristimulus values C_i using Equation (7.17) in Theorem 7.3. The result, while straightforward, is sufficiently important to state as a theorem.

Theorem 7.8 Let $[\mathbf{C}] = [C(\lambda)]$ be a color in $\mathcal{C}$, where $C(\lambda)$ is an arbitrary representative of the class. Let $[\mathbf{P}_i], i = 1, 2, 3$ be three primaries with corresponding color matching functions $\overline{p}_i(\lambda)$. Then the tristimulus values with respect to the primaries $[\mathbf{P}_i]$ are given by

$$C_i = \int_{\mathcal{V}} C(\lambda)\overline{p}_i(\lambda)\,d\lambda, \qquad i = 1, 2, 3. \tag{7.32}$$

Proof: We know that $C(\lambda)$ can be written as $C(\lambda) = C_1(\lambda) - C_2(\lambda)$ for some (nonunique) $C_1(\lambda), C_2(\lambda) \in \mathcal{P}$. From Theorem 7.3, we have

$$C_{1i} = \int_{\mathcal{V}} C_1(\lambda)\overline{p}_i(\lambda)\,d\lambda, \qquad i = 1, 2, 3$$

$$C_{2i} = \int_{\mathcal{V}} C_2(\lambda)\overline{p}_i(\lambda)\,d\lambda, \qquad i = 1, 2, 3.$$

Then

$$[\mathbf{C}] = [\mathbf{C}_1] - [\mathbf{C}_2] = \sum_{i=1}^{3}(C_{1i} - C_{2i})[\mathbf{P}_i], \tag{7.33}$$

so that

$$\begin{aligned} C_i = C_{1i} - C_{2i} &= \int_{\mathcal{V}} (C_1(\lambda) - C_2(\lambda))\overline{p}_i(\lambda)\,d\lambda \\ &= \int_{\mathcal{V}} C(\lambda)\overline{p}_i(\lambda)\,d\lambda, \qquad i = 1, 2, 3. \end{aligned} \tag{7.34}$$

□

It is convenient to represent a set of tristimulus values with respect to a given basis as a column vector. Let $\mathcal{B} = \{[\mathbf{P}_1], [\mathbf{P}_2], [\mathbf{P}_3]\}$ denote a set of primaries that form a basis for $\mathcal{C}$. Then, if $[\mathbf{C}] = \sum_{i=1}^{3} C_i[\mathbf{P}_i]$, we denote the set of tristimulus values as

$$\mathbf{C}_{\mathcal{B}} = \begin{bmatrix} C_1 \\ C_2 \\ C_3 \end{bmatrix}. \tag{7.35}$$

7.4.6 Transformation of Primaries

Since primaries form a basis for the color vector space, the change of representation from one set of primaries to another is a change of basis operation. Let $\mathcal{B} = \{[\mathbf{P}_1], [\mathbf{P}_2], [\mathbf{P}_3]\}$ be a set of primaries, and let $\tilde{\mathcal{B}} = \{[\tilde{\mathbf{P}}_1], [\tilde{\mathbf{P}}_2], [\tilde{\mathbf{P}}_3]\}$ be a different set of primaries. We have two representations of a color $[\mathbf{C}]$, namely

$$[\mathbf{C}] = \sum_{j=1}^{3} C_j[\mathbf{P}_j] = \sum_{k=1}^{3} \tilde{C}_k[\tilde{\mathbf{P}}_k]. \tag{7.36}$$

We can express the primaries $[\mathbf{P}_j]$ in terms of the $[\tilde{\mathbf{P}}_k]$ as

$$[\mathbf{P}_j] = \sum_{k=1}^{3} a_{kj}[\tilde{\mathbf{P}}_k], \quad j = 1, 2, 3, \tag{7.37}$$

i.e. the a_{kj} are the tristimulus values of the color $[\mathbf{P}_j]$ with respect to the primaries $[\tilde{\mathbf{P}}_k], k = 1, 2, 3$. This allows us to transform the representations of arbitrary colors. Specifically

$$\begin{aligned}
[\mathbf{C}] &= \sum_{j=1}^{3} C_j[\mathbf{P}_j] \\
&= \sum_{j=1}^{3} C_j \sum_{k=1}^{3} a_{kj}[\tilde{\mathbf{P}}_k] \\
&= \sum_{k=1}^{3} \left(\sum_{j=1}^{3} C_j a_{kj} \right) [\tilde{\mathbf{P}}_k],
\end{aligned} \tag{7.38}$$

from which we identify

$$\tilde{C}_k = \sum_{j=1}^{3} C_j a_{kj} = \sum_{j=1}^{3} a_{kj} C_j, \qquad k = 1, 2, 3. \tag{7.39}$$

This can be written in matrix form as

$$\begin{bmatrix} \tilde{C}_1 \\ \tilde{C}_2 \\ \tilde{C}_3 \end{bmatrix} = \begin{bmatrix} a_{11} & a_{12} & a_{13} \\ a_{21} & a_{22} & a_{23} \\ a_{31} & a_{32} & a_{33} \end{bmatrix} \begin{bmatrix} C_1 \\ C_2 \\ C_3 \end{bmatrix} \tag{7.40}$$

or $\mathbf{C}_{\tilde{B}} = \mathbf{A}\mathbf{C}_{B}$. Furthermore, it is clear that $\mathbf{C}_{B} = \mathbf{A}^{-1}\mathbf{C}_{\tilde{B}}$. To explicitly identify the original and target bases in this transformation, we adopt the notation [Dubois (2010)]

$$\mathbf{C}_{\tilde{B}} = \mathbf{A}_{B\to\tilde{B}}\mathbf{C}_{B}. \tag{7.41}$$

Using this notation, we also identify that $(\mathbf{A}_{B\to\tilde{B}})^{-1} = \mathbf{A}_{\tilde{B}\to B}$.

This matrix operation can also be used to transform the color matching functions for the primaries $[\mathbf{P}_j]$ to color matching functions for the primaries $[\tilde{\mathbf{P}}_k]$, recognizing that color matching functions specify tristimulus values of monochromatic lights for each λ:

$$\begin{bmatrix} \bar{\tilde{p}}_1(\lambda) \\ \bar{\tilde{p}}_2(\lambda) \\ \bar{\tilde{p}}_3(\lambda) \end{bmatrix} = \mathbf{A}_{B\to\tilde{B}} \begin{bmatrix} \bar{p}_1(\lambda) \\ \bar{p}_2(\lambda) \\ \bar{p}_3(\lambda) \end{bmatrix}. \tag{7.42}$$

The relationship between primaries can also be written in matrix form as

$$\begin{bmatrix} [\mathbf{P}_1] \\ [\mathbf{P}_2] \\ [\mathbf{P}_3] \end{bmatrix} = \mathbf{A}^T_{B\to\tilde{B}} \begin{bmatrix} [\tilde{\mathbf{P}}_1] \\ [\tilde{\mathbf{P}}_2] \\ [\tilde{\mathbf{P}}_3] \end{bmatrix}. \tag{7.43}$$

Note that Equation (7.43) is not a conventional real matrix operation. Rather, a real matrix multiplies a column matrix of abstract elements of the color space C, which is

Table 7.1 Conversions of tristimulus values, color matching functions and primaries between two sets of primaries defined by bases $\mathcal{B}$ and $\tilde{\mathcal{B}}$.

1.	$\begin{bmatrix}\tilde{C}_1\\ \tilde{C}_2\\ \tilde{C}_3\end{bmatrix} = \mathbf{A}_{\mathcal{B}\to\tilde{\mathcal{B}}}\begin{bmatrix}C_1\\ C_2\\ C_3\end{bmatrix}$	4.	$\begin{bmatrix}C_1\\ C_2\\ C_3\end{bmatrix} = \mathbf{A}^{-1}_{\mathcal{B}\to\tilde{\mathcal{B}}}\begin{bmatrix}\tilde{C}_1\\ \tilde{C}_2\\ \tilde{C}_3\end{bmatrix}$
2.	$\begin{bmatrix}\tilde{\bar{p}}_1(\lambda)\\ \tilde{\bar{p}}_2(\lambda)\\ \tilde{\bar{p}}_3(\lambda)\end{bmatrix} = \mathbf{A}_{\mathcal{B}\to\tilde{\mathcal{B}}}\begin{bmatrix}\bar{p}_1(\lambda)\\ \bar{p}_2(\lambda)\\ \bar{p}_3(\lambda)\end{bmatrix}$	5.	$\begin{bmatrix}\bar{p}_1(\lambda)\\ \bar{p}_2(\lambda)\\ \bar{p}_3(\lambda)\end{bmatrix} = \mathbf{A}^{-1}_{\mathcal{B}\to\tilde{\mathcal{B}}}\begin{bmatrix}\tilde{\bar{p}}_1(\lambda)\\ \tilde{\bar{p}}_2(\lambda)\\ \tilde{\bar{p}}_3(\lambda)\end{bmatrix}$
3.	$\begin{bmatrix}[\tilde{\mathbf{P}}_1]\\ [\tilde{\mathbf{P}}_2]\\ [\tilde{\mathbf{P}}_3]\end{bmatrix} = \mathbf{A}^{-T}_{\mathcal{B}\to\tilde{\mathcal{B}}}\begin{bmatrix}[\mathbf{P}_1]\\ [\mathbf{P}_2]\\ [\mathbf{P}_3]\end{bmatrix}$	6.	$\begin{bmatrix}[\mathbf{P}_1]\\ [\mathbf{P}_2]\\ [\mathbf{P}_3]\end{bmatrix} = \mathbf{A}^{T}_{\mathcal{B}\to\tilde{\mathcal{B}}}\begin{bmatrix}[\tilde{\mathbf{P}}_1]\\ [\tilde{\mathbf{P}}_2]\\ [\tilde{\mathbf{P}}_3]\end{bmatrix}$

well defined by the properties of a vector space, to yield another column matrix of elements of $\mathcal{C}$.

There are six transformations just defined, counting both directions, and they are often confused. Table 7.1 summarizes these six transformations, all defined in terms of the matrix $\mathbf{A}_{\mathcal{B}\to\tilde{\mathcal{B}}}$.

7.4.7 The CIE Standard Observer

As we have seen, the trichromatic color vector space is determined by the generalized metameric equivalence $\boxed{=}$, which is in turn determined by three linearly independent continuous functions $\bar{p}_i(\lambda)$ corresponding to three primaries $[\mathbf{P}_i]$, $i = 1, 2, 3$. Any three primaries can be used, since we can convert to any other primaries as shown in the previous section. In principle, each observer has their own color space, which would have to be determined by experimentally matching all monochromatic lights with a linear combination of three independent primary lights. This is not practical, nor necessary, since most trichromatic human observers have very similar color matching functions. Thus a standardized set of color matching functions can be selected to represent this entire population.

Standardization in the fields of photometry and colorimetry is handled by the Commission international de l'éclairage or CIE (in English, known as the International Illumination Commission). Various color spaces have been standardized, but the most important one used for most color imaging work is known as the CIE 1931 standard colorimetric observer. Based on experimental results by Wright with ten observers and Guild with seven observers, the CIE defined this color space (see [Wright (2007)] for a description by Wright of the process that led to the establishment of the CIE 1931 standard). The CIE provided two sets of color matching functions (related by a nonsingular transformation) and corresponding primaries. The primaries used in most color imaging applications are defined in terms of these CIE 1931 primaries.

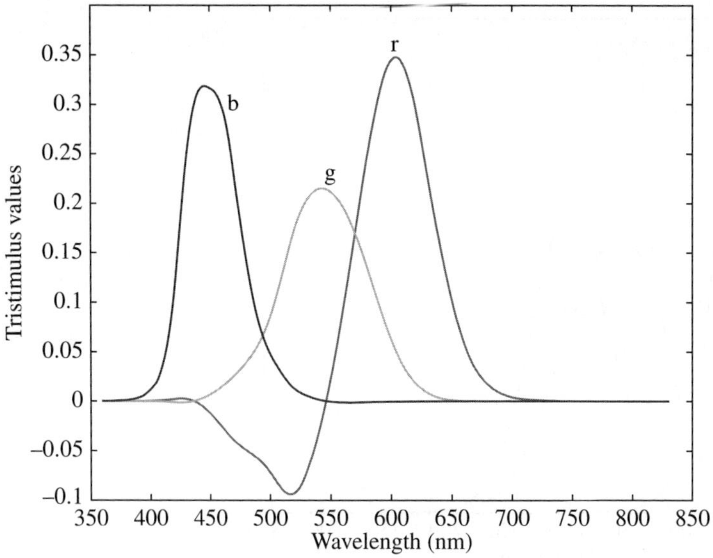

Figure 7.4 Color matching functions of the CIE 1931 RGB primaries: monochromatic lights at wavelengths 700.0 nm, 546.1 nm and 435.8 nm respectively. (*See color plate section for the color representation of this figure.*)

The first set of primaries used in CIE 1931 consists of monochromatic red, green and blue lights, specifically

$$R_{31}(\lambda) = k_R\delta(\lambda - 700.0)$$
$$G_{31}(\lambda) = k_G\delta(\lambda - 546.1)$$
$$B_{31}(\lambda) = k_B\delta(\lambda - 435.8)$$

with the reference white being equal-energy white. The resulting color matching functions are referred to as $\overline{r}_{31}(\lambda)$, $\overline{g}_{31}(\lambda)$ and $\overline{b}_{31}(\lambda)$, see Figure 7.4. The basis is referred to here as $\mathcal{RGB}31 = \{[R_{31}(\lambda)], [G_{31}(\lambda)], [B_{31}(\lambda)]\}$.

The second and most important set of primaries are known as $[\mathbf{X}]$, $[\mathbf{Y}]$ and $[\mathbf{Z}]$; most color coordinate systems in use today are defined in terms of XYZ tristimulus values. We denote this basis by

$$\mathcal{XYZ} = \{[\mathbf{X}], [\mathbf{Y}], [\mathbf{Z}]\}, \tag{7.44}$$

and the canonical representation of a color $[\mathbf{C}]$ is

$$[\mathbf{C}] = C_X[\mathbf{X}] + C_Y[\mathbf{Y}] + C_Z[\mathbf{Z}]. \tag{7.45}$$

The primaries $[\mathbf{X}]$, $[\mathbf{Y}]$ and $[\mathbf{Z}]$ do not correspond to physically realizable colors. They are defined by a nonsingular transformation of $[\mathbf{R}_{31}]$, $[\mathbf{G}_{31}]$ and $[\mathbf{B}_{31}]$. These primaries were selected to have certain useful properties: the color-matching functions

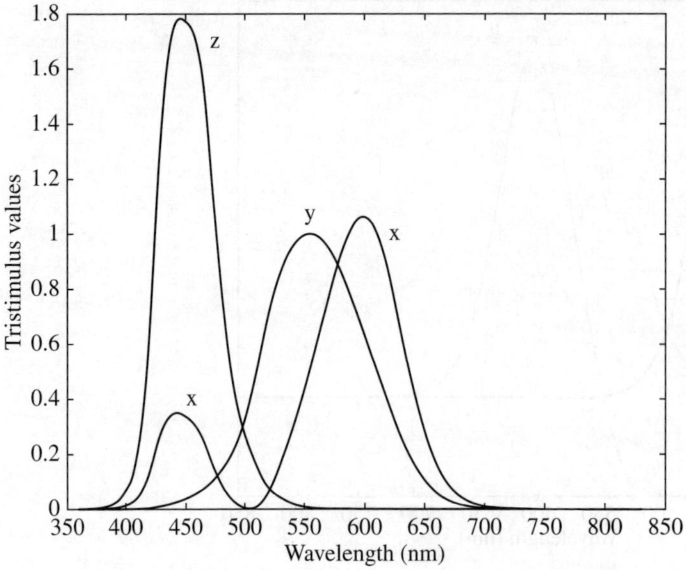

Figure 7.5 Color matching functions of the CIE 1931 XYZ primaries.

are positive everywhere, and the Y color-matching function is equal to the relative luminous efficiency curve of photometry (introduced in Section 7.5.2).

The transformation relating the $\mathcal{XYZ}$ basis to the $\mathcal{RGB}31$ basis is determined by the matrix

$$\mathbf{A}_{\mathcal{RGB}31\to\mathcal{XYZ}} = \begin{bmatrix} 2.768892 & 1.751748 & 1.130160 \\ 1.000000 & 4.590700 & 0.060100 \\ 0.000000 & 0.056508 & 5.594292 \end{bmatrix}. \tag{7.46}$$

We can of course use this matrix to transform primaries, tristimulus values and color-matching functions in either direction using the formulas from Table 7.1.

The CIE 1931 XYZ color-matching functions are illustrated in Figure 7.5. They are standardized by the CIE and widely available. These functions are provided for values of λ ranging from 360 nm to 830 nm at 1 nm increments with seven digits of precision, and are assumed to be linearly interpolated within the 1 nm intervals. In this way, the color-matching functions of the CIE 1931 standard observer are continuous and specified with high precision, allowing reproducible results. A good source for these and many other functions related to color vision is the Colour & Vision Research Laboratory (CVRL) website (www.cvrl.org). The curves plotted here use data from this website.

7.4.8 Specification of Primaries

Based on the above discussions, we can immediately identify four methods that are frequently used to identify a set of primaries $[\mathbf{P}_i]$, $i = 1, 2, 3$.

1) The representation of each of the primaries as a linear combination of the XYZ primaries is specified directly, as in Equation (7.43), where $[\mathbf{X}]$, $[\mathbf{Y}]$ and $[\mathbf{Z}]$ take the role

of the $[\tilde{\mathbf{P}}_i]$. In other words, the matrix $\mathbf{A}^T_{\mathcal{B}\to\mathcal{XYZ}}$ is specified. For example, the 1976 CIE Uniform Chromaticity Scale (UCS) primaries are given by (case 6 of Table 7.1)

$$\begin{aligned} [\mathbf{U}'] &= 2.25[\mathbf{X}] + 2.25[\mathbf{Z}] \\ [\mathbf{V}'] &= [\mathbf{Y}] - 2[\mathbf{Z}] \\ [\mathbf{W}'] &= 3[\mathbf{Z}] \end{aligned} \tag{7.47}$$

which is specified by the matrix

$$\mathbf{A}_{\mathcal{UCS}76\to\mathcal{XYZ}} = \begin{bmatrix} 2.25 & 0 & 0 \\ 0 & 1 & 0 \\ 2.25 & -2 & 3 \end{bmatrix}. \tag{7.48}$$

2) The matrix equation to calculate the tristimulus values of an arbitrary color with respect to the primaries $[\mathbf{P}_i]$ as a function of the tristimulus values for the XYZ primaries is specified directly, i.e., the matrix $\mathbf{A}_{\mathcal{XYZ}\to\mathcal{B}} = \mathbf{A}^{-1}_{\mathcal{B}\to\mathcal{XYZ}}$ is specified. For the same example of the 1976 UCS primaries with $\mathbf{A}$ as above, we have (case 4. of Table 7.1)

$$\begin{aligned} \begin{bmatrix} C_{U'} \\ C_{V'} \\ C_{W'} \end{bmatrix} &= \mathbf{A}_{\mathcal{XYZ}\to\mathcal{UCS}76} \begin{bmatrix} C_X \\ C_Y \\ C_Z \end{bmatrix} \\ &= \begin{bmatrix} \frac{4}{9} & 0 & 0 \\ 0 & 1 & 0 \\ -\frac{1}{3} & \frac{2}{3} & \frac{1}{3} \end{bmatrix} \begin{bmatrix} C_X \\ C_Y \\ C_Z \end{bmatrix}. \end{aligned} \tag{7.49}$$

Here, $\mathbf{A}_{\mathcal{XYZ}\to\mathcal{UCS}76} = \mathbf{A}^{-1}_{\mathcal{UCS}76\to\mathcal{XYZ}}$ in 1) above.

3) The spectral densities $P_i(\lambda)$ of one member of each of the equivalence classes $[\mathbf{P}_i]$ is provided. For example, this could be the spectral density of the light emitted from the red, green, and blue phosphors of a CRT display. In this case, the XYZ tristimulus values of each of the primaries can be computed using Equation (7.32) with the known XYZ color matching functions of Figure 7.5. The power spectral densities of the phosphors in a typical CRT display were shown in Figure 7.1(c). If these are normalized so that their sum matches the reference white $[\mathbf{D}_{65}]$ and the integral of Equation (7.32) is approximated by a sum of products spaced every 2 nm, we find

$$\begin{aligned} \begin{bmatrix} [\mathbf{R}_C] \\ [\mathbf{G}_C] \\ [\mathbf{B}_C] \end{bmatrix} &= \mathbf{A}^T_{\mathcal{RGBCRT}\to\mathcal{XYZ}} \begin{bmatrix} [\mathbf{X}] \\ [\mathbf{Y}] \\ [\mathbf{Z}] \end{bmatrix} \\ &= \begin{bmatrix} 0.4641 & 0.2596 & 0.0358 \\ 0.3056 & 0.6594 & 0.1423 \\ 0.1808 & 0.0810 & 0.9107 \end{bmatrix} \begin{bmatrix} [\mathbf{X}] \\ [\mathbf{Y}] \\ [\mathbf{Z}] \end{bmatrix}. \end{aligned} \tag{7.50}$$

Using formula 5 from Table 7.1, we find the color matching functions for these CRT display primaries,

$$\begin{bmatrix} \overline{p}_{RC}(\lambda) \\ \overline{p}_{GC}(\lambda) \\ \overline{p}_{BC}(\lambda) \end{bmatrix} = \mathbf{A}^{-1}_{\mathcal{RGBCRT}\to\mathcal{XYZ}} \begin{bmatrix} \overline{p}_X(\lambda) \\ \overline{p}_Y(\lambda) \\ \overline{p}_Z(\lambda) \end{bmatrix} \tag{7.51}$$

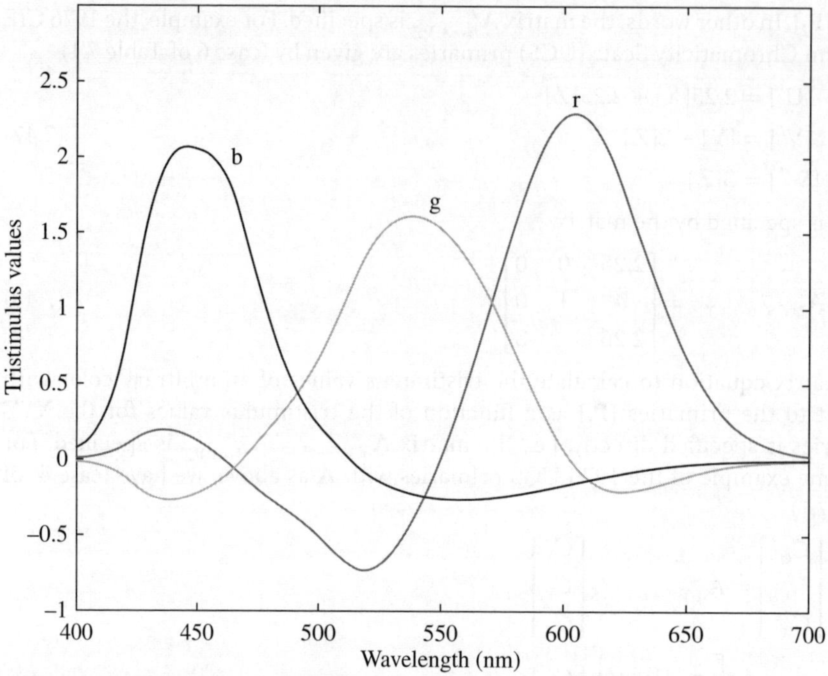

Figure 7.6 Color matching functions of a typical CRT RGB display (EIZO) with the red, green and blue spectral densities given in Figure 7.1(c). (*See color plate section for the color representation of this figure.*)

as shown in Figure 7.6. These color-matching functions would be used by a color camera for example to compute tristimulus values to accurately reproduce a color with spectral density $C(\lambda)$ on the CRT display.

4) A set of three color-matching functions $\overline{p}_i(\lambda)$, assumed to be a linear combination of $\overline{x}(\lambda)$, $\overline{y}(\lambda)$ and $\overline{z}(\lambda)$, is provided. An example would be the spectral sensitivities of the L, M and S type cone photoreceptors in the human retina. Given these, we must find the matrix $\mathbf{A}^{-1}$ relating the two sets of color matching functions, which is unique and easily determined using standard methods of linear algebra. If these color matching functions do not lie in the subspace of $C^{(0)}(\mathcal{V})$ spanned by $\overline{x}(\lambda)$, $\overline{y}(\lambda)$, and $\overline{z}(\lambda)$, then they define a different color space than the CIE 1931 standard observer, e.g., the color space of a digital camera or of a specific human observer.

7.4.9 Physically Realizable Colors

We have defined a physically realizable color $[\mathbf{C}]$ to be an element of the color vector space that we can physically synthesize, because at least one element of the equivalence class is non-negative on $\mathcal{V}$. All the spectra of Figure 7.1 correspond to physically realizable colors, while the primaries $[\mathbf{X}]$, $[\mathbf{Y}]$, and $[\mathbf{Z}]$ of the CIE XYZ basis do not. In this section, we examine the form of the set $\mathcal{C}_R$ of physically realizable colors, as well as the issue of additive combination of such colors, as occurs in a color display device.

Theorem 7.9 The set of physically realizable colors $\mathcal{C}_R$ is a convex subset of $\mathcal{C}$ that has the form of a convex cone.

Proof: Let $[\mathbf{C}_1], [\mathbf{C}_2] \in \mathcal{C}_R$. We need to show that $\alpha[\mathbf{C}_1] + (1 - \alpha)[\mathbf{C}_2] \in \mathcal{C}_R$ for all $0 \leq \alpha \leq 1$. Let $C_1(\lambda)$ and $C_2(\lambda)$ be elements of the equivalence classes $[\mathbf{C}_1]$ and $[\mathbf{C}_2]$ that are members of $\mathcal{P}$. It follows that $\alpha C_1(\lambda) + (1 - \alpha)C_2(\lambda) \in \mathcal{P}$ for any $\alpha \in [0, 1]$, and since this is a member of the equivalence class $\alpha[\mathbf{C}_1] + (1 - \alpha)[\mathbf{C}_2]$, it follows that $\alpha[\mathbf{C}_1] + (1 - \alpha)[\mathbf{C}_2] \in \mathcal{C}_R$. The set $\mathcal{C}_R$ is a convex cone if $[\mathbf{C}] \in \mathcal{C}_R$ implies that $\alpha[\mathbf{C}] \in \mathcal{C}_R$ for $\alpha \geq 0$. This clearly holds, since if $C(\lambda) \in \mathcal{P}$, then $\alpha C(\lambda) \in \mathcal{P}$ for $\alpha \geq 0$. □

It can be shown that the set $\mathcal{C}_R$ is the convex hull of the set of monochromatic lights [Dubois (2010)]. For the CIE standard observer, its boundary consists of the monochromatic lights, as well as purples, which are an additive mixture of the extreme colors red and blue. Figure 7.7 illustrates the set $\mathcal{C}_R$ in the XYZ coordinate system. In particular, we see that $[\mathbf{X}]$, $[\mathbf{Y}]$, and $[\mathbf{Z}]$ (not visible in this view) are outside of $\mathcal{C}_R$, as we have stated.

Physically realizable colors are particularly important, as these are the elements of the color vector space that we can physically synthesize and view, for example on a display device such as a CRT or LCD screen. These devices typically have three basic colors called display primaries that can be directly displayed, typically red, green, and blue. By taking a weighted sum of these display primaries, of course with positive weights, other colors can be generated. However, not all physically realizable colors can be generated in this way. By a judicious choice of the display primaries, we can attempt to maximize the portion of the cone of physically realizable colors that can be synthesized on the display. The following theorem specifies which colors can be synthesized by a positive additive mixture of physically realizable colors.

Theorem 7.10 Let $[\mathbf{P}_1]$, $[\mathbf{P}_2]$ and $[\mathbf{P}_3]$ be three physically realizable and linearly independent colors. The set $\mathcal{Q}_\mathcal{B}$ of all colors that can be formed as linear combinations of

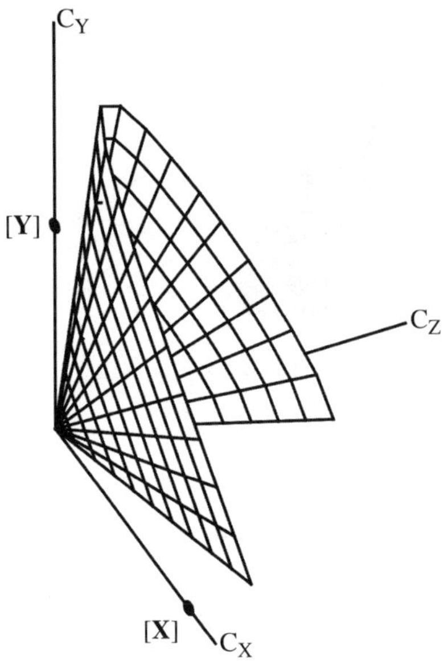

Figure 7.7 Illustration of the cone of physically realizable colors in the CIE 1931 *XYZ* coordinate system. Its boundary is the set of monochromatic lights, on the curved surface toward the top, and the plane of purples on the bottom (not shown). All physically realizable colors can be formed as a convex combination of two elements on the curved surface, which is why the set is closed off by the plane containing the extreme colors on the bottom.

$[\mathbf{P}_1]$, $[\mathbf{P}_2]$ and $[\mathbf{P}_3]$ with non-negative coefficients is a convex cone with a triangular cross section. It is the convex hull of the three half lines $\alpha_1[\mathbf{P}_1]$, $\alpha_2[\mathbf{P}_2]$ and $\alpha_3[\mathbf{P}_3]$ with $\alpha_i \geq 0$, for $i = 1, 2, 3$.

Proof: Since $[\mathbf{P}_1]$, $[\mathbf{P}_2]$ and $[\mathbf{P}_3]$ are linearly independent, they form a basis of C. Now

$$\mathcal{Q}_B = \{\alpha_1[\mathbf{P}_1] + \alpha_2[\mathbf{P}_2] + \alpha_3[\mathbf{P}_3] \mid \alpha_i \geq 0, i = 1, 2, 3\}. \tag{7.52}$$

This set is easily seen to satisfy the definition of a convex cone. It contains the half lines $\alpha_i[\mathbf{P}_i]$ with $\alpha_1 \geq 0$, $i = 1, 2, 3$, and thus any element of the convex hull of these lines. Conversely, any element of the convex hull of the half lines belongs to $\mathcal{Q}_B$. This cone is bounded by the three planes containing the pairs of lines $\{\alpha_1[\mathbf{P}_1], \alpha_2[\mathbf{P}_2]\}$, $\{\alpha_1[\mathbf{P}_1], \alpha_3[\mathbf{P}_3]\}$ and $\{\alpha_2[\mathbf{P}_2], \alpha_3[\mathbf{P}_3]\}$ respectively, and thus has a triangular cross section. (The reader should fill in the details of this proof.) □

Figure 7.8(a) illustrates this triangular cone in the *XYZ* coordinate system for the CRT RGB primaries with color matching functions given in Figure 7.6. Of course, it lies entirely within the cone C_R. The goal in choosing additive display primaries is to fill as much of C_R as possible with the cone of additively reproducible colors. It is clear from the illustration why red, green, and blue are used as primaries in televisions and computer monitors as these lie near the extremities of the cone C_R. In practice, the weights are between 0 and 1, so the reproducible colors are limited to a finite region, as shown in Figure 7.8(b).

This discussion is easily extended to additive combination with positive coefficients of more than three base colors. In this case the set of physically reproducible colors lies in a cone with a polygonal cross section, where each vertex of the polygon is specified by one of the given colors. However, it may be that some of the base colors lie within the polygon, for example if white is used as a base color to get a brighter display. If more

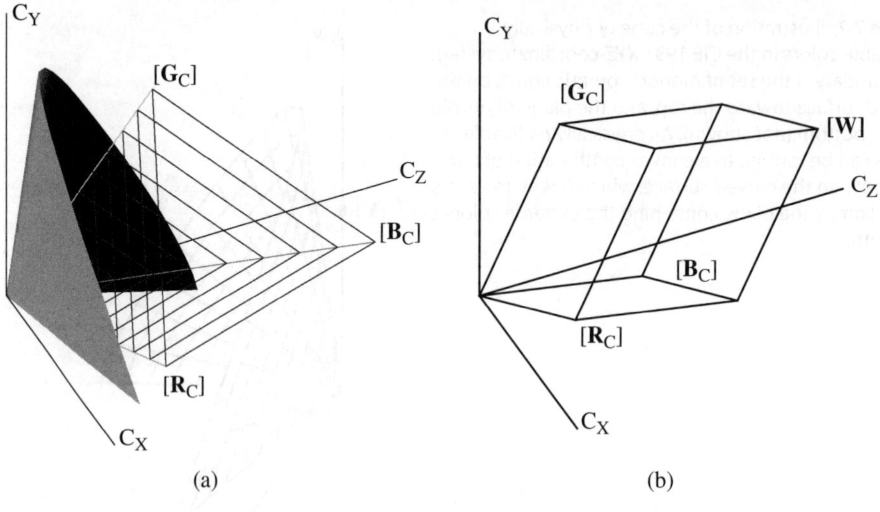

Figure 7.8 (a) Illustration of the cone of reproducible colors with red, green, and blue display primaries, in the *XYZ* color coordinate system. (b) The set of reproducible colors with red, green, and blue coefficients between 0 and 1. (*See color plate section for the color representation of this figure.*)

than three base colors are used in a display, the coefficients to synthesize a particular color may not be unique.

7.5 Color Coordinate Systems

7.5.1 Introduction

In this book, we make a clear distinction between a color space and a color coordinate system. A color space, as we have defined it, is associated with a particular observer, group of observers, or even a device like a color camera. It is specified by a set of three linearly independent color matching functions. A set of color coordinates serves to uniquely identify a specific color within the color space. In many cases, these color coordinates may only be meaningfully defined for elements of the set C_R of physically realizable colors.

Given a set of primaries, the three tristimulus values uniquely specify an element of the color space and form our basic set of color coordinates. Different sets of primaries lead to different color coordinates, e.g., RGB, XYZ, etc., which can be converted to other sets of primaries by a linear transformation. We will only consider trichromatic color spaces here. Any one-to-one transformation from tristimulus values to some subset of $\mathbb{R}^3$ can define a set of color coordinates. Many different color coordinate systems are used in imaging systems and image processing (they are sometimes called color spaces, e.g., CIELAB color space, but we will not use that designation for the reasons given above).

There are several motivations for the use of coordinate systems other than tristimulus values (i.e. that involve a nonlinear transformation of tristimulus values with respect to a certain set of primaries). In some cases, we would like color coordinates that describe particular perceptual attributes of a color, such as brightness, hue, and saturation. We may wish to provide a more perceptually uniform representation of colors and color differences. There is a highly variable, nonlinear relationship between Euclidean distance of sets of tristimulus values and perceptual color difference. Some color coordinate systems, such as CIELAB, are far more perceptually uniform. Finally, some color coordinate systems are tied to color display systems such as CRT displays with gamma correction. This section provides an overview of some color coordinate systems used in image processing.

7.5.2 Luminance and Chromaticity

The first set of coordinates that we introduce consists of *relative luminance*, a measure of relative brightness, and two *chromaticity* coordinates that specify the hue/saturation aspect of a color. These coordinates are defined for colors in C_R. If two lights have equal luminance, they appear to be equally bright to a viewer when viewed in the same context, independently of their chromatic attributes. Luminance is a standardized *quantitative* measure, whereas brightness is used as a *qualitative* descriptor of this familiar concept. Although it may be difficult to judge if, say, a red light and a blue light have equal brightness when viewing them side by side, this judgement is easier if they are viewed in alternation one after the other. At a low frequency of switching, we can see the display flipping red, blue, red, blue,.... However, as the switching frequency increases

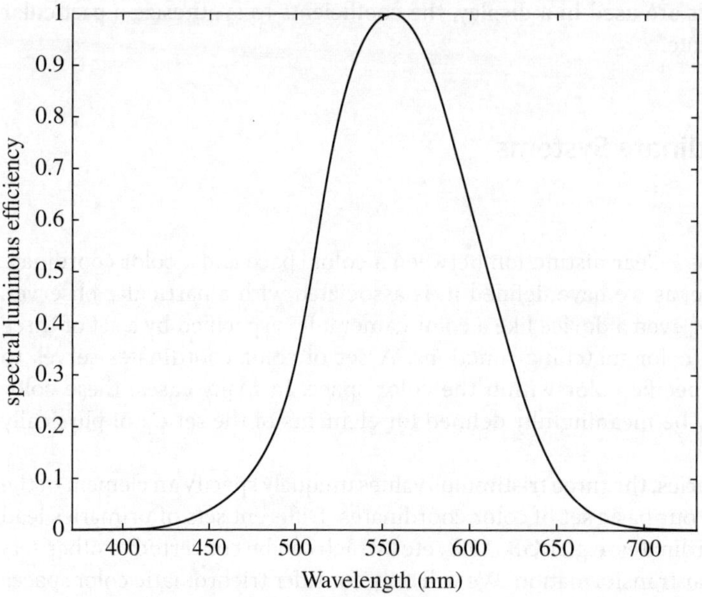

Figure 7.9 CIE 1924 relative luminous efficiency $V(\lambda)$.

and passes a certain limit, the two colors merge into one, which flickers if they have different brightness. The intensity of one of the lights can be adjusted until the flickering disappears. At this point, the two lights have equal perceptual brightness. This brightness depends on the power density spectrum of the light. A light with a spectrum concentrated near 550 nm appears much brighter than a light of equal total power with a spectrum concentrated near 700 nm. This property is captured by the *relative luminous efficiency* curve $V(\lambda)$, shown in Figure 7.9. The curve tells us that a monochromatic light at wavelength λ_0 with power density spectrum $\delta(\lambda - \lambda_0)$ appears equally as bright as a monochromatic light with power density spectrum $V(\lambda_0)\delta(\lambda - \lambda_{max})$, where $\lambda_{max} \approx 555$ nm. The relative luminous efficiency curve was standardized by the CIE in 1924 and is specified from 360 nm to 830 nm in intervals of 1 nm.

For an arbitrary colored light with radiance power density spectrum $C(\lambda)$, the absolute luminance is defined as

$$C_L = K_m \int C(\lambda)V(\lambda)\,d\lambda \tag{7.53}$$

where $K_m = 683.002$ lm W−1 (lumens per Watt) is a constant that determines the units of luminance. This formula can be developed using axioms of brightness matching, as was done previously to develop the formula for tristimulus values (see Dubois (2010) for details). Thus, two different lights with power density spectra $C_1(\lambda)$ and $C_2(\lambda)$ will appear equally bright in the same context if $C_{1L} = C_{2L}$, although they may look different in their chromatic attributes. Of course, we expect that if $C_1(\lambda) \equiv C_2(\lambda)$, then $C_{1L} = C_{2L}$. Indeed, as mentioned previously, the Y tristimulus value C_Y of the XYZ primary system was chosen to be proportional to the luminance, i.e. $C_Y \propto C_L$ and $\overline{p}_Y(\lambda) \propto V(\lambda)$, so this property clearly holds. According to the definition of luminance, if the power density spectrum is multiplied by a constant α, say $C_1(\lambda) = \alpha C(\lambda)$, then the luminance is also multiplied by α, $C_{1L} = \alpha C_L$; the light appears brighter but its chromatic aspect remains unchanged. Luminance is also additive: if $C(\lambda) = C_1(\lambda) + C_2(\lambda)$, then $C_L = C_{1L} + C_{2L}$. From this, we conclude that if

$$C(\lambda) \boxed{\equiv} C_1 P_1(\lambda) + C_2 P_2(\lambda) + C_3 P_3(\lambda)$$

then $C_L = C_1 P_{1L} + C_2 P_{2L} + C_3 P_{3L}$, where $P_{iL} = K_m \int P_i(\lambda) V(\lambda)\, d\lambda$ is the luminance of the primary $[\mathbf{P}_i]$, $i = 1, 2, 3$. In image processing, we usually work with normalized quantities, assuming that the luminance of reference white is unity. We refer to this normalized quantity as *relative luminance*. Since this normalization applies to the Y tristimulus value, we denote relative luminance by C_Y. Since we are rarely concerned with absolute luminance in this book, the term luminance will generally refer to relative luminance.

In terms of the tristimulus values, if $[\mathbf{C}] = C_1[\mathbf{P}_1] + C_2[\mathbf{P}_2] + C_3[\mathbf{P}_3]$, then the tristimulus values of $\alpha[\mathbf{C}]$ are αC_1, αC_2 and αC_3. This corresponds to moving along a line passing through the origin and through the point (C_1, C_2, C_3) in a 3D color coordinate system, as illustrated in Figure 7.10 for RGB primaries. (Note: although the RGB axes are drawn as orthogonal in Figure 7.10, we have not established any (nor does there exist any) meaningful concept of orthogonality in the color vector space. The axes are drawn as orthogonal only for convenience. Since lines and planes remain lines and planes under any change of basis, the arguments made in this section hold independently of the existence of orthogonality.) The chromatic properties of the light are determined by a specification of this line, while luminance is proportional to the distance from the origin. One way of specifying the line is by giving the coordinates of its intersection with the plane $C_1 + C_2 + C_3 = 1$, i.e. by finding γ such that the tristimulus values of $\gamma[\mathbf{C}] = \gamma C_1[\mathbf{P}_1] + \gamma C_2[\mathbf{P}_2] + \gamma C_3[\mathbf{P}_3]$ lie on the said plane. It follows that $\gamma C_1 + \gamma C_2 + \gamma C_3 = 1$,

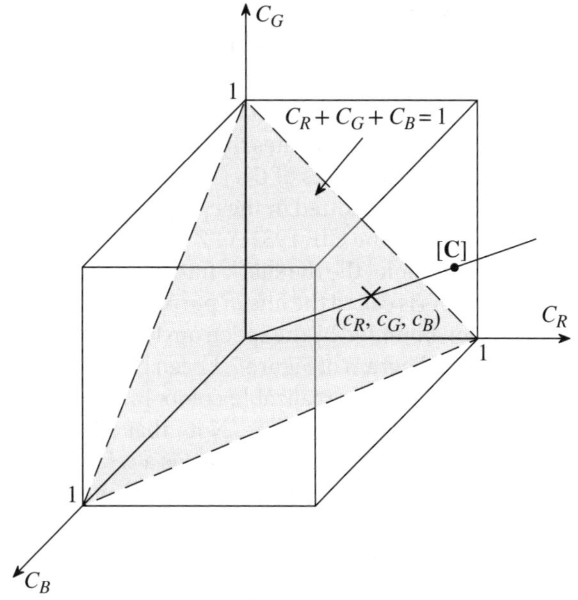

Figure 7.10 Illustration of the concept of chromaticity in RGB coordinates. The triangle with vertices $(1, 0, 0)$, $(0, 1, 0)$ and $(0, 0, 1)$ and equation $C_R + C_G + C_B = 1$ is known as the Maxwell triangle. The intersection of the line through a color $[\mathbf{C}]$ and the Maxwell triangle has coordinates given by the chromaticities (c_R, c_G, c_B) of the given color. The cube shown here corresponds to the parallelepiped in Figure 7.8(b) but with the R, G, and B tristimulus values drawn as orthogonal axes.

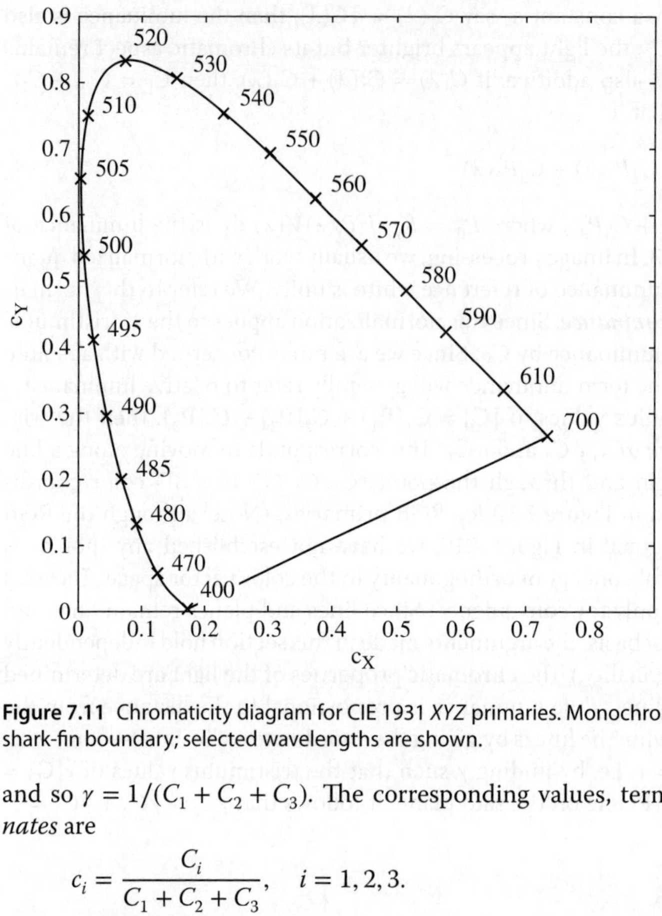

Figure 7.11 Chromaticity diagram for CIE 1931 *XYZ* primaries. Monochromatic colors lie on the shark-fin boundary; selected wavelengths are shown.

and so $\gamma = 1/(C_1 + C_2 + C_3)$. The corresponding values, termed *chromaticity coordinates* are

$$c_i = \frac{C_i}{C_1 + C_2 + C_3}, \quad i = 1, 2, 3. \tag{7.54}$$

They are dependent since $c_3 = 1 - c_1 - c_2$, so only two need be specified. Note that we systematically denote tristimulus values with uppercase letters and chromaticity coordinates with (the same) lowercase letters. Thus, for a given set of primaries, (C_Y, c_1, c_2) form a set of color coordinates that uniquely specify the color, although these are usually only defined for elements of C_R.

The set of colors plotted in the c_1c_2 plane is termed a chromaticity diagram, as shown in Figure 7.11 for the CIE 1931 *XYZ* primaries. The chromaticities of the spectral colors $[\delta(\lambda - \lambda_0)]$ lie along the the shark-fin-shaped curve. The straight line joining the extreme spectral colors is called the line of purples. All physical colors have chromaticities that lie in the region bounded by the spectrum locus and the line of purples. The boundary of the chromaticity diagram of Figure 7.11 can be seen to be the intersection of the boundary of the cone of physically realizable colors in Figure 7.7 with the plane $C_X + C_Y + C_Z = 1$, projected on the C_X–C_Y plane. Note that for a normalized set of primaries, such that $[\mathbf{P}_1] + [\mathbf{P}_2] + [\mathbf{P}_3] = [\mathbf{W}]$ where $[\mathbf{W}]$ is a reference white, then $W_1 = W_2 = W_3 = 1$ and thus $w_1 = w_2 = w_3 = \frac{1}{3}$, i.e. the chromaticity coordinates of reference white are all $\frac{1}{3}$.

For luminance and two chromaticities to form a valid set of coordinates, we must be able to uniquely recover C_1, C_2, and C_3 from C_Y, c_1, and c_2 for the given primaries. This is indeed the case, as summarized in the following theorem.

Theorem 7.11 For a given set of primaries $[\mathbf{P}_1]$, $[\mathbf{P}_2]$, and $[\mathbf{P}_3]$ with relative luminances P_{1Y}, P_{2Y}, P_{3Y}, the luminance and two chromaticities defined by

$$C_Y = C_1 P_{1Y} + C_2 P_{2Y} + C_3 P_{3Y} \tag{7.55}$$

$$c_i = \frac{C_i}{C_1 + C_2 + C_3}, \quad i = 1, 2 \tag{7.56}$$

form a valid set of color coordinates for elements of $\mathcal{C}_\mathrm{R}$.

Proof: Given C_1, C_2, and C_3 for a color in $\mathcal{C}_\mathrm{R}$, then C_Y, c_1, and c_2 are uniquely defined. We must show that, given $C_Y \geq 0$ and c_1, c_2 within the shark-fin region that is the intersection of $\mathcal{C}_\mathrm{R}$ with the plane $C_1 + C_2 + C_3 = 1$, projected on the $C_1 - C_2$ plane, we can uniquely recover C_1, C_2, and C_3. Now $c_3 = 1 - c_1 - c_2$.

Taking $C_Y = C_1 P_{1Y} + C_2 P_{2Y} + C_3 P_{3Y}$ and dividing by $C_1 + C_2 + C_3$, we obtain

$$\frac{C_Y}{C_1 + C_2 + C_3} = c_1 P_{1Y} + c_2 P_{2Y} + c_3 P_{3Y}.$$

Multiplying by C_i for $i = 1, 2, 3$ yields

$$C_Y c_i = C_i (c_1 P_{1Y} + c_2 P_{2Y} + c_3 P_{3Y})$$

with the unique solution

$$C_i = \frac{C_Y c_i}{c_1 P_{1Y} + c_2 P_{2Y} + c_3 P_{3Y}}, \quad i = 1, 2, 3. \tag{7.57}$$

□

The inverse formula is particularly simple for the $XY\,Z$ primaries. Although, it was not previously stated, another criterion in the definition of $[\mathbf{X}]$, $[\mathbf{Y}]$, and $[\mathbf{Z}]$ was that their relative luminances are given by $X_Y = 0$, $Y_Y = 1$, and $Z_Y = 0$. Thus, given the luminance C_Y and the chromaticities c_X, c_Y, the XYZ tristimulus values are given by

$$C_X = \frac{C_Y c_X}{c_Y} \tag{7.58}$$

$$C_Y = C_Y (\text{trivially}) \tag{7.59}$$

$$C_Z = \frac{C_Y c_Z}{c_Y}. \tag{7.60}$$

Chromaticity diagrams are often used to visualize the range of hue and saturation that can be achieved in a given imaging system. As we have seen, in an additive color reproduction system such as a CRT or LCD display, colors are synthesized by additively mixing a positive amount of three (or more) basic colors. Not all physically realizable colors can be synthesized in this way; only colors that lie in a cone with triangular cross section can be synthesized (Theorem 7.10). In a chromaticity diagram, the colors that can be synthesized lie in a triangle as stated in the following theorem.

Theorem 7.12 On a chromaticity diagram, the chromaticities of all colors that are realized by the sum of a positive quantity of three physically realizable colors lie within a triangle whose vertices are the chromaticities of these three colors.

Proof: Let $\mathcal{B} = \{[\mathbf{P}_1], [\mathbf{P}_2], [\mathbf{P}_3]\}$ be the basis used to form the chromaticity diagram, usually $\mathcal{XYZ}$. Let [**A**], [**B**], and [**C**] be three physically realizable colors that are used to additively synthesize other colors, e.g., [**R**], [**G**], and [**B**] of an RGB display device. In the color coordinate system of tristimulus values with respect to the primaries of $\mathcal{B}$, the colors that can be synthesized as a positive linear combination of [**A**], [**B**], and [**C**] lie within a cone with triangular cross section and edges given by the straight lines $\alpha_1[\mathbf{A}]$, $\alpha_2[\mathbf{B}]$, and $\alpha_3[\mathbf{C}]$, $\alpha_i \geq 0$ (Theorem 7.10). The chromaticities lie on the intersection of this cone with the plane $C_1 + C_2 + C_3 = 1$, which is a triangle. In a chromaticity diagram, they lie on the projection of this triangle on the $C_1 - C_2$ plane, which is also a triangle, whose vertices are the chromaticities of [**A**], [**B**], and [**C**] respectively. □

This theorem can also be proved by directly showing that the chromaticities of all colors that are realized by the sum of a positive quantity of *two* physically realizable colors lie on the straight line joining the chromaticities of the two colors. Then, adding this sum to a positive quantity of the third color will yield chromaticities that lie within the stated triangle. The formal proof is left as an exercise for the reader, or it can be found in Dubois (2010).

Figure 7.12 illustrates this result for the red, green, and blue display primaries of a typical RGB display on an *xy* chromaticity diagram. The subset of all possible colors that can be reproduced on this display have chromaticities that lie within the indicated triangle. This expains why red, green, and blue are used as display primaries, since the extremities of the shark fin of all physically realizable colors have these hues.

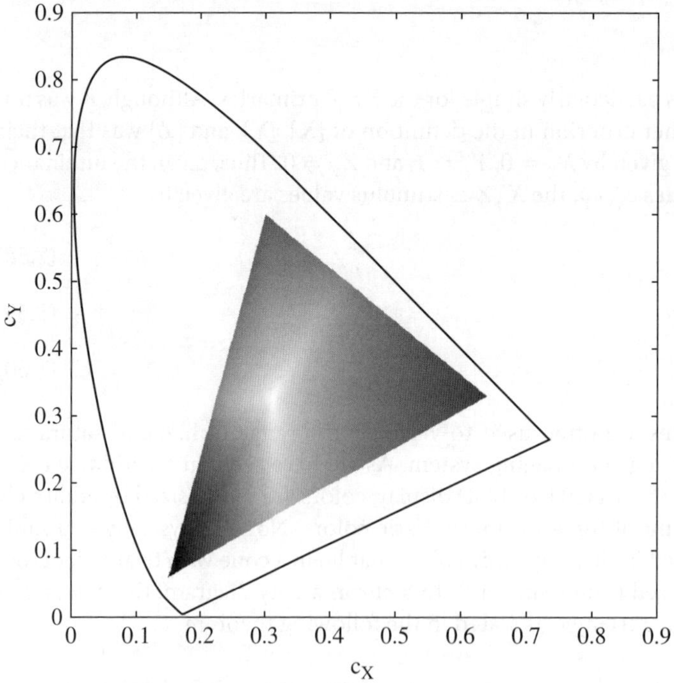

Figure 7.12 Chromaticities of colors that can be reproduced by addition of positive quantities of typical RGB display primaries. (*See color plate section for the color representation of this figure.*)

7.5.3 Linear Color Representations

A linear color representation is one where the color coordinates are tristimulus values with respect to a specific set of primaries. As we have seen, all such representations are related by a linear transformation. We have already described several linear coordinate systems for CIE 1931 color space, namely XYZ, RGB31, and 1976 UCS. In this section we discuss only one linear representation for this color space, namely the one given by the ITU-R Rec. BT.709/sRGB primaries. Others will be introduced later as needed, and in the problems.

Many sets of red-green-blue (RGB) primaries have been standardized over the years. A number of these are described in Poynton (2012). The most widely used set at this time is specified in the ITU-R Rec BT.709 standard for HDTV as well as the sRBG standard used in computing and defined in IEC 61966-2-1. These primaries are representative of the phosphors used in modern CRTs and most other displays emulate them in at least one operating mode. They are considered to form a basis for the 1931 CIE space and are defined in terms of their 1931 XYZ chromaticities and a given reference white. This section will serve to illustrate many of the calculations done in colorimetry to specify a given linear coordinate system. These primaries can be denoted as $[\mathbf{R}_{709}]$, $[\mathbf{G}_{709}]$ and $[\mathbf{B}_{709}]$, and the basis is referred to as $\mathcal{R}709$. For this section, we use the simplified notation $[\mathbf{R}] = [\mathbf{R}_{709}]$, $[\mathbf{G}] = [\mathbf{G}_{709}]$, $[\mathbf{B}] = [\mathbf{B}_{709}]$. The chromaticities of the primaries and reference white are specified in Table 7.2. Of course the Z-chromaticities do not *need* to be provided in the table, since the sum of the values in each column is 1. Slightly different values for the transformation matrices shown below can be found in the literature, generally due to slightly different values of the xy chomaticities of $[\mathbf{D}_{65}]$. We use here the official CIE values rounded to four decimal digits, and so obtain the same matrices as found in IEC 61966-2-1. We use $[\mathbf{D}_{65}]$ as reference white here, but $[\mathbf{D}_{50}]$ is also used in some applications.

D_{65} is a reference white specified by the CIE and is meant to be typical of daylight. A representative power spectral density is shown in Figure 7.1(d). In this section, we use the simplified notation $[\mathbf{D}] = [\mathbf{D}_{65}]$. Normalization is provided by setting the relative luminance of reference white to unity, $D_Y = 1$. It follows that the XYZ tristimulus values of reference white are given by

$$D_X = \frac{\mathrm{d}_X}{\mathrm{d}_Y} = 0.9505$$

$$D_Y = 1$$

$$D_Z = \frac{\mathrm{d}_Z}{\mathrm{d}_Y} = 1.0891.$$

Table 7.2 XYZ chromaticities of ITU-R Rec. 709 red, green and blue primaries and reference white D_{65}.

	Red	Green	Blue	White, D_{65}
x	0.640	0.300	0.150	0.3127
y	0.330	0.600	0.060	0.3290
z	0.030	0.100	0.790	0.3583

To find the XYZ tristimulus values of the red, green and blue primaries, we need to find their relative luminances R_Y, G_Y and B_Y. These are determined through the constraint $[\mathbf{R}] + [\mathbf{G}] + [\mathbf{B}] = [\mathbf{D}]$. This equation implies the matrix equation

$$\begin{bmatrix} R_X \\ R_Y \\ R_Z \end{bmatrix} + \begin{bmatrix} G_X \\ G_Y \\ G_Z \end{bmatrix} + \begin{bmatrix} B_X \\ B_Y \\ B_Z \end{bmatrix} = \begin{bmatrix} D_X \\ D_Y \\ D_Z \end{bmatrix}. \tag{7.61}$$

Using $R_X = R_Y \frac{r_X}{r_Y}$ and so on (Equations (7.58), (7.59), (7.60)) in the above equation with the chromaticities from Table 7.2 gives

$$\begin{bmatrix} 1.\overline{93} & 0.5 & 2.5 \\ 1.0 & 1.0 & 1.0 \\ 0.\overline{09} & 0.1\overline{6} & 13.1\overline{6} \end{bmatrix} \begin{bmatrix} R_Y \\ G_Y \\ B_Y \end{bmatrix} = \begin{bmatrix} 0.9505 \\ 1.0 \\ 1.0891 \end{bmatrix}. \tag{7.62}$$

Solving this matrix equation, we find

$$\begin{bmatrix} R_Y \\ G_Y \\ B_Y \end{bmatrix} = \begin{bmatrix} 0.2126 \\ 0.7152 \\ 0.0722 \end{bmatrix} \tag{7.63}$$

and substituting into the expressions $R_X = R_Y \frac{r_X}{r_Y}$ etc., we find

$$\begin{bmatrix} [\mathbf{R}] \\ [\mathbf{G}] \\ [\mathbf{B}] \end{bmatrix} = \underbrace{\begin{bmatrix} 0.4124 & 0.2126 & 0.0193 \\ 0.3576 & 0.7152 & 0.1192 \\ 0.1805 & 0.0722 & 0.9505 \end{bmatrix}}_{\mathbf{A}^T_{R709\rightarrow XYZ}} \begin{bmatrix} [\mathbf{X}] \\ [\mathbf{Y}] \\ [\mathbf{Z}] \end{bmatrix}. \tag{7.64}$$

To convert tristimulus values between the two sets of primaries, suppose that

$$\begin{aligned} [C] &= C_{\mathrm{R}}[\mathbf{R}] + C_{\mathrm{G}}[\mathbf{G}] + C_{\mathrm{B}}[\mathbf{B}] \\ &= C_X[\mathbf{X}] + C_Y[\mathbf{Y}] + C_Z[\mathbf{Z}]. \end{aligned}$$

Then, referring to Table 7.1,

$$\begin{bmatrix} C_X \\ C_Y \\ C_Z \end{bmatrix} = \underbrace{\begin{bmatrix} 0.4124 & 0.3576 & 0.1805 \\ 0.2126 & 0.7152 & 0.0722 \\ 0.0193 & 0.1192 & 0.9505 \end{bmatrix}}_{\mathbf{A}_{R709\rightarrow XYZ}} \begin{bmatrix} C_{\mathrm{R}} \\ C_{\mathrm{G}} \\ C_{\mathrm{B}} \end{bmatrix} \tag{7.65}$$

and

$$\begin{bmatrix} C_{\mathrm{R}} \\ C_{\mathrm{G}} \\ C_{\mathrm{B}} \end{bmatrix} = \underbrace{\begin{bmatrix} 3.2406 & -1.5372 & -0.4986 \\ -0.9689 & 1.8758 & 0.0415 \\ 0.0557 & -0.2040 & 1.0570 \end{bmatrix}}_{\mathbf{A}^{-1}_{R709\rightarrow XYZ}} \begin{bmatrix} C_X \\ C_Y \\ C_Z \end{bmatrix}. \tag{7.66}$$

Note that the matrix of Equation 7.66 is obtained by inverting the matrix of Equation 7.65, as shown, i.e. rounded to four digits after the decimal, and rounding the result to four digits after the decimal.

The color matching functions, obtained by transforming the XYZ color matching functions using $\mathbf{A}^{-1}_{R709\to\mathcal{XYZ}}$ are very similar to those shown in Figure 7.6.

7.5.4 Perceptually Uniform Color Coordinates

The color coordinate systems introduced so far, namely tristimulus values with respect to given basis and luminance/chromaticity coordinates, can precisely specify a color in the given color space. However, they have the disadvantage that equal differences in the color coordinates can correspond to widely different perceptual differences. This is a problem if we are trying to quantify the magnitude of the difference between two colors from a perceptual point of view.

Many attempts have been made to establish a perceptually uniform color coordinate system, and this work is ongoing. We describe here one very popular coordinate system, namely CIELAB, that is widely used in many applications, including image processing. This color coordinate system is based on XYZ tristimulus values, and requires specification of a reference white. To a first approximation, equal Euclidean differences in CIELAB coordinates correspond to equal perceptual differences.

The CIELAB coordinates are denoted L^*, a^*, b^*. We adapt the notation used for tristimulus values in this book to denote the CIELAB values as well. Thus, assume that $\mathbf{C}_{\mathcal{XYZ}} = [\; C_X \; C_Y \; C_Z \;]^T$ is the tristimulus vector for a color $[\mathbf{C}]$ in the $\mathcal{XYZ}$ basis, and that $\mathbf{W}_{\mathcal{XYZ}} = [\; W_X \; W_Y \; W_Z \;]^T$ is the selected reference white (e.g., $[\mathbf{D}_{65}]$). Define the nonlinear real-valued function $f(x)$ on $[0,1]$,

$$f(x) = \begin{cases} \frac{1}{3}\left(\frac{29}{6}\right)^2 x + \frac{4}{29} & \text{if } 0 \le x \le \left(\frac{6}{29}\right)^3 \\ x^{1/3} & \text{if } \left(\frac{6}{29}\right)^3 \le x \le 1. \end{cases} \tag{7.67}$$

Note that $f(x)$ has been chosen to be one-to-one, monotonically increasing, and continuous with a continuous first derivative on $[0,1]$ that is finite at the origin. For physically realizable colors, we assume $0 \le C_i \le W_i$, for $i \in \{X, Y, Z\}$ and CIELAB coordinates are defined for colors in the set $\{[0, W_X] \times [0, W_Y] \times [0, W_Z]\} \cap \mathcal{C}_{\mathrm{R}}$.

Then the CIELAB coordinates of $[\mathbf{C}]$ are given by

$$\begin{aligned} C_{L^*} &= 116 f(C_Y/W_Y) - 16 \\ C_{a^*} &= 500(f(C_X/W_X) - f(C_Y/W_Y)) \\ C_{b^*} &= 200(f(C_Y/W_Y) - f(C_Z/W_Z)). \end{aligned} \tag{7.68}$$

We note that for any gray color $[\mathbf{G}_\alpha] = \alpha[\mathbf{W}]$ for $0 \le \alpha \le 1$, $G_{\alpha,a^*} = G_{\alpha,b^*} = 0$ and only G_{α,L^*} is nonzero, with values ranging from 0 to 100. C_{L^*}, a perceptually uniform measure of brightness, is referred to as *lightness*. Although defined in terms of the cube root function in Equation 7.67, the overall function for C_{L^*} is very close to the simple power law $100(C_Y/W_Y)^{0.42}$, but is better behaved near the origin, having a finite derivative [Poynton and Funt (2014)].

A measure of the perceptual difference between two colors $[\mathbf{C}_1]$ and $[\mathbf{C}_2]$ is

$$\Delta^*_{ab}([\mathbf{C}_1],[\mathbf{C}_2]) = \sqrt{(C_{1,L^*} - C_{2,L^*})^2 + (C_{1,a^*} - C_{2,a^*})^2 + (C_{1,b^*} - C_{2,b^*})^2}. \tag{7.69}$$

This formula is by no means perfect and more elaborate color difference formulas have been developed. See Witt (2007) for a discussion of efforts in this direction.

Figure 7.13 shows some examples of color step patterns with linearly changing tristimulus values on the left, and linearly changing CIELAB coordinates on the right. The steps of constant change in CIELAB appear more uniform than the steps of constant change in tristimulus values. These images should be viewed on an sRGB monitor with gamma of 2.2 and suitable magnification for best effect.

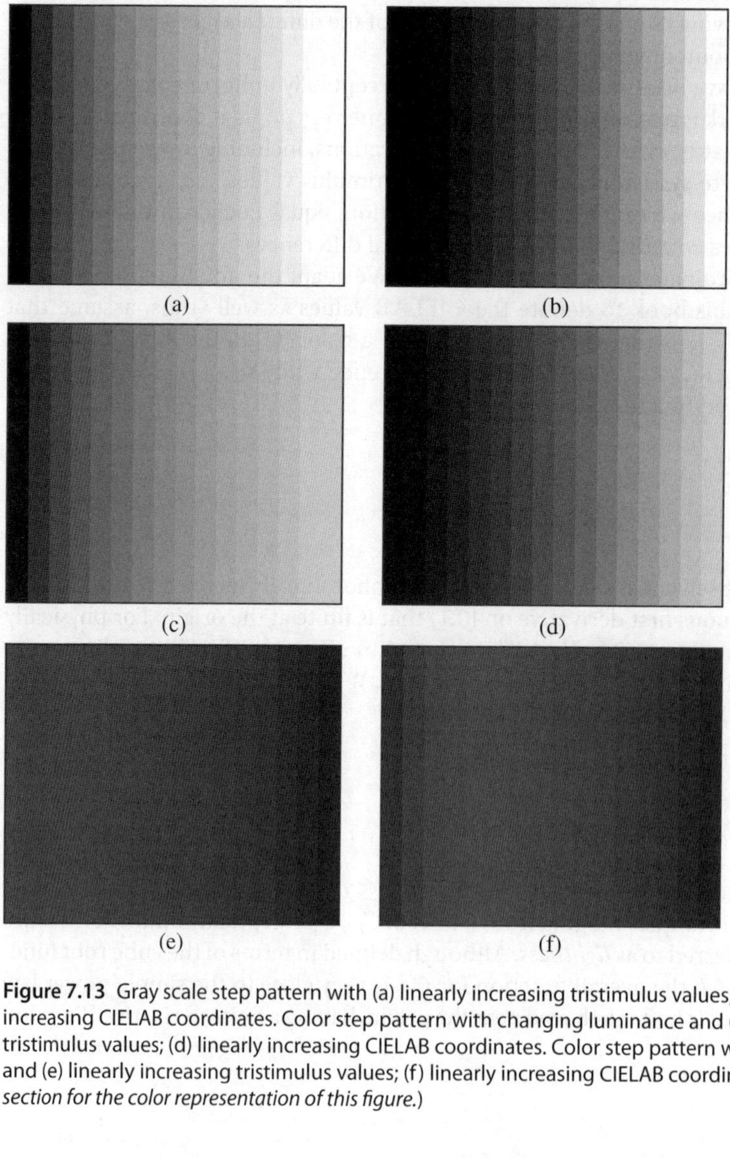

Figure 7.13 Gray scale step pattern with (a) linearly increasing tristimulus values; (b) linearly increasing CIELAB coordinates. Color step pattern with changing luminance and (c) linearly increasing tristimulus values; (d) linearly increasing CIELAB coordinates. Color step pattern with fixed luminance and (e) linearly increasing tristimulus values; (f) linearly increasing CIELAB coordinates. (*See color plate section for the color representation of this figure.*)

7.5.5 Display Referred Coordinates

One of the most common color coordinate systems in imaging is the sRGB representation. This nonlinear coordinate system is historically related to cathode ray tube (CRT) display devices, now obsolete. The light output of a CRT is related to the voltage applied approximately by a power law

$$\text{intensity} = \text{voltage}^{\gamma}.$$

This applies to each of the RGB components of a display device, so that if the voltages applied are C'_R, C'_G and C'_B, the tristimulus values of the color emitted are $C_R = g(C'_R)$, $C_G = g(C'_G)$ and $C_B = g(C'_B)$, where $g(\cdot)$ is an invertible pointwise nonlinearity such as the power law mentioned above. Several such nonlinearities have been standardized, such as the ones used in the sRGB and Rec. 709 standards mentioned previously. For example, the power-law nonlinearity given in sRGB is

$$g_{\text{sRGB}}(x) = \begin{cases} \frac{x}{12.92} & 0 \le x \le 0.04045 \\ \left(\frac{x+0.055}{1.055}\right)^{2.4} & 0.04045 \le x \le 1. \end{cases} \tag{7.70}$$

Note that the function has a linear segment near the origin. The function is monotonically increasing, continuous with continuous derivative on [0,1]. The inverse function is given by

$$g^{-1}_{\text{sRGB}}(x) = \begin{cases} 12.92x & 0 \le x \le 0.0031308 \\ 1.055x^{0.41666} - 0.055 & 0.0031308 \le x \le 1. \end{cases} \tag{7.71}$$

Thus, sRGB coordinates are obtained by taking the Rec. 709/sRGB tristimulus values as defined in Section 7.5.3 and applying the sRGB gamma correction function g^{-1}_{sRGB} independently to each of the three tristimulus values. These gamma-corrected coordinates are usually denoted with primes: $C'_i = g^{-1}_{\text{sRGB}}(C_i)$ for $i =$R, G, B. This transformation is one-to-one on the subset of $\mathcal{C}_R$ given by $0 \le C_i \le 1$ for $i =$R, G, B and so is a valid set of coordinate for the set of displayable colors, the region shown in Figure 7.8(b). It is a well behaved transformation, since both g_{sRGB} and g^{-1}_{sRGB} are continuous, monotonic functions with continuous first derivative that is finite at the origin.

It should be noted that the sRGB gamma function above closely approximates a pure power law with a gamma value of 2.2 (not 2.4), due to the linear segment near the origin [Poynton and Funt (2014)]. Thus, the sRGB gamma correction is very near to the inverse function for a display with a gamma of 2.2. If RGB tristimulus values are passed through $g^{-1}(\cdot)$, referred to as gamma correction, they can be applied directly to a CRT display to produce the correct tristimulus values. Today, as CRT displays are disappearing, other displays such as LCD displays apply a gamma function electronically for compatibility, often selectable under user control. The gamma-corrected sRGB coordinates are widely used, with most digital color images implicitly or explicitly assumed to be in this representation. Although this representation is not as perceptually uniform as CIELAB, it is much more so than the linear tristimulus representation, and is sufficient for many situations. This is a good reason to keep gamma-corrected values for color image representation even though CRTs are obsolete. See Poynton and Funt (2014) for an excellent discussion of many of these issues. Various transformations of the gamma-corrected

RGB values are widely used (or have been used over the years) in video and digital photography. See Poynton's book [Poynton (2012)] for an excellent treatment of these nonlinear coordinate systems.

7.5.6 Luma-Color-Difference Representation

The luma-color-difference representation (and its variants) is one of the most widely used color coordinate systems in image compression and transmission. It is used in standard television, as well as in digital compression schemes like JPEG and MPEG. It is based on gamma-corrected RGB values as seen in the previous section. In this color system, the luma carries the brightness information and the color differences carry the chromatic information. It is a nonlinear coordinate system, so again these coordinates are *not* tristimulus values, although they are related to tristimulus values by a reversible nonlinear transformation. One of the advantages of this coordinate system is that the color difference components can be subsampled with respect to the luma component, say by a factor of two in the horizontal and vertical dimension, with little impact on the perceived quality of the color image. This will be discussed in the next chapter. The transformation to this coordinate system is defined by ITU-R Rec. 601

$$\begin{aligned} C'_Y &= 0.299C'_\mathrm{R} + 0.587C'_\mathrm{G} + 0.114C'_\mathrm{B} \\ C_{P_\mathrm{B}} &= 0.564(C'_\mathrm{B} - C'_Y) \\ C_{P_\mathrm{R}} &= 0.713(C'_\mathrm{R} - C'_Y) \end{aligned}$$

where C'_R, C'_G, C'_B are gamma-corrected RGB values, assumed to lie between 0 and 1. It follows that $0 \leq C'_Y \leq 1$ and $-0.5 \leq C_{P_\mathrm{B}}, C_{P_\mathrm{R}} \leq 0.5$. This can be expressed in matrix form as

$$\begin{bmatrix} C'_Y \\ C_{P_\mathrm{B}} \\ C_{P_\mathrm{R}} \end{bmatrix} = \begin{bmatrix} 0.299 & 0.587 & 0.114 \\ -0.169 & -0.331 & 0.5 \\ 0.5 & -0.419 & -0.081 \end{bmatrix} \begin{bmatrix} C'_\mathrm{R} \\ C'_\mathrm{G} \\ C'_\mathrm{B} \end{bmatrix}.$$

The quantity C'_Y is referred to as *luma*, although it is often incorrectly called luminance in the literature. For a gray color, with $C_\mathrm{R} = C_\mathrm{G} = C_\mathrm{B}$, it follows that $C'_Y = C'_\mathrm{R} = C'_\mathrm{G} = C'_\mathrm{B}$, and $C_{P_\mathrm{B}} = C_{P_\mathrm{R}} = 0$.

Note that there is a closely related luma-color-difference coordinate system in wide use referred to as $Y'C_\mathrm{B}C_\mathrm{R}$, where the coordinates of a color **C** would be denoted C'_Y, C_{C_B}, and C_{C_R}, and there are several variations on these. Also the term $Y'UV$ is often used, ambiguously, to refer to these color coordinate systems. See Poynton (2012) for a detailed treatment of these issues.

Problems

1 The web color *goldenrod* that we will denote [**Q**] is specified by the RGB values 218, 165, 32, on a scale from 0 to 255. Thus they can be assumed to be $Q'_\mathrm{R} = 0.8549$, $Q'_\mathrm{G} = 0.6471$, $Q'_\mathrm{B} = 0.1255$ on a scale from 0 to 1. We assume that these are gamma-corrected values, according to the sRGB gamma law (Section 7.5.5), and that the primaries are the Rec. 709/sRGB primaries, normalized with respect to reference white D_{65} (Section 7.5.3). The goal of this problem is to determine

representations of this color in other color coordinate representations. Determine the following (show your work):

(a) The tristimulus values Q_R, Q_G, Q_B in the Rec. 709/sRGB color representation (Section 7.5.3).
(b) The luminance Q_L and the chromaticities q_R, q_G, q_B in the Rec. 709/sRGB representation.
(c) The XYZ tristimulus values Q_X, Q_Y, Q_Z and the corresponding chromaticities q_X, q_Y, q_Z (Section 7.4.7).
(d) The 1976 $U'V'W'$ tristimulus values $Q_{U'}$, $Q_{V'}$, $Q_{W'}$ and the corresponding chromaticities $q_{U'}$, $q_{V'}$, $q_{W'}$ (Section 7.4.8).
(e) The CIELAB coordinates Q_{L^*}, Q_{a^*}, Q_{b^*} (Section 7.5.4).
(f) The Luma and color differences $Q_{Y'}$, Q_{P_B}, Q_{P_R} (Section 7.5.6).

* You can visualize this color in any Windows program that lets you specify the RGB values of a color. For example, in Microsoft Word, draw a shape like a rectangle and set the fill color using "More Colors – Custom" and enter the gamma-corrected red, green and blue values in the boxes.

2 As stated at the end of Section 7.4.8, the spectral absorption curves of the three types of cone photoreceptors in the human retina should be a linear combination of any set of three color-matching functions. The outputs of these receptors can be considered to be tristimulus values with respect to some set of primaries that we will call **[L]**, **[M]**, **[S]** (which stand for long, medium and short). It has been found that the tristimulus values with respect to these primaries for a color **[C]**, denoted $\mathbf{C}_l$, C_m, C_s, can be obtained from the XYZ tristimulus values by

$$\begin{bmatrix} \mathbf{C}_l \\ C_m \\ C_s \end{bmatrix} = \begin{bmatrix} 0.4002 & 0.7076 & -0.0808 \\ -0.2263 & 1.1653 & 0.0457 \\ 0.0 & 0.0 & 0.9182 \end{bmatrix} \begin{bmatrix} C_X \\ C_Y \\ C_Z \end{bmatrix}$$

(a) Determine and plot the color matching functions for the LMS primaries, denoted $\bar{l}(\lambda)$, $\overline{m}(\lambda)$, $\bar{s}(\lambda)$. The data for the xyz color-matching functions are given on the CVRL website (www.cvrl.org).
(b) Express the primaries **[L]**, **[M]**, **[S]** in terms of the primaries **[X]**, **[Y]**, **[Z]**. What color is **[L]** + **[M]** + **[S]**?
(c) Why are these primaries called **[L]**, **[M]** and **[S]**?
(d) Determine the LMS tristimulus values of the color *goldenrod* of Problem 1. Can you give a physical interpretation (in terms of your eye) of these tristimulus values?

3 The Bayer color sampling strategy induces a new set of color signals from the original RGB values (assume Rec. 709) as follows:

$$\begin{bmatrix} f_L \\ f_{C1} \\ f_{C2} \end{bmatrix} = \begin{bmatrix} 0.25 & 0.5 & 0.25 \\ -0.25 & 0.5 & -0.25 \\ -0.25 & 0.0 & 0.25 \end{bmatrix} \begin{bmatrix} f_R \\ f_G \\ f_B \end{bmatrix}.$$

These can be considered to be tristimulus values with respect to a new set of primaries denoted **[L]**, **[C1]**, **[C2]**.

(a) Determine and plot the color matching functions for the LC1C2 primaries, denoted $\bar{l}(\lambda)$, $\overline{c1}(\lambda)$, $\overline{c2}(\lambda)$. The data for the XYZ color-matching functions are given on the CVRL website (www.cvrl.org).
(b) Express the primaries [**L**], [**C1**], [**C2**] in terms of the primaries [**R**], [**G**], [**B**] and in terms of the primaries [**X**], [**Y**], [**Z**].
(c) Determine the LC1C2 tristimulus values of the color *goldenrod* of Problem 1 and of the reference white D_{65}.

4 The recommendation 709 RGB primaries can be expressed in terms of the CIE *XYZ* primaries by

$$\begin{bmatrix}[\mathbf{R}]\\ [\mathbf{G}]\\ [\mathbf{B}]\end{bmatrix} = \begin{bmatrix}0.4125 & 0.2127 & 0.0193\\ 0.3576 & 0.7152 & 0.1192\\ 0.1804 & 0.0722 & 0.9502\end{bmatrix}\begin{bmatrix}[\mathbf{X}]\\ [\mathbf{Y}]\\ [\mathbf{Z}]\end{bmatrix}.$$

Consider the cyan, magenta and yellow (CMY) primaries used in printing. These are given by [**C**] = [**B**] + [**G**], [**M**] = [**R**] + [**B**] and [**YE**] = [**R**] + [**G**].
(a) Determine the *tristimulus values* of [**C**], [**M**] and [**YE**] with respect to the *XYZ* primaries. Compute the *XYZ chromaticities* of [**C**], [**M**] and [**YE**] and plot them on an *xy* chromaticity diagram. Comment on the suitability of cyan, magenta and yellow as primaries for an additive color display device like a cathode ray tube (CRT).
(b) Suppose that [**C**], [**M**] and [**YE**], as in part (a), are taken as primaries in a color system. Determine the tristimulus values of a monochromatic light $\delta(\lambda - 510\text{ nm})$ with respect to these primaries. You will need to use the *XYZ* color matching functions. Carefully explain all steps. Can the given light be *physically* synthesized as a sum of a positive quantity of the [**C**], [**M**] and [**YE**] primaries?
(c) Determine the tristimulus values of the color *goldenrod* of question 1 with respect to the [**C**], [**M**] and [**YE**] primaries. Can this color be *physically* synthesized as a sum of a positive quantity of the [**C**], [**M**] and [**YE**] primaries?

5 The EBU (European Broadcasting Union) primaries, have the following specification

	Red	Green	Blue	White D_{65}
x	0.640	0.290	0.150	0.3127
y	0.330	0.600	0.060	0.3290
z	0.030	0.110	0.790	0.3582

Assume that the reference white has unit luminance $D_L = 1.0$ and that [**R**] + [**G**] + [**B**] = [$\mathbf{D}_{65}$].
(a) Find the *XYZ* tristimulus values of the reference white [$\mathbf{D}_{65}$], i.e. D_X, D_Y and D_Z.
(b) Using [**R**] + [**G**] + [**B**] = [$\mathbf{D}_{65}$], determine the luminances of the three primaries, R_L, G_L and B_L.

(c) Now find the XYZ tristimulus values of the three primaries, i.e. if $[\mathbf{R}] = R_X[\mathbf{X}] + R_Y[\mathbf{Y}] + R_Z[\mathbf{Z}]$, find R_X, R_Y, R_Z, and similarly for $[\mathbf{G}]$ and $[\mathbf{B}]$.

(d) If an arbitrary color $[Q]$ is written

$$[\mathbf{Q}] = Q_X[\mathbf{X}] + Q_Y[\mathbf{Y}] + Q_Z[\mathbf{Z}] = Q_{\mathrm{R}}[\mathbf{R}] + Q_{\mathrm{G}}[\mathbf{G}] + Q_{\mathrm{B}}[\mathbf{B}]$$

determine the matrix relations to find Q_X, Q_Y, Q_Z from Q_{R}, Q_{G}, Q_{B} and vice versa.

(e) Plot an xy chromaticity diagram showing the triangles of chromaticities reproducible with the EBU RGB primaries.

(f) Compute and plot the color matching functions for the EBU RGB primaries by transforming the XYZ color matching functions using the results of (d).

(g) For the three spectral densities $Q_1(\lambda)$, $Q_2(\lambda)$ and $Q_3(\lambda)$ in the following table, compute their chromaticities and plot them on an xy chromaticity diagram. Would they make good primaries for a physical color image synthesis system such as a CRT? Explain.

Q_1	Q_2	Q_3	λ
.19	.00	.60	400 nm
.20	.00	.63	
.20	.00	.64	
.20	.00	.63	
.20	.00	.62	
.20	.02	.59	450 nm
.20	.06	.53	
.19	.19	.43	
.18	.31	.31	
.16	.43	.20	
.13	.52	.10	500 nm
.08	.61	.05	
.06	.67	.02	
.04	.69	.01	
.03	.69	.00	
.04	.67	.00	550 nm
.08	.64	.00	
.14	.60	.00	
.22	.55	.00	
.32	.49	.00	
.41	.43	.00	600 nm
.50	.38	.00	
.56	.33	.00	
.63	.28	.00	

Q_1	Q_2	Q_3	λ
.67	.25	.00	
.71	.23	.00	650 nm
.75	.21	.00	
.77	.20	.00	
.79	.19	.00	
.80	.19	.00	
.81	.18	.00	700 nm

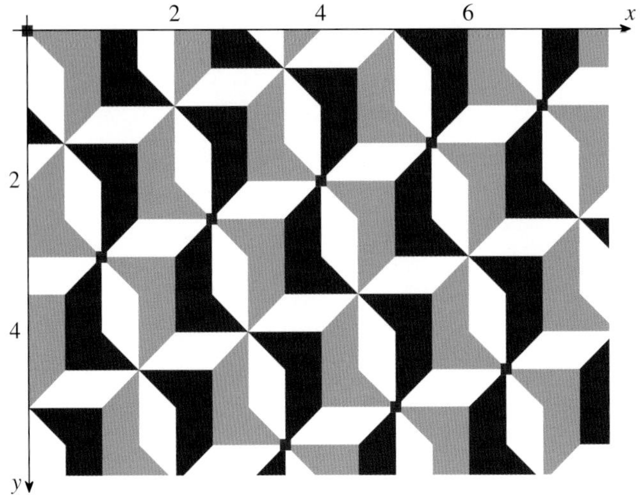

Figure 4.1 Portion of a periodic image representative of a wallpaper pattern. The periodicity is given by the lattice Γ shown, which is a sublattice of the underlying lattice Λ (not shown).

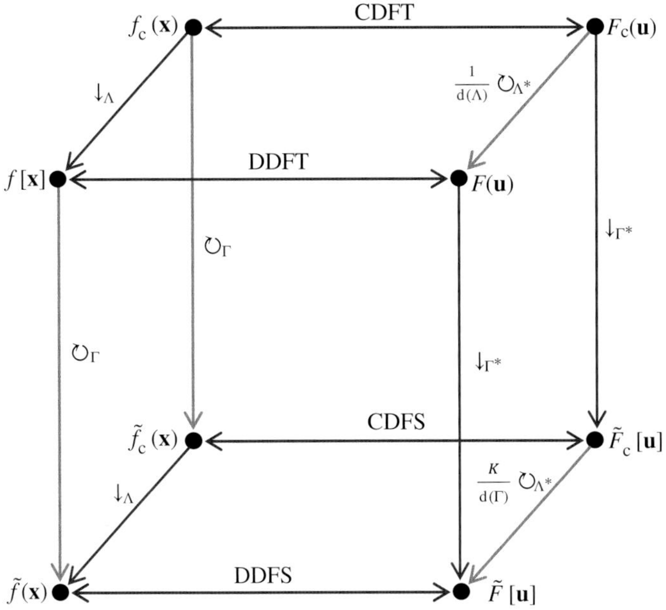

Figure 6.4 Fourier–Poisson cube illustrating Fourier transform relationships for sampling and periodization of multidimensional signals.

Multidimensional Signal and Color Image Processing Using Lattices, First Edition. Eric Dubois.

Companion website: www.wiley.com/go/Dubois/multiSP

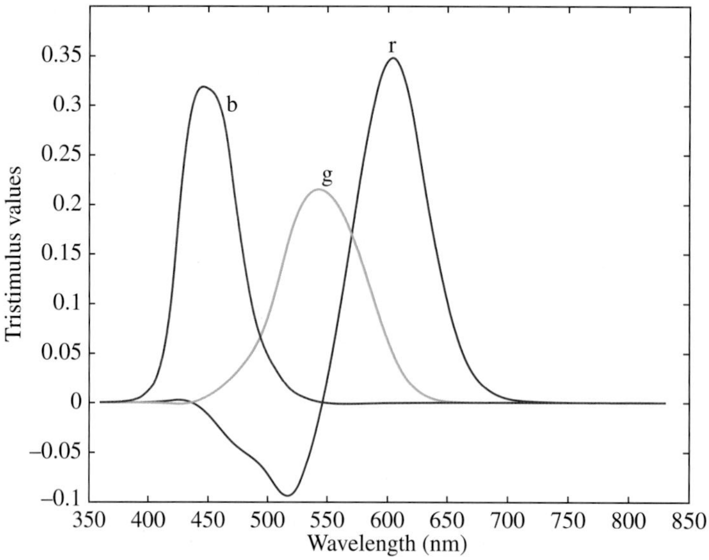

Figure 7.4 Color matching functions of the CIE 1931 RGB primaries: monochromatic lights at wavelengths 700.0 nm, 546.1 nm and 435.8 nm respectively.

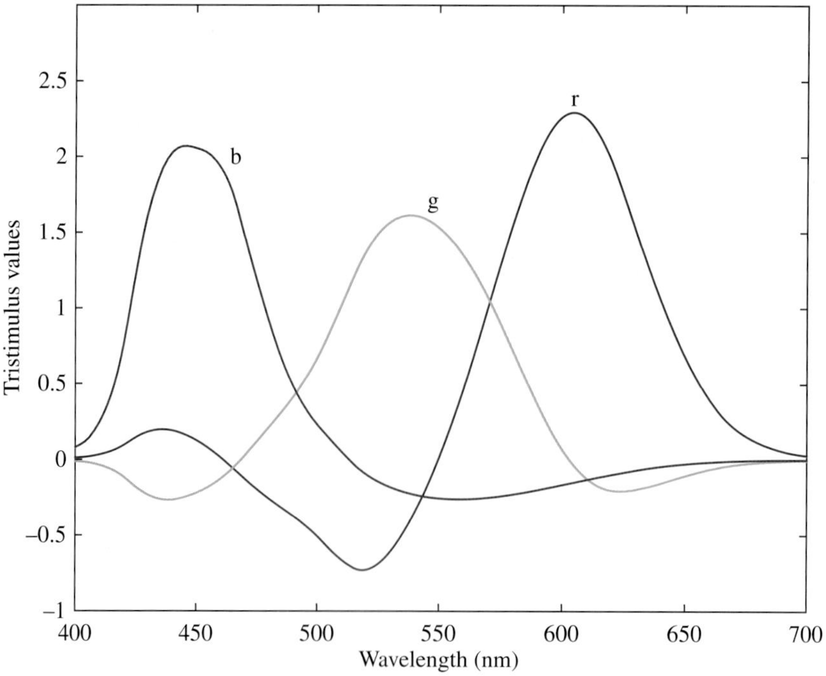

Figure 7.6 Color matching functions of a typical CRT RGB display (EIZO) with the red, green and blue spectral densities given in Figure 7.1(c).

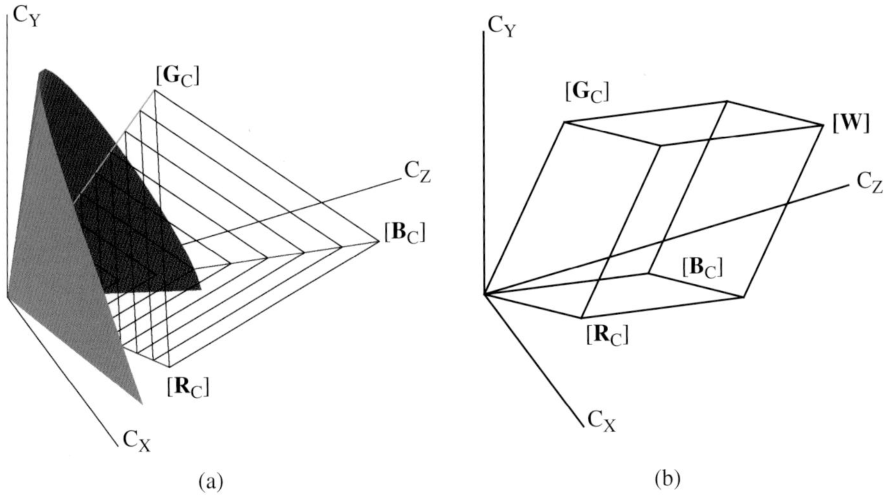

Figure 7.8 (a) Illustration of the cone of reproducible colors with red, green, and blue display primaries, in the *XYZ* color coordinate system. (b) The set of reproducible colors with red, green, and blue coefficients between 0 and 1.

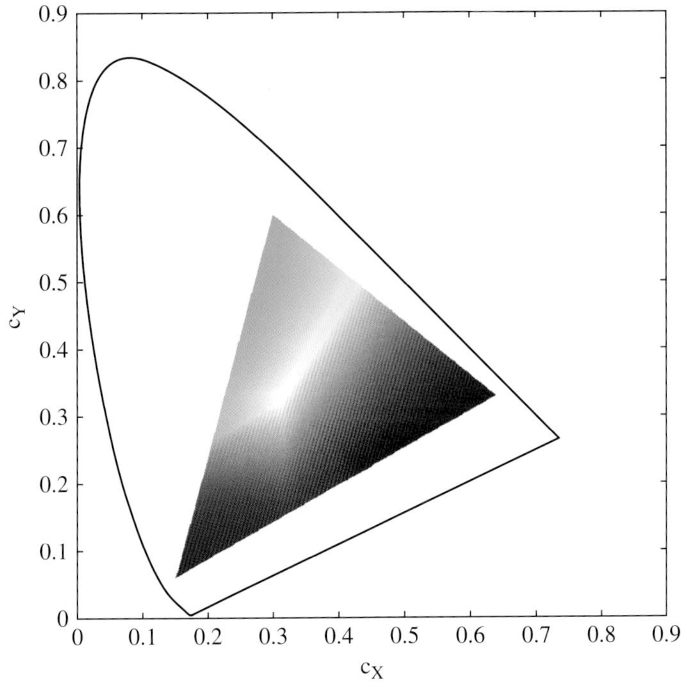

Figure 7.12 Chromaticities of colors that can be reproduced by addition of positive quantities of typical RGB display primaries.

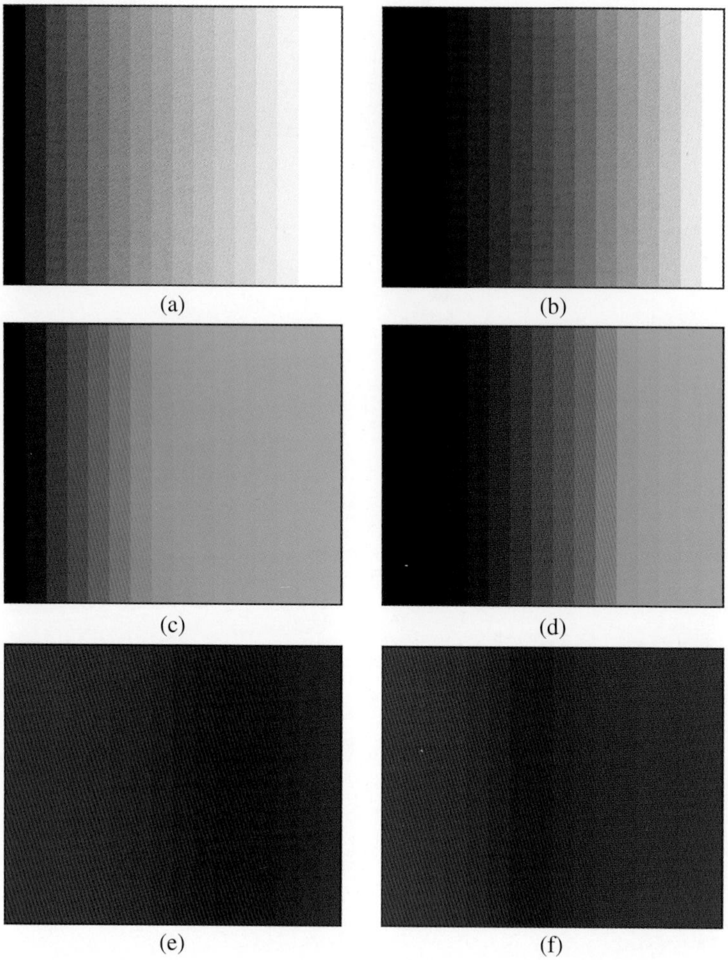

Figure 7.13 Gray scale step pattern with (a) linearly increasing tristimulus values; (b) linearly increasing CIELAB coordinates. Color step pattern with changing luminance and (c) linearly increasing tristimulus values; (d) linearly increasing CIELAB coordinates. Color step pattern with fixed luminance and (e) linearly increasing tristimulus values; (f) linearly increasing CIELAB coordinates.

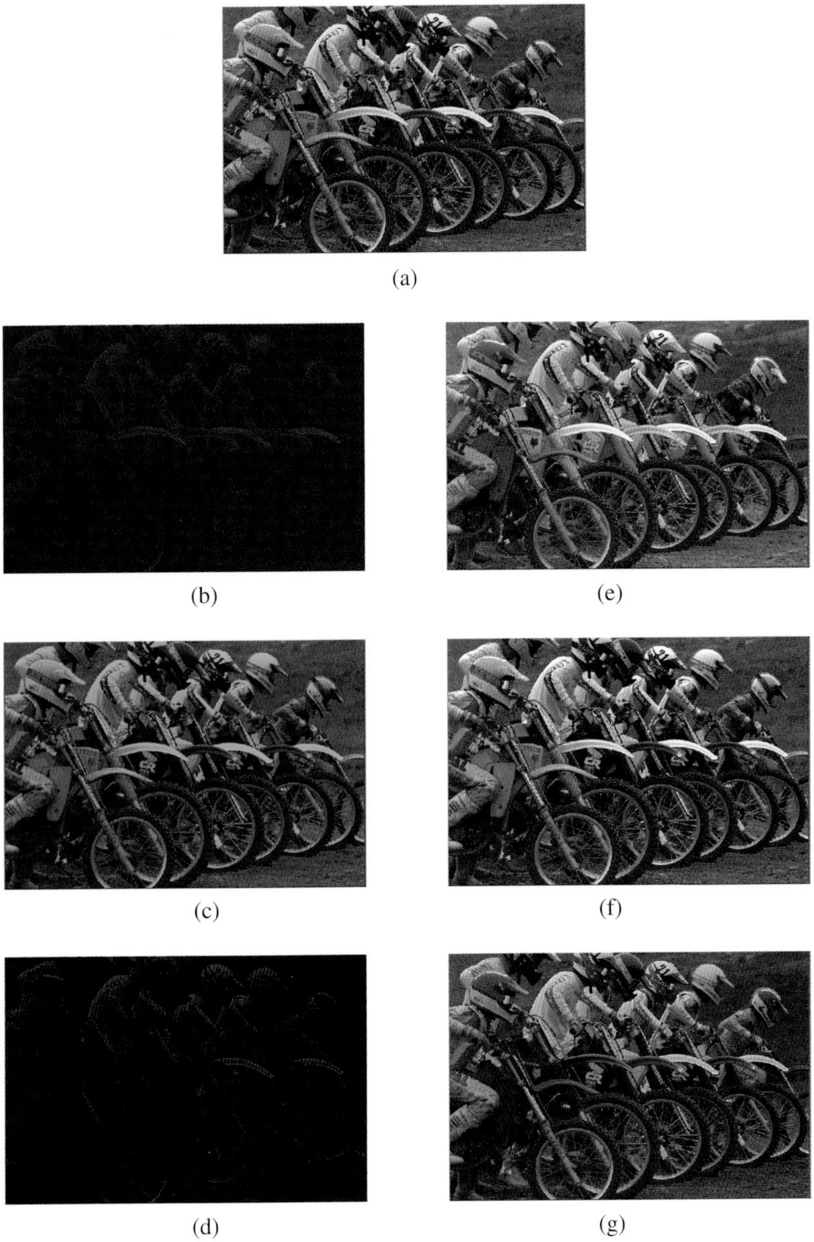

(a) (b) (c) (d) (e) (f) (g)

Figure 8.1 Example of a real color image: (a) its sRGB components shown as color images: (b) red; (c) green; (d) blue; the tristimulus images shown as gray-scale images: (e) red; (f) green; (g) blue.

Figure 8.2 Example of a real color sinusoidal signal.

(a)

(b)

Figure 8.5 Motorcycles image with different types of noise added to the NTSC YIQ components. (a) Low-pass noise, with greater noise level in the luma component. (b) High-pass noise, with greater noise level in the chroma components.

(a)

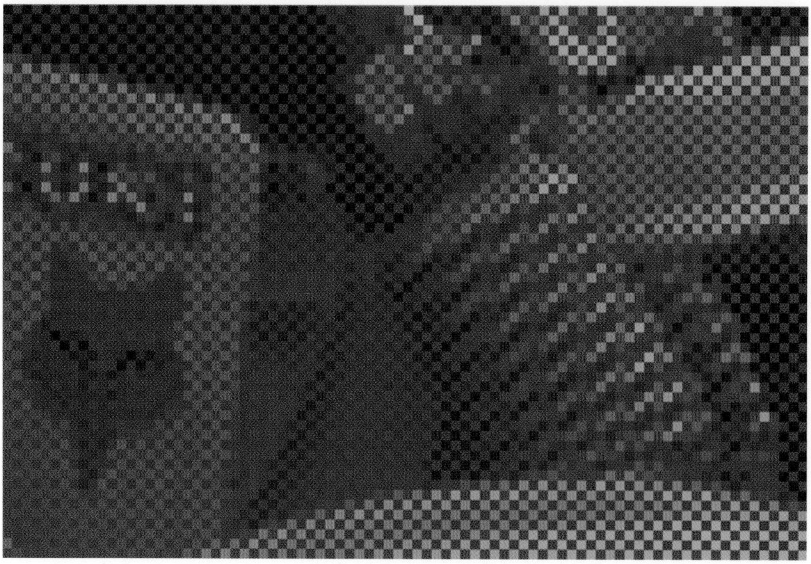

(b)

Figure 8.7 Example of a Bayer sampled color image. (a) Full image. (b) Sub-image magnified by a factor of eight.

Figure 8.8 Bayer sampled image of Figure 8.7 reconstructed by separate low-pass filtering of the three channels of $\mathbf{C}_{d,\mathcal{L}CC}[\mathbf{x}]$, followed by conversion to RGB.

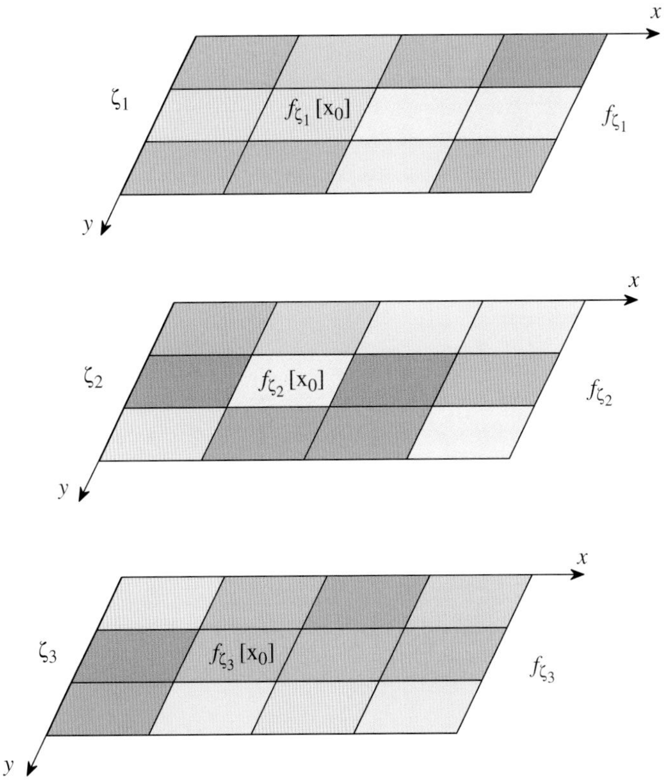

Figure 9.1 Illustration of three realizations of a two-dimensional color random field f_{ζ_i}, $i = 1, 2, 3$, identifying the three realizations of the random vector corresponding to $\mathbf{x} = \mathbf{x}_0$.

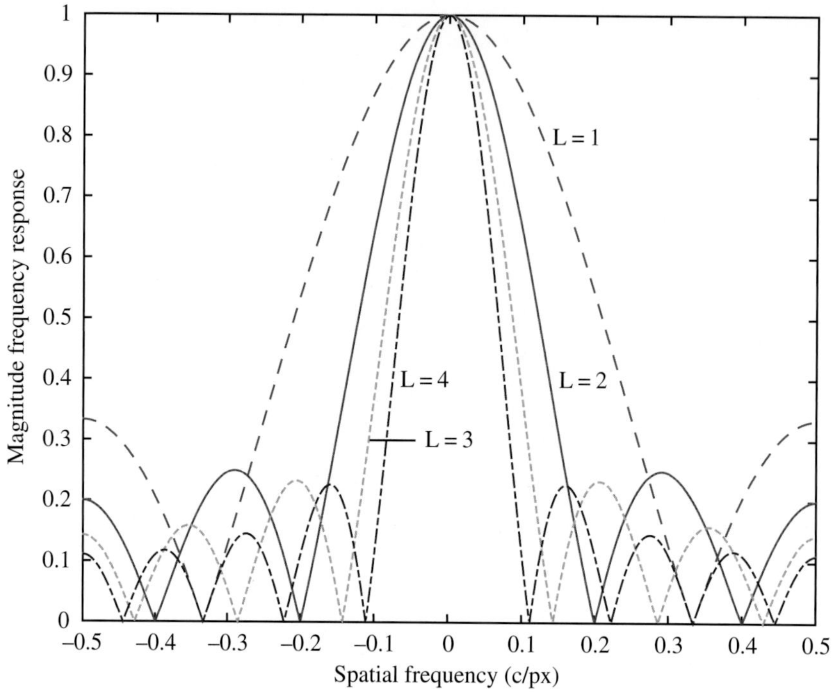

Figure 10.1 Profile $|H(u,0)|$ of the frequency response of the moving average filter for $L = 1, 2, 3, 4$.

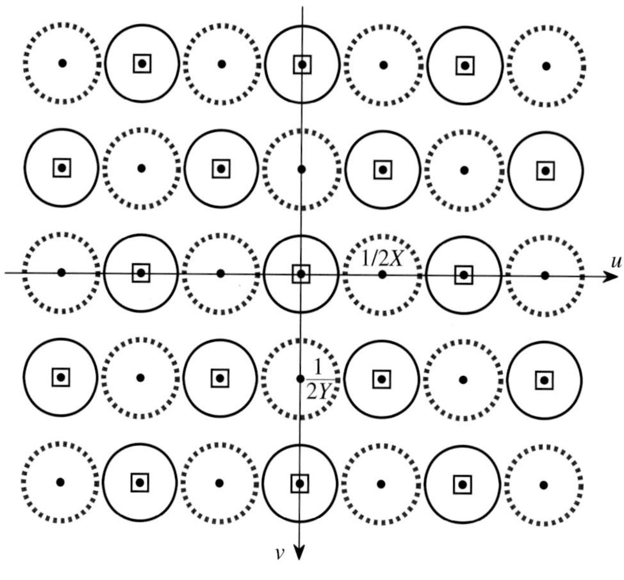

Figure 11.2 Illustration of reciprocal lattices Λ^* (•) and Γ^* (□) of Example 11.1, where $\Gamma^* \subset \Lambda^*$. Solid circles represent spectral repeats on the points of Γ^*. Dashed circles represent spectral repeats on the points of $\Lambda^* \backslash \Gamma^*$ that are to be removed in the upsampling process.

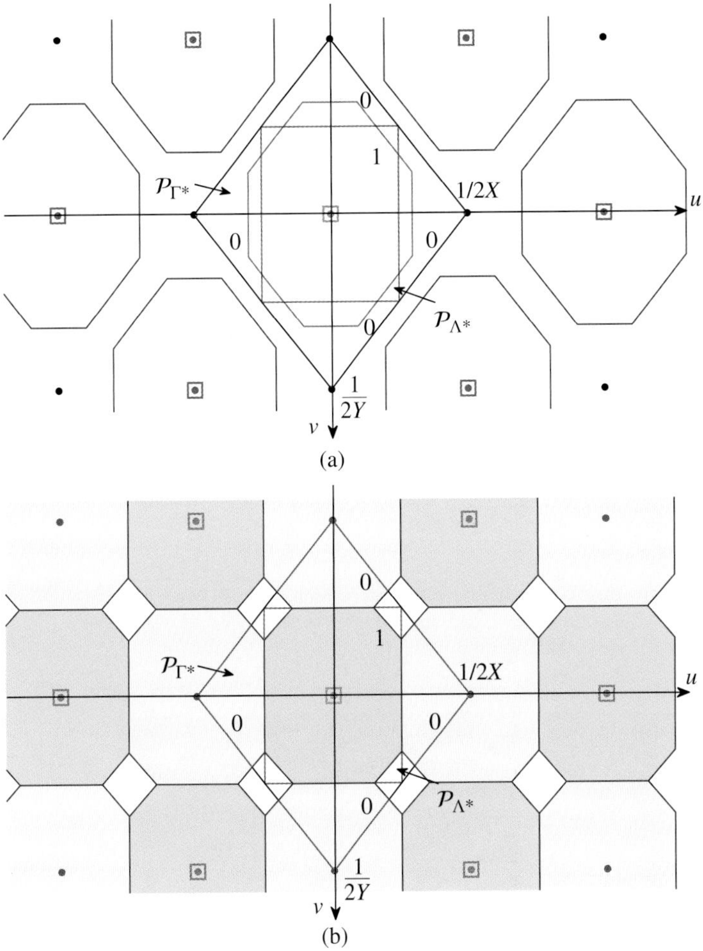

Figure 11.9 Illustration of a downsampling system using the lattices of Example 11.1. (a) Selected points of the reciprocal lattices Λ^* (•) and Γ^* (□) are shown. The Fourier transform of the original signal defined on Γ is limited to the octagonal region shown, replicated at points of Γ^*. (b) Output of the downsampling system. The original filtered replicas on Γ^* are shown in green and the additional replicas on points of $\Lambda^*\backslash\Gamma^*$ are shown in yellow.

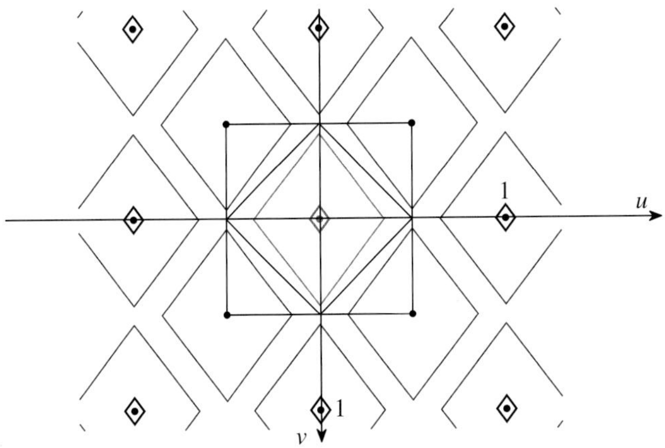

Figure 11.13 Frequency domain situation for upconversion from Λ_1 to Λ_3 in Example 11.3. The reciprocal lattices Λ_1^* (•) and Λ_3^* (◇) and their unit cells are shown. The input signal has a Fourier transform confined to the diamond-shaped region. The interpolation filter, defined on Λ_3, should keep the shaded replicas and reject the unshaded replicas.

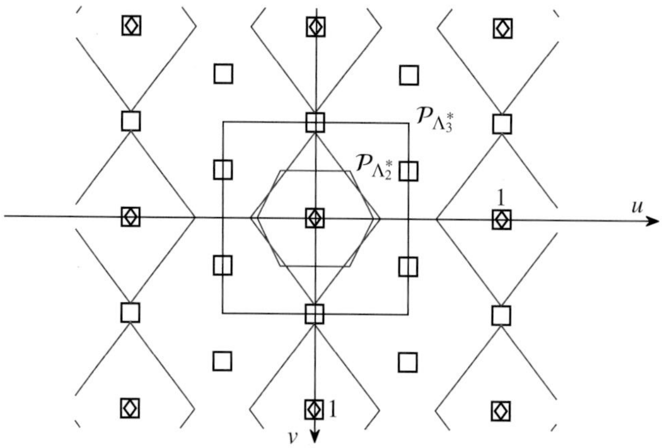

Figure 11.14 Frequency domain situation for downconversion from Λ_3 to Λ_2 in Example 11.3. The reciprocal lattices Λ_2^* (□) and Λ_3^* (◇) and their unit cells are shown. The anti-aliasing prefilter has passband equal to the hexagonal unit cell of Λ_2^*.

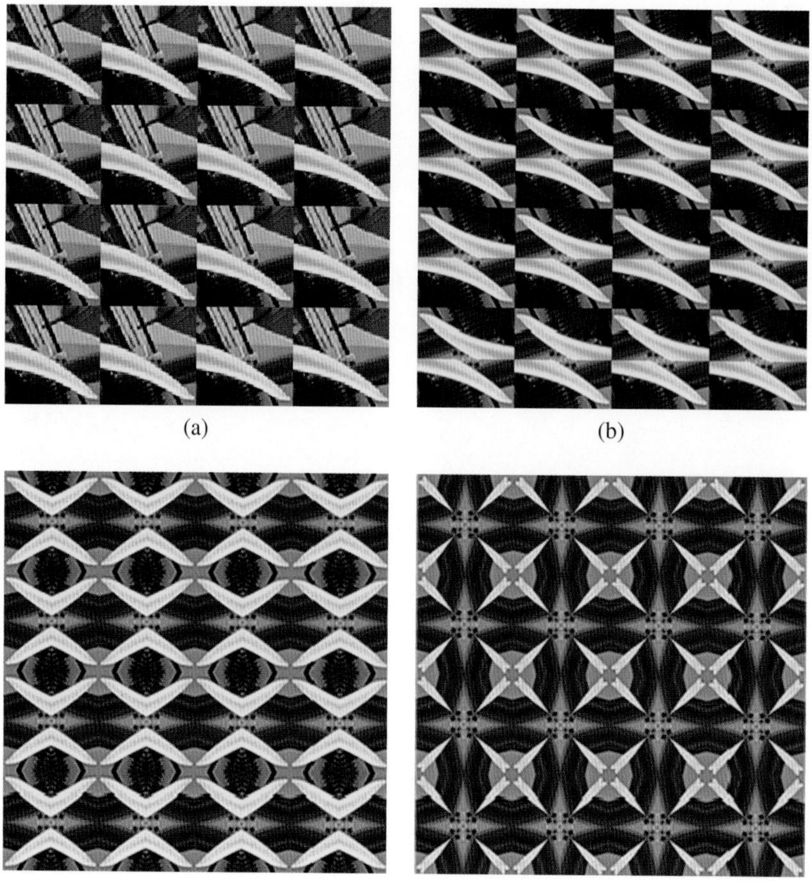

(a) (b) (c) (d)

Figure 12.6 Periodic extension of image blocks with (a) no symmetry, (b) inversion symmetry, (c) quadrantal symmetry, (d) eight-fold symmetry.

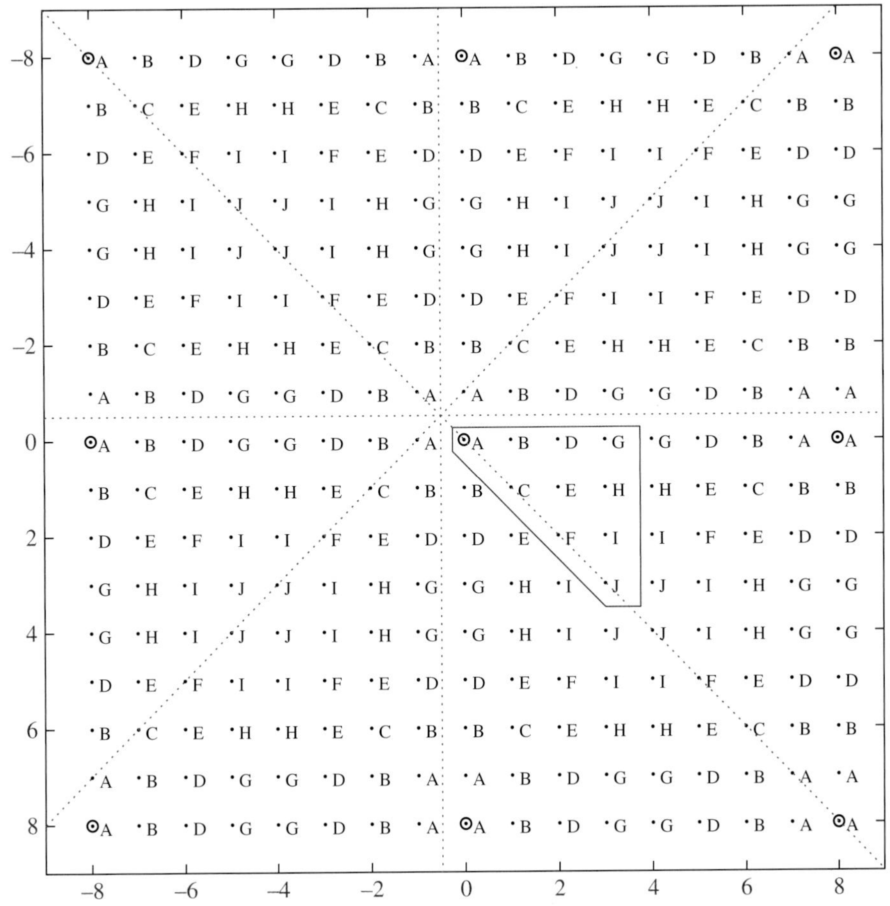

Figure 12.7 Illustration of orbits and orbit representatives for the square lattice $\Lambda = \mathbb{Z}^2$ (.) with eight-fold symmetry about the point $[-0.5\ -0.5\ ]^T$ and periodicity lattice $\Gamma = (8\mathbb{Z})^2$ (⊙). A set $\mathcal{B}_{\mathcal{G}}$ of orbit representatives A–J is indicated within the polygonal region. This signal is invariant to reflections about the four axes shown and to rotations of multiples of 90° about the point $[-0.5\ -0.5\ ]^T$.

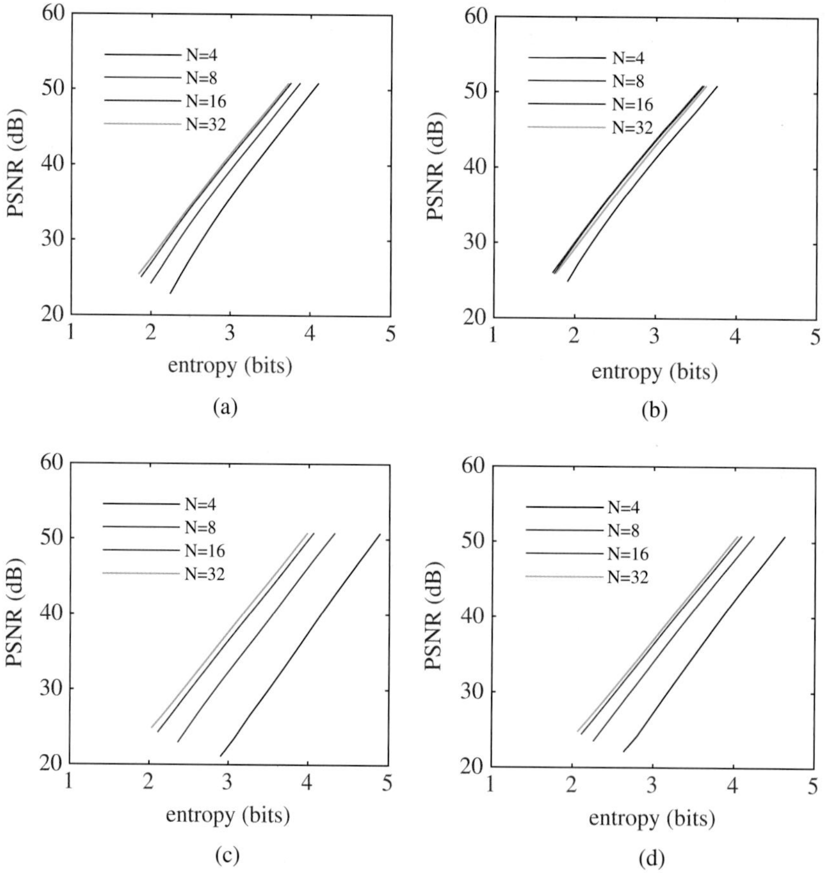

Figure 12.9 Curves of PSNR versus sample entropy over the Kodak dataset for different blocksize N_1. (a) DCT-1, grayscale images; (b) DCT-2, grayscale images; (c) DCT-1, Bayer CFA images; (d) DCT-2, Bayer CFA images.

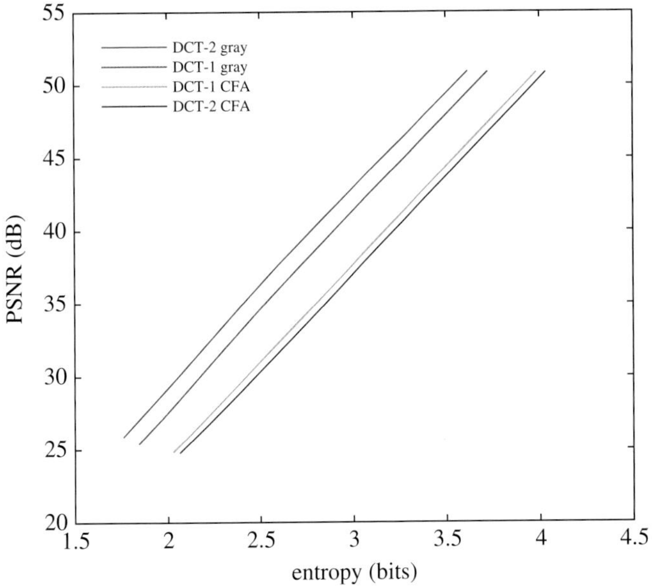

Figure 12.10 Curves of PSNR versus sample entropy over the Kodak dataset for DCT-1 and DCT-2 basis, applied to grayscale images and Bayer CFA images. The legend identifies the curves in decreasing order of PSNR.

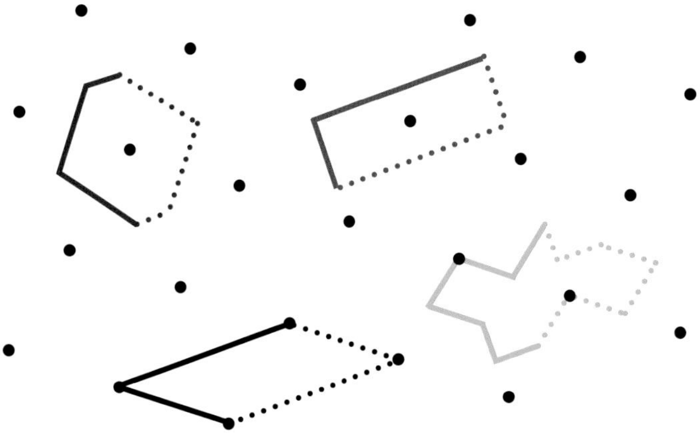

Figure 13.2 Portion of the lattice of Figure 13.1 with different possible unit cells. Clockwise from bottom left: fundamental parallelepiped, Voronoi cell, rectangular unit cell, irregular unit cell. All unit cells have the same area, $0.87X^2$. Solid borders are part of the unit cells, dashed borders are not. Axes are not shown, so that in each case the origin can be chosen as the one lattice point contained in the unit cell.

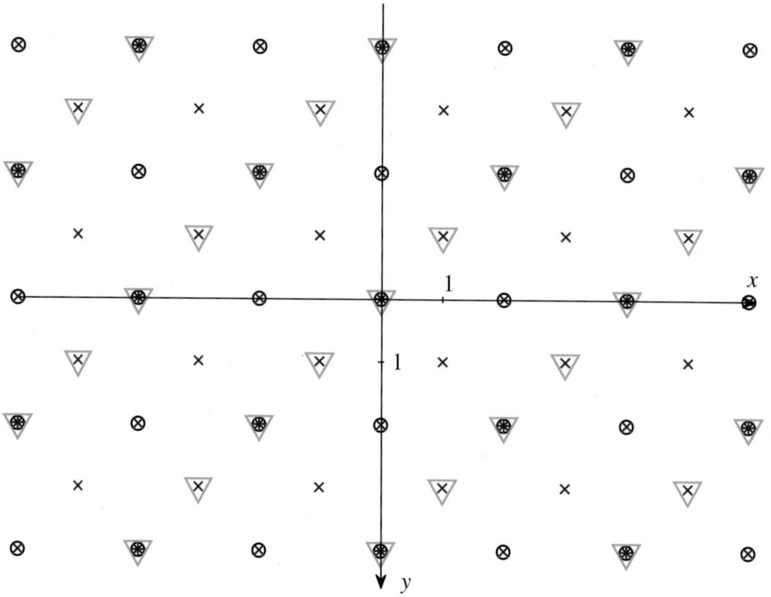

Figure 13.6 Intersection and sum of lattices. Λ_1: O; Λ_2: △; $\Lambda_1 \cap \Lambda_2$: +; $\Lambda_1 + \Lambda_2$: ×.

8

Processing of Color Signals

8.1 Introduction

This chapter presents the theory of the representation and processing of color signals, namely still and time-varying color images. A color image is a function of two (spatial) or three (spatiotemporal) independent variables with values in a color space. We introduce the ideas of signals, linear shift-invariant systems and Fourier transforms, in the continuous domain and in the sampled domain. This chapter extends the material in Chapters 2–6 from scalar valued signals to vector valued signals. It can also be viewed as a revision and extension of Chapter 6 of Dubois (2010).

8.2 Continuous-Domain Systems for Color Images

8.2.1 Continuous-Domain Color Signals

A continuous-space color image is a function that assigns an element of a given color space $\mathcal{C}$ to each spatial coordinate (x, y) within a given image window. This image will be denoted $[\mathbf{C}](x, y)$. A time-varying color image would be of the form $[\mathbf{C}](x, y, t)$. As before, we use a vector notation for the independent variables, and write $[\mathbf{C}](\mathbf{x})$ for either still or time-varying images, where $\mathbf{x}$ denotes either (x, y) or (x, y, t) according to context. The corresponding domain is denoted $\mathbb{R}^D$, where $D = 2$ or $D = 3$. If we choose a basis $\mathcal{B} = \{[\mathbf{P}_1], [\mathbf{P}_2], [\mathbf{P}_3]\}$ for $\mathcal{C}$, then the color signal can be written

$$[\mathbf{C}](\mathbf{x}) = C_1(\mathbf{x})[\mathbf{P}_1] + C_2(\mathbf{x})[\mathbf{P}_2] + C_3(\mathbf{x})[\mathbf{P}_3] \tag{8.1}$$

so that the color image is represented by three scalar-valued images in this basis. It may be convenient to represent the color signal as a column matrix of scalar signals with respect to the given basis:

$$\mathbf{C}_{\mathcal{B}}(\mathbf{x}) = \begin{bmatrix} C_1(\mathbf{x}) \\ C_2(\mathbf{x}) \\ C_3(\mathbf{x}) \end{bmatrix}. \tag{8.2}$$

Multidimensional Signal and Color Image Processing Using Lattices, First Edition. Eric Dubois.

Companion website: www.wiley.com/go/Dubois/multiSP

Of course, the representation can be changed to any other basis for the color space using Equation (7.41) at each point $\mathbf{x}$, i.e. $\mathbf{C}_{\tilde{\mathcal{B}}}(\mathbf{x}) = \mathbf{A}_{\mathcal{B}\to\tilde{\mathcal{B}}}\mathbf{C}_{\mathcal{B}}(\mathbf{x})$. We consider here three-dimensional color spaces, corresponding to trichromatic human vision, but the ideas can easily be adapted to any other dimension for the color space, for example to deal with the signals from multispectral or hyperspectral cameras. A physically realizable color image is one such that $[\mathbf{C}](\mathbf{x}) \in \mathcal{C}_{\mathrm{R}}$ for all $\mathbf{x}$. Such an image would normally correspond to a (nonunique) non-negative spectral density $C(\mathbf{x}, \lambda)$, projected onto the given color space using the appropriate color matching functions, again independently at each point $\mathbf{x}$.

We can also represent a color image using one of the nonlinear coordinate systems introduced in Section 7.5. For example, we can represent an image in CIELAB coordinates, and denote it as

$$\mathbf{C}_{\mathrm{LAB}}(\mathbf{x}) = \begin{bmatrix} C_{\mathrm{L}^*}(\mathbf{x}) \\ C_{a^*}(\mathbf{x}) \\ C_{b^*}(\mathbf{x}) \end{bmatrix}. \tag{8.3}$$

We use roman text font in the subscript to denote a nonlinear coordinate system (e.g., LAB), as opposed to a script font for a basis in a linear tristimulus representation. In this chapter, we denote a change of coordinate function as $\xi_{\mathrm{rep1}\to\mathrm{rep2}}$. Note that this does not change the color, it just changes its representation. Thus, we would denote the change of representation from XYZ tristimulus values to LAB coordinates by

$$\mathbf{C}_{\mathrm{LAB}}(\mathbf{x}) = \xi_{\mathcal{XYZ}\to\mathrm{LAB}}(\mathbf{C}_{\mathcal{XYZ}}(\mathbf{x})), \tag{8.4}$$

where the function $\xi_{\mathcal{XYZ}\to\mathrm{LAB}}$ implements Equation (7.68) independently at each point $\mathbf{x}$. For a change of primaries, the function ξ is a matrix multiplication, so that $\xi_{\mathcal{B}\to\tilde{\mathcal{B}}}(\mathbf{C}_{\mathcal{B}}(\mathbf{x})) = \mathbf{A}_{\mathcal{B}\to\tilde{\mathcal{B}}}\mathbf{C}_{\mathcal{B}}(\mathbf{x})$.

The set of color images forms a vector space $\mathcal{U}$ under the obvious definitions of pointwise addition and scalar multiplication at each $\mathbf{x}$ using the vector space operations of the color space $\mathcal{C}$, as long as the set is closed under these operations. As an example, we could have

$$\mathcal{U} = \{[\mathbf{C}](\mathbf{x}) \mid |C_i(\mathbf{x})| < b \text{ for all } \mathbf{x};\ i = 1, 2, 3;\ b < \infty\}, \tag{8.5}$$

a space of bounded color images. The vector space $\mathcal{U}$ is of course infinite dimensional. Strictly speaking, $[\mathbf{C}](\mathbf{x})$ is the value of the signal at coordinate $\mathbf{x}$ and is an element of $\mathcal{C}$. However, following conventional usage and to avoid introduction of an additional symbol (e.g., $[\vec{\mathbf{C}}]$) for elements of $\mathcal{U}$, we will also use $[\mathbf{C}](\mathbf{x})$ to denote an element of $\mathcal{U}$ where it is clear from context what is meant. We write $[\mathbf{C}](\mathbf{x})|_{\mathbf{x}=\mathbf{s}}$ to explicitly denote the color signal at a particular spatiotemporal coordinate.

Figure 8.1(a) shows an example of a typical color image, which we denote $[\mathbf{C}](\mathbf{x})$. Assuming the sRGB basis, which we denote $\{[\mathbf{R}], [\mathbf{G}], [\mathbf{B}]\}$, the color image is represented as $[\mathbf{C}](\mathbf{x}) = C_{\mathrm{R}}(\mathbf{x})[\mathbf{R}] + C_{G}(\mathbf{x})[\mathbf{G}] + C_{B}(\mathbf{x})[\mathbf{B}]$. Figures 8.1(b)–(d) illustrate the three color images $C_{\mathrm{R}}(\mathbf{x})[\mathbf{R}]$, $C_{G}(\mathbf{x})[\mathbf{G}]$, and $C_{B}(\mathbf{x})[\mathbf{B}]$ respectively. Each of the three tristimulus values can be treated as a scalar-valued image and visualized as a gray-scale image. Figures 8.1 (e)–(g) show $C_{\mathrm{R}}(\mathbf{x})$, $C_{G}(\mathbf{x})$, and $C_{B}(\mathbf{x})$, respectively, in this way. All of the displayed image components maintain the original gamma correction.

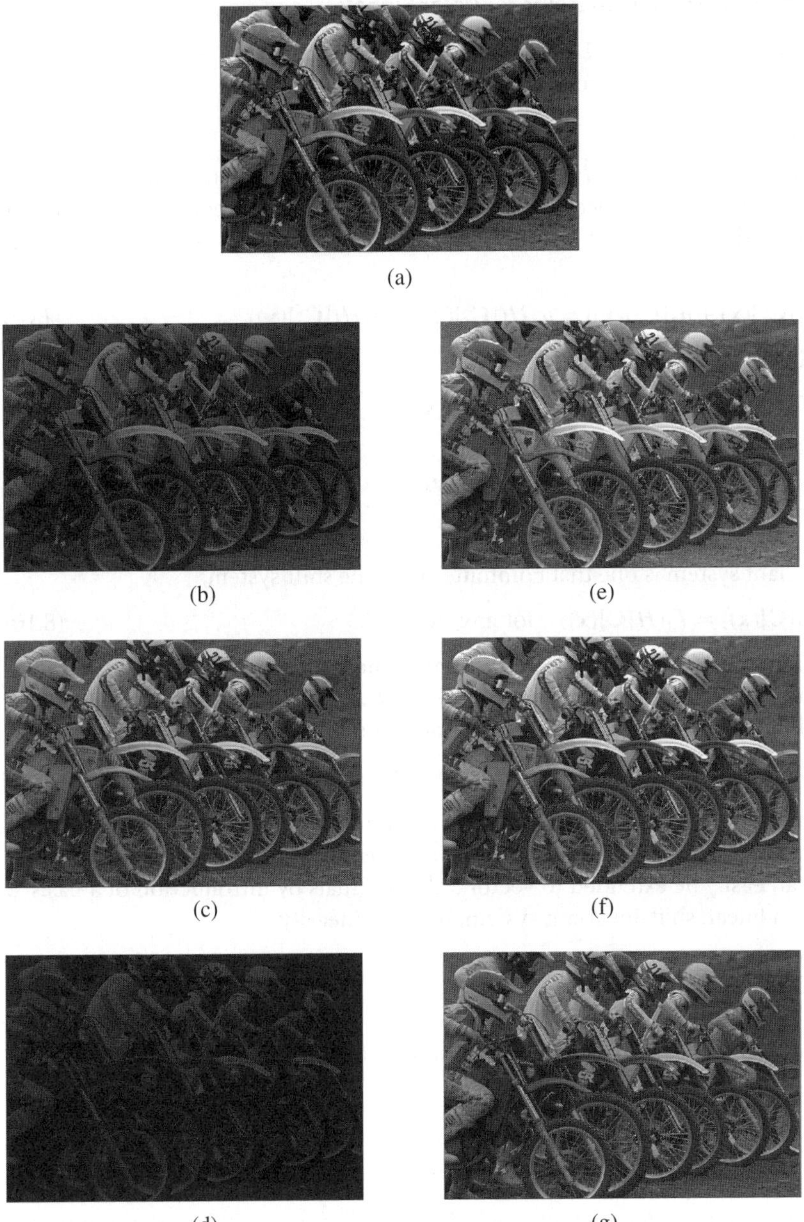

Figure 8.1 (a) Example of a real color image. Its sRGB components shown as color images: (b) red; (c) green; (d) blue; the tristimulus images shown as gray-scale images: (e) red; (f) green; (g) blue. (*See color plate section for the color representation of this figure.*)

8.2.2 Continuous-Domain Systems for Color Signals

A *system* is any operator $\mathcal{H} : \mathcal{U} \rightarrow \mathcal{U}$ (or more generally $\mathcal{H} : \mathcal{U}_1 \rightarrow \mathcal{U}_2$) that takes a color image as input and produces another color image as output:

$$[\mathbf{Q}](\mathbf{x}) = \mathcal{H}([\mathbf{C}](\mathbf{x})). \tag{8.6}$$

When needed, we will write $(\mathcal{H}[\mathbf{C}])(\mathbf{x})$ to denote $\mathcal{H}([\mathbf{C}](\mathbf{x}))$ at coordinate $\mathbf{x}$. We are mainly concerned with two specific classes of systems, linear systems and memoryless systems, and perhaps combinations of these. As in all previous cases, a *linear system* is one that satisfies

$$\mathcal{H}(\alpha_1[\mathbf{C}_1](\mathbf{x}) + \alpha_2[\mathbf{C}_2](\mathbf{x})) = \alpha_1\mathcal{H}([\mathbf{C}_1](\mathbf{x})) + \alpha_2\mathcal{H}([\mathbf{C}_2](\mathbf{x})). \tag{8.7}$$

A memoryless system is one for which

$$(\mathcal{H}[\mathbf{C}])(\mathbf{x})|_{\mathbf{x}=\mathbf{s}} \quad \text{depends only on} \quad [\mathbf{C}](\mathbf{x})|_{\mathbf{x}=\mathbf{s}} \tag{8.8}$$

for all $\mathbf{s}$.

A simple linear system is the translation system

$$\mathcal{T}_\mathbf{d} : (\mathcal{T}_\mathbf{d}[\mathbf{C}])(\mathbf{x}) = [\mathbf{C}](\mathbf{x} - \mathbf{d}). \tag{8.9}$$

A shift-invariant system is one that commutes with the shift system,

$$\mathcal{H}(\mathcal{T}_\mathbf{d}[\mathbf{C}](\mathbf{x})) = \mathcal{T}_\mathbf{d}(\mathcal{H}[\mathbf{C}](\mathbf{x})) \quad \text{for any} \;\; \mathbf{d} \in \mathbb{R}^D. \tag{8.10}$$

We will mainly consider shift-invariant systems. A memoryless shift-invariant system is characterized by an arbitrary mapping from $\mathcal{C}$ into $\mathcal{C}$, which is applied independently at each coordinate $\mathbf{x}$. A simple example of a memoryless shift-invariant system is given by

$$\mathcal{H}[\mathbf{C}](\mathbf{x}) = \sqrt{C_1(\mathbf{x})}[\mathbf{P}_1] + \sqrt{C_2(\mathbf{x})}[\mathbf{P}_2] + \sqrt{C_3(\mathbf{x})}[\mathbf{P}_3]. \tag{8.11}$$

As in the usual scalar case, linear shift-invariant systems for color signals are characterized by their impulse response. The approach for scalar systems presented in Chapter 2 can easily be extended to vector-valued signals by introduction of a basis $\mathcal{B}$ for $\mathcal{C}$. If $\mathcal{H}$ is a linear, shift-invariant system, then by linearity

$$\mathcal{H}([\mathbf{C}](\mathbf{x})) = \sum_{i=1}^{3} \mathcal{H}(C_i(\mathbf{x})[\mathbf{P}_i]). \tag{8.12}$$

Let the input signal be a Dirac delta with values in the one dimensional subspace of $\mathcal{C}$ spanned by $[\mathbf{P}_i]$, i.e. $\delta(\mathbf{x})[\mathbf{P}_i]$, and define

$$\mathcal{H}(\delta(\mathbf{x})[\mathbf{P}_i]) = \sum_{k=1}^{3} H_{ki}(\mathbf{x})[\mathbf{P}_k]. \tag{8.13}$$

Here it is important to note that although the input lies in span$([\mathbf{P}_i])$, the output can be anywhere in the color space, except in special circumstances (although this special case tends to be the rule in the literature). Since the $H_{ki}(\mathbf{x})$ are tristimulus values, following our notation for color spaces we denote these signals with uppercase letters. Thus, we will need a different notation in this chapter for Fourier transforms than used in Chapter 2. Following the same development as in Section 2.5.4, it follows that

$$\mathcal{H}(C_i(\mathbf{x})[\mathbf{P}_i]) = \sum_{k=1}^{3} (H_{ki} * C_i)(\mathbf{x})[\mathbf{P}_k], \tag{8.14}$$

where the symbol $*$ denotes scalar convolution as defined in Chapter 2, and so

$$\mathcal{H}([\mathbf{C}](\mathbf{x})) = \sum_{k=1}^{3}\left(\sum_{i=1}^{3}(H_{ki} * C_i)(\mathbf{x})\right)[\mathbf{P}_k]. \tag{8.15}$$

Thus, if $[\mathbf{Q}](\mathbf{x}) = \mathcal{H}([\mathbf{C}](\mathbf{x}))$, then the components of $[\mathbf{Q}](\mathbf{x})$ in the basis $\mathcal{B}$ are

$$Q_k(\mathbf{x}) = \sum_{i=1}^{3}(H_{ki} * C_i)(\mathbf{x}). \tag{8.16}$$

We see that given a basis, a general linear shift-invariant system for color signals is characterized by *nine* scalar-valued impulse response functions, which we express in matrix form as

$$\mathbf{H}_{\mathcal{B}}(\mathbf{x}) = \begin{bmatrix} H_{11}(\mathbf{x}) & H_{12}(\mathbf{x}) & H_{13}(\mathbf{x}) \\ H_{21}(\mathbf{x}) & H_{22}(\mathbf{x}) & H_{23}(\mathbf{x}) \\ H_{31}(\mathbf{x}) & H_{32}(\mathbf{x}) & H_{33}(\mathbf{x}) \end{bmatrix}. \tag{8.17}$$

We denote the response of the system in matrix-vector notation by

$$\mathbf{Q}_{\mathcal{B}}(\mathbf{x}) = (\mathbf{H}_{\mathcal{B}} \circledast \mathbf{C}_{\mathcal{B}})(\mathbf{x}). \tag{8.18}$$

This is simply a short-form notation to express the components of Equation (8.16). In most practical applications of color image filtering, three components with respect to some basis are filtered independently. This corresponds to a diagonal impulse response matrix in the given basis.

In some analyses, several different bases may be involved and we wish to represent the system with respect to any of these bases. In this case, we will refer to the component impulse responses as $H_{\mathcal{B},ki}$ to explicitly identify the basis.

Theorem 8.1 Suppose that $\mathcal{B}$ and $\widetilde{\mathcal{B}}$ are two bases for $\mathcal{C}$, related by the change of basis matrix $\mathbf{A}_{\mathcal{B}\to\widetilde{\mathcal{B}}}$, so that $\mathbf{C}_{\widetilde{\mathcal{B}}} = \mathbf{A}_{\mathcal{B}\to\widetilde{\mathcal{B}}}\mathbf{C}_{\mathcal{B}}$. Then

$$\mathbf{H}_{\widetilde{\mathcal{B}}}(\mathbf{x}) = \mathbf{A}_{\mathcal{B}\to\widetilde{\mathcal{B}}}\mathbf{H}_{\mathcal{B}}(\mathbf{x})\mathbf{A}^{-1}_{\mathcal{B}\to\widetilde{\mathcal{B}}}. \tag{8.19}$$

Proof: If $\mathbf{Q}_{\mathcal{B}}(\mathbf{x}) = (\mathbf{H}_{\mathcal{B}} \circledast \mathbf{C}_{\mathcal{B}})(\mathbf{x})$, then

$$Q_{\mathcal{B},k}(\mathbf{x}) = \sum_{i=1}^{3}(H_{\mathcal{B},ki} * C_{\mathcal{B},i})(\mathbf{x}). \tag{8.20}$$

In terms of the basis $\widetilde{\mathcal{B}}$,

$$Q_{\widetilde{\mathcal{B}},\ell}(\mathbf{x}) = \sum_{k=1}^{3}\mathbf{A}_{\mathcal{B}\to\widetilde{\mathcal{B}},lk}Q_{\mathcal{B},k}(\mathbf{x}), \tag{8.21}$$

while

$$C_{\mathcal{B},i}(\mathbf{x}) = \sum_{m=1}^{3}\mathbf{A}_{\widetilde{\mathcal{B}}\to\mathcal{B},im}C_{\widetilde{\mathcal{B}},m}(\mathbf{x}). \tag{8.22}$$

Thus

$$Q_{\widetilde{\mathcal{B}},\ell}(\mathbf{x}) = \sum_{k=1}^{3}\mathbf{A}_{\mathcal{B}\to\widetilde{\mathcal{B}},lk}\sum_{i=1}^{3}\left(H_{\mathcal{B},ki} * \sum_{m=1}^{3}\mathbf{A}_{\widetilde{\mathcal{B}}\to\mathcal{B},im}C_{\widetilde{\mathcal{B}},m}\right)(\mathbf{x}). \tag{8.23}$$

Rearranging, and using the linearity of scalar convolution,

$$Q_{\widetilde{B},\ell}(\mathbf{x}) = \sum_{m=1}^{3}\sum_{i=1}^{3}\sum_{k=1}^{3}(\mathbf{A}_{B\to\widetilde{B},lk}H_{B,ki}\mathbf{A}_{\widetilde{B}\to B,im} * C_{\widetilde{B},m})(\mathbf{x}), \tag{8.24}$$

from which we identify

$$H_{\widetilde{B},\ell m}(\mathbf{x}) = \sum_{i=1}^{3}\sum_{k=1}^{3}\mathbf{A}_{B\to\widetilde{B},lk}H_{B,ki}(\mathbf{x})\mathbf{A}_{\widetilde{B}\to B,im}. \tag{8.25}$$

Thus

$$\mathbf{H}_{\widetilde{B}}(\mathbf{x}) = \mathbf{A}_{B\to\widetilde{B}}\mathbf{H}_{B}(\mathbf{x})\mathbf{A}_{\widetilde{B}\to B} = \mathbf{A}_{B\to\widetilde{B}}\mathbf{H}_{B}(\mathbf{x})\mathbf{A}^{-1}_{B\to\widetilde{B}}. \tag{8.26}$$

□

In particular, if an impulse response matrix is diagonal with respect to one basis for color space, it will not in general be diagonal with respect to a different basis.

8.2.3 Frequency Response and Fourier Transform

A sinusoidal color signal of frequency $\mathbf{u}$ is defined as

$$[\mathbf{C}](\mathbf{x}) = [\mathbf{A}]\cos(2\pi\mathbf{u}\cdot\mathbf{x} + \phi). \tag{8.27}$$

As in scalar signals and systems, it is very convenient to introduce complex exponential sinusoidal signals. In order to do this, we must introduce the *complexification* of the real vector space C [Shaw (1982), Wonham (1985)]. Specifically, we define $C_{\mathbb{C}}$ as the set of formal sums

$$C_{\mathbb{C}} = \{[\mathbf{C}_1] + j[\mathbf{C}_2] \mid [\mathbf{C}_1], [\mathbf{C}_2] \in C\}, \tag{8.28}$$

where j is the imaginary unit. Vector addition and scalar multiplication with elements of $\mathbb{C}$ are done in the obvious fashion. Note that $\dim(C_{\mathbb{C}}) = \dim(C)$ and any basis for C is also a basis for $C_{\mathbb{C}}$. We can now define the complex exponential sinusoidal signals:

$$\begin{aligned}[\mathbf{C}](\mathbf{x}) &= [\mathbf{A}]\exp(j2\pi\mathbf{u}\cdot\mathbf{x}) \\ &= \sum_{i=1}^{3} A_i \exp(\mathrm{j}2\pi\mathbf{u}\cdot\mathbf{x})[\mathbf{P}_i]\end{aligned} \tag{8.29}$$

expressed with respect to the basis B, where $[\mathbf{A}]$ is an arbitrary element of $C_{\mathbb{C}}$. Real sinusoidal color signals can be constructed as linear combinations of complex exponential sinusoidal signals. Figure 8.2 shows the real sinusoidal signal

$$[\mathbf{C}](\mathbf{x}) = [\mathbf{A}_1] + [\mathbf{A}_2]\cos(2\pi\mathbf{u}\cdot\mathbf{x}) \tag{8.30}$$

where $\mathbf{A}_{1,sRGB} = \begin{bmatrix}0.5 & 0.35 & 0.5\end{bmatrix}^T$, $\mathbf{A}_{2,sRGB} = \begin{bmatrix}-0.5 & 0.35 & -0.5\end{bmatrix}^T$ and $\mathbf{u} = (-20, 5)$ c/ph. This can be written as the sum of three complex exponentials,

$$[\mathbf{C}](\mathbf{x}) = [\mathbf{A}_1]\exp(\mathrm{j}2\pi\mathbf{0}\cdot\mathbf{x}) + 0.5[\mathbf{A}_2]\exp(\mathrm{j}2\pi\mathbf{u}\cdot\mathbf{x}) + 0.5[\mathbf{A}_2]\exp(\mathrm{j}2\pi(-\mathbf{u})\cdot\mathbf{x}). \tag{8.31}$$

Figure 8.2 Example of a real color sinusoidal signal. (*See color plate section for the color representation of this figure.*)

Note that sRGB gamma correction has been applied for the display of this sinusoidal image.

If we apply the complex sinusoidal signal $[\mathbf{C}](\mathbf{x})$ of Equation (8.29) as input to an LSI system, then using Equation (8.15), we obtain as output

$$\begin{aligned}[\mathbf{Q}](\mathbf{x}) &= \sum_{k=1}^{3}\left(\sum_{i=1}^{3}\int_{\mathbb{R}^D} H_{ki}(\mathbf{s})A_i \exp(\mathrm{j}2\pi\mathbf{u}\cdot(\mathbf{x}-\mathbf{s}))\,\mathrm{d}\mathbf{s}\right)[\mathbf{P}_k] \\ &= \exp(\mathrm{j}2\pi\mathbf{u}\cdot\mathbf{x})\sum_{k=1}^{3}\left(\sum_{i=1}^{3}A_i\int_{\mathbb{R}^D} H_{ki}(\mathbf{s})\exp(-\mathrm{j}2\pi\mathbf{u}\cdot\mathbf{s})\,\mathrm{d}\mathbf{s}\right)[\mathbf{P}_k]. \end{aligned} \tag{8.32}$$

We identify

$$\widehat{H}_{ki}(\mathbf{u}) = \int_{\mathbb{R}^D} H_{ki}(\mathbf{s})\exp(-\mathrm{j}2\pi\mathbf{u}\cdot\mathbf{s})\,\mathrm{d}\mathbf{s} \tag{8.33}$$

as the standard multidimensional Fourier transform of the scalar signal $H_{ki}(\mathbf{s})$. Thus

$$[\mathbf{Q}](\mathbf{x}) = \left(\sum_{k=1}^{3}\left(\sum_{i=1}^{3}A_i\widehat{H}_{ki}(\mathbf{u})\right)[\mathbf{P}_k]\right)\exp(\mathrm{j}2\pi\mathbf{u}\cdot\mathbf{x}). \tag{8.34}$$

Since we use uppercase letters to denote tristimulus values, we adopt here the alternate standard notation for Fourier transforms using the circumflex (hat). Using the matrix notation

$$\begin{aligned}\mathbf{Q}_{\mathcal{B}}(\mathbf{x}) &= \begin{bmatrix}\widehat{H}_{11}(\mathbf{u}) & \widehat{H}_{12}(\mathbf{u}) & \widehat{H}_{13}(\mathbf{u})\\ \widehat{H}_{21}(\mathbf{u}) & \widehat{H}_{22}(\mathbf{u}) & \widehat{H}_{23}(\mathbf{u})\\ \widehat{H}_{31}(\mathbf{u}) & \widehat{H}_{32}(\mathbf{u}) & \widehat{H}_{33}(\mathbf{u})\end{bmatrix}\mathbf{A}_{\mathcal{B}}\exp(\mathrm{j}2\pi\mathbf{u}\cdot\mathbf{x}) \\ &= \widehat{\mathbf{H}}_{\mathcal{B}}(\mathbf{u})\mathbf{A}_{\mathcal{B}}\exp(\mathrm{j}2\pi\mathbf{u}\cdot\mathbf{x}), \end{aligned} \tag{8.35}$$

which defines the matrix $\hat{\mathbf{H}}_{\mathcal{B}}(\mathbf{u})$, following our convention, $\mathbf{A}_{\mathcal{B}}$ is the column matrix of tristimulus values of $[\mathbf{A}]$ with respect to the basis $\mathcal{B}$.

Referring back to the general case of Equation (8.16), and applying the standard multidimensional convolution theorem,

$$\hat{Q}_k(\mathbf{u}) = \sum_{i=1}^{3} \hat{H}_{ki}(\mathbf{u})\hat{C}_i(\mathbf{u}). \tag{8.36}$$

In matrix form we can write

$$\hat{\mathbf{Q}}_{\mathcal{B}}(\mathbf{u}) = \hat{\mathbf{H}}_{\mathcal{B}}(\mathbf{u})\hat{\mathbf{C}}_{\mathcal{B}}(\mathbf{u}) \tag{8.37}$$

where

$$\hat{\mathbf{Q}}_{\mathcal{B}}(\mathbf{u}) = \begin{bmatrix} \hat{Q}_1(\mathbf{u}) \\ \hat{Q}_2(\mathbf{u}) \\ \hat{Q}_3(\mathbf{u}) \end{bmatrix} \qquad \hat{\mathbf{C}}_{\mathcal{B}}(\mathbf{u}) = \begin{bmatrix} \hat{C}_1(\mathbf{u}) \\ \hat{C}_2(\mathbf{u}) \\ \hat{C}_3(\mathbf{u}) \end{bmatrix} \tag{8.38}$$

and $\hat{C}_i(\mathbf{u})$ is the multidimensional Fourier transform of $C_i(\mathbf{x})$,

$$\hat{C}_i(\mathbf{u}) = \int_{\mathbb{R}^D} C_i(\mathbf{x}) \exp(-\mathrm{j}2\pi\mathbf{u}\cdot\mathbf{x})\, \mathrm{d}\mathbf{x} \tag{8.39}$$

and similarly for $\hat{Q}_i(\mathbf{u})$. We can consider $\hat{\mathbf{C}}_{\mathcal{B}}(\mathbf{u})$ to be the *basis-dependent* form of the Fourier transform of the color signal $\mathbf{C}_{\mathcal{B}}(\mathbf{x})$. We could formally express the *basis-independent* version as

$$[\hat{\mathbf{C}}](\mathbf{u}) = \int_{\mathbb{R}^D} [\mathbf{C}](\mathbf{x}) \exp(-\mathrm{j}2\pi\mathbf{u}\cdot\mathbf{x})\, \mathrm{d}\mathbf{x} \tag{8.40}$$

where we can interpret $[\hat{\mathbf{C}}](\mathbf{u})$ as a mapping from the frequency domain $\mathbb{R}^D$ to $\mathcal{C}_{\mathbb{C}}$. In practice, we would introduce a basis to actually compute Equation (8.40).

As already mentioned, if we change the basis of the color space, say from $\mathcal{B}$ to $\tilde{\mathcal{B}}$, then both the space-domain and the frequency-domain representations are transformed using the appropriate change-of-basis matrix:

$$\begin{aligned} \mathbf{C}_{\tilde{\mathcal{B}}}(\mathbf{x}) &= \mathbf{A}_{\mathcal{B}\to\tilde{\mathcal{B}}}\mathbf{C}_{\mathcal{B}}(\mathbf{x}), \\ \hat{\mathbf{C}}_{\tilde{\mathcal{B}}}(\mathbf{u}) &= \mathbf{A}_{\mathcal{B}\to\tilde{\mathcal{B}}}\hat{\mathbf{C}}_{\mathcal{B}}(\mathbf{u}). \end{aligned} \tag{8.41}$$

If we apply this to Equation (8.37), we find

$$\hat{\mathbf{Q}}_{\tilde{\mathcal{B}}}(\mathbf{u}) = \mathbf{A}_{\mathcal{B}\to\tilde{\mathcal{B}}}\hat{\mathbf{H}}_{\mathcal{B}}(\mathbf{u})\mathbf{A}^{-1}_{\mathcal{B}\to\tilde{\mathcal{B}}}\hat{\mathbf{C}}_{\tilde{\mathcal{B}}}(\mathbf{u}) \tag{8.42}$$

from which we conclude that

$$\hat{\mathbf{H}}_{\tilde{\mathcal{B}}}(\mathbf{u}) = \mathbf{A}_{\mathcal{B}\to\tilde{\mathcal{B}}}\hat{\mathbf{H}}_{\mathcal{B}}(\mathbf{u})\mathbf{A}^{-1}_{\mathcal{B}\to\tilde{\mathcal{B}}}. \tag{8.43}$$

This corresponds to Equation (8.19) in the frequency domain, and requires a much simpler proof than the one given in the proof of Theorem 8.1. We conclude that all transfer matrices for a given LSI color system with respect to different bases are equivalent and related by a similarity transformation as in Equation (8.43).

Poirson and Wandell [Poirson and Wandell (1993)], [Poirson and Wandell (1996)] found bases (for specific observers) such that the first stages of the human visual system can be approximated by a linear system with a diagonal transfer matrix in the given bases. Based on that work, Zhang and Wandell defined the spatial CIELAB (S-CIELAB) model [Zhang and Wandell (1997)]. This is intended to model the initial stages of the human visual system and was specifically developed in the context of an error criterion for color images that takes into account the spatial frequency response of human vision. We present it here as a good example of a continuous-space linear system for color images. In a specific basis, referred to as an *opponent-color* basis, the transfer matrix of the linear system is diagonal. The linear system is followed by the conversion to nonlinear CIELAB coordinates; we will introduce this aspect later. Assume that an image is represented in the CIE 1931 color space. Let us refer to the opponent-color basis defined by Zhang and Wandell as $\mathcal{OZW} = \{[\mathbf{O}_1], [\mathbf{O}_2], [\mathbf{O}_3]\}$. This basis is specified by the transformation matrix

$$\mathbf{A}_{\mathcal{XYZ}\rightarrow\mathcal{OZW}} = \begin{bmatrix} 0.279 & 0.720 & -0.107 \\ -0.449 & 0.290 & -0.077 \\ 0.086 & -0.590 & 0.501 \end{bmatrix}. \tag{8.44}$$

In this basis, the transfer matrix $\widehat{\mathbf{H}}_{\mathcal{OZW}}(\mathbf{u})$ is diagonal. In the S-CIELAB model, each term is circularly symmetric and formed as a weighted sum of Gaussian filters, where the parameters of the filters were obtained empirically to best fit the Poirson and Wandell data. Reproducing the model of Zhang and Wandell (1997) in the present notation,

$$H_{ll}(x, y) = \sum_{i=1}^{I_l} w_{il} k_{il} \exp(-(x^2 + y^2)/\sigma_{il}^2), \quad l = 1, 2, 3. \tag{8.45}$$

The scale factors k_{il} are chosen so that the corresponding Gaussian filters have a DC gain (response at $(u, v) = (0, 0)$) of 1.0. The w_{il} are scaled such that the overall filter has a DC gain of unity, $\widehat{H}_{ll}(0, 0) = 1.0$.

Using the fact that the two-dimensional Fourier transform of the Gaussian impulse response is given by

$$\exp(-(x^2 + y^2)/\sigma_{il}^2) \overset{\text{CDFT}}{\longleftrightarrow} \pi\sigma_{il}^2 \exp(-\pi^2(u^2 + v^2)\sigma_{il}^2), \tag{8.46}$$

we see that the DC gain is 1.0 if $k_{il} = 1/(\pi\sigma_{il}^2)$. Thus, in the frequency domain, the filter responses are

$$\widehat{H}_{ll}(u, v) = \sum_{i=1}^{I_l} w_{il} \exp(-\pi^2(u^2 + v^2)\sigma_{il}^2), \tag{8.47}$$

and of course $\widehat{H}_{lm}(u, v) = 0$ for $l \neq m$. In any other basis, the transfer matrix is not diagonal, and can be obtained using Equation (8.43).

The empirical parameters from Zhang and Wandell (1997) are shown in Table 8.1, along with the above derived k_{il}. The weights are different than in Zhang and Wandell (1997) as they have been scaled to sum to 1.0, as is done in the widely available software that accompanies Zhang and Wandell (1997) (see Zhang (1998)). Note that $I_1 = 3$, $I_2 = I_3 = 2$, and the unit of distance for x, y and σ_{il} is one degree of visual angle. The viewing distance is required to convert this to distance on the screen in a measure such as pixels or picture height. See Johnson and Fairchild (2003) for an illustration of this conversion in the context of the present model. (Note that the σ parameters given in Johnson and Fairchild (2003) are not the same as the ones in Zhang (1998).)

Table 8.1 Parameters of the filters used in the S-CIELAB model.

Component l	Term i	w_{il}	σ_{il}	k_{il}
1	1	1.0033	0.0283	397.4
	2	0.1144	0.1330	17.99
	3	−0.1176	4.336	0.0169
2	1	0.6167	0.0392	207.1
	2	0.3833	0.4940	1.304
3	1	0.5681	0.0536	110.8
	2	0.4319	0.3860	2.136

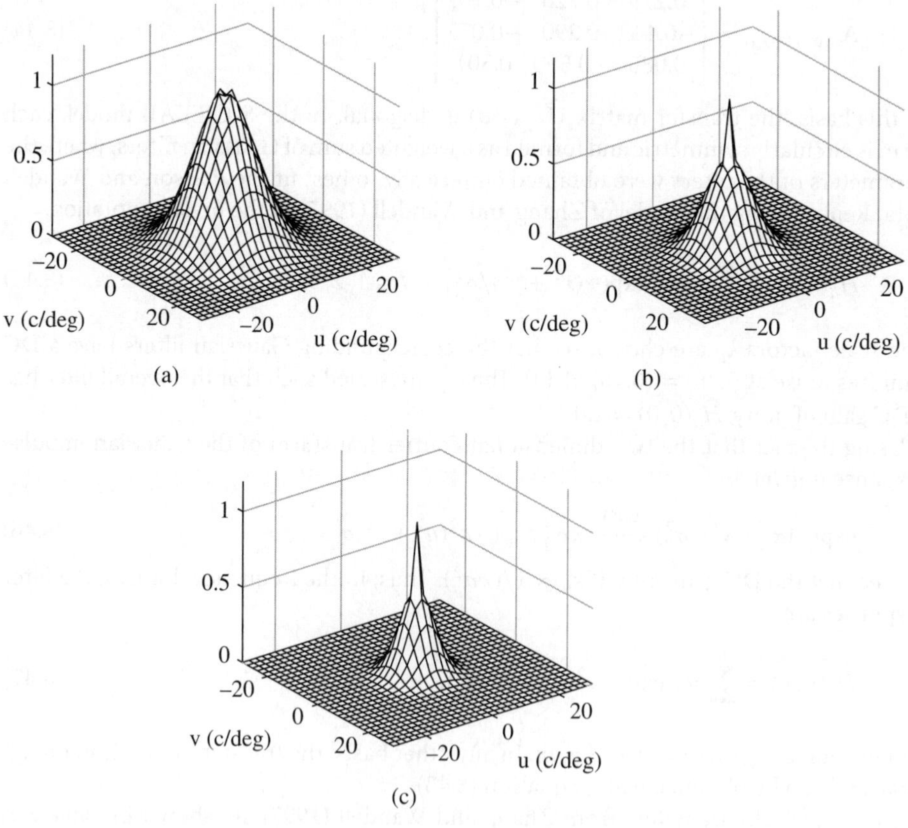

Figure 8.3 Perspective view of the frequency response of the three continuous-domain filters of Zhang and Wandell in the range from 0 to 30 c/deg. (a) $\widehat{H}_{11}(u.v)$. (b) $\widehat{H}_{22}(u.v)$. (c) $\widehat{H}_{33}(u.v)$.

Figure 8.3 shows a perspective view of the frequency response of the three filters described above, plotted in the range from 0 to 30 cycles per degree (c/deg). We observe that these filters have a circularly symmetric freguency response, with successively smaller bandwidths. The filter $\widehat{H}_{11}(u, v)$ has a peak at a radial frequency of about 1 c/deg.

8.3 Discrete-Domain Color Images

Although real-world physical color images such as the images on the surface of a display or on a sensor are continuous-domain signals, they must be converted to a discrete-domain (sampled) representation for digital processing, storage and transmission. We first discuss the discrete-domain representation and processing of color images, and then address the sampling and reconstruction of continuous-domain signals. This is a straightforward extension of the material in Chapter 3 and Chapter 6.

Discrete-domain images are defined with the use of lattices. We consider first the case where the color signal is defined at all points of a lattice $\Lambda \subset \mathbb{R}^D$. The thornier (but very widely used) situation when different components with respect to some color coordinate system are defined on different sampling structures will follow.

8.3.1 Color Signals With All Components on a Single Lattice

We consider the case here where the complete color vector is given at each point of a lattice Λ. Again, a common notation can be used for both still and time-varying imagery using a vector notation for the independent variables. A color image is denoted $[\mathbf{C}][\mathbf{x}]$, $\mathbf{x} \in \Lambda$. Note that square brackets enclosing an uppercase bold symbol generally indicates an element of the color space, while square brackets around a lowercase bold symbol indicates the independent variable in a discrete-domain signal. While not ideal, this should be clear from the context. In most cases (but certainly not all), the lattice Λ is the square lattice, and using the pixel spacing, 1 px, as the unit of distance, then $\Lambda = \mathbb{Z}^2$.

The theory and development are very similar to that for continuous-domain images. We again assume that the collection of color images of interest belongs to a vector space that is referred to as a signal space, e.g.,

$$\mathcal{V}_d = \{[\mathbf{C}][\mathbf{x}] \mid \mathbf{x} \in \Lambda;\ |C_i[\mathbf{x}]| < b,\ i = 1,2,3;\ b < \infty\}, \tag{8.48}$$

where $C_i[\mathbf{x}]$ are tristimulus values with respect to any basis for $\mathcal{C}$. A system is any operator $\mathcal{H} : \mathcal{V}_d \to \mathcal{V}_d$, and the definitions of linear and memoryless systems are unchanged from Equation (8.7) and Equation (8.8). Similarly, the translation system has the same definition as in Equation (8.9) with the proviso that $\mathbf{d} \in \Lambda$, and shift invariance is given by Equation (8.10) with $\mathbf{d} \in \Lambda$.

The characterization of a linear shift-invariant system follows the same path, but is in fact simpler since we do not need Dirac deltas. As before, we define the *unit sample* function $\delta_\Lambda[\mathbf{x}]$ on a lattice Λ by

$$\delta_\Lambda[\mathbf{x}] = \begin{cases} 1 & \mathbf{x} = \mathbf{0} \\ 0 & \mathbf{x} \in \Lambda\backslash\mathbf{0}, \end{cases} \tag{8.49}$$

where $\Lambda\backslash\mathbf{0}$ denotes all points of Λ except $\mathbf{0}$. An arbitrary color signal can be expressed

$$\begin{aligned}[\mathbf{C}][\mathbf{x}] &= \sum_{\mathbf{s}\in\Lambda} [\mathbf{C}][\mathbf{s}]\delta_\Lambda[\mathbf{x}-\mathbf{s}] \\ &= \sum_{i=1}^{3}\sum_{\mathbf{s}\in\Lambda} C_i[\mathbf{s}]\delta_\Lambda[\mathbf{x}-\mathbf{s}][\mathbf{P}_i],\end{aligned} \tag{8.50}$$

which is a superposition of color signals of the form $\delta_\Lambda[\mathbf{x}-\mathbf{s}][\mathbf{P}_i]$ with weights $C_i[\mathbf{s}]$. Applying linearity and shift invariance

$$[\mathbf{Q}][\mathbf{x}] = \mathcal{H}[\mathbf{C}][\mathbf{x}] = \sum_{i=1}^{3}\sum_{\mathbf{s}\in\Lambda} C_i[\mathbf{s}]\mathcal{H}(\delta_\Lambda[\mathbf{x}-\mathbf{s}][\mathbf{P}_i])$$
$$= \sum_{i=1}^{3}\sum_{\mathbf{s}\in\Lambda} C_i[\mathbf{s}]\mathcal{T}_{\mathbf{s}}\mathcal{H}(\delta_\Lambda[\mathbf{x}][\mathbf{P}_i]). \tag{8.51}$$

Here, $\mathcal{H}(\delta_\Lambda[\mathbf{x}][\mathbf{P}_i])$ is the response of the LSI system to a unit sample in the one-dimensional subspace of $\mathcal{C}$ spanned by $[\mathbf{P}_i]$, and is a general color signal (not necessarily confined to that subspace) that can be expressed in terms of the basis as

$$\mathcal{H}(\delta_\Lambda[\mathbf{x}][\mathbf{P}_i]) = \sum_{k=1}^{3} H_{ki}[\mathbf{x}][\mathbf{P}_k]. \tag{8.52}$$

With this definition of the H_{ki}, we have

$$[\mathbf{Q}][\mathbf{x}] = \sum_{i=1}^{3}\sum_{k=1}^{3}\left(\sum_{\mathbf{s}\in\Lambda} C_i[\mathbf{s}]H_{ki}[\mathbf{x}-\mathbf{s}]\right)[\mathbf{P}_k]$$
$$= \sum_{k=1}^{3}\left(\sum_{i=1}^{3}(C_i * H_{ki})[\mathbf{s}]\right)[\mathbf{P}_k]$$
$$= \sum_{k=1}^{3}\left(\sum_{i=1}^{3}(H_{ki} * C_i)[\mathbf{s}]\right)[\mathbf{P}_k] \tag{8.53}$$

which is formally the same as Equation (8.15), but now $H_{ki} * C_i$ denotes discrete-domain multidimensional convolution on the lattice Λ as defined in Chapter 3, which is commutative. So once again, a linear shift-invariant discrete domain system for color signals is fully characterized by nine scalar-valued unit sample responses. To describe a discrete-domain system for color signals in terms of coordinates, we adopt the same short-form notation as in the continuous-domain case, as given in Equation (8.18), $\mathbf{Q}_\mathcal{B}[\mathbf{x}] = (\mathbf{H}_\mathcal{B} \circledast \mathbf{C}_\mathcal{B})[\mathbf{x}]$.

A complex exponential signal on a lattice Λ again has the same definition as in the continuous domain case, as given in Equation (8.29). However, as in the scalar case, there is an important difference. In the continuous domain, each different frequency vector $\mathbf{u} \in \mathbb{R}^D$ corresponds to a different sinusoidal signal, with increasing frequency as $|\mathbf{u}|$ increases. In the discrete domain,

$$\exp(\mathrm{j}2\pi\mathbf{u}\cdot\mathbf{x}) = \exp(\mathrm{j}2\pi(\mathbf{u}+\mathbf{r})\cdot\mathbf{x}), \qquad \mathbf{r} \in \Lambda^* \tag{8.54}$$

where Λ^* is the reciprocal lattice. Thus all frequencies in the set $\{\mathbf{u}+\mathbf{r} \mid \mathbf{r} \in \Lambda^*\}$ correspond to the same sinusoidal signal. It follows that we only need to consider frequencies in one unit cell of Λ^*.

If we apply a sinusoidal signal $\sum_{i=1}^{3} A_i \exp(\mathrm{j}2\pi\mathbf{u}\cdot\mathbf{x})[\mathbf{P}_i]$ as input to a linear shift-invariant system, we obtain

$$[\mathbf{Q}][\mathbf{x}] = \sum_{k=1}^{3}\left(\sum_{i=1}^{3}\sum_{\mathbf{s}\in\Lambda} H_{ki}[\mathbf{s}]A_i \exp(\mathrm{j}2\pi\mathbf{u}\cdot(\mathbf{x}-\mathbf{s}))\right)[\mathbf{P}_k]$$
$$= \exp(\mathrm{j}2\pi\mathbf{u}\cdot\mathbf{x})\sum_{k=1}^{3}\left(\sum_{i=1}^{3} A_i \sum_{\mathbf{s}\in\Lambda} H_{ki}[\mathbf{s}]\exp(-\mathrm{j}2\pi\mathbf{u}\cdot\mathbf{s})\right)[\mathbf{P}_k]. \tag{8.55}$$

We identify

$$\widehat{H}_{ki}(\mathbf{u}) = \sum_{\mathbf{s}\in\Lambda} H_{ki}[\mathbf{s}] \exp(-j2\pi\mathbf{u}\cdot\mathbf{s}) \tag{8.56}$$

as the discrete-domain Fourier transform of H_{ki} on the lattice Λ. Note that $\widehat{H}_{ki}(\mathbf{u}) = \widehat{H}_{ki}(\mathbf{u}+\mathbf{r})$ for all $\mathbf{r} \in \Lambda^*$, so that $\widehat{H}_{ki}(\mathbf{u})$ is periodic. We need only specify it on one period, which is a unit cell of the reciprocal lattice Λ^*. Thus

$$[\mathbf{Q}][\mathbf{x}] = \left(\sum_{k=1}^{3}\sum_{i=1}^{3} A_i \widehat{H}_{ki}(\mathbf{u})[\mathbf{P}_k]\right) \exp(j2\pi\mathbf{u}\cdot\mathbf{x}) \tag{8.57}$$

and formally we can use Equation (8.35).

These developments lead us to define the Fourier transform for a general color signal defined on a lattice Λ as

$$[\widehat{\mathbf{C}}](\mathbf{u}) = \sum_{\mathbf{x}\in\Lambda} [\mathbf{C}][\mathbf{x}] \exp(-j2\pi\mathbf{u}\cdot\mathbf{x}) \tag{8.58}$$

which can be implemented using any desired basis. All the relations for change of basis on the Fourier transform and the transfer matrix given in Equation (8.41)–(8.43) apply unchanged.

A system for linear processing of color images is specified by the choice of the nine unit-sample responses with respect to some basis. These may be determined by some optimization process, or chosen heuristically. In almost all cases in practice, three filters are specified and applied independently to three components in some linear or nonlinear coordinate system. Thus, all methods discussed in previous chapters can be applied to the three components to generate a color image processing system.

8.3.1.1 Sampling a Continuous-Domain Color Signal Using a Single Lattice

Let $[\mathbf{C}_c](\mathbf{x})$ be a continuous-domain color signal. This can be converted to a discrete-domain color signal by sampling on a lattice Λ: $[\mathbf{C}_d][\mathbf{x}] = [\mathbf{C}_c](\mathbf{x})$, $\mathbf{x} \in \Lambda$. This is simply achieved by sampling the three components with respect to a given basis, e.g.,

$$C_{d,i}[\mathbf{x}] = C_{c,i}(\mathbf{x}), \qquad \mathbf{x}\in\Lambda, \quad i = 1,2,3. \tag{8.59}$$

For example, an RGB camera (not using a color filter array) would sample R, G, and B components at each point of the lattice Λ. Thus, all the analysis of Chapter 6 would be applied separately to each component and thus to the vector color signal. We can conclude that if the continuous-domain Fourier transform of $[\mathbf{C}_c](\mathbf{x})$ is $[\widehat{\mathbf{C}}_c](\mathbf{u})$, then the discrete-domain Fourier transform of the signal sampled on Λ is

$$[\widehat{\mathbf{C}}_d](\mathbf{u}) = \frac{1}{d(\Lambda)} \sum_{\mathbf{r}\in\Lambda^*} [\widehat{\mathbf{C}}_c](\mathbf{u}+\mathbf{r}) \tag{8.60}$$

since the sampling formula of Equation 6.7 applies to the three component signals with respect to any basis. Thus all considerations of prefiltering, sampling and reconstruction given in Chapter 6 can be applied directly to any set of three components of a color signal with respect to the corresponding basis.

8.3.1.2 S-CIELAB Error Criterion

It is a difficult and unsolved problem to numerically measure the magnitude of the difference between two color images in a perceptually consistent fashion, and this is the subject of ongoing research at the time of writing. One error criterion specifically designed for still color images that accounts for the spatial frequency response of human vision as well as the perceptual nonuniformity of color space is the S-CIELAB criterion. This is often presented along with the root mean squared error (or CPSNR) on gamma-corrected sRGB coordinates, which does not account for the spatial frequency response of human vision and only roughly accounts for perceptual nonuniformity.

Assume that we wish to measure the perceptual difference between a discrete-domain reference image $[\mathbf{C}_r][\mathbf{x}]$ and a degraded or processed image $[\mathbf{C}_p][\mathbf{x}]$. To apply the S-CIELAB error criterion, the images $[\mathbf{C}_r][\mathbf{x}]$ and $[\mathbf{C}_p][\mathbf{x}]$ are represented in the linear OZW coordinates described in Section 8.2.3. How precisely this is done depends on the coordinates initially used to represent these images (often gamma-corrected sRGB) as well as the color space used (the original authors considered several besides CIE1931). If the original images are represented in a coordinate system 'rep', then this change of coordinates is implemented by the function $\xi_{\text{rep}\to\mathcal{OZW}}$. These images are then filtered with a discrete-domain version of the system described in Section 8.2.3; the transfer matrix is denoted $\mathbf{H}_{ZW,\mathcal{OZW}}[\mathbf{x}]$ and is diagonal; the subscript portion ZW refers to Zhang and Wandell filters. To convert that continuous-domain system to the discrete domain, we must express one degree of visual angle in pixels (px) and then sample the impulse responses of the three channels, although some adjustments are needed. The final step is to convert the coordinates to CIELAB and compute a root mean-square error over a set of pixels $\mathcal{L}$ in the image window. The set $\mathcal{L}$ may be the entire image, or it may exclude a boundary region so that edge effects due to the spatial filtering do not influence the error measure. These steps can all be summarized in a single expression for the S-CIELAB error Δ_S,

$$\Delta_S([\mathbf{C}_r](\mathbf{x}),[\mathbf{C}_p](\mathbf{x})) = \left(\frac{1}{|\mathcal{L}|} \sum_{\mathbf{x}\in\mathcal{L}} \|\xi_{\mathcal{OZW}\to\text{LAB}}(\mathbf{H}_{ZW,\mathcal{OZW}} \circledast (\xi_{\text{rep}\to\mathcal{OZW}}(\mathbf{C}_{r,\text{rep}})))(\mathbf{x}) - \xi_{\mathcal{OZW}\to\text{LAB}}(\mathbf{H}_{ZW,\mathcal{OZW}} \circledast (\xi_{\text{rep}\to\mathcal{OZW}}(\mathbf{C}_{p,\text{rep}})))(\mathbf{x})\|_2^2 \right)^{0.5}, \tag{8.61}$$

where $\|\cdot\|_2^2$ denotes a squared Euclidean norm in CIELAB coordinates. Rather than simply taking the average of the Δ_{ab}^{*2} error between the filtered images, more sophisticated analyses of the error image can be carried out to take other phenomena into account such as spatial masking (see Johnson and Fairchild (2003) and Zhang and Wandell (1998) for examples).

We present a detailed example of computation of the S-CIELAB error, which simultaneously illustrates several aspects of a discrete-domain color image processing system. To convert the continuous-domain filters of Equation (8.45) to discrete-domain filters, we must determine the number of px per degree of visual angle. If the viewing distance is h px, then from simple trigonometry, the visual angle θ_{px} subtended by one pixel is given by

$$\begin{aligned}\theta_{\text{px}} &= 2\tan^{-1}(1/2h) \quad \text{rad} \\ &= (360/\pi)\tan^{-1}(1/2h) \quad \text{deg}.\end{aligned} \tag{8.62}$$

For the author's current viewing setup, 1 ph = 1050 px and $h = 2.5$ ph = 2625 px. Thus, $\theta_{\text{px}} = 0.0218$ deg, and so the number of pixels per degree subtended at the eye is PPD = $1/\theta_{\text{px}} = 45.8$ px/deg. This parameter would be an input to any S-CIELAB calculator. If we maintain degrees subtended at the eye as the spatial unit of distance, and assume a square lattice, then the sampling lattice is $\theta_{\mathrm{px}}\mathbb{Z}^2$, and the reciprocal lattice is $\frac{1}{\theta_{\mathrm{px}}}\mathbb{Z}^2$ with spatial frequencies in cycles per degree (c/deg).

If we sample an impulse response at a high enough sampling density that the aliasing is negligible, then the frequency response of the discrete domain filter is equal to the frequency response of the continuous domain filter multiplied by $1/d(\Lambda)$ (see Equation (6.8)). Since we want to maintain a DC gain of 1.0, we should multiply the impulse response of the continuous domain filters by $d(\Lambda)$ when sampling them. We also need to specify a finite spatial extent of these filters. We take the usual approach of taking an odd integer just above one degree of visual angle, in this case 47 px, so that the region of support of the three filters is $\mathcal{A} = [-23, 23] \times [-23, 23]$. Since some of the component filters extend beyond this region, we introduce a gain factor of g_ℓ to maintain an overall DC gain of 1.0. The filters are circularly symmetric and are the sum of two or three separable filters, which can be used to reduce computational complexity if required. In conclusion, the three discrete-domain filters that comprise $\mathbf{H}_{ZW,\mathcal{O}ZW}(\mathbf{x})$ are given by

$$H_{\ell\ell}[m,n] = g_\ell \theta_{\mathrm{px}}^2 \sum_{i=1}^{I_\ell} w_{il} k_{il} \exp\left(-\frac{(m^2+n^2)\theta_{\mathrm{px}}^2}{\sigma_{i\ell}^2}\right), \qquad (m,n) \in \mathcal{A}. \tag{8.63}$$

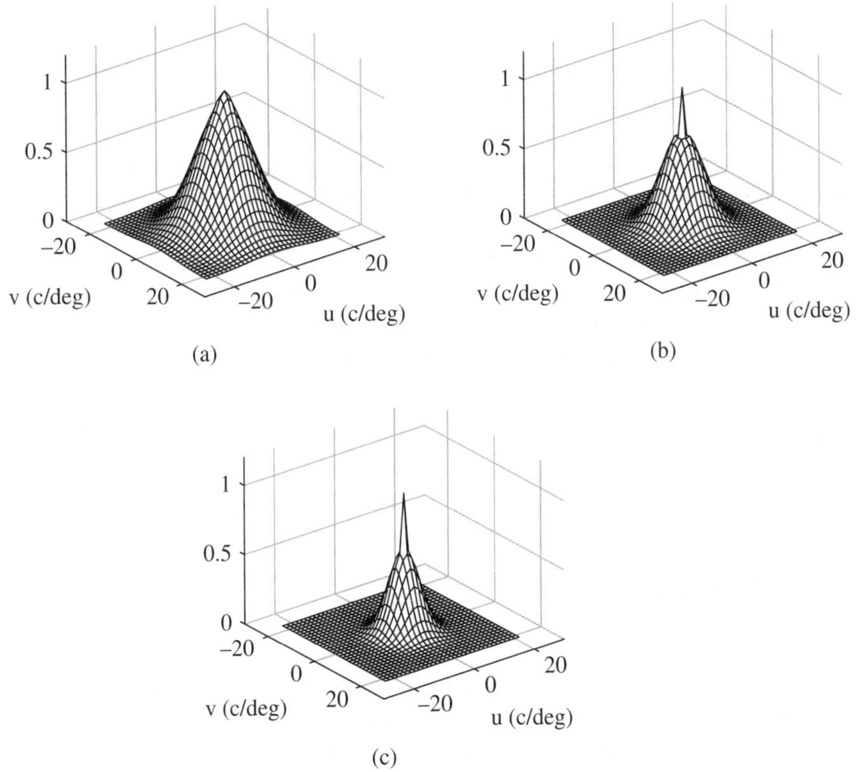

Figure 8.4 Perspective view of the frequency response of the three discrete-domain filters of Zhang and Wandell in the range from 0 to 22.9 c/deg. (a) $\widehat{H}_{11}(u,v)$. (b) $\widehat{H}_{22}(u,v)$. (c) $\widehat{H}_{33}(u,v)$.

Figure 8.4 shows the 2D frequency response of these discrete-domain filters over one unit cell of Λ^*, $[-22.9\text{c/deg}, 22.9\text{c/deg}]^2$. We see that these reponses look very similar to the responses of the continuous-domain filters in Figure 8.3. The filter for channel 1 is missing the small dip at DC and has not completely decayed to zero at the boundary of the unit cell of Λ^*. We also see a small difference in the response of H_{22} at the transition between the two Gaussians. These differences in the shape can be reduced by increasing the extent of the filters, but this would have little effect on the S-CIELAB computation, so we retain the usual extent of one degree of visual angle as used in the literature. To correct the effect that the response of channel 1 has not fully decayed to zero at the boundary of the unit cell of Λ^*, we can increase the sampling density by upsampling the input image, but again this would not have much impact on the S-CIELAB computation. With these filters specifying $\mathbf{H}_{ZW,OZW}$, computation of the S-CIELAB error is a straightforward implementation of Equation (8.61). We illustrate the computation of the S-CIELAB error and comparison with the conventional color root mean square error (CRMSE) and corresponding color peak signal-to-noise ratio (CPSNR) with an example.

Example 8.1
Compute the S-CIELAB error and compare with the CRMSE and CPSNR for two noisy images with different noise characteristics such as those shown in Figure 8.5 (a) and (b).

Solution:
The two images in Figure 8.5 have noise with different characteristics added. The original sRGB image is the motorcycles image of Figure 8.1(a). It is converted to the NTSC YIQ luma-chroma representation. Gaussian noise is added to each of the three components. Specifically, in Figure 8.5(a), low-pass noise obtained by filtering white Gaussian noise with a Gaussian low-pass filter h_{LP} with parameter $r_0 = 2$ px is added to each component. The noise added to the luma component has greater variance than the noise added to the chroma components. In Figure 8.5(b), high-pass noise obtained by filtering white Gaussian noise with a filter having unit-sample response $\delta[\mathbf{x}] - 0.5h_{\mathrm{LP}}[\mathbf{x}]$ is added to the components. In this case the noise added to the luma component has lower variance than the noise added to the chroma components. The noisy YIQ images are then converted back to sRGB format as shown in Figure 8.5. Let us refer to these two gamma-corrected sRGB images as $\mathbf{C}_1[\mathbf{x}]$ and $\mathbf{C}_2[\mathbf{x}]$ and the original image as $\mathbf{C}[\mathbf{x}]$.

The CRMSE is computed on the gamma-corrected sRGB images as follows:

$$\text{CRMSE} = \left(\frac{1}{|\mathcal{L}|} \sum_{\mathbf{x}\in\mathcal{L}} \|\mathbf{C}[\mathbf{x}] - \mathbf{C}_i[\mathbf{x}]\|^2 \right)^{0.5}, \tag{8.64}$$

and the CPSNR is obtained by

$$\text{CPSNR} = 20 * \log_{10}(1/\text{CMSE}). \tag{8.65}$$

Note that signal components are assumed to lie in the range [0,1]. The S-CIELAB error $\Delta_S(\mathbf{C}, \mathbf{C}_i)$ is computed using Equation (8.61). We find it more meaningful to compare these values on a logarithmic scale. We define an S-CIELAB peak signal-to-noise ratio as

$$\text{S-CIELAB-PSNR} = 20 * \log_{10}(100/\Delta_S). \tag{8.66}$$

We use 100 in the numerator since the range of the L^* component is [0,100].

(a)

(b)

Figure 8.5 Motorcycles image with different types of noise added to the NTSC YIQ components. (a) Low-pass noise, with greater noise level in the luma component. (b) High-pass noise, with greater noise level in the chroma components. (*See color plate section for the color representation of this figure.*)

For one instance of these two noisy images, the ones shown in Figure 8.5, we find the CRMSE and the CPSNR of the two images to be essentially the same, CRMSE = 0.07 and CPSNR = 22.8 dB. However, it is clear that image $\mathbf{C}_2[\mathbf{x}]$ is far superior to image $\mathbf{C}_1[\mathbf{x}]$, showing these measures to be totally inadequate in this circumstance. On the other hand, the S-CIELAB error measures are very different: $\Delta_S(\mathbf{C}, \mathbf{C}_1) = 6.79$, S-CIELAB-PSNR$_1$ = 23.36 dB for image 1 and $\Delta_S(\mathbf{C}, \mathbf{C}_2) = 2.47$, S-CIELAB-PSNR$_2$ = 32.13 dB for image 2. Thus, the S-CIELAB measures show a large difference, in keeping with our visual assessment. Note that CPSNR and S-CIELAB-PSNR values cannot be meaningfully compared with each other. □

8.3.2 Color Signals With Different Components on Different Sampling Structures

It is common practice to sample color signals with different components (according to some color coordinate system) defined on different sampling structures; in fact it is almost universally applied at some point in the imaging chain. The most notable cases are: image acquisition using color filter arrays (CFAs), such as the Bayer array where red, green and blue components are sampled on different structures; image displays using color pixel arrays such as LCD RGB displays; luma-chroma representations, where chroma is downsampled with respect to luma. Although the latter application usually applies to nonlinear color coordinates, we will discuss the theory here mainly for linear color representations.

Suppose that a continuous-domain signal is defined with respect to a specific basis $\mathcal{B} = \{[\mathbf{P}_1], [\mathbf{P}_2], [\mathbf{P}_3]\}$,

$$[\mathbf{C}_c](\mathbf{x}) = C_{c1}(\mathbf{x})[\mathbf{P}_1] + C_{c2}(\mathbf{x})[\mathbf{P}_2] + C_{c3}(\mathbf{x})[\mathbf{P}_3], \quad \mathbf{x} \in \mathbb{R}^D. \tag{8.67}$$

We assume that the three components $C_{ci}(\mathbf{x}), i = 1, 2, 3$ are represented on three sampling structures Ψ_i, which are all subsets of a common lattice Λ. Although the Ψ_i may not necessarily be lattices (if they do not contain the origin), they are generally shifted versions of a sublattice of Λ, or perhaps the union of several shifted sublattices. Thus, the sampled components are $C_{di}[\mathbf{x}], \mathbf{x} \in \Psi_i$. Each component can be treated separately using standard scalar signal processing; this is the conventional way to do things. However, it may be advantageous to take a holistic view as in previous sections. It is often problematic that not all components are available at a given sample location. However, we can define a color signal on the super-lattice Λ,

$$[\mathbf{C}_d][\mathbf{x}] = \sum_{i=1}^{3} \gamma_i C_{di}[\mathbf{x}][\mathbf{P}_i], \quad \mathbf{x} \in \Lambda, \tag{8.68}$$

where we set $C_{di}[\mathbf{x}] = 0$ for $\mathbf{x} \in \Lambda \backslash \Psi_i$. Thus, at any point of Λ, there could be anywhere from 0 to 3 nonzero components. The coefficients γ_i are required to maintain the same average ratios of the components and are inversely related to the relative sampling densities. An explicit formula will be given shortly.

The Fourier transform of $[\mathbf{C}_d][\mathbf{x}]$ is given by

$$\begin{aligned}[\hat{\mathbf{C}}_d](\mathbf{u}) &= \sum_{i=1}^{3} \gamma_i \left(\sum_{\mathbf{x}\in\Lambda} C_{di}[\mathbf{x}] \exp(-j2\pi\mathbf{u}\cdot\mathbf{x}) \right) [\mathbf{P}_i] \\ &= \sum_{i=1}^{3} \gamma_i \left(\sum_{\mathbf{x}\in\Psi_i} C_{di}[\mathbf{x}] \exp(-j2\pi\mathbf{u}\cdot\mathbf{x}) \right) [\mathbf{P}_i]. \end{aligned} \tag{8.69}$$

Although $[\hat{\mathbf{C}}_d](\mathbf{u})$ is periodic, with one period being a unit cell of Λ^*, individual components $\hat{C}_{di}(\mathbf{u})$ may have a smaller period, and thus several copies of their basic spectrum lie in one unit cell of Λ^*.

A common task and preoccupation is to relate the sampled signal C_d defined in Equation (8.68) to one in which all components are defined at *every* point of the same lattice Λ. This can be studied by adapting the methods of Dubois (2009) developed for color filter arrays to the present vector space representation. We assume that the sampling scheme is periodic and that the periodicity is given by a sublattice Γ of Λ. This is illustrated in Figure 8.6 for a Bayer color filter array sampling scheme. A color image with this sampling scheme is shown in Figure 8.7; (a) shows the full image and (b) shows a subimage expanded with pixel replication. Note that γ_i is not included in this image as this would result in overflow. The signal of Equation 8.68 is not meant to be displayed. The basic lattice Λ is a rectangular lattice with equal horizontal and vertical spacing X. To simplify notation, we take X as the unit of length called the pixel height (px), so that $X = 1$ px. Then Λ is the integer Cartesian lattice $\Lambda = \mathbb{Z}^2$. In the Bayer structure, one period of the sampling structure is replicated on points of the sublattice $\Gamma = (2\mathbb{Z})^2$. We will use the notation introduced in Chapter 4 for discrete-domain periodic signals. The number of sample points in one period is given by the index of Γ in Λ, denoted $K = \mathrm{d}(\Gamma)/\mathrm{d}(\Lambda)$, where $K = 4$ for the Bayer structure. The lattice Λ can be partitioned into a union of cosets of the sublattice Γ in Λ. For any $\mathbf{x} \in \Lambda$, $\mathbf{x} + \Gamma$ is a coset; cosets are either identical or disjoint and there are K distinct cosets. We choose one element from each coset as a *coset representative* denoted $\mathbf{b}_k, k = 0, \dots, K-1$. (Note that to maintain consistency with Chapter 4, we index the cosets from 0 to $K-1$, whereas in Dubois (2009) and Dubois (2010) they are indexed from 1 to K.) It follows that $\Lambda = \cup_{k=0}^{K-1}(\mathbf{b}_k + \Gamma)$. For the Bayer structure of Figure 8.6, a suitable set of coset representatives is $\mathbf{b}_0 = [0\ \ 0]^T, \mathbf{b}_1 = [1\ \ 0]^T, \mathbf{b}_2 = [0\ \ 1]^T, \mathbf{b}_3 = [1\ \ 1]^T$. For convenience, we arrange the coset representatives into a $2 \times K$ matrix $\mathbf{B} = [\mathbf{b}_0\ \mathbf{b}_1 \cdots \mathbf{b}_{K-1}]$.

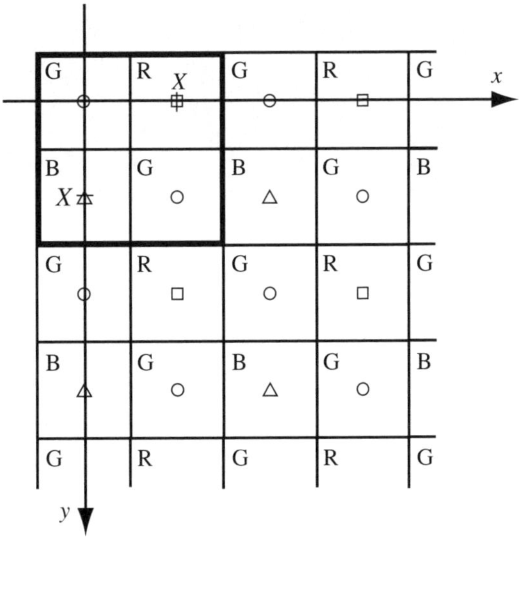

Figure 8.6 Illustration of the Bayer color filter array showing the sampling structures for the R, G, and B components.

(a)

(b)

Figure 8.7 Example of a Bayer sampled color image. (a) Full image. (b) Sub-image magnified by a factor of eight. (*See color plate section for the color representation of this figure.*)

The key concept here is that each component C_{di} is defined on a selected subset of the set of K cosets. Specifically, the sampling structure Ψ_i of component i is given by

$$\Psi_i = \bigcup_{k \in \mathcal{J}_i} (\mathbf{b}_k + \Gamma) \tag{8.70}$$

where the set $\mathcal{J}_i \subset \{0, 1, \ldots, K-1\}$ identifies which cosets form Ψ_i. For the Bayer structure, we have $\mathcal{J}_R = \{1\}$, $\mathcal{J}_G = \{0, 3\}$ and $\mathcal{J}_B = \{2\}$, giving $\Psi_R = (\mathbf{b}_1 + \Gamma)$, $\Psi_G = (\mathbf{b}_0 + \Gamma) \cup (\mathbf{b}_3 + \Gamma)$ and $\Psi_B = (\mathbf{b}_2 + \Gamma)$. Although the indexing of cosets and the choice of coset representatives from each coset is arbitrary, we choose coset representatives near the origin, and choose $\mathbf{b}_0 = 0$ by convention. The assignment of cosets to sampling structures can be summarized with the $K \times 3$ matrix $\mathbf{J}$ defined by

$$J_{li} = \begin{cases} 1 & l-1 \in \mathcal{J}_i \\ 0 & \text{otherwise.} \end{cases} \qquad l = 1, \ldots, K; i = 1, 2, 3. \tag{8.71}$$

With this formalism, we define the sampling functions

$$\tilde{m}_i[\mathbf{x}] = \begin{cases} 1 & \mathbf{x} \in \Psi_i \\ 0 & \mathbf{x} \in \Lambda \backslash \Psi_i. \end{cases} \tag{8.72}$$

This can be used to relate the Fourier transform of $[\mathbf{C}_d][\mathbf{x}]$ to the Fourier transform of a hypothetical signal $[\mathbf{C}][\mathbf{x}]$ in which all components are defined on all of Λ. Specifically, we have

$$C_{di}[\mathbf{x}] = C_i[\mathbf{x}]\tilde{m}_i[\mathbf{x}]. \tag{8.73}$$

Now each of the $\tilde{m}_i[\mathbf{x}]$ is a periodic function on Λ, with $\tilde{m}_i[\mathbf{x}+\mathbf{y}] = \tilde{m}_i[\mathbf{x}]$ for all $\mathbf{y} \in \Gamma$, and so can be expressed using the discrete-domain Fourier series, see Section 4.4. Since $\Gamma \subset \Lambda$, the inverse relation for reciprocal lattices is $\Lambda^* \subset \Gamma^*$ with $d(\Lambda^*)/d(\Gamma^*) = K$. We let $\{\mathbf{d}_0, \ldots, \mathbf{d}_{K-1}\}$ be a suitable set of coset representatives for Λ^* in Γ^*, summarized as before in the $2 \times K$ matrix $\mathbf{D} = [\mathbf{d}_0 \cdots \mathbf{d}_{K-1}]$. Again, we let $\mathbf{d}_0 = 0$. Recall that the reciprocal lattices are in the frequency domain so that the $\mathbf{d}_i$ are frequency vectors.

Continuing to use the formulation of Dubois (2009) (with modified indexing), the Fourier series representation of $\tilde{m}_i[\mathbf{x}]$ is given by

$$\begin{aligned} \tilde{m}_i[\mathbf{x}] &= \sum_{k=0}^{K-1} \widetilde{M}_i[\mathbf{d}_k] \exp(\mathrm{j}2\pi\mathbf{x}\cdot\mathbf{d}_k) \\ &= \sum_{k=0}^{K-1} M_{ki} \exp(\mathrm{j}2\pi\mathbf{x}\cdot\mathbf{d}_k), \end{aligned} \tag{8.74}$$

where

$$\begin{aligned} M_{ki} &= \frac{1}{K}\sum_{l=0}^{K-1} \tilde{m}_i(\mathbf{b}_l)\exp(-\mathrm{j}2\pi\mathbf{b}_l\cdot\mathbf{d}_k) \\ &= \frac{1}{K}\sum_{l\in\mathcal{J}_i} \exp(-\mathrm{j}2\pi\mathbf{b}_l\cdot\mathbf{d}_k). \end{aligned} \tag{8.75}$$

The previously defined matrix $\mathbf{J}$ can equivalently be defined by $J_{li} = m_i(\mathbf{b}_{l-1})$. We can then express the matrix $\mathbf{M}$ as

$$\mathbf{M} = \frac{1}{K}[\exp(-\mathrm{j}2\pi\mathbf{D}^T\mathbf{B})]\mathbf{J} \tag{8.76}$$

where the exponential is carried out term by term for each element of the matrix (i.e. it is not the matrix exponential). We can now express the sampled color image as

$$[\mathbf{C}_d][\mathbf{x}] = \sum_{i=1}^{3} \gamma_i C_i[\mathbf{x}] \sum_{k=0}^{K-1} M_{ki}\exp(\mathrm{j}2\pi\mathbf{x}\cdot\mathbf{d}_k)[\mathbf{P}_i]$$

$$= \sum_{k=0}^{K-1} \left(\sum_{i=1}^{3} \gamma_i M_{ki} C_i[\mathbf{x}][\mathbf{P}_i] \right) \exp(j2\pi \mathbf{x} \cdot \mathbf{d}_k)$$

$$= \sum_{k=0}^{K-1} [\mathbf{F}_k][\mathbf{x}] \exp(j2\pi \mathbf{x} \cdot \mathbf{d}_k), \tag{8.77}$$

where the new color signals $[\mathbf{F}_k][\mathbf{x}]$ are defined as

$$[\mathbf{F}_k][\mathbf{x}] = \sum_{i=1}^{3} \gamma_i M_{ki} C_i[\mathbf{x}][\mathbf{P}_i]. \tag{8.78}$$

Taking the Fourier transform, and using the modulation property,

$$[\widehat{\mathbf{C}}_d](\mathbf{u}) = \sum_{k=0}^{K-1} [\widehat{\mathbf{F}}_k](\mathbf{u} - \mathbf{d}_k). \tag{8.79}$$

Thus, we can view the sampled signal as the sum of the baseband component $[\mathbf{F}_0][\mathbf{x}]$ at zero frequency, and each of the other $[\mathbf{F}_k][\mathbf{x}]$ shifted (modulated) to the non-zero frequencies $\mathbf{d}_k$. Note that since $\mathbf{d}_0 = 0$, then $M_{0i} = |\mathcal{J}_i|/K$. Thus, irrespective of the sampling scheme used, the baseband component $[\mathbf{F}_0][\mathbf{x}] = [\mathbf{C}][\mathbf{x}]$ if

$$\gamma_i = \frac{1}{M_{1i}} = \frac{K}{|\mathcal{J}_i|}, \tag{8.80}$$

where $|\mathcal{J}_i|$ denotes the number of elements in the set $\mathcal{J}_i$. We will systematically use this choice for the γ_i.

Dubois (2009) develops this representation for scalar CFA signals for a number of different examples of color filter arrays. Each of these can be extended to the vector representation presented here. For this chapter, only the Bayer structure is examined in detail. The basis used for this structure is RGB, $[\mathbf{P}_1] = [\mathbf{R}]$, $[\mathbf{P}_2] = [\mathbf{G}]$, $[\mathbf{P}_3] = [\mathbf{B}]$. It does not matter which RGB basis is used (as long as it is known) and we can safely assume that it is the Rec. 709/sRGB basis of Section 7.5.3. For the Bayer structure, $\Lambda^* = \mathbb{Z}^2$ and $\Gamma^* = \left(\frac{1}{2}\mathbb{Z}\right)^2$, and suitable choices for the matrices defined above are:

$$\mathbf{B} = \begin{bmatrix} 0 & 1 & 0 & 1 \\ 0 & 0 & 1 & 1 \end{bmatrix}$$

$$\mathbf{D} = \begin{bmatrix} 0 & \frac{1}{2} & \frac{1}{2} & 0 \\ 0 & \frac{1}{2} & 0 & \frac{1}{2} \end{bmatrix}$$

$$\mathbf{J} = \begin{bmatrix} 0 & 1 & 0 \\ 1 & 0 & 0 \\ 0 & 0 & 1 \\ 0 & 1 & 0 \end{bmatrix}$$

$$\mathbf{M} = \frac{1}{4} \begin{bmatrix} 1 & 2 & 1 \\ -1 & 2 & -1 \\ -1 & 0 & 1 \\ 1 & 0 & -1 \end{bmatrix} \tag{8.81}$$

We can identify the baseband component as

$$[\mathbf{F}_0][\mathbf{x}] = \frac{1}{4}\gamma_1 C_1[\mathbf{x}][\mathbf{P}_1] + \frac{1}{2}\gamma_2 C_2[\mathbf{x}][\mathbf{P}_2] + \frac{1}{4}\gamma_3 C_3[\mathbf{x}][\mathbf{P}_3]. \quad (8.82)$$

The baseband component will be equal to the desired color signal $[\mathbf{C}](\mathbf{x})$ if $\gamma_1 = 4$, $\gamma_2 = 2$ and $\gamma_3 = 4$, in agreement with Equation (8.80). With these values, the remaining three transformed signals are

$$\begin{aligned}
[\mathbf{F}_1][\mathbf{x}] &= -C_1[\mathbf{x}][\mathbf{P}_1] + C_2[\mathbf{x}][\mathbf{P}_2] - C_3[\mathbf{x}][\mathbf{P}_3] \\
[\mathbf{F}_2][\mathbf{x}] &= -C_1[\mathbf{x}][\mathbf{P}_1] + C_3[\mathbf{x}][\mathbf{P}_3] \\
[\mathbf{F}_3][\mathbf{x}] &= C_1[\mathbf{x}][\mathbf{P}_1] - C_3[\mathbf{x}][\mathbf{P}_3] = -[\mathbf{F}_2][\mathbf{x}]
\end{aligned} \quad (8.83)$$

and in the frequency domain, we have

$$[\widehat{\mathbf{C}}_d](u, v) = [\widehat{\mathbf{F}}_0](u, v) + [\widehat{\mathbf{F}}_1](u - \tfrac{1}{2}, v - \tfrac{1}{2}) + [\widehat{\mathbf{F}}_2](u - \tfrac{1}{2}, v) - [\widehat{\mathbf{F}}_2](u, v - \tfrac{1}{2}). \quad (8.84)$$

This representation suggests a frequency-domain approach to recover $[\mathbf{C}][\mathbf{x}]$ (i.e. $[\mathbf{F}_0][\mathbf{x}]$) from the subsampled signal $[\mathbf{C}_d][\mathbf{x}]$, namely to isolate and remove the components at the frequencies $(\frac{1}{2}, 0)$, $(0, \frac{1}{2})$ and $(\frac{1}{2}, \frac{1}{2})$ with suitable filters. Using linear filters on the lattice Λ, the filters would be characterized by the 3×3 transfer matrices, as defined in Section 8.3.1. Doing separate processing on the $[\mathbf{P}_1]$, $[\mathbf{P}_2]$ and $[\mathbf{P}_3]$ (i.e. RGB) components would not work well. This simply amounts to separate interpolation of the subsampled RGB components, a method known to give relatively poor results [Gunturk *et al.* (2005)]. An alternative approach to full vector processing is to perform a change of basis so that separate processing of the transformed coordinates is effective. We also note that the component $[\mathbf{F}_2][\mathbf{x}]$ appears separately at frequencies $(\frac{1}{2}, 0)$ and $(0, \frac{1}{2})$, interacting differently with the neighboring components $[\mathbf{F}_0][\mathbf{x}]$ and $[\mathbf{F}_1][\mathbf{x}]$ located at the frequencies $(0, 0)$ and $(\frac{1}{2}, \frac{1}{2})$ respectively. This suggests the use of a locally adaptive filtering that weights more heavily the component suffering the least from crosstalk. This has been found to be a very effective approach in scalar demosaicking [Dubois (2005), Leung *et al.* (2011)].

One interesting basis to consider for independent processing of components arises in the frequency-domain analysis of a CFA signal, obtained by multiplexing the red, green and blue samples defined on Ψ_1, Ψ_2 and Ψ_3 into a single scalar signal defined on $\Lambda = \Psi_1 \cup \Psi_2 \cup \Psi_3$,

$$C_{\mathrm{CFA}}[\mathbf{x}] = C_1[\mathbf{x}]\tilde{m}_1[\mathbf{x}] + C_2[\mathbf{x}]\tilde{m}_2[\mathbf{x}] + C_3[\mathbf{x}]\tilde{m}_3[\mathbf{x}]. \quad (8.85)$$

Using the Fourier series representation of the $\tilde{m}_i[\mathbf{x}]$ of Equation (8.74) with $K = 4$, we find

$$C_{\mathrm{CFA}}[\mathbf{x}] = \sum_{k=0}^{3} \left(\sum_{i=1}^{3} M_{ki} C_i[\mathbf{x}] \right) \exp(\mathrm{j}2\pi\mathbf{x} \cdot \mathbf{d}_k), \quad (8.86)$$

where $\mathbf{M}$ is given for the Bayer pattern in Equation (8.81). Since the last row of $\mathbf{M}$ is the negative of the third row, we define only three new signals

$$\begin{bmatrix} C_{\mathrm{L}}[\mathbf{x}] \\ C_{\mathrm{CH1}}[\mathbf{x}] \\ C_{\mathrm{CH2}}[\mathbf{x}] \end{bmatrix} = \underbrace{\begin{bmatrix} \frac{1}{4} & \frac{1}{2} & \frac{1}{4} \\ -\frac{1}{4} & \frac{1}{2} & -\frac{1}{4} \\ -\frac{1}{4} & 0 & \frac{1}{4} \end{bmatrix}}_{\mathbf{A}_{\mathcal{RGB}\to\mathcal{LCC}}} \begin{bmatrix} C_1[\mathbf{x}] \\ C_2[\mathbf{x}] \\ C_3[\mathbf{x}] \end{bmatrix}. \quad (8.87)$$

With this definition, we obtain

$$C_{\mathrm{CFA}}[\mathbf{x}] = C_{\mathrm{L}}[\mathbf{x}] + C_{\mathrm{CH1}}[\mathbf{x}]\exp(\mathrm{j}2\pi\mathbf{x}\cdot\mathbf{d}_1) + C_{\mathrm{CH2}}[\mathbf{x}](\exp(\mathrm{j}2\pi\mathbf{x}\cdot\mathbf{d}_2) - \exp(\mathrm{j}2\pi\mathbf{x}\cdot\mathbf{d}_3)) \tag{8.88}$$

with the corresponding frequency domain representation

$$\hat{C}_{\mathrm{CFA}}(u,v) = \hat{C}_{\mathrm{L}}(u,v) + \hat{C}_{\mathrm{CH1}}(u-\tfrac{1}{2}, v-\tfrac{1}{2}) + \hat{C}_{\mathrm{CH2}}(u-\tfrac{1}{2}, v) - \hat{C}_{\mathrm{CH2}}(u, v-\tfrac{1}{2}). \tag{8.89}$$

The CFA signal is seen to be represented as the frequency-domain multiplexing of the luma component C_{L} at baseband and two chroma components C_{CH1} centered at frequency $(\frac{1}{2},\frac{1}{2})$ and C_{CH2} centered at frequencies $(\frac{1}{2},0)$ and $(0,\frac{1}{2})$. The luma-chroma demultiplexing method [Leung *et al.* (2011)] involves extracting C_{L}, C_{CH1}, and C_{CH2} from the CFA signal using spatial filtering and demodulation to baseband, and then using the inverse of Equation (8.87) to recover C_1, C_2, and C_3.

In this process, three new signals have been defined in Equation (8.87), which are a linear transformation of the original tristimulus values. Thus, these new values can be considered to be tristimulus values with respect to a new basis. As per Table 7.1, this defines the new primaries forming the basis $\mathcal{LCC} = \{[\mathbf{L}], [\mathbf{CH}_1], [\mathbf{CH}_2]\}$,

$$\begin{bmatrix}[\mathbf{L}]\\ [\mathbf{CH}_1]\\ [\mathbf{CH}_2]\end{bmatrix} = \underbrace{\begin{bmatrix}1 & 1 & 1\\ -1 & 1 & -1\\ -2 & 0 & 2\end{bmatrix}}_{\mathbf{A}^{-T}_{\mathcal{RGB}\to\mathcal{LCC}}}\begin{bmatrix}[\mathbf{P}_1]\\ [\mathbf{P}_2]\\ [\mathbf{P}_3]\end{bmatrix}. \tag{8.90}$$

We observe that the basis vector $[\mathbf{L}] = [\mathbf{P}_1] + [\mathbf{P}_2] + [\mathbf{P}_3]$ is in fact reference white, and that for any gray-scale image in which $C_1[\mathbf{x}] = C_2[\mathbf{x}] = C_3[\mathbf{x}]$, the two components $C_{\mathrm{CH1}}[\mathbf{x}]$ and $C_{\mathrm{CH2}}[\mathbf{x}]$ are zero.

Applying the transformation of Equation (8.87) to the color signals $[\mathbf{F}_0][\mathbf{x}]$, $[\mathbf{F}_1][\mathbf{x}]$ and $[\mathbf{F}_2][\mathbf{x}]$ defined in Equation (8.82) and Equation (8.83), we obtain

$$\mathbf{F}_{0,\mathcal{LCC}}[\mathbf{x}] = \begin{bmatrix}\frac{1}{4}C_1[\mathbf{x}] + \frac{1}{2}C_2[\mathbf{x}] + \frac{1}{4}C_3[\mathbf{x}]\\ -\frac{1}{4}C_1[\mathbf{x}] + \frac{1}{2}C_2[\mathbf{x}] - \frac{1}{4}C_3[\mathbf{x}]\\ -\frac{1}{4}C_1[\mathbf{x}] + \frac{1}{4}C_3[\mathbf{x}]\end{bmatrix} = \begin{bmatrix}C_{\mathrm{L}}[\mathbf{x}]\\ C_{\mathrm{CH1}}[\mathbf{x}]\\ C_{\mathrm{CH2}}[\mathbf{x}]\end{bmatrix}$$

$$\mathbf{F}_{1,\mathcal{LCC}}[\mathbf{x}] = \begin{bmatrix}-\frac{1}{4}C_1[\mathbf{x}] + \frac{1}{2}C_2[\mathbf{x}] - \frac{1}{4}C_3[\mathbf{x}]\\ \frac{1}{4}C_1[\mathbf{x}] + \frac{1}{2}C_2[\mathbf{x}] + \frac{1}{4}C_3[\mathbf{x}]\\ \frac{1}{4}C_1[\mathbf{x}] - \frac{1}{4}C_3[\mathbf{x}]\end{bmatrix} = \begin{bmatrix}C_{\mathrm{CH1}}[\mathbf{x}]\\ C_{\mathrm{L}}[\mathbf{x}]\\ -C_{\mathrm{CH2}}[\mathbf{x}]\end{bmatrix}$$

$$\mathbf{F}_{2,\mathcal{LCC}}[\mathbf{x}] = \begin{bmatrix}-\frac{1}{4}C_1[\mathbf{x}] + \frac{1}{4}C_3[\mathbf{x}]\\ \frac{1}{4}C_1[\mathbf{x}] - \frac{1}{4}C_3[\mathbf{x}]\\ \frac{1}{4}C_1[\mathbf{x}] + \frac{1}{4}C_3[\mathbf{x}]\end{bmatrix} = \begin{bmatrix}C_{\mathrm{CH2}}[\mathbf{x}]\\ -C_{\mathrm{CH2}}[\mathbf{x}]\\ \frac{1}{2}(C_{\mathrm{L}}[\mathbf{x}] - C_{\mathrm{CH1}}[\mathbf{x}])\end{bmatrix} = -\mathbf{F}_{3,\mathcal{LCC}}[\mathbf{x}] \tag{8.91}$$

In this basis, $[\mathbf{C}_d][\mathbf{x}]$ in Equation (8.77) can be written out explicitly as

$$\mathbf{C}_{d,\mathcal{LCC}}[\mathbf{x}] = \begin{bmatrix} C_{\mathrm{L}}[\mathbf{x}] + C_{\mathrm{CH1}}[\mathbf{x}]\exp(\mathrm{j}2\pi\mathbf{x}\cdot\mathbf{d}_1) + C_{\mathrm{CH2}}[\mathbf{x}] \\ (\exp(\mathrm{j}2\pi\mathbf{x}\cdot\mathbf{d}_2) - \exp(\mathrm{j}2\pi\mathbf{x}\cdot\mathbf{d}_3)) \\ C_{\mathrm{CH1}}[\mathbf{x}] + C_{\mathrm{L}}[\mathbf{x}]\exp(\mathrm{j}2\pi\mathbf{x}\cdot\mathbf{d}_1) - C_{\mathrm{CH2}}[\mathbf{x}] \\ (\exp(\mathrm{j}2\pi\mathbf{x}\cdot\mathbf{d}_2) - \exp(\mathrm{j}2\pi\mathbf{x}\cdot\mathbf{d}_3)) \\ C_{\mathrm{CH2}}[\mathbf{x}] - C_{\mathrm{CH2}}[\mathbf{x}]\exp(\mathrm{j}2\pi\mathbf{x}\cdot\mathbf{d}_1) + \frac{1}{2}(C_{\mathrm{L}}[\mathbf{x}] - C_{\mathrm{CH1}}[\mathbf{x}]) \\ (\exp(\mathrm{j}2\pi\mathbf{x}\cdot\mathbf{d}_2) - \exp(\mathrm{j}2\pi\mathbf{x}\cdot\mathbf{d}_3)) \end{bmatrix} \quad (8.92)$$

Here, the $[\mathbf{L}]$ component is the conventional CFA signal. Given the definition of the $\mathbf{d}_i$ for the Bayer structure, it is straightforward to show that $C_{d,CH1}[\mathbf{x}] = C_{d,L}[\mathbf{x}]\exp(2\pi(\mathbf{x}\cdot\mathbf{d}_1))$, and so the $[\mathbf{CH1}]$ component provides the same information as the $[\mathbf{L}]$ component. However, the $[\mathbf{CH2}]$ component depends only on $C_1[\mathbf{x}]$ and $C_3[\mathbf{x}]$ and could perhaps provide additional information about the original components, although this has not been established. In any case, we can recover an estimate of the original full color image by using separate filters on the L, CH1 and CH2 components of $\mathbf{C}_{d,\mathcal{LCC}}[\mathbf{x}]$ to suppress the high frequency components in the vicinty of the frequencies $\mathbf{d}_1$, $\mathbf{d}_2$ and $\mathbf{d}_3$. When applied to the CFA image of Figure 8.7, we obtain the result shown in Figure 8.8. Although this is not intended to be a state-of-the-art method, it does give a 2.2 dB increase in CPSNR compared to separate interpolation of the R, G, and B components. A key observation is that we can obtain a good estimate of the full color image using only the $[\mathbf{L}]$ component in the $\mathcal{LCC}$ basis and that the CH2 component is modulated at both frequencies $\mathbf{d}_2$ and

Figure 8.8 Bayer sampled image of Figure 8.7 reconstructed by separate low-pass filtering of the three channels of $\mathbf{C}_{d,\mathcal{LCC}}[\mathbf{x}]$, followed by conversion to RGB. (*See color plate section for the color representation of this figure.*)

$\mathbf{d}_3$. Either one can be used to reconstruct the full color image and we can locally choose which one gives the best result [Dubois (2005)].

8.4 Color Mosaic Displays

Most color image display devices, including CRT, LCD, plasma, etc., are based on a mosaic of color elements. A model and analysis for such displays has been presented by Farrell *et al.* (2008). In such models, it is assumed that the display screen is partitioned into nonoverlapping regions. One of a finite number of base colors is displayed in each region; there are usually three base colors (red, green and blue) but recent work has considered more than three to expand the gamut of displayable colors or to increase brightness (e.g., Kutas *et al.* (2006)). Using more than three base colors can allow a larger portion of the cone of physically realizable colors to be displayable by additive synthesis on a given device, as discussed in Section 7.4.9. We adapt our preceding development to such displays to describe the continuous-space color image on the screen. For the purpose of this chapter, we assume three base colors (red, green and blue, denoted P_1, P_2, P_3) but the development can be extended to more than three base colors in a straightforward fashion. Assume that the three base colors at full intensity emit light with spectral density $P_i(\lambda)$, $i = 1, 2, 3$, with a typical example shown in Figure 7.1(c).

According to the linear additive model, the displayed continuous-space image (expressed as spectral radiant exitance) is given by

$$C_o(\mathbf{x}, \lambda) = \sum_{i=1}^{3} \sum_{\mathbf{s}\in\Psi_i} C_{di}[\mathbf{s}]a(\mathbf{x} - \mathbf{s})P_i(\lambda). \tag{8.93}$$

We assume that a specific color space is used to represent the displayed color image, usually the 1931 CIE standard observer. In this case, the spectral densities $P_i(\lambda)$ are projected to the color space in the usual fashion to obtain $[\mathbf{P}_i]$, $i = 1, 2, 3$. Then, the displayed color signal is given by

$$[\mathbf{C}_o](\mathbf{x}) = \sum_{i=1}^{3} \sum_{\mathbf{s}\in\Psi_i} C_{di}[\mathbf{s}]a(\mathbf{x} - \mathbf{s})[\mathbf{P}_i]. \tag{8.94}$$

In this model, we assume that the set of display elements lie on a lattice Λ that is partitioned as the union of three subsets, $\Lambda = \cup\Psi_i$, where the Ψ_i are cosets of a sublattice Γ in Λ. The continuous space function $a(\mathbf{x})$ is the display aperture. In typical mosaic displays such as an LCD display, shifted versions of $a(\mathbf{x})$ on Λ do not overlap, i.e. $a(\mathbf{x} - \mathbf{s}_1) \cdot a(\mathbf{x} - \mathbf{s}_2) = 0$ for $\mathbf{s}_1, \mathbf{s}_2 \in \Lambda$, $\mathbf{s}_1 \neq \mathbf{s}_2$. A common example of display mosaic is illustrated in Figure 8.9(a), corresponding to the rectangular display aperture function in Figure 8.9(b). A popular alternative aperture is the chevron of Figure 8.9(c), which can tile the display screen according to the same arrangement as Figure 8.9(a). See Farrell *et al.* (2008) for close-up photographs of display mosaics of each type.

We can directly compute the Fourier transform of the displayed signal of Equation (8.94),

$$[\hat{\mathbf{C}}_o](\mathbf{u}) = \sum_{i=1}^{3} \sum_{\mathbf{s}\in\Psi_i} C_{di}[\mathbf{s}] \int_{\mathbb{R}^2} a(\mathbf{x} - \mathbf{s}) \exp(-\mathrm{j}2\pi\mathbf{u} \cdot \mathbf{x})\, \mathrm{d}\mathbf{x}[\mathbf{P}_i]. \tag{8.95}$$

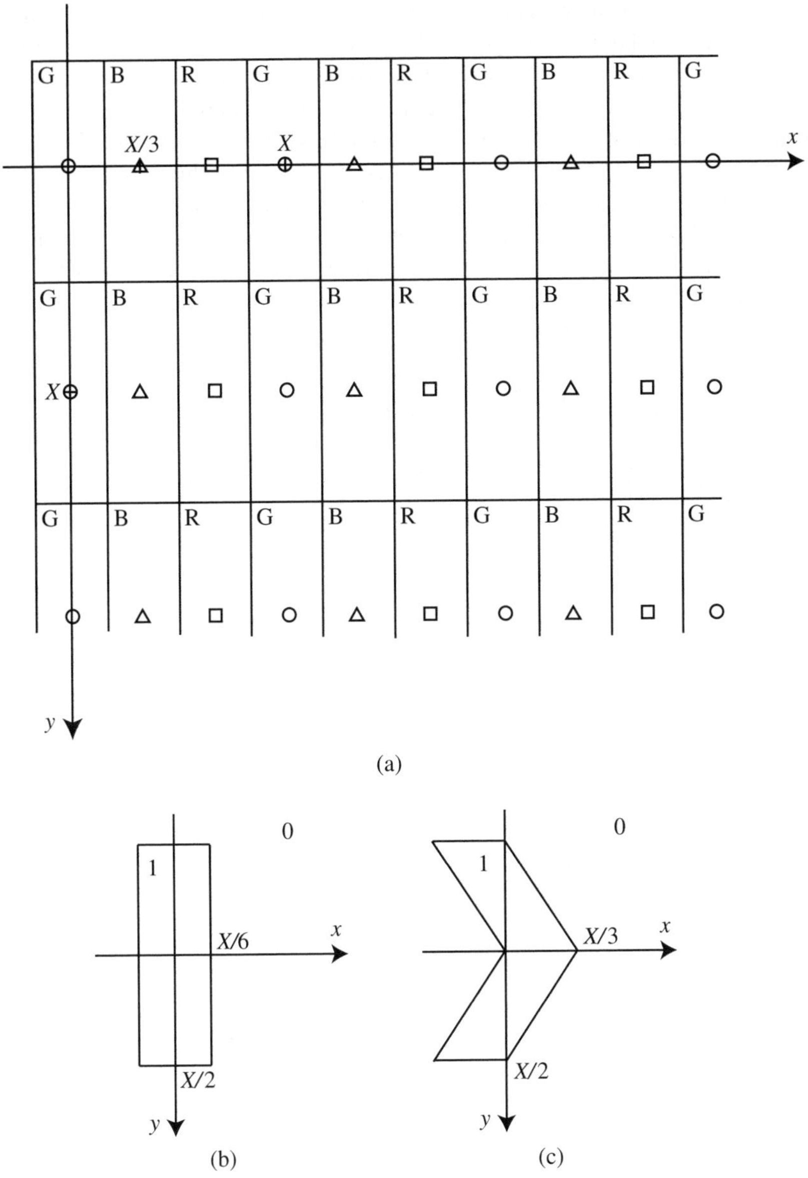

Figure 8.9 (a) Example of a typical display mosaic with a rectangular display aperture. (b) Rectangular aperture function. (c) Chevron display aperture.

Using the shift property of the continuous space Fourier transform,

$$\int_{\mathbb{R}^2} a(\mathbf{x}-\mathbf{s}) \exp(-\mathrm{j}2\pi\mathbf{u}\cdot\mathbf{x})\,\mathrm{d}\mathbf{x} = \hat{a}(\mathbf{u}) \exp(-\mathrm{j}2\pi\mathbf{u}\cdot\mathbf{x}) \tag{8.96}$$

so that

$$[\widehat{\mathbf{C}}_o](\mathbf{u}) = \hat{a}(\mathbf{u}) \sum_{i=1}^{3} \left(\sum_{\mathbf{s}\in\Psi_i} C_{di}[\mathbf{s}] \exp(-\mathrm{j}2\pi\mathbf{u}\cdot\mathbf{x}) \right) [\mathbf{P}_i]. \tag{8.97}$$

We see that the latter term is essentially the same as Equation (8.69) with $\gamma_i = 1$, and thus the analysis of the preceding section can be applied directly. Then, this periodic Fourier transform is multiplied by the aperiodic $\hat{a}(\mathbf{u})$ to get the Fourier transform of the displayed signal. This can be coupled with a model of the human visual system that accounts for the frequency response of the color channels, such as the S-CIELAB model to optimize the display mosaic, as well as any processing to choose the $C_{di}(\mathbf{x})$ from the stored color image to be displayed (see Xu *et al.* (2008) for an example).

To illustrate these ideas, in the following we identify all the parameters and matrices associated with the display mosaic of Figure 8.9(a) in order to determine the Fourier transform in Equation (8.97). By inspection of Figure 8.9(a), we identify the following, using the notation of the preceding section.

$$\Gamma = \text{LAT}\left(\begin{bmatrix} X & 0 \\ 0 & X \end{bmatrix}\right) \qquad \Lambda = \left(\begin{bmatrix} \frac{X}{3} & 0 \\ 0 & X \end{bmatrix}\right)$$

$$\Gamma^* = \text{LAT}\left(\begin{bmatrix} \frac{1}{X} & 0 \\ 0 & \frac{1}{X} \end{bmatrix}\right) \qquad \Lambda^* = \text{LAT}\left(\begin{bmatrix} \frac{3}{X} & 0 \\ 0 & \frac{1}{X} \end{bmatrix}\right)$$

$$K = 3$$

$$\mathbf{B} = \begin{bmatrix} 0 & \frac{X}{3} & \frac{2X}{3} \\ 0 & 0 & 0 \end{bmatrix} \qquad \mathbf{D} = \begin{bmatrix} 0 & \frac{1}{X} & -\frac{1}{X} \\ 0 & 0 & 0 \end{bmatrix}$$

$$\mathbf{J} = \begin{bmatrix} 1 & 0 \\ 0 & 1 \end{bmatrix}$$

$$\mathbf{M} = \frac{1}{3}\begin{bmatrix} 1 & 1 & 1 \\ 1 & e^{-j2\pi/3} & e^{j2\pi/3} \\ 1 & e^{j2\pi/3} & e^{-j2\pi/3} \end{bmatrix}.$$

Finally, the three new signals defined are

$$\begin{aligned} [\mathbf{F}_1][\mathbf{x}] &= \frac{1}{3}(C_{d1}[\mathbf{x}][\mathbf{P}_1] + C_{d2}[\mathbf{x}][\mathbf{P}_2] + C_{d3}[\mathbf{x}][\mathbf{P}_3]) \\ [\mathbf{F}_2][\mathbf{x}] &= \frac{1}{3}(C_{d1}[\mathbf{x}][\mathbf{P}_1] + e^{-j2\pi/3}C_{d2}[\mathbf{x}][\mathbf{P}_2] + e^{j2\pi/3}C_{d3}[\mathbf{x}][\mathbf{P}_3]) \\ [\mathbf{F}_3][\mathbf{x}] &= \frac{1}{3}(C_{d1}[\mathbf{x}][\mathbf{P}_1] + e^{j2\pi/3}C_{d2}[\mathbf{x}][\mathbf{P}_2] + e^{-j2\pi/3}C_{d3}[\mathbf{x}][\mathbf{P}_3]) \end{aligned}$$

and the desired Fourier transform is

$$[\hat{\mathbf{C}}_o](u,v) = \hat{a}(u,v)([\hat{\mathbf{F}}_1](u,v) + [\hat{\mathbf{F}}_2](u - \frac{1}{X}, v) + [\hat{\mathbf{F}}_3](u + \frac{1}{X}, v). \tag{8.98}$$

Note that if $C_{di}[\mathbf{x}]$ is equal to the desired signal on Ψ_i, the baseband component is essentially the desired color signal, scaled by a factor of one third, since two out of three samples for each component are zero. The components at $\frac{1}{X}$ and $-\frac{1}{X}$ are due to the display mosaic. Higher spectral repeats are attenuated by a combination of the display aperture $\hat{a}(u,v)$ and the human visual system. The display aperture Fourier transform is easily found using standard methods, either analytically as in Chapter 2 or using an approximation with the discrete Fourier transform. The overall perceptual effect of the display process can be modeled with the S-CIELAB model discussed previously and used to effectively optimize the system. Several investigations have been carried out along these lines (Klompenhouwer and de Haan (2003), Platt (2000), Hirakawa and Wolfe (2007) and Zheng (2014) to name just a few).

9

Random Field Models

9.1 Introduction

Many phenomena in imaging systems are random in nature and are most appropriately addressed using probabilistic techniques. The most notable example is random noise that may corrupt an image. However, many image structures and patterns, such as textures, can also be considered to be random phenomena. Essentially, anything that varies in a fashion that is not perfectly predictable can be modeled with probabilistic methods. Besides image structure and noise, this can include random variations due to scene lighting, object surface orientation, motion, etc. To correctly model these quantities, we must extend the techniques of probability and stochastic processes to random functions of two or more independent variables. We call such multidimensional random processes *random fields*. In this chapter we present a first introduction to some aspects of random field modeling of images. We assume a basic familiarity with topics in probability such as random variables, probability distributions, expectation, conditional distributions, etc. There are numerous textbooks covering these topics; some examples that are particularly relevant to signal and image processing are Priestley (1981), Papoulis and Unnikrishna Pillai (2002), Stark and Woods (2002) and Leon-Garcia (2008).

Random field models have many applications in image processing. One of the most fundamental of these is image estimation: to estimate an unknown image (or selected image samples) given some related observations. A basic example is image restoration. We assume that there is an underlying ideal image that we cannot observe directly, but we can observe a related image, say one obtained by blurring and adding noise to the ideal image. Given a model for the blurring process, and random field models of the ideal image and the noise, we can attempt to optimally estimate the ideal image according to some criterion; this process is generally called image restoration. Related problems are image prediction, for example to predict an image frame from an observed previous frame, or image interpolation, say to estimate an image frame given some previous and future image frames. These operations occur frequently in image compression systems. We can also try to estimate other unobservable image functions from observed images. Common examples include motion, displacement and range images. Another frequent application is to estimate image regions according to some classification scheme, for example regions corresponding to different objects, textures, motion,

Multidimensional Signal and Color Image Processing Using Lattices, First Edition. Eric Dubois.

Companion website: www.wiley.com/go/Dubois/multiSP

etc. These operations are closely related to pattern recognition and computer vision. A final application we can mention, related to computer graphics, is the synthesis of images with certain stochastic properties, for example textures.

In this chapter, some basic concepts and properties of random field models of scalar and color images are introduced, which could then be applied to various image estimation problems.

9.2 What is a Random Field?

A random field is the extension of the random process or stochastic process to functions of several independent variables. Other treatments can be found in chapter 7 of Woods (2006), Fieguth (2011) and in Fieguth and Zhang (2005). We assume that there is an underlying probability sample space Ω with outcomes ζ. Each outcome from the sample space corresponds to a multidimensional function f defined on domain D. As in the case of deterministic signals, we can have continuous-domain random fields and discrete-domain random fields. Although we will discuss both to some extent, our main focus will be on discrete-domain random fields. The signal value can be real

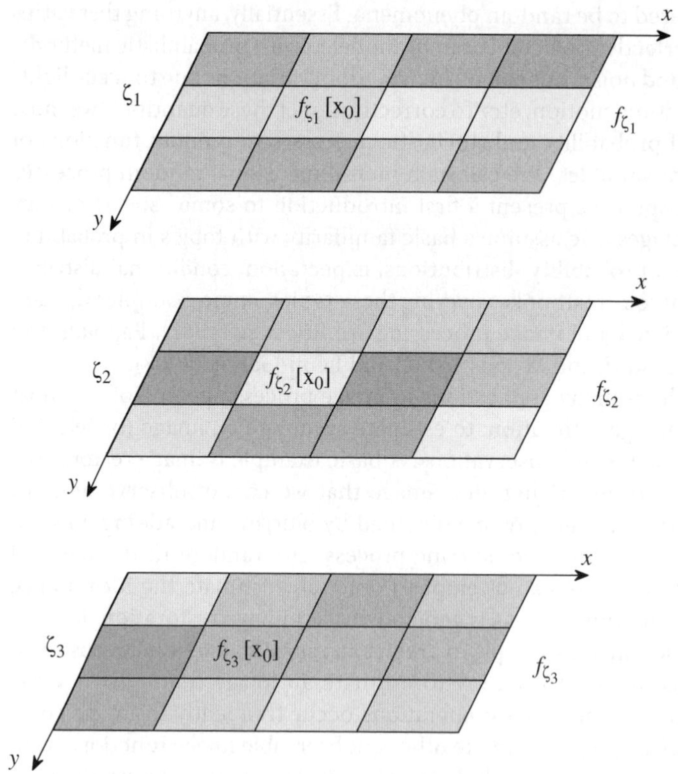

Figure 9.1 Illustration of three realizations of a two-dimensional color random field f_{ζ_i}, $i = 1, 2, 3$, identifying the three realizations of the random vector corresponding to $\mathbf{x} = \mathbf{x}_0$. *(See color plate section for the color representation of this figure.)*

as in the case of gray-scale images, or belong to a vector space as in the case of color images. Thus, to each $\zeta \in \Omega$ there corresponds the signal $f_\zeta : \mathcal{D} \to \mathcal{R}$, and to each $\mathbf{x} \in \mathcal{D}$ there corresponds a random variable or random vector $f(\mathbf{x}) \in \mathcal{R}$ (or $f[\mathbf{x}]$ in the discrete-domain case; we will simply use parentheses to enclose the independent variable when it can be either continuous or discrete). Thus, for example, a scalar two-dimensional discrete-domain random field f with finite support is a collection of random variables

$$f = \{f[\mathbf{x}], \mathbf{x} \in \mathcal{L}\} \tag{9.1}$$

where $\mathcal{L}$ is a finite subset (of size L) of a lattice Λ, usually a rectangular window. For a color image, it would be a collection of random vectors chosen from the color vector space $\mathcal{C}$. This concept is illustrated in Figure 9.1, where $\Omega = \{\zeta_1, \zeta_2, \zeta_3, \dots\}$. The figure shows a portion of three realizations of the color random field, and also illustrates one random vector corresponding to $\mathbf{x} = \mathbf{x}_0$.

The complete characterization of a random field is provided by the joint probability distribution of the pixel values, $P_f(f(\mathbf{x}_1), \dots, f(\mathbf{x}_N))$ for any set of locations $\mathbf{x}_1, \dots \mathbf{x}_N \in \mathcal{D}$ and for any integer N (up to L in the bounded discrete-domain case). For a typical discrete domain image, L is a number in the millions, so we are talking of a joint distribution of millions of random variables. For the continuous-domain case, the number of positions is unbounded. Clearly a model is needed!

9.3 Image Moments

Although the joint probability distribution of all sets of sample values is required to completely characterize the random field, we can do many things knowing a few low-order moments like the mean and autocorrelation, especially in the case of wide-sense stationary (or homogeneous) random fields. These moments are introduced here, along with some important properties.

Recall that a *random variable* is a function that assigns a real number to each outcome ζ from the sample space. An example is the image value (say luminance or a tristimulus value) at a particular location in the realization of the random field, $f_\zeta(\mathbf{x}_0)$ for any $\mathbf{x}_0 \in \mathcal{D}$. If Z is a continuous random variable with possible values $z \in \mathbb{R}$, and we know the probability density $p_Z(z)$, then the expected value, or expectation, of Z is denoted $E\{Z\} = \int z p_Z(z)\, dz$. If Z is a discrete random variable with values $z_1, z_2, \dots$, with discrete probabilities $p_Z(z_k)$, then the expectation is given by $E\{Z\} = \sum z_k p_Z(z_k)$. Refer to the basic references on probability for more detailed and rigorous discussions.

9.3.1 Mean, Autocorrelation, Autocovariance

We first define the mean, autocorrelation, and autocovariance for a general real scalar random field with domain $\mathcal{D}$, whether continuous space or discrete space:

Mean: $\mu_f(\mathbf{x}) = E\{f(\mathbf{x})\}$ for all $\mathbf{x} \in \mathcal{D}$
Autocorrelation: $R_f(\mathbf{x}_1, \mathbf{x}_2) = E\{f(\mathbf{x}_1)f(\mathbf{x}_2)\}$ for all $\mathbf{x}_1, \mathbf{x}_2 \in \mathcal{D}$
Autocovariance: $K_f(\mathbf{x}_1, \mathbf{x}_2) = E\{(f(\mathbf{x}_1) - \mu_f(\mathbf{x}_1))(f(\mathbf{x}_2) - \mu_f(\mathbf{x}_2))\}$.

Note that $K_f(\mathbf{x}_1, \mathbf{x}_2) = R_f(\mathbf{x}_1, \mathbf{x}_2) - \mu_f(\mathbf{x}_1)\mu_f(\mathbf{x}_2)$. We see that the mean function has the same domain as the image, whereas the autocorrelation and autocovariance functions have domain $\mathcal{D} \times \mathcal{D}$. In specific applications, these averages may depend on where the points are located. An example is a database of ID photos for a specific population, say students at a certain university. The means may be quite different in the background region of the photos than in the foreground. However, in many applications of interest, there is no reason to believe that the means are position dependent, for example the ensemble of all television images that may appear on the TV screen during its lifetime. If these first and second order moments are independent of position, we call the random field *wide-sense stationary* (WSS) or *homogeneous*. It follows that in this case

1) $\mu_f(\mathbf{x}) = \mu_f$, independently of position $\mathbf{x}$
2) $R_f(\mathbf{x}_1 + \mathbf{w}, \mathbf{x}_2 + \mathbf{w}) = R_f(\mathbf{x}_1, \mathbf{x}_2)$, independently of $\mathbf{w}$, for $\mathbf{x}_1, \mathbf{x}_2, \mathbf{w} \in \mathcal{D}$.

In the WSS case, we change the notation slightly and write

$$R_f(\mathbf{x}) = E\{f(\mathbf{w}+\mathbf{x})f(\mathbf{w})\} \tag{9.2}$$

$$K_f(\mathbf{x}) = E\{(f(\mathbf{w}+\mathbf{x}) - \mu_f)(f(\mathbf{w}) - \mu_f)\}. \tag{9.3}$$

Here, $\mathbf{w}, \mathbf{x} \in \mathcal{D}$, but the result does not depend on $\mathbf{w}$. In this WSS case, the domain of the autocorrelation and autocovariance functions is the same as the domain of the image. We note that $R_f(\mathbf{0}) = E\{|f(\mathbf{w})|^2\}$, which can be interpreted as the average power of the signal. Also, $K_f(\mathbf{0}) = E\{|f(\mathbf{w}) - \mu_f|^2\}$, which is the variance of f, denoted σ_f^2. If $\mu_f = 0$, these two values are the same.

The mean and autocorrelation can be estimated from a large collection of images that are typical of a given application. Such a collection is called a training set. Alternatively, they can be taken to be analytically defined functions with parameters chosen to best represent a given dataset.

Example 9.1 Discrete-space WSS Gaussian white noise

The mean and autocorrelation of white noise $v[\mathbf{x}]$ are given by $\mu_v = 0$ and $R_v[\mathbf{x}] = \sigma_v^2\delta[\mathbf{x}]$. Thus, any two samples of white noise at different spatial locations are uncorrelated. Since they are Gaussian random variables, they are also independent. The probability density function for an arbitrary sample is the well-known form

$$p_v(v[\mathbf{x}]) = \frac{1}{\sqrt{2\pi}\sigma_v}\exp(-v[\mathbf{x}]^2/(2\sigma_v^2)). \tag{9.4}$$

Because of independence, the joint probability density of any two samples at locations $\mathbf{x}_1$ and $\mathbf{x}_2$ is given by

$$p_v(v[\mathbf{x}_1], v[\mathbf{x}_2]) = \frac{1}{2\pi\sigma_v^2}\exp(-(v[\mathbf{x}_1]^2 + v[\mathbf{x}_2]^2)/(2\sigma_v^2)). \tag{9.5}$$

Indeed, we can write the joint probability density of an arbitrary number of samples,

$$p_v(v[\mathbf{x}_1], v[\mathbf{x}_2], \dots, v[\mathbf{x}_L]) = \frac{1}{(2\pi\sigma_v^2)^{L/2}}\exp\left(-\frac{1}{2\sigma_v^2}\sum_{k=1}^{L} v[\mathbf{x}_k]^2\right). \tag{9.6}$$

□

White noise can also be defined in the continuous-domain case, with $R_v(\mathbf{x}) = c\delta(\mathbf{x})$, where now $\delta(\mathbf{x})$ denotes the Dirac delta. However, care must be taken with this model, as this is a non-physically realizable signal with infinite power.

Example 9.2 Separable exponential autocorrelation
In this analytically defined model, we let $\mu_f = 0$ and we assume that the spatial autocorrelation function is separable in x and y. In the continuous domain case,

$$R_f(x, y) = \sigma_f^2 \exp(-\gamma_1 |x|) \exp(-\gamma_2 |y|) \tag{9.7}$$

where $\gamma_1 > 0$ and $\gamma_2 > 0$.

In the discrete-domain case, on a square lattice with spacing X,

$$R_f[mX, nX] = \sigma_f^2 \rho_1^{|m|} \rho_2^{|n|} \tag{9.8}$$

where $|\rho_1| < 1$ and $|\rho_2| < 1$. In this case, we note that

$$\sigma_f^2 = R_f[0, 0], \qquad \rho_1 = \frac{R_f[X, 0]}{R_f[0, 0]}, \qquad \rho_2 = \frac{R_f[0, X]}{R_f[0, 0]}. \tag{9.9}$$

Here, ρ_1 and ρ_2 are the first-order horizontal and vertical correlation coefficients, respectively. The separable autocorrelation is a commonly applied model when images are assumed to be samples of a WSS random field. □

In the case of a vector-valued random field representing a color image, at each point of the domain we have a random vector $\mathbf{C}_\mathcal{B}(\mathbf{x})$ composed of three tristimulus values with respect to a given basis (set of primaries) $\mathcal{B}$ of the color space. In this case, we define first and second order moments as follows:

Mean: $\boldsymbol{\mu}_{C_\mathcal{B}}(\mathbf{x}) = E\{\mathbf{C}_\mathcal{B}(\mathbf{x})\}$ for all $\mathbf{x} \in \mathcal{D}$
Autocorrelation matrix: $\mathbf{R}_{C_\mathcal{B}}(\mathbf{x}_1, \mathbf{x}_2) = E\{\mathbf{C}_\mathcal{B}(\mathbf{x}_1)\mathbf{C}_\mathcal{B}^T(\mathbf{x}_2)\}$ for all $\mathbf{x}_1, \mathbf{x}_2 \in \mathcal{D}$
Autocovariance matrix: $\mathbf{K}_{C_\mathcal{B}}(\mathbf{x}_1, \mathbf{x}_2) = E\{(\mathbf{C}_\mathcal{B}(\mathbf{x}_1) - \boldsymbol{\mu}_{C_\mathcal{B}}(\mathbf{x}_1))(\mathbf{C}_\mathcal{B}(\mathbf{x}_2) - \boldsymbol{\mu}_{C_\mathcal{B}}(\mathbf{x}_2))^T\}$.

These expressions extend the definitions in Brillinger (2001) for vector-valued time series to the case of vector-valued random fields. For the WSS case, we have

1) $\boldsymbol{\mu}_{C_\mathcal{B}}(\mathbf{x}) = \boldsymbol{\mu}_{C_\mathcal{B}}$, independently of position $\mathbf{x}$
2) $\mathbf{R}_{C_\mathcal{B}}(\mathbf{x}_1 + \mathbf{w}, \mathbf{x}_2 + \mathbf{w}) = \mathbf{R}_{C_\mathcal{B}}(\mathbf{x}_1, \mathbf{x}_2)$, independently of $\mathbf{w}$, for $\mathbf{x}_1, \mathbf{x}_2, \mathbf{w} \in \mathcal{D}$.

Thus, similar to the scalar case, we denote the matrix-valued autocorrelation and autocovariance functions in the WSS case by

$$\mathbf{R}_{C_\mathcal{B}}(\mathbf{x}) = E\{\mathbf{C}_\mathcal{B}(\mathbf{w} + \mathbf{x})\mathbf{C}_\mathcal{B}^T(\mathbf{w})\} \tag{9.10}$$

$$\mathbf{K}_{C_\mathcal{B}}(\mathbf{x}) = E\{(\mathbf{C}_\mathcal{B}(\mathbf{w} + \mathbf{x}) - \boldsymbol{\mu}_{C_\mathcal{B}})(\mathbf{C}_\mathcal{B}(\mathbf{w}) - \boldsymbol{\mu}_{C_\mathcal{B}})^T\} \tag{9.11}$$

for all $\mathbf{x}, \mathbf{w} \in \mathcal{D}$, where these expressions are independent of $\mathbf{w}$. The mean vector, and the correlation and covariance matrices as defined depend on the basis. However, if we know these functions for a given basis, we can convert them to another basis using the change of basis matrix. Suppose that $\mathcal{B}$ and $\widetilde{\mathcal{B}}$ are two bases for the color space $\mathcal{C}$, related by the change of basis matrix $\mathbf{A}_{\mathcal{B}\to\widetilde{\mathcal{B}}}$. Then

$$\boldsymbol{\mu}_{C_{\widetilde{\mathcal{B}}}} = \mathbf{A}_{\mathcal{B}\to\widetilde{\mathcal{B}}} \boldsymbol{\mu}_{C_\mathcal{B}} \tag{9.12}$$

$$\begin{aligned} \mathbf{R}_{C_{\widetilde{\mathcal{B}}}}(\mathbf{x}) &= E\{\mathbf{A}_{\mathcal{B}\to\widetilde{\mathcal{B}}}\mathbf{C}_\mathcal{B}(\mathbf{w} + \mathbf{x})\mathbf{C}_\mathcal{B}^T(\mathbf{w})\mathbf{A}_{\mathcal{B}\to\widetilde{\mathcal{B}}}^T\} \\ &= \mathbf{A}_{\mathcal{B}\to\widetilde{\mathcal{B}}}\mathbf{R}_{C_\mathcal{B}}(\mathbf{x})\mathbf{A}_{\mathcal{B}\to\widetilde{\mathcal{B}}}^T. \end{aligned} \tag{9.13}$$

Similarly,

$$\mathbf{K}_{C_{\widetilde{B}}}(\mathbf{x}) = \mathbf{A}_{B\to\widetilde{B}}\mathbf{K}_{C_B}(\mathbf{x})\mathbf{A}^T_{B\to\widetilde{B}}. \tag{9.14}$$

We have defined these functions for the three-dimensional color space of human vision, but the definitions extend in a straightforward fashion to other dimensions of color space and to other vector-valued observations.

9.3.2 Properties of the Autocorrelation Function

We state here a few properties of the autocorrelation function of a real WSS scalar random field f. Of course, the properties apply equally well to the autocovariance function, which is the autocorrelation of the WSS random field $f - \mu_f$. These properties apply to both continuous domain and discrete domain WSS random fields.

1) $R_f(\mathbf{x})$ is even, i.e. $R_f(-\mathbf{x}) = R_f(\mathbf{x})$. This follows directly from the WSS property:

$$\begin{aligned} R_f(-\mathbf{x}) &= E\{f(\mathbf{w}-\mathbf{x})f(\mathbf{w})\} = E\{f(\mathbf{w})f(\mathbf{w}+\mathbf{x})\} \\ &= E\{f(\mathbf{w}+\mathbf{x})f(\mathbf{w})\} = R_f(\mathbf{x}). \end{aligned}$$

2) The autocorrelation function is maximum at the origin, $|R_f(\mathbf{x})| \le R_f(\mathbf{0})$ for all $\mathbf{x} \in \mathcal{D}$.

$$\begin{aligned} R_f^2(\mathbf{x}) &= (E\{f(\mathbf{w}+\mathbf{x})f(\mathbf{w})\})^2 \\ &\le E\{(f(\mathbf{w}+\mathbf{x}))^2\}E\{(f(\mathbf{w}))^2\} \\ &= R_f(\mathbf{0})R_f(\mathbf{0}). \end{aligned}$$

This follows from the standard property that $(E\{YZ\})^2 \le E\{Y^2\}E\{Z^2\}$ for any two jointly distributed random variables Y and Z (an instance of the Cauchy–Schwartz inequality).

3) The autocorrelation function is positive semi-definite; for any points $\mathbf{x}_1, \mathbf{x}_2, \dots, \mathbf{x}_n \in \mathcal{D}$ and any real numbers $z_1, z_2, \dots, z_n$,

$$\sum_{r=1}^{n}\sum_{s=1}^{n} z_r R_f(\mathbf{x}_r - \mathbf{x}_s) z_s \ge 0. \tag{9.15}$$

To show this [Priestley (1981)], for any real numbers $z_1, z_2, \dots, z_n$, define the random variable $W = \sum_{r=1}^{n} z_r f(\mathbf{x}_r)$. Then

$$\begin{aligned} 0 \le E\{W^2\} &= E\left\{\sum_{r=1}^{n} z_r f(\mathbf{x}_r)\sum_{s=1}^{n} z_s f(\mathbf{x}_s)\right\} \\ &= \sum_{r=1}^{n}\sum_{s=1}^{n} z_r E\{f(\mathbf{x}_r)f(\mathbf{x}_s)\} z_s \\ &= \sum_{r=1}^{n}\sum_{s=1}^{n} z_r R_f(\mathbf{x}_r - \mathbf{x}_s) z_s. \end{aligned}$$

These properties show that we cannot select an arbitrary function to model an autocorrelation of a random field. Any such model must satisfy the properties above.

For a vector-valued color signal, property (1) becomes $\mathbf{R}_{C_B}(-\mathbf{x}) = \mathbf{R}^T_{C_B}(\mathbf{x})$, while properties (2) and (3) apply to the diagonal elements of $\mathbf{R}_{C_B}(\mathbf{x})$.

9.3.3 Cross-Correlation

It frequently occurs that we consider several random fields simultaneously, for example an input image, a noise field, and a corrupted image. The complete joint specification of two random fields f_1 and f_2 would be provided by all possible joint probability distributions of the form $P_{f_1 f_2}(f_1(\mathbf{x}_{11}), \dots, f_1(\mathbf{x}_{1N}), f_2(\mathbf{x}_{21}), \dots, f_2(\mathbf{x}_{2M}))$ for arbitrary N, M and choice of $N+M$ points. Again, our goal is to find a model from which we can deduce all these joint distributions. As in the case of a single random field, we can often get much information from low-order moments, specifically cross-correlation and cross-covariance, defined as follows for scalar random fields:

Definitions:
Cross-correlation: $R_{fg}(\mathbf{x}_1, \mathbf{x}_2) = E\{f(\mathbf{x}_1)g(\mathbf{x}_2)\}$ for all $\mathbf{x}_1, \mathbf{x}_2 \in \mathcal{D}$
Cross-covariance: $K_{fg}(\mathbf{x}_1, \mathbf{x}_2) = E\{(f(\mathbf{x}_1) - \mu_f(\mathbf{x}_1))(g(\mathbf{x}_2) - \mu_g(\mathbf{x}_2))\}$.
The random fields are said to be uncorrelated if $K_{fg}(\mathbf{x}_1, \mathbf{x}_2) = 0$ for all $\mathbf{x}_1, \mathbf{x}_2 \in \mathcal{D}$.

Two random fields are jointly wide-sense stationary if both are individually wide-sense stationary, and if the cross-correlation depends only on $\mathbf{x}_1 - \mathbf{x}_2$. In this case, we define

$$R_{fg}(\mathbf{x}) = E\{f(\mathbf{w} + \mathbf{x})g(\mathbf{w})\}$$
$$K_{fg}(\mathbf{x}) = E\{(f(\mathbf{w} + \mathbf{x}) - \mu_f)(g(\mathbf{w}) - \mu_g)\}.$$

where both expressions are independent of $\mathbf{w}$. The extension to vector-valued color signals is straightforward.

Example 9.3
Suppose that we observe two views f_1 and f_2 of the same scene, where one is shifted by an amount $\mathbf{d} \in \mathcal{D}$ and scaled by α with respect to the other, and both are corrupted by additive white noise, where the noise fields are uncorrelated with each other and with the image. Thus,

$$f_1 = g + v_1$$
$$f_2 = \alpha \mathcal{T}_{\mathbf{d}} g + v_2,$$

where $R_{v_1 v_2} = R_{v_1 g} = R_{v_2 g} = 0$, $E\{v_1\} = E\{v_2\} = 0$, and $E\{v_1^2\} = E\{v_2^2\} = \sigma_v^2$. We wish to find the cross-correlation $R_{f_2 f_1}$.

Solution:

$$\begin{aligned} R_{f_2 f_1}(\mathbf{x}) &= E\{f_2(\mathbf{w} + \mathbf{x})f_1(\mathbf{w})\} \\ &= E\{(\alpha g(\mathbf{w} + \mathbf{x} - \mathbf{d}) + v_2(\mathbf{w} + \mathbf{x}))(g(\mathbf{w}) + v_1(\mathbf{w}))\} \\ &= \alpha R_g(\mathbf{x} - \mathbf{d}). \end{aligned}$$

Note that $R_{f_2 f_1}$ attains its maximum value at $\mathbf{x} = \mathbf{d}$, so that we can estimate $\mathbf{d}$ by first estimating $R_{f_2 f_1}$ and then finding the location of its maximum. □

9.4 Power Density Spectrum

As we have seen, for WSS random fields the autocorrelation function is defined on the same domain as the image. We define the *power density spectrum* (also called power spectral density by some authors) as the Fourier transform of the autocorrelation, whether in the discrete or continuous-domain case. We explain shortly the significance of the term power density spectrum. The autocorrelation is recovered with the inverse Fourier transform. These pairs of relations are as follows for discrete-domain and continuous-domain real scalar WSS random fields.

Discrete domain

$$S_f(\mathbf{u}) = \sum_{\mathbf{x}\in\Lambda} R_f[\mathbf{x}] \exp(-\mathrm{j}2\pi\mathbf{u}\cdot\mathbf{x}) \tag{9.16}$$

$$R_f[\mathbf{x}] = d(\Lambda)\int_{\mathcal{P}_{\Lambda^*}} S_f(\mathbf{u}) \exp(\mathrm{j}2\pi\mathbf{u}\cdot\mathbf{x})d\mathbf{u}. \tag{9.17}$$

Continuous domain

$$S_f(\mathbf{u}) = \int_{\mathbb{R}^D} R_f(\mathbf{x}) \exp(-\mathrm{j}2\pi\mathbf{u}\cdot\mathbf{x})\, d\mathbf{x} \tag{9.18}$$

$$R_f(\mathbf{x}) = \int_{\mathbb{R}^D} S_f(\mathbf{u}) \exp(\mathrm{j}2\pi\mathbf{u}\cdot\mathbf{x})d\mathbf{u} \tag{9.19}$$

We assume that the autocorrelation is such that these expressions are well defined, for example that the autocorrelation is absolutely summable in the discrete-domain case and absolutely integrable in the continuous-domain case.

9.4.1 Properties of the Power Density Spectrum

The power density spectrum has a number of properties that follow from the properties of the autocorrelation.

1) $S_f(\mathbf{u})$ is real. This follows from the fact that $R_f(\mathbf{x})$ is real and even.
2) $S_f(\mathbf{u}) \geq 0$ for all $\mathbf{u}$. We will give an argument for this shortly.
3) $R_f[\mathbf{0}] = E\{|f[\mathbf{x}]|^2\} = d(\Lambda)\int_{\mathcal{P}_{\Lambda^*}} S_f(\mathbf{u})d\mathbf{u}$ in the discrete domain case, and $\int_{\mathbb{R}^D} S_f(\mathbf{u})\, d\mathbf{u}$ in the continuous domain case.

Example 9.4 Discrete space white noise
We found that $R_v[\mathbf{x}] = \sigma_v^2\delta[\mathbf{x}]$. Taking the Fourier transform, we find that $S_v(\mathbf{u}) = \sigma_v^2$. In the continuous case, we also find that the power density spectrum of white noise is a constant. □

Example 9.5 Continuous-domain separable exponential autocorrelation
We had that

$$R_f(x, y) = \sigma_f^2 \exp(-\gamma_1|x|)\exp(-\gamma_2|y|).$$

The power-density spectrum is also separable. It is a straightforward exercise to compute the Fourier transform:

$$S_f(u, v) = \sigma_f^2 \frac{2\gamma_1}{\gamma_1^2 + 4\pi^2u^2}\frac{2\gamma_2}{\gamma_2^2 + 4\pi^2v^2}.$$

□

Example 9.6 Discrete-domain separable exponential autocorrelation
Here, we recall

$$R_f[mX, nX] = \sigma_f^2 \rho_1^{|m|} \rho_2^{|n|}.$$

Again, the power density spectrum is separable and is obtained by taking the discrete-domain Fourier transform. Each term is the sum of two geometric series for positive and negative values of the independent variable.

$$S_f(u, v) = \sigma_f^2 \frac{1 - \rho_1^2}{(1 + \rho_1^2) - 2\rho_1 \cos(2\pi uX)} \frac{1 - \rho_2^2}{(1 + \rho_2^2) - 2\rho_2 \cos(2\pi vX)}.$$

□

We see that all these examples satisfy the properties (1)–(3) above.

9.4.2 Cross Spectrum

In a similar fashion, we define the cross spectrum of two jointly WSS random fields to be the Fourier transform of the cross-correlation function:

$$S_{fg}(\mathbf{u}) = \sum_{\mathbf{x} \in \Lambda} R_{fg}[\mathbf{x}] \exp(-\mathrm{j}2\pi \mathbf{u} \cdot \mathbf{x}) \qquad \text{discrete domain;} \tag{9.20}$$

$$S_{fg}(\mathbf{u}) = \int_{\mathbb{R}^D} R_{fg}(\mathbf{x}) \exp(-\mathrm{j}2\pi \mathbf{u} \cdot \mathbf{x})\, d\mathbf{x} \qquad \text{continuous domain.} \tag{9.21}$$

Note that the cross spectrum is not necessarily real, even if the underlying random fields are real.

9.4.3 Spectral Density Matrix

We can obtain spectral densities for a color signal by taking the term-by-term Fourier transform of the matrix autocorrelation function $\mathbf{R}_{C_B}(\mathbf{x})$. We denote this as the *spectral density matrix* $\mathbf{S}_{C_B}(\mathbf{u})$, where the diagonal terms represent the power density spectra of the tristimulus signals and the off-diagonal terms are the cross spectra of different tristimulus signals. The spectral analysis of vector-valued one-dimensional time series is treated in depth in Brillinger (2001) and Priestley (1981). As in the case of the matrix-valued autocorrelation function, the spectral density matrix depends on the basis of the color space. If $\mathcal{B}$ and $\widetilde{\mathcal{B}}$ are two bases for the color space $\mathcal{C}$, related by the change of basis matrix $\mathbf{A}_{\mathcal{B}\to\widetilde{\mathcal{B}}}$, then using Equation 9.13 and linearity of the Fourier transform, it follows that

$$\mathbf{S}_{C_{\widetilde{\mathcal{B}}}}(\mathbf{u}) = \mathbf{A}_{\mathcal{B}\to\widetilde{\mathcal{B}}} \mathbf{S}_{C_{\mathcal{B}}}(\mathbf{u}) \mathbf{A}^T_{\mathcal{B}\to\widetilde{\mathcal{B}}}. \tag{9.22}$$

We provide the following important property of the spectral density matrix.

Theorem 9.1 The spectral density matrix $\mathbf{S}_C(\mathbf{u})$ of a WSS vector-valued random field is Hermitian and positive semidefinite for each $\mathbf{u} \in \mathcal{D}$.

Proof: For the matrix autocorrelation function, we have

$$\begin{aligned} R_{C,ji}(\mathbf{x}) &= E\{C_j(\mathbf{w}+\mathbf{x})C_i(\mathbf{w})\} \\ &= E\{C_i(\mathbf{w}-\mathbf{x})C_j(\mathbf{w})\} \\ &= R_{C,ij}(-\mathbf{x}) \end{aligned} \tag{9.23}$$

or in other words

$$\mathbf{R}_C^T(\mathbf{x}) = \mathbf{R}_C(-\mathbf{x}). \tag{9.24}$$

It follows that $S_{C,ji}(\mathbf{u}) = S^*_{C,ij}(\mathbf{u})$, and thus $\mathbf{S}_C^T(\mathbf{u}) = \mathbf{S}_C^*(\mathbf{u})$, and conjugating,

$$\mathbf{S}_C^H(\mathbf{u}) = \mathbf{S}_C(\mathbf{u}), \qquad \text{for all } \mathbf{u} \in \mathcal{D}. \tag{9.25}$$

For a complex matrix $\mathbf{A}$, the superscript H denotes the Hermitian transpose, $\mathbf{A}^H = (\mathbf{A}^*)^T$, and a complex matrix is called Hermitian if $\mathbf{A}^H = \mathbf{A}$. We have used the fact that the Fourier transform of $f(-\mathbf{x})$ is $F(-\mathbf{u})$ and if f is real, this is $F^*(\mathbf{u})$.

To show that $\mathbf{S}(\mathbf{u})$ is positive semidefinite for every $\mathbf{u}$, let $\mathbf{a} = [a_1, a_2, a_3]^T$ be a column matrix of arbitrary real constants, and form the scalar process

$$f(\mathbf{x}) = \mathbf{a}^T\mathbf{C}(\mathbf{x}). \tag{9.26}$$

The autocorrelation function of f is

$$\begin{aligned} R_f(\mathbf{x}) &= E\{\mathbf{a}^T\mathbf{C}(\mathbf{w}+\mathbf{x})\mathbf{C}^T(\mathbf{w})\mathbf{a}\} \\ &= \mathbf{a}^T\mathbf{R}_C(\mathbf{x})\mathbf{a} \end{aligned} \tag{9.27}$$

and the power density spectrum of f is

$$S_f(\mathbf{u}) = \mathbf{a}^T\mathbf{S}_C(\mathbf{u})\mathbf{a}. \tag{9.28}$$

Since $S_f(\mathbf{u}) \geq 0$ by property (2) of the power density spectrum, $\mathbf{S}_C(\mathbf{u})$ must be positive semidefinite for each $\mathbf{u}$. □

9.5 Filtering and Sampling of WSS Random Fields

The realizations of a random field can be subjected to filtering and sampling operations, as done for deterministic signals. In general, the result is another random field. In the case of WSS random fields passed through linear, shift-invariant filters, the resulting random field is also WSS and we can determine the autocorrelation and power density spectrum. Similarly, when applying sampling operations, we can characterize the moments of the resulting random field. A few results along these lines are presented in this section, first for scalar random fields and then for vector-valued random fields.

9.5.1 LSI Filtering of a Scalar WSS Random Field

Theorem 9.2 If a real, scalar WSS random field f with zero mean and power density spectrum $S_f(\mathbf{u})$ is passed through an LSI filter with frequency response $H(\mathbf{u})$, then the output g of the filter is also a WSS random field with zero mean and power density spectrum

$$S_g(\mathbf{u}) = |H(\mathbf{u})|^2 S_f(\mathbf{u}). \tag{9.29}$$

We will prove this result for a discrete-domain random field and filter. The continuous-domain case is done in a similar fashion and is left as an exercise.

Proof: Assume the random field and filter are defined on a lattice Λ. The LSI filter has unit-sample response $h[\mathbf{x}]$, which is the inverse Fourier transform of $H(\mathbf{u})$. The output of the filter is $g[\mathbf{x}] = \sum_{\mathbf{s}\in\Lambda} h[\mathbf{s}]f[\mathbf{x}-\mathbf{s}]$. By linearity of the expectation operator,

$$E\{g[\mathbf{x}]\} = \sum_{\mathbf{s}\in\Lambda} h[\mathbf{s}]E\{f[\mathbf{x}-\mathbf{s}]\} = 0,$$

so the output process is zero mean. The autocorrelation function of the output signal is given by

$$\begin{aligned}
R_g[\mathbf{w}+\mathbf{x},\mathbf{w}] &= E\{g[\mathbf{w}+\mathbf{x}]g[\mathbf{w}]\} \\
&= E\left\{\sum_{\mathbf{s}_1\in\Lambda} h[\mathbf{s}_1]f[\mathbf{w}+\mathbf{x}-\mathbf{s}_1]\sum_{\mathbf{s}_2\in\Lambda} h[\mathbf{s}_2]f[\mathbf{w}-\mathbf{s}_2]\right\} \\
&= \sum_{\mathbf{s}_1\in\Lambda}\sum_{\mathbf{s}_2\in\Lambda} h[\mathbf{s}_1]h[\mathbf{s}_2]E\{f[\mathbf{w}+\mathbf{x}-\mathbf{s}_1]f[\mathbf{w}-\mathbf{s}_2]\} \\
&= \sum_{\mathbf{s}_1\in\Lambda}\sum_{\mathbf{s}_2\in\Lambda} h[\mathbf{s}_1]h[\mathbf{s}_2]R_f[\mathbf{x}-\mathbf{s}_1+\mathbf{s}_2],
\end{aligned}$$

which is independent of $\mathbf{w}$ and thus can be denoted $R_g[\mathbf{x}]$. Thus, the output of the LSI system is a zero-mean, wide-sense stationary random field.

Let $q = h * R_f$. Then

$$\begin{aligned}
R_g[\mathbf{x}] &= \sum_{\mathbf{s}_2\in\Lambda} h[\mathbf{s}_2]\sum_{\mathbf{s}_1\in\Lambda} h[\mathbf{s}_1]R_f[\mathbf{x}+\mathbf{s}_2-\mathbf{s}_1] \\
&= \sum_{\mathbf{s}_2\in\Lambda} h[\mathbf{s}_2]q[\mathbf{x}+\mathbf{s}_2].
\end{aligned}$$

Define $h_R[\mathbf{x}] = h[-\mathbf{x}]$, and make the change of variables $\mathbf{s} = -\mathbf{s}_2$. Then

$$R_g[\mathbf{x}] = \sum_{\mathbf{s}\in\Lambda} h_R[\mathbf{s}]q[\mathbf{x}-\mathbf{s}],$$

i.e. $R_g = h_R * q = h_R * h * R_f$. Taking Fourier transforms

$$\begin{aligned}
S_g(\mathbf{u}) &= H_R(\mathbf{u})Q(\mathbf{u}) \\
&= H_R(\mathbf{u})H(\mathbf{u})S_f(\mathbf{u}).
\end{aligned}$$

Finally,

$$\begin{aligned}
H_R(\mathbf{u}) &= \sum_{\mathbf{x}\in\Lambda} h[-\mathbf{x}]\exp(-\mathrm{j}2\pi\mathbf{u}\cdot\mathbf{x}) \\
&= \sum_{\mathbf{s}\in\Lambda} h[\mathbf{s}]\exp(-\mathrm{j}2\pi\mathbf{u}\cdot(-\mathbf{s})) \qquad (\mathbf{s}=-\mathbf{x}) \\
&= \left[\sum_{\mathbf{s}\in\Lambda} h[\mathbf{s}]\exp(-\mathrm{j}2\pi\mathbf{u}\cdot\mathbf{s})\right]^* \\
&= H^*(\mathbf{u}),
\end{aligned}$$

and $S_g(\mathbf{u}) = |H(\mathbf{u})|^2 S_f(\mathbf{u})$. Note that for the result to hold, we must assume that $h[\mathbf{x}]$ is *real*. □

We can also determine the cross-correlation and cross-power density spectrum of f and $g = h * f$.

Theorem 9.3 Define $R_{gf}[\mathbf{x}] = E\{g[\mathbf{w} + \mathbf{x}]f[\mathbf{w}]\}$ (which is independent of $\mathbf{w}$), and let $S_{gf}(\mathbf{u})$ be the cross-power density spectrum, i.e. the discrete-space Fourier transform of R_{gf} on the lattice Λ. Then $S_{gf}(\mathbf{u}) = H(\mathbf{u})S_f(\mathbf{u})$.

Proof:

$$\begin{aligned} E\{g[\mathbf{w}+\mathbf{x}]f[\mathbf{w}]\} &= E\left\{\sum_{\mathbf{s}\in\Lambda} h[\mathbf{s}]f[\mathbf{w}+\mathbf{x}-\mathbf{s}]f[\mathbf{w}]\right\} \\ &= \sum_{\mathbf{s}\in\Lambda} h[\mathbf{s}]E\{f[\mathbf{w}+\mathbf{x}-\mathbf{s}]f[\mathbf{w}]\} \\ &= \sum_{\mathbf{s}\in\Lambda} h[\mathbf{s}]R_f[\mathbf{x}-\mathbf{s}], \end{aligned} \tag{9.30}$$

which is independent of $\mathbf{w}$ and can be denoted $R_{gf}(\mathbf{x})$. Thus f and g are jointly WSS, and $R_{gf} = h * R_f$. Taking Fourier transforms, $S_{gf}(\mathbf{u}) = H(\mathbf{u})S_f(\mathbf{u})$. □

9.5.2 Why is $S_f(u)$ Called a Power Density Spectrum?

Given these results, we can now explain why $S_f(\mathbf{u})$ is called a power density spectrum, and why $S_f(\mathbf{u}) \geq 0$.

In one dimension, suppose we pass a WSS random field f with power density spectrum $S_f(u)$ through the band-pass filter shown in Figure 9.2. Then the power density spectrum of the output g is

$$S_g(u) = \begin{cases} S_f(u) & u_0 \leq u \leq u_0 + \Delta u \\ 0 & \text{otherwise} \end{cases}.$$

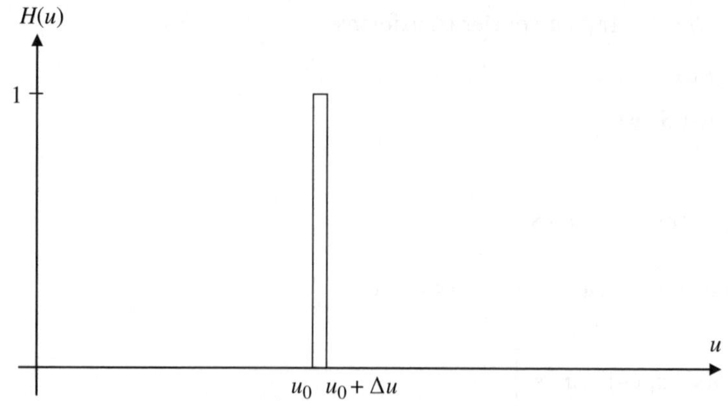

Figure 9.2 Frequency response of a narrowband band-pass filter in one dimension.

In the continuous-domain case, the power of g is $E\{|g[x]|^2\} = \int_{\mathbb{R}} S_g(u)\,du \approx S_f(u_0)\Delta u$. In other words, the power of the signal f in the frequency band $[u, u+\Delta u]$ is $S_f(u)\Delta u$ and so the *power density* per unit of frequency is $S_f(u)$. This is easily extended to two or more dimensions, by using a band-pass filter with a small rectangular passband.

In the discrete-domain case, the power of g is $E\{|g[x]|^2\} = \mathrm{d}(\Lambda)\int S_g(u)\,du \approx \mathrm{d}(\Lambda)S_f(u_0)\Delta u$. Here, the power of the signal f in the frequency band $[u, u+\Delta u]$ is $\mathrm{d}(\Lambda)S_f(u)\Delta u$ and so the *power density* per unit of frequency is $\mathrm{d}(\Lambda)S_f(u)$. Again, this is easily extended to two or more dimensions.

It is also clear that since the power of g must be positive for any u_0 and any Δu, that $S_f(u_0)$ must be positive for any u_0.

9.5.3 LSI Filtering of a WSS Color Random Field

Suppose that realizations of a discrete-domain WSS color random field are passed through the LSI color system with transfer matrix $\mathbf{H}_{\mathcal{B}}[\mathbf{x}]$, and corresponding frequency response matrix $\widehat{\mathbf{H}}_{\mathcal{B}}(\mathbf{u})$ for some basis $\mathcal{B}$. If the input to the system is $\mathbf{C}_{\mathcal{B}}[\mathbf{x}]$ and the output is $\mathbf{Q}_{\mathcal{B}}[\mathbf{x}]$, then, as shown in Chapter 8, $\mathbf{Q}_{\mathcal{B}}[\mathbf{x}] = (\mathbf{H}_{\mathcal{B}} \circledast \mathbf{C}_{\mathcal{B}})[\mathbf{x}]$, where explicitly

$$Q_k[\mathbf{x}] = \sum_{i=1}^{3} (H_{ki} * C_i)[\mathbf{x}]. \tag{9.31}$$

The basis is assumed to be $\mathcal{B}$ throughout this development and so will be omitted to lighten the notation. We wish to obtain the matrix autocorrelation function and the spectral density matrix of the random field $\mathbf{Q}$ when the input random field is a WSS zero-mean process with matrix autocorrelation function $\mathbf{R}_C[\mathbf{x}]$ and corresponding spectral density matrix $\mathbf{S}_C(\mathbf{u})$.

From the linearity of expectation, $E\{Q_k[\mathbf{x}]\} = 0$ and

$$\begin{aligned}
R_{Q,kl}[\mathbf{w}+\mathbf{x}, \mathbf{w}] &= E\left\{Q_k[\mathbf{w}+\mathbf{x}]Q_\ell[\mathbf{w}]\right\} \\
&= E\left\{\sum_{i=1}^{3}(H_{ki} * C_i)[\mathbf{w}+\mathbf{x}]\sum_{j=1}^{3}(H_{\ell j} * C_j)[\mathbf{w}]\right\} \\
&= E\left\{\sum_{i=1}^{3}\sum_{\mathbf{s}_1\in\Lambda} H_{ki}[\mathbf{s}_1]C_i[\mathbf{w}+\mathbf{x}-\mathbf{s}_1]\sum_{j=1}^{3}\sum_{\mathbf{s}_2\in\Lambda} H_{\ell j}[\mathbf{s}_2]C_j[\mathbf{w}-\mathbf{s}_2]\right\} \\
&= \sum_{i=1}^{3}\sum_{j=1}^{3}\sum_{\mathbf{s}_1\in\Lambda}\sum_{\mathbf{s}_2\in\Lambda} H_{ki}[\mathbf{s}_1]H_{\ell j}[\mathbf{s}_2]R_{C_iC_j}[\mathbf{x}-\mathbf{s}_1+\mathbf{s}_2]
\end{aligned} \tag{9.32}$$

which is independent of $\mathbf{w}$ for all k, l, so $\mathbf{Q}$ is WSS. Defining $H_{R,ki}[\mathbf{x}] = H_{ki}[-\mathbf{x}]$ and following the development of Theorem 9.2,

$$R_{Q,k\ell}[\mathbf{x}] = \sum_{i=1}^{3}\sum_{j=1}^{3}(H_{ki} * H_{R,\ell j} * R_{C,ij})[\mathbf{x}], \tag{9.33}$$

and taking the Fourier transform

$$S_{Q,k\ell}(\mathbf{u}) = \sum_{i=1}^{3}\sum_{j=1}^{3}\widehat{H}_{ki}(\mathbf{u})\widehat{H}^*_{\ell j}(\mathbf{u})S_{C,ij}(\mathbf{u}). \tag{9.34}$$

In matrix form, this is written

$$\mathbf{S}_Q(\mathbf{u}) = \widehat{\mathbf{H}}(\mathbf{u})\mathbf{S}_C(\mathbf{u})\widehat{\mathbf{H}}^H(\mathbf{u}). \tag{9.35}$$

In the common special case where $\widehat{\mathbf{H}}(\mathbf{u})$ is diagonal, i.e. the three components are filtered independently, then

$$S_{Q,k\ell}(\mathbf{u}) = \widehat{H}_{kk}(\mathbf{u})\widehat{H}^*_{\ell\ell}(\mathbf{u})S_{C,k\ell}(\mathbf{u}), \tag{9.36}$$

and in particular

$$S_{Q,kk}(\mathbf{u}) = |\widehat{H}_{kk}(\mathbf{u})|^2 S_{C,kk}(\mathbf{u}). \tag{9.37}$$

9.5.4 Sampling of a WSS Continuous-Domain Random Field

Let $\mathbf{C}_a(\mathbf{x})$ be a realization of a WSS continuous-domain random field with matrix-valued autocorrelation function $\mathbf{R}_{C_a}(\mathbf{x})$ and corresponding spectral density matrix $\mathbf{S}_{C_a}(\mathbf{u})$, with respect to some arbitrary given basis for the color space. Suppose that we sample the components of $\mathbf{C}_a(\mathbf{x})$ on a lattice Λ to obtain

$$\mathbf{C}[\mathbf{x}] = \mathbf{C}_a(\mathbf{x}), \quad \mathbf{x} \in \Lambda. \tag{9.38}$$

Then, the sampled functions $\mathbf{C}[\mathbf{x}]$ are realizations of a discrete-domain random field satisfying the following properties.

Theorem 9.4 If we sample a WSS continuous-domain random field on a lattice Λ, the resulting discrete-domain random field is WSS, with matrix-valued autocorrelation function and spectral density matrix given by

$$\mathbf{R}_C[\mathbf{x}] = \mathbf{R}_{C_a}(\mathbf{x}), \quad \mathbf{x} \in \Lambda \tag{9.39}$$

$$\mathbf{S}_C(\mathbf{u}) = \frac{1}{\mathrm{d}(\Lambda)} \sum_{\mathbf{r}\in\Lambda^*} \mathbf{S}_{C_a}(\mathbf{u} - \mathbf{r}) \tag{9.40}$$

respectively.

Proof: For $\mathbf{w}, \mathbf{x} \in \Lambda$, we have

$$\mu[\mathbf{x}] = E\{\mathbf{C}[\mathbf{x}]\} = E\{\mathbf{C}_a(\mathbf{x})\} = \mu_{C_a}, \quad \text{independently of } \mathbf{x} \tag{9.41}$$

and

$$\begin{aligned}\mathbf{R}_C[\mathbf{w}+\mathbf{x}, \mathbf{w}] &= E\{\mathbf{C}[\mathbf{w}+\mathbf{x}]\mathbf{C}^T[\mathbf{w}]\} \\ &= E\{\mathbf{C}_a(\mathbf{w}+\mathbf{x})\mathbf{C}_a^T(\mathbf{w})\} \\ &= \mathbf{R}_{C_a}(\mathbf{x}), \quad \text{independently of } \mathbf{w}.\end{aligned} \tag{9.42}$$

Thus, $\mathbf{C}[\mathbf{x}]$ are realizations of a WSS discrete-domain random field with mean $\mu_C = \mu_{C_a}$ and matrix autocorrelation function

$$\mathbf{R}_C[\mathbf{x}] = \mathbf{R}_{C_a}(\mathbf{x}), \quad \mathbf{x} \in \Lambda. \tag{9.43}$$

Since each element of the matrix $\mathbf{R}_C[\mathbf{x}]$ is the sampled version on Λ of the corresponding element of $\mathbf{R}_{C_a}(\mathbf{x})$, it follows from Equation 6.7 that

$$\mathbf{S}_C(\mathbf{u}) = \frac{1}{\mathrm{d}(\Lambda)} \sum_{\mathbf{r}\in\Lambda^*} \mathbf{S}_{C_a}(\mathbf{u} - \mathbf{r}), \tag{9.44}$$

where the sum is applied independently to each element of the matrix. □

9.6 Estimation of the Spectral Density Matrix

In this section we briefly describe how one could get an estimate of the spectral density matrix for a vector-valued discrete-domain WSS random field. Estimation of the power density spectrum of a scalar random field follows as a simple special case. Spectral estimation has been studied at great length for one-dimensional signals and there exists a wide variety of methods ranging from taking Fourier transforms of sample data to parametric modeling methods such as auto-regressive (AR) modeling, etc. Here, we present a direct extension of the method of averaging modified periodograms, which has been shown to be an unbiased and consistent estimator of the power density spectrum. Analysis of these methods is quite involved and we simply motivate, describe and illustrate the method. More details for the one-dimensional case can be found in references such as Priestley (1981) and Brillinger (2001).

Let $\mathbf{C}_{\mathcal{B}}[\mathbf{x}]$ be a realization of a WSS discrete-domain vector-valued random field defined on a lattice Λ. We assume a fixed basis $\mathcal{B}$ throughout this section and so will suppress the subscript $\mathcal{B}$ (the symbol $\mathcal{B}$ will be used for another purpose). We also assume a zero-mean random field, since a nonzero mean implies a Dirac delta in the power density spectrum at $\mathbf{u} = 0$. For non-zero-mean processes, we start by subtracting an estimate of the mean. Let Γ be a sublattice of Λ (typically rectangular) and let $\mathcal{B}$ be a set of K coset representatives, typically in a rectangular unit cell of Γ. In other words, $\mathcal{B}$ denotes a rectangular block of the data consisting of K samples. We then define a window function $w[\mathbf{x}]$ on Λ that is only nonzero on $\mathcal{B}$. The windowed data is $w[\mathbf{x}]\mathbf{C}[\mathbf{x}]$, i.e. each component of $\mathbf{C}[\mathbf{x}]$ is multiplied by the same scalar valued window function. The purpose of the window is to limit the data to a finite region. If $w[\mathbf{x}] = 1$ for all $\mathbf{x} \in \mathcal{B}$, we call it a rectangular window. The performance of the spectral estimator can be improved by using a tapered window. Windows will also be seen in the context of FIR filter design in Section 10.5.1.

Let $\hat{\mathbf{V}}(\mathbf{u})$ be the discrete-domain Fourier transform of $w[\mathbf{x}]\mathbf{C}[\mathbf{x}]$,

$$\hat{\mathbf{V}}(\mathbf{u}) = \sum_{\mathbf{x}\in\mathcal{B}} w[\mathbf{x}]\mathbf{C}[\mathbf{x}] \exp(-\mathrm{j}2\pi\mathbf{u}\cdot\mathbf{x}). \tag{9.45}$$

The proposed estimator of $\mathbf{S}_C(\mathbf{u})$ based on the data in $\mathcal{B}$ is

$$\hat{\mathbf{Q}}(\mathbf{u}) = \frac{1}{K}\hat{\mathbf{V}}(\mathbf{u})\hat{\mathbf{V}}^H(\mathbf{u}). \tag{9.46}$$

The elements of $\hat{\mathbf{Q}}(\mathbf{u})$ are referred to as *second order periodograms* [Brillinger (2001)]. The motivation for this estimate is that $\hat{\mathbf{Q}}(\mathbf{u})$ is the Fourier transform of the windowed sample autocorrelation

$$\mathbf{Q}[\mathbf{x}] = \frac{1}{K}\sum_{\mathbf{z}\in\mathcal{B}} w[\mathbf{z}+\mathbf{x}]\mathbf{C}[\mathbf{z}+\mathbf{x}]\mathbf{C}^T[\mathbf{z}]w[\mathbf{z}]. \tag{9.47}$$

The proof of this result is left as an exercise.

We can find the expected value of these two estimates. For a WSS random field,

$$\begin{aligned} E\{\mathbf{Q}[\mathbf{x}]\} &= \frac{1}{K}\mathbf{R}_C[\mathbf{x}]\sum_{\mathbf{z}\in\mathcal{B}} w[\mathbf{z}+\mathbf{x}]w[\mathbf{z}] \\ &= \frac{1}{K}\mathbf{R}_C[\mathbf{x}]R_w[\mathbf{x}], \end{aligned} \tag{9.48}$$

and

$$E\{\widehat{\mathbf{Q}}(\mathbf{u})\} = \frac{1}{K}\sum_{\mathbf{x}\in\Lambda} \mathbf{R}_C[\mathbf{x}]R_w[\mathbf{x}]\exp(-\mathrm{j}2\pi\mathbf{u}\cdot\mathbf{x})$$
$$= \frac{\mathrm{d}(\Lambda)}{K}\int_{\mathcal{P}_{\Lambda^*}} \mathbf{S}_C(\mathbf{r})R_w(\mathbf{u}-\mathbf{r})\,\mathrm{d}\mathbf{r}. \tag{9.49}$$

Thus, the expected value of our spectral estimate is equal to the true spectral density matrix convolved with a scaled version of the Fourier transform of the window autocorrelation. If $(\mathrm{d}(\Lambda)/K)R_w(\mathbf{u})$ approaches $\delta(\mathbf{u})$ when K is large, the estimate is asymptotically unbiased. We note that $R_w(\mathbf{u}) = |W(\mathbf{u})|^2$ and that if $(\mathrm{d}(\Lambda)/K)R_w(\mathbf{u}) \to \delta(\mathbf{u})$, then

$$\frac{\mathrm{d}(\Lambda)}{K}\int_{\mathcal{P}_{\Lambda^*}} R_w(\mathbf{u})\,\mathrm{d}\mathbf{u} = \frac{\mathrm{d}(\Lambda)}{K}\int_{\mathcal{P}_{\Lambda^*}} |W(\mathbf{u})|^2\,\mathrm{d}\mathbf{u} \to 1, \tag{9.50}$$

and thus

$$\frac{1}{K}\sum_{\mathbf{x}\in\Lambda} |w[\mathbf{x}]|^2 \to 1. \tag{9.51}$$

Thus, the window should be normalized such that $\sum_{\mathbf{x}\in\Lambda} w^2[\mathbf{x}] = K$ (which automatically holds for the rectangular window).

Although the estimate is asymptotically unbiased, the variance of the estimate increases with K, resulting in a very poor estimate. The standard approach to overcome this problem is to segment the data sample into intermediate sized blocks, e.g., 64 in each dimension. These blocks can overlap by up to 50% in each dimension without compromising independence of the estimates. If we assume that there are L such data blocks, denoted $\mathbf{C}^{(i)}$, shifted to the origin, then

$$\widehat{\mathbf{V}}^{(i)}(\mathbf{u}) = \sum_{\mathbf{x}\in B} w[\mathbf{x}]\mathbf{C}^{(i)}[\mathbf{x}]\exp(-\mathrm{j}2\pi\mathbf{u}\cdot\mathbf{x}) \tag{9.52}$$

$$\widehat{\mathbf{Q}}^{(i)}(\mathbf{u}) = \frac{1}{K}\widehat{\mathbf{V}}^{(i)}(\mathbf{u})\widehat{\mathbf{V}}^{(i)H}(\mathbf{u}) \tag{9.53}$$

and our final spectral estimate of $\mathbf{S}_C(\mathbf{u})$ is

$$\frac{1}{L}\sum_{i=1}^{L} \widehat{\mathbf{Q}}^{(i)}(\mathbf{u}). \tag{9.54}$$

We note that the frequency variable $\mathbf{u}$ is continuous that we only need to consider one unit cell of Λ^*, since the spectral density matrix is periodic with this unit cell as the period. Since the windowed data $w[\mathbf{x}]\mathbf{C}^{(i)}[\mathbf{x}]$ is space limited to B, we can exactly compute samples of $\widehat{\mathbf{V}}^{(i)}(\mathbf{u})$ using the periodic discrete-domain Fourier transform on Γ (i.e. the multidimensional DFT). We usually choose the data block size in each dimension to be a power of two (e.g., 64) so that we can compute the DFT efficiently using the FFT algorithm.

We show a few examples of this estimation procedure for still color images. In all the examples presented here, we use a block size of 64×64 ($K = 64^2 = 4096$), a block overlap of 25% in each dimension, and a separable four-term Blackman–Harris window. Harris (1978) gives a comprehensive overview of windows for spectral analysis and identifies the four-term Blackman–Harris window as a good window. Figure 9.3 shows the magnitude (on a dB scale) of the six distinct components of the estimated

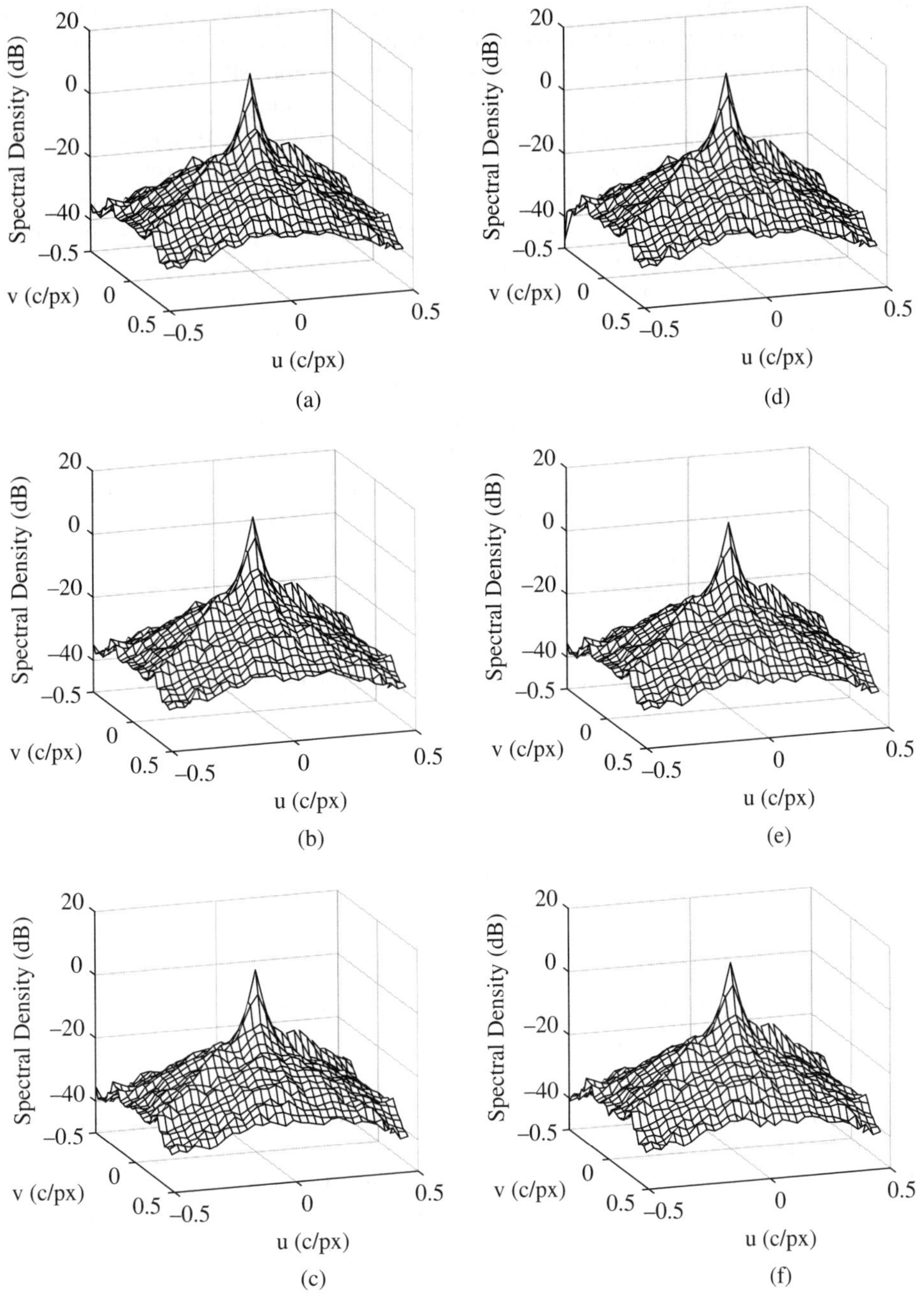

Figure 9.3 Elements of spectral density matrix in dB for Kodak motorcycles image (#5) using sRGB basis of color space. (a) $\widehat{Q}_{RR}(\mathbf{u})$. (b) $\widehat{Q}_{GG}(\mathbf{u})$. (c) $\widehat{Q}_{BB}(\mathbf{u})$. (d) $\widehat{Q}_{RG}(\mathbf{u})$. (e) $\widehat{Q}_{RB}(\mathbf{u})$. (f) $\widehat{Q}_{GB}(\mathbf{u})$.

spectral density matrix for a typical sRGB color image (Kodak image 5, motorcycles, size 768×512). The estimated spectral densities are the average of 160 modified periodograms for these parameters and image size. In this example, sRGB gamma has been applied to reverse the gamma correction and convert the image to linear sRGB color coordinates. Observing these spectra, we see that the power density spectra of the R, G, and B components in the left column are all quite similar and the cross spectra are similar in magnitude, showing the high correlation between the R, G, and B components.

Figure 9.4 shows the same spectral density estimates with respect to the OZW opponent basis of Zhang and Wandell, as specified in Equation 8.44. The component O_1 is a brightness component while O_2 and O_3 represent opponent chromatic components. We see that O_1 has significantly higher power than O_2 and O_3, and the cross spectra are in turn significantly lower, showing reduced correlation between the components.

We conclude with an illustration of the effect of filtering on the spectral density matrix. We consider a filter transfer matrix that is diagonal in the OZW basis. Specifically, in this example system, the O_1 component is left untouched while the O_2 and O_3 components each undergo an identical low-pass filtering operation with the filter whose frequency response $H_L(\mathbf{u})$ is illustrated in Figure 10.17(c). Thus, the transfer matrix of this filter is

$$\widehat{\mathbf{H}}_{\mathcal{OZW}}(\mathbf{u}) = \begin{bmatrix} 1 & 0 & 0 \\ 0 & H_L(\mathbf{u}) & 0 \\ 0 & 0 & H_L(\mathbf{u}) \end{bmatrix}. \tag{9.55}$$

We can (if desired) obtain the filter transfer matrix directly in the sRGB basis using Equation (8.43)

$$\widehat{\mathbf{H}}_{s\mathcal{RGB}}(\mathbf{u}) = \mathbf{A}_{\mathcal{OZW}\to s\mathcal{RGB}}\widehat{\mathbf{H}}(\mathbf{u})\mathbf{A}^{-1}_{\mathcal{OZW}\to s\mathcal{RGB}}. \tag{9.56}$$

Now, if the realizations $\mathbf{C}_{\mathcal{OZW}}[\mathbf{x}]$ of a WSS random field with spectral density matrix $\mathbf{S}_{C,\mathcal{OZW}}(\mathbf{u})$ are filtered with the above system, to obtain $\mathbf{Q}_{\mathcal{OZW}}[\mathbf{x}]$, from Equation (9.35) we find that

$$\begin{aligned} \mathbf{S}_{Q,\mathcal{OZW}}(\mathbf{u}) &= \widehat{\mathbf{H}}_{\mathcal{OZW}}(\mathbf{u})\mathbf{S}_{C,\mathcal{OZW}}(\mathbf{u})\widehat{\mathbf{H}}^H_{\mathcal{OZW}}(\mathbf{u}) \\ &= \begin{bmatrix} S_{11}(\mathbf{u}) & H_L^*(\mathbf{u})S_{12}(\mathbf{u}) & H_L^*(\mathbf{u})S_{13}(\mathbf{u}) \\ H_L(\mathbf{u})S_{21}(\mathbf{u}) & |H_L(\mathbf{u})|^2 S_{22}(\mathbf{u}) & |H_L(\mathbf{u})|^2 S_{23}(\mathbf{u}) \\ H_L(\mathbf{u})S_{31}(\mathbf{u}) & |H_L(\mathbf{u})|^2 S_{32}(\mathbf{u}) & |H_L(\mathbf{u})|^2 S_{33}(\mathbf{u}) \end{bmatrix} \end{aligned} \tag{9.57}$$

where for simplicity of notation, we write $\mathbf{S}_{C,\mathcal{OZW}}(\mathbf{u}) = [S_{ij}(\mathbf{u})]$.

Figure 9.5 shows an estimate of the six distinct spectral densities in the OZW basis for the motorcycles image filtered with the above system. As predicted by Equation (9.57), the power density spectrum of the O_1 component is unchange, while the high frequencies of the O_2 and O_2 component are significantly attenuated. We see that the high frequencies of the cross spectra are also significantly attenuated, especially the O_2O_3 cross spectrum. Figure 9.6 shows the estimate of the six distinct spectral densities in the sRGB basis. We see that these spectral densities have been modified from the original spectral densities shown in Figure 9.3, although the exact nature of the modification is harder to interpret in this basis.

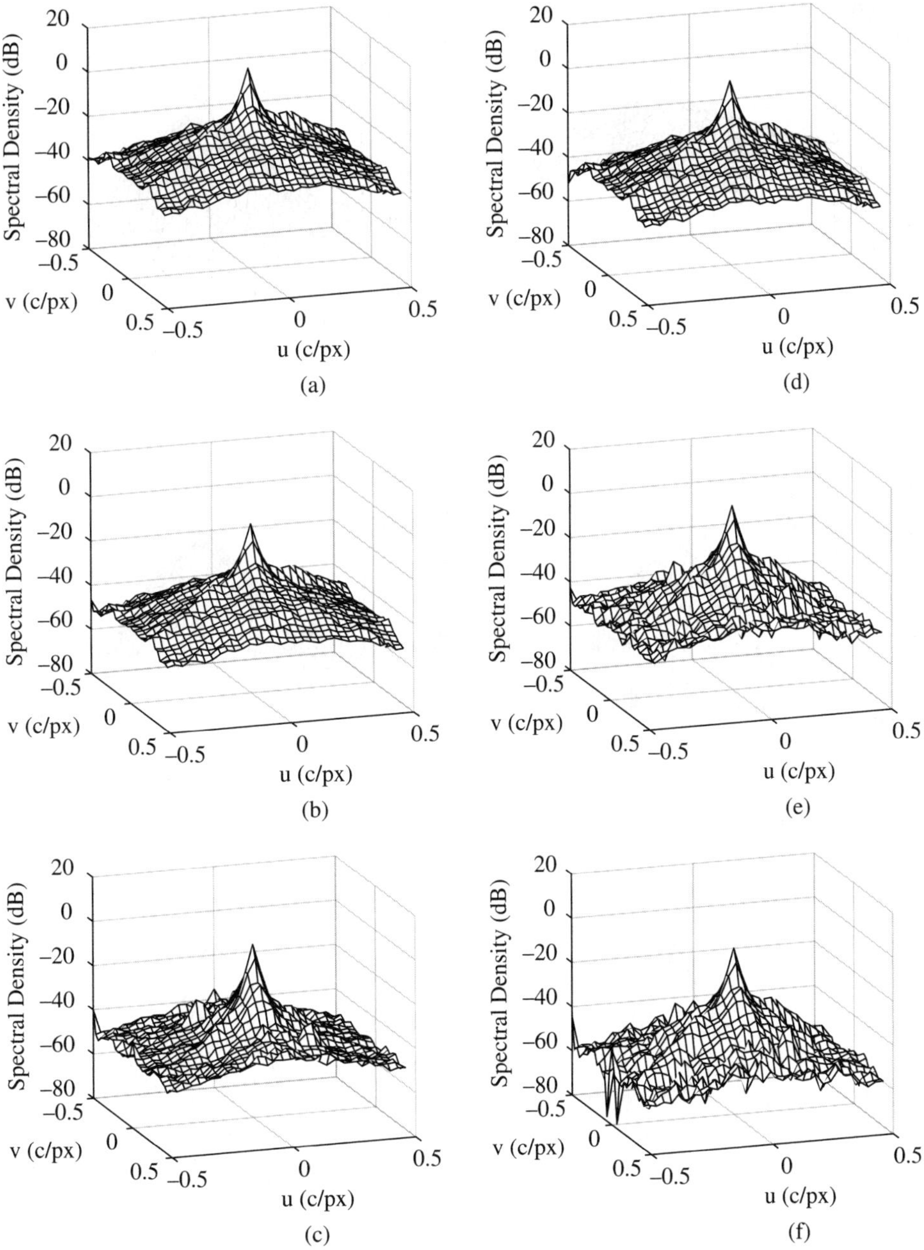

Figure 9.4 Elements of spectral density matrix in dB for Kodak motorcycles image (#5) using the Zhang and Wandell OZW opponent basis of color space. (a) $\widehat{Q}_{OZW,11}(\mathbf{u})$. (b) $\widehat{Q}_{OZW,22}(\mathbf{u})$. (c) $\widehat{Q}_{OZW,33}(\mathbf{u})$. (d) $\widehat{Q}_{OZW,12}(\mathbf{u})$. (e) $\widehat{Q}_{OZW,13}(\mathbf{u})$. (f) $\widehat{Q}_{OZW,23}(\mathbf{u})$.

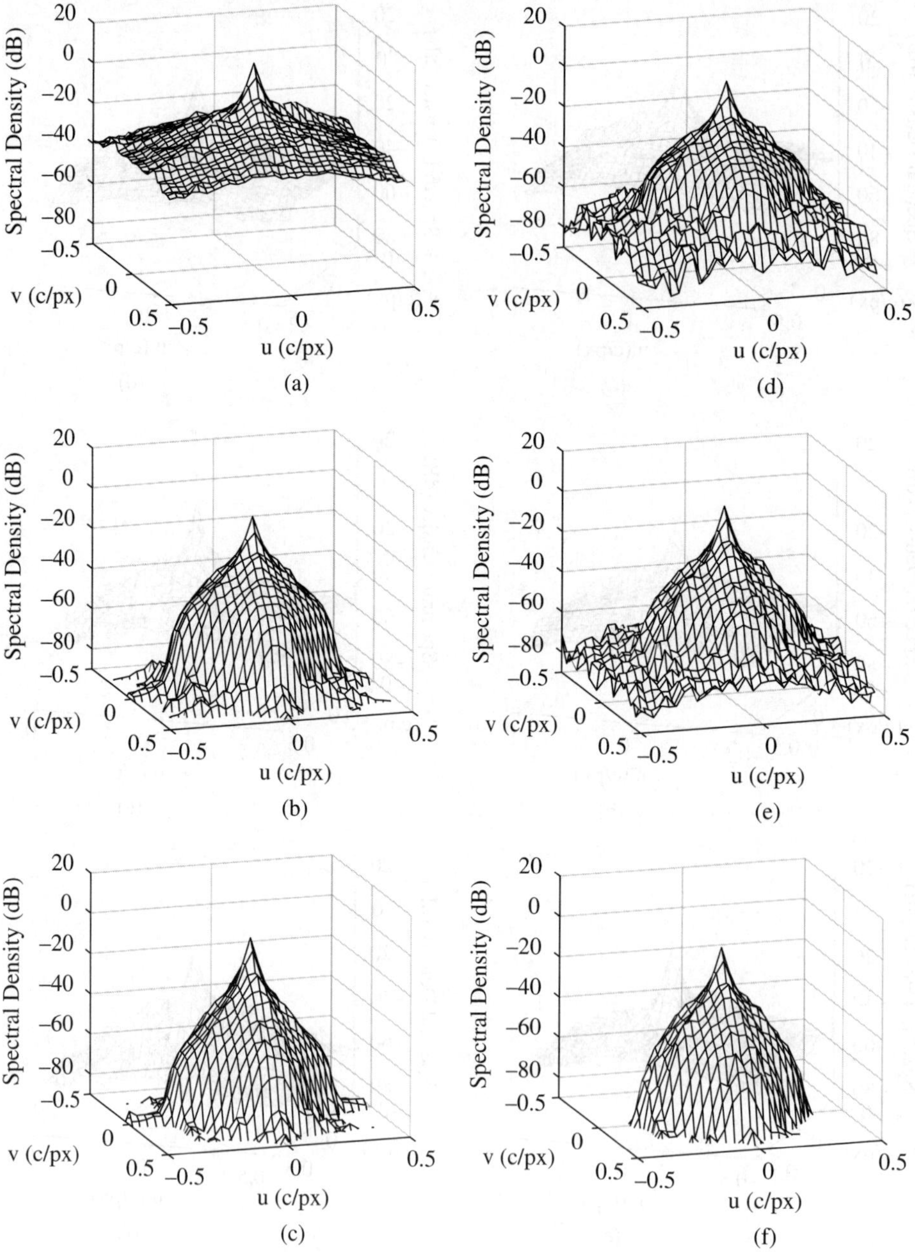

Figure 9.5 Elements of spectral density matrix in dB for Kodak motorcycles image (#5) using the Zhang and Wandell OZW opponent basis of color space, where the components O_2 and O_3 have undergone a low-pass filtering operation. (a) $\hat{Q}_{OZW,11}(\mathbf{u})$. (b) $\hat{Q}_{OZW,22}(\mathbf{u})$. (c) $\hat{Q}_{OZW,33}(\mathbf{u})$. (d) $\hat{Q}_{OZW,12}(\mathbf{u})$. (e) $\hat{Q}_{OZW,13}(\mathbf{u})$. (f) $\hat{Q}_{OZW,23}(\mathbf{u})$.

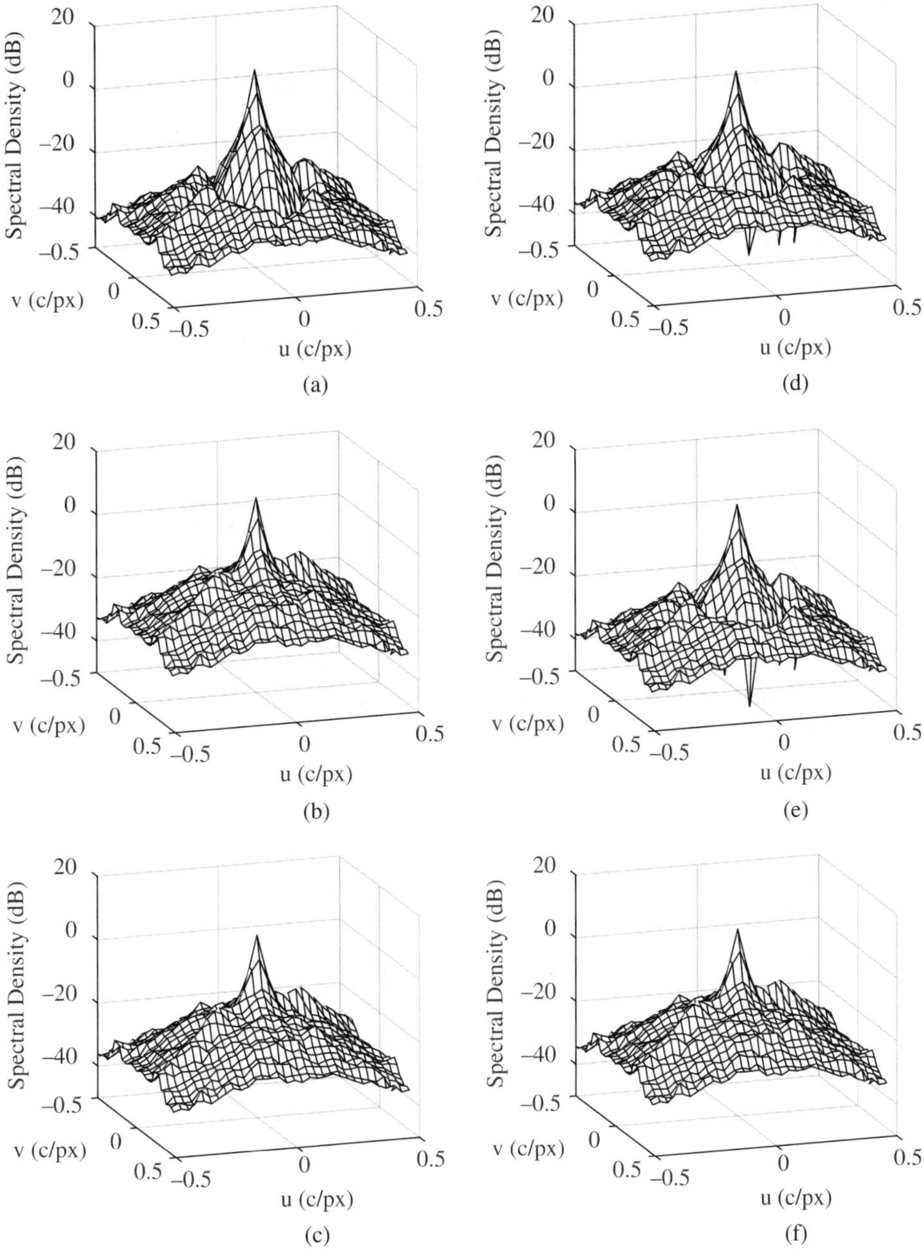

Figure 9.6 Elements of spectral density matrix in dB for Kodak motorcycles image (#5) using sRGB basis of color space after filtering. (a) $\widehat{Q}_{RR}(\mathbf{u})$. (b) $\widehat{Q}_{GG}(\mathbf{u})$. (c) $\widehat{Q}_{BB}(\mathbf{u})$. (d) $\widehat{Q}_{RG}(\mathbf{u})$. (e) $\widehat{Q}_{RB}(\mathbf{u})$. (f) $\widehat{Q}_{GB}(\mathbf{u})$.

Problems

1 Continuous-domain separable exponential autocorrelation. Show that if

$$R_f(x,y) = \sigma_f^2 \exp(-\gamma_1|x|)\exp(-\gamma_2|y|),$$

then the power-density spectrum is given by

$$S_f(u,v) = \sigma_f^2 \frac{2\gamma_1}{\gamma_1^2 + 4\pi^2u^2} \frac{2\gamma_2}{\gamma_2^2 + 4\pi^2v^2}.$$

2 Discrete-domain separable exponential autocorrelation. Show that if

$$R_f[mX, nX] = \sigma_f^2 \rho_1^{|m|} \rho_2^{|n|},$$

then the power density spectrum is separable and is given by

$$S_f(u,v) = \sigma_f^2 \frac{1-\rho_1^2}{(1+\rho_1^2) - 2\rho_1\cos(2\pi uX)} \frac{1-\rho_2^2}{(1+\rho_2^2) - 2\rho_2\cos(2\pi vX)}.$$

Each term is the sum of two geometric series for positive and negative values of the independent variable.

3 Prove that the spectral density matrix estimate $\hat{\mathbf{Q}}(\mathbf{u})$ of Equation 9.46 is equal to the discrete-domain Fourier transform of the windowed sample autocorrelation $\mathbf{Q}[\mathbf{x}]$ given in Equation 9.47.

10

Analysis and Design of Multidimensional FIR Filters

10.1 Introduction

This chapter presents and analyzes a number of examples of multidimensional discrete-domain FIR filters that are often used, along with some issues of design. As discussed in Section 3.8, an FIR filter has a unit-sample response that is zero outside of a finite set of sample locations $\mathcal{A}$. The frequency response is given by the finite sum

$$H(\mathbf{u}) = \sum_{\mathbf{x}\in\mathcal{A}} h[\mathbf{x}] \exp(-\mathrm{j}2\pi\mathbf{u}\cdot\mathbf{x}). \tag{10.1}$$

It is periodic, with one period given by any unit cell $\mathcal{P}^*$ of the reciprocal lattice Λ^*. Filter design often involves finding the filter unit-sample response such that the frequency response meets certain requirements or specifications. We first start with a few *ad hoc* designs that are widely used.

10.2 Moving Average Filters

The moving average filter simply takes the average of the pixel values over the region $\mathcal{A}$ shifted to the output point $\mathbf{x}$,

$$g[\mathbf{x}] = \frac{1}{|\mathcal{A}|} \sum_{\mathbf{s}\in\mathcal{A}} f[\mathbf{x}-\mathbf{s}], \tag{10.2}$$

where $|\mathcal{A}|$ denotes the number of elements in the set $\mathcal{A}$. This corresponds to an LSI filter with unit sample response

$$h_{\mathrm{MA}}[\mathbf{x}] = \begin{cases} \frac{1}{|\mathcal{A}|} & \text{if } \mathbf{x}\in\mathcal{A}, \\ 0 & \text{if } \mathbf{x}\in\Lambda\backslash\mathcal{A}. \end{cases} \tag{10.3}$$

Its frequency response is

$$H_{\mathrm{MA}}(\mathbf{u}) = \frac{1}{|\mathcal{A}|} \sum_{\mathbf{x}\in\mathcal{A}} \exp(-\mathrm{j}2\pi\mathbf{u}\cdot\mathbf{x}). \tag{10.4}$$

It is clear that $H_{\mathrm{MA}}(0) = 1$, i.e. the DC gain is unity.

Multidimensional Signal and Color Image Processing Using Lattices, First Edition. Eric Dubois.

Companion website: www.wiley.com/go/Dubois/multiSP

Consider the most common usage of this filter: Λ is a rectangular 2D lattice LAT $\left(\left[\begin{smallmatrix} X & 0 \\ 0 & Y \end{smallmatrix}\right]\right)$ and $\mathcal{A}$ is a square region centered at the origin,

$$\mathcal{A} = \{(n_1X, n_2Y) \mid -L \leq n_1 \leq L, -L \leq n_2 \leq L\}, \tag{10.5}$$

with $|\mathcal{A}| = (2L+1)^2$. In this case we can write

$$g[x,y] = \frac{1}{(2L+1)^2} \sum_{n_1=-L}^{L} \sum_{n_2=-L}^{L} f[x - n_1X, y - n_2Y]. \tag{10.6}$$

This filter is a crude low-pass filter. Its frequency response is

$$\begin{aligned} H(u,v) &= \frac{1}{(2L+1)^2} \sum_{n_1=-L}^{L} \sum_{n_2=-L}^{L} \exp(-\mathrm{j}2\pi(un_1X + vn_2Y)) \\ &= \frac{1}{2L+1} \sum_{n_1=-L}^{L} \exp(-\mathrm{j}2\pi un_1X) \frac{1}{2L+1} \sum_{n_2=-L}^{L} \exp(-\mathrm{j}2\pi vn_2Y). \end{aligned} \tag{10.7}$$

Each component sum has the same form; it is a geometric series familiar from 1D signals and systems. Evaluating gives

$$H(u,v) = \frac{1}{(2L+1)^2} \frac{\sin(\pi u(2L+1)X)\sin(\pi v(2L+1)Y)}{\sin(\pi uX)\sin(\pi vY)}. \tag{10.8}$$

Figure 10.1 shows the profile $|H(u,0)|$ of this response for several values of L. We see that after the frequency response goes to zero, it rebounds to a fairly significant value, which may not be the desired behavior for a low-pass filter.

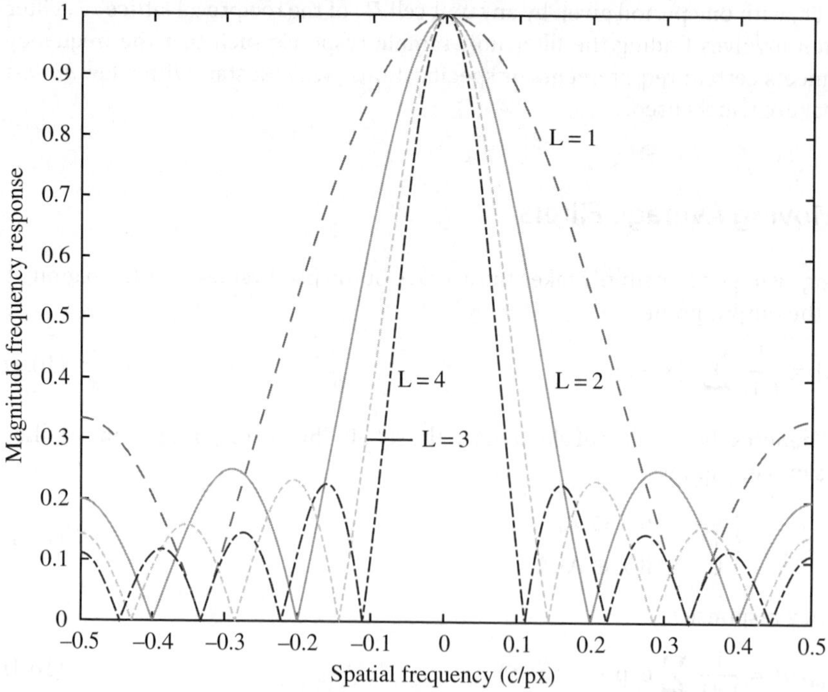

Figure 10.1 Profile $|H(u,0)|$ of the frequency response of the moving average filter for $L = 1, 2, 3, 4$. *(See color plate section for the color representation of this figure.)*

10.3 Gaussian Filters

Gaussian filters are very popular in image processing. In continuous space, the frequency response of a filter with a Gaussian impulse response is also Gaussian (Table 2.2). The Gaussian has the property of being simultaneously concentrated in space and frequency and having no oscillations. In discrete space, the unit sample response of an FIR Gaussian filter has the form

$$h_G[\mathbf{x}] = c\exp(-\|\mathbf{x}\|^2/2r^2), \quad \mathbf{x} \in \mathcal{A}, \tag{10.9}$$

where r is a parameter that determines the spread of the unit sample response. The factor 2 in the denominator is included here for consistency with the Gaussian probability density function; it is not always used in the literature. The region $\mathcal{A}$ is normally chosen large enough that samples outside it are negligibly small, which is not difficult since the Gaussian decays rapidly beyond the distance r. In the common case of a square 2D lattice

$$h_G[n_1X, n_2X] = c\exp(-(n_1^2 + n_2^2)X^2/2r^2), -L \le n_1, n_2 \le L. \tag{10.10}$$

A Gaussian filter is a type of low-pass filter, and since we typically want a DC gain of 1.0, we choose the constant c so that $H(\mathbf{0}) = \sum_{\mathbf{x}\in\mathcal{A}} h_G[\mathbf{x}] = 1$, i.e.

$$c = \frac{1}{\sum_{\mathbf{x}\in\mathcal{A}} \exp(-\|\mathbf{x}\|^2/2r^2)}. \tag{10.11}$$

Since the frequency response is periodic in the frequency domain, it cannot be a Gaussian as in the continuous case. However, if r is such that the Fourier transform of the continuous Gaussian is *mostly* confined to a unit cell of Λ^*, then using Equation (6.8), we can conclude that the frequency response of $h_G[\mathbf{x}]$ within a unit cell will be approximately Gaussian:

$$H(u, v) \approx \frac{c}{d(\Lambda)} 2\pi r^2 \exp(-2\pi^2(u^2 + v^2)r^2), \quad (u, v) \in \mathcal{P}^*. \tag{10.12}$$

Note that a DC gain of 1 implies that $c \approx d(\Lambda)/2\pi r^2$. Let u_c be the (radial) frequency at which the frequency response has dropped to $1/\sqrt{2}$ of its DC value (i.e. −3 dB). Then $\exp(-2\pi^2 u_c^2 r^2) = 2^{-1/2}$, or $2\pi^2 u_c^2 r^2 = 0.5 \ln 2$ and

$$u_c = \frac{\sqrt{\ln 2}}{2\pi r} \approx \frac{0.1325}{r}. \tag{10.13}$$

If we are given u_c, then we should choose $r = 0.1325/u_c$.

Example 10.1
Design a 2D low-pass filter on a square lattice with a 3 dB bandwidth of $1/8X$ c/ph (or equivalently 0.125 c/px) in both the horizontal and the vertical directions (3 dB bandwidth corresponds to an attenuation of 0.707).

Solution:
For a moving average filter, referring to Figure 10.1, to meet the specification we need $L = 2$. For a Gaussian filter, according to Equation (10.13), we should choose $r = 0.1325/(1/8X) = 1.06X$, and using Equation (10.11) to have a DC gain of 1.0, we

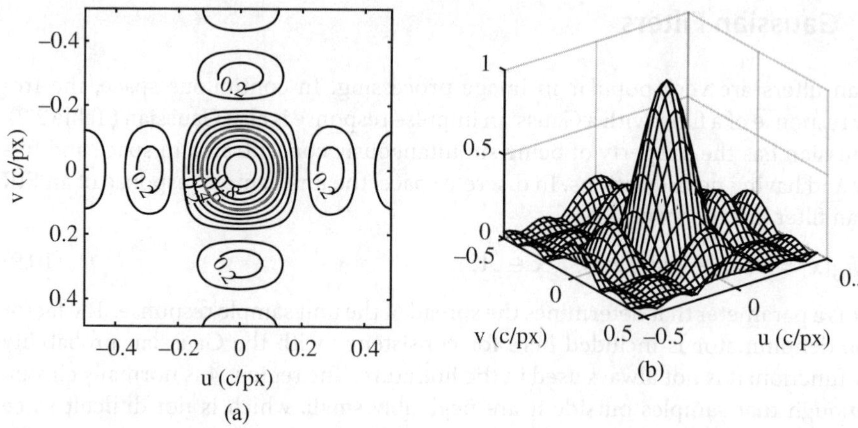

Figure 10.2 Frequency response of a moving average filter with $L = 2$. (a) Contour plot, contour spacing 0.1. (b) Perspective plot.

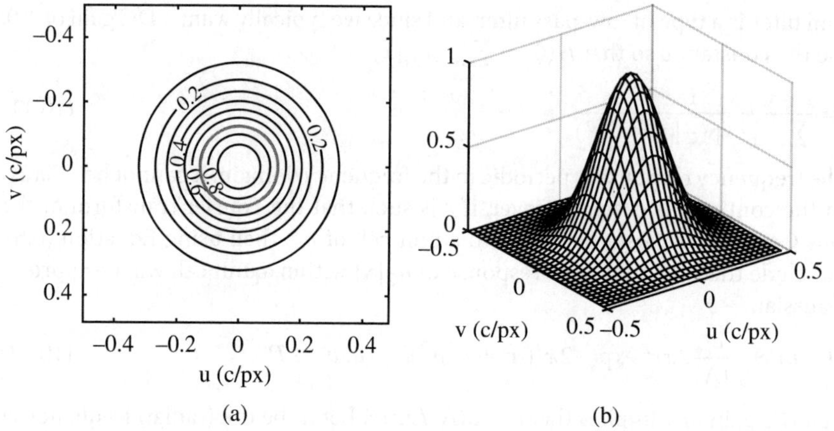

Figure 10.3 Frequency response of a Gaussian filter with $r = 1.06X$ and $c = 0.1416$. (a) Contour plot, contour spacing 0.1. (b) Perspective plot.

find $c = 0.1416$. Figure 10.2 shows the contour and perspective plots of the resulting moving average filter, and Figure 10.3 shows the response of the Gaussian filter. The desired 3 dB passband is shown on both contour plots. Figure 10.4 shows the result of applying each of these two filters to the Barbara image of Figure 3.12(a). Note that although the filters have the same target 3 dB bandwidth, the moving average filter has too much attenuation at the desired cutoff frequency and has introduced greater blurring into the image than the Gaussian filter (see for example the stripes on the headscarf). At the same time, the moving average filter has passed significantly more high frequency energy which could potentially cause problems (see for example the tops of the knees). □

(a)

(b)

Figure 10.4 Result of filtering the Barbara image with the filters of Example 10.1. (a) Moving average filter. (b) Gaussian filter.

10.4 Band-pass and Band-stop Filters

The moving-average and Gaussian filters we have just seen are low-pass filters. It is possible to generate band-pass and band-stop filters from these using the modulation property of the Fourier transform. Assume that $h_{\mathrm{LP}}[\mathbf{x}]$ is the unit sample response of a low-pass filter, which has a DC gain of about 1.0. Then

$$h_{\mathrm{BP}}[\mathbf{x}] = h_{\mathrm{LP}}[\mathbf{x}]\exp(\mathrm{j}2\pi\mathbf{u}_0 \cdot \mathbf{x}) \tag{10.14}$$

is the unit-sample response of a band-pass filter with center frequency $\mathbf{u}_0$. From Property 3.3 in Table 3.1,

$$H_{\mathrm{BP}}(\mathbf{u}) = H_{\mathrm{LP}}(\mathbf{u} - \mathbf{u}_0). \tag{10.15}$$

Note that, in general, $h_{\mathrm{BP}}[\mathbf{x}]$ as defined above will be complex, except for special frequencies on the boundary of a unit cell of Λ^*, where $2\mathbf{u}_0 \in \Lambda^*$ (and so, $-\mathbf{u}_0$ and $\mathbf{u}_0$ represent the same discrete-domain frequency on the lattice Λ).

A real band-pass filter will have a frequency response that is symmetric about zero, with center frequencies $\pm\mathbf{u}_0$. Such a filter can be obtained by

$$\begin{aligned} h_{\mathrm{BP}}[\mathbf{x}] &= 2h_{\mathrm{LP}}[\mathbf{x}]\cos(2\pi\mathbf{u}_0 \cdot \mathbf{x}) \\ &= h_{\mathrm{LP}}[\mathbf{x}](\exp(\mathrm{j}2\pi\mathbf{u}_0 \cdot \mathbf{x}) + \exp(-\mathrm{j}2\pi\mathbf{u}_0 \cdot \mathbf{x})) \end{aligned} \tag{10.16}$$

$$H_{\mathrm{BP}}(\mathbf{u}) = H_{\mathrm{LP}}(\mathbf{u} - \mathbf{u}_0) + H_{\mathrm{LP}}(\mathbf{u} + \mathbf{u}_0). \tag{10.17}$$

Note that in both cases discussed above, the frequency response retains its periodicity with respect to the reciprocal lattice.

A band-stop filter can be obtained from a band-pass filter by simply subtracting the unit-sample response of the band-pass filter from that of an all-pass filter $h_{AP}[\mathbf{x}] = \delta[\mathbf{x}]$. Thus

$$h_{\mathrm{BS}}[\mathbf{x}] = \delta[\mathbf{x}] - h_{\mathrm{BP}}[\mathbf{x}] \tag{10.18}$$

$$H_{\mathrm{BS}}(\mathbf{u}) = 1 - H_{\mathrm{BP}}(\mathbf{u}). \tag{10.19}$$

Example 10.2

Design a real Gaussian band-stop filter with center frequencies $\pm\mathbf{u}_0$ and 3 dB bandwidth $u_c = 15$ c/ph for a 512×512 image on a square sampling lattice. Use it to remove the specific frequency $\mathbf{u}_0 = (150, -86)$ c/ph from the image Barbara.

Solution:

From the specification, $r = 0.1325/u_c = 0.00883$ ph. Using a filter size of 25×25, to get a DC gain of unity, we need $c = 0.00787$. Note that using the approximation $c \approx \mathrm{d}(\Lambda)/2\pi r^2$, we obtain $c = 0.00778$. With x and y in units of picture heights, the desired unit-sample response is

$$\begin{aligned} h_{\mathrm{BS}}[\mathbf{x}] &= \delta[\mathbf{x}] - 2c\cos(2\pi(u_0 x + v_0 y))\exp\left(-\frac{(x^2+y^2)}{2r^2}\right) \\ &= \delta[\mathbf{x}] - 0.0157\cos(2\pi(150x - 86y))\exp\left(\frac{-(x^2+y^2)}{0.0001561}\right). \end{aligned}$$

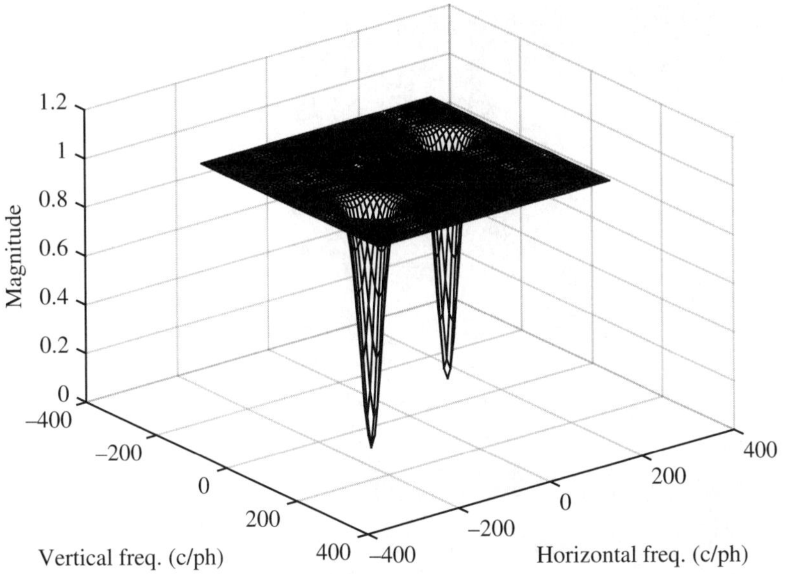

Figure 10.5 Frequency response of a Gaussian band-stop filter with center frequency (150, −86) c/ph.

A perspective plot of the frequency response is shown in Figure 10.5. The filtered Barbara image is shown in Figure 10.6. A close up of the original and filtered image are shown in Figure 10.7. □

Example 10.3

Design a band-stop filter to attenuate the halftone pattern in the image of Figure 10.8, which has been scanned from a newspaper.

Solution:

This is an open-ended question with no unique solution. We observe that the halftone pattern is a periodic pattern that lies on a hexagonal lattice. Thus we can surmise that in the frequency domain, the halftone interference is concentrated on the points of the reciprocal lattice, and we can attenuate it with a multi-band-stop filter with stop bands located on points of the reciprocal lattice. Figure 10.9 shows an estimate of the power spectral density of the halftone image, which is sampled on the square lattice $\Lambda = \mathbb{Z}^2$. Spatial frequencies are given in units of c/px. By examining this contour plot, we observe that the interference is indeed concentrated on the points of a hexagonal lattice Γ^* in the frequency domain; specifically, we graphically estimate that

$$\Gamma^* = \mathrm{LAT}\left(\begin{bmatrix} 0.1474 & 0.1614 \\ 0.1614 & -0.1474 \end{bmatrix}\right) = \mathrm{LAT}\left(\begin{bmatrix} \mathbf{v}_1 & \mathbf{v}_2 \end{bmatrix}\right). \tag{10.20}$$

We note that this hexagonal lattice is slightly rotated, since the newspaper clipping was not perfectly horizontal when it was scanned. For the band-stop filter, we will take a low-pass Gaussian filter with a parameter r_0 (to be optimized), modulate it to selected mirror frequency pairs with a cosine function as before and subtract these from an all-pass filter to get the desired multi-band-stop filter. Several values of the parameter r_0

Figure 10.6 Result of filtering the Barbara image with the Gaussian band-stop filter of Example 10.2. Observe the area on the right knee.

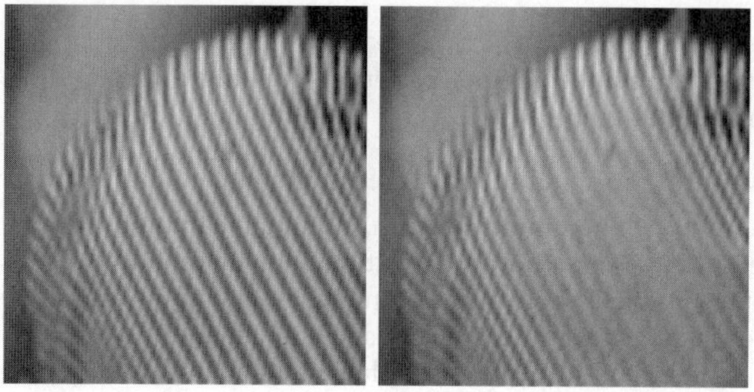

Figure 10.7 Result of filtering the Barbara image with the Gaussian band-stop filter: closeup of her right knee. Left is the original, right is filtered.

were tried and the value $r_0 = 3$ was found to give a suitable result, with a corresponding gain of $c = 0.0178$ to give unit DC gain for a 17×17 filter.

In this example, we will remove all the halftone components within the unit cell of Λ. These are located at 10 mirror frequency pairs that we see by inspection are $\{\pm\mathbf{v}_1, \pm\mathbf{v}_2, \pm(\mathbf{v}_1 + \mathbf{v}_2), \pm(\mathbf{v}_1 - \mathbf{v}_2), \pm 2\mathbf{v}_1, \pm 2\mathbf{v}_2, \pm(2\mathbf{v}_1 + \mathbf{v}_2), \pm(2\mathbf{v}_1 - \mathbf{v}_2), \pm(\mathbf{v}_1 + 2\mathbf{v}_2), \pm(-\mathbf{v}_1 + 2\mathbf{v}_2)\}$. Let us label these frequencies $\pm\mathbf{u}_{m,i}$, $i = 1, 10$ in the order given, as illustrated in Figure 10.10. We see from the spectral plots that the components at

Figure 10.8 Halftone image of the author scanned from a 1962 newspaper clipping.

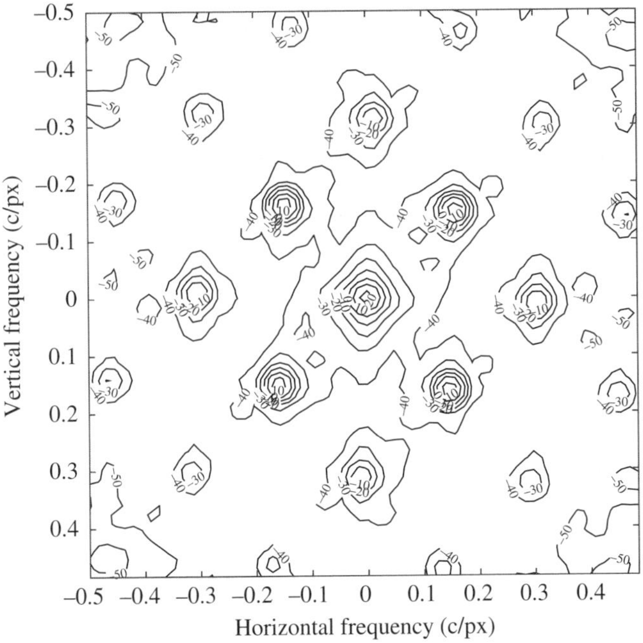

Figure 10.9 Contour plot of the power density spectrum estimate of the halftone image of Figure 10.8

$\pm\mathbf{u}_{m,i}$, $i = 1, 6$ are well isolated and can be suppressed by the low-pass Gaussian filter modulated to these frequencies. However, due to the essentially random relationship between the sampling lattice Λ and the halftone lattice Γ, components at $\mathbf{u}_{m,7}$ and $-\mathbf{u}_{m,9}$ overlap, as do the components at $\mathbf{u}_{m,8}$ and $\mathbf{u}_{m,10}$ (recall the pattern of Figure 10.10 is repeated periodically). As a result, the Gaussian filters modulated to these frequencies

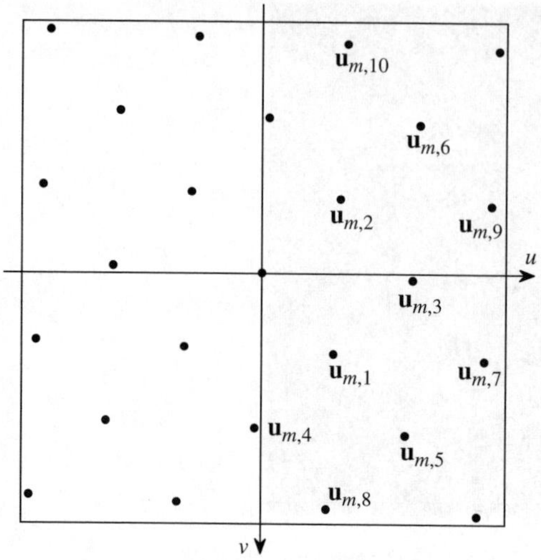

Figure 10.10 Labelled points $\mathbf{u}_{m,i}$ of the halftone interference pattern within one unit cell of the reciprocal lattice Λ^*.

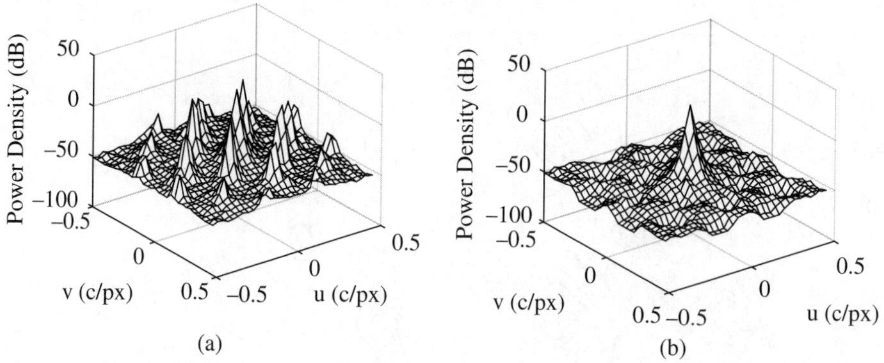

Figure 10.11 (a) Perspective view of the power density spectrum of the input halftone image. (b) Perspective view of the power density spectrum of the output image filtered by the multi-band-stop filter.

will overlap and have a peak gain greater than 1.0. Thus we need to adjust the gain of these filters to have a peak frequency response gain of 1.0 at the modulation frequency, so that the band-stop filter will effectively remove that component. For $r_0 = 3$ the uncorrected gain is found to be $g = 1.459$ and so we should divide by this factor. The resulting multi-band-stop filter has unit sample response

$$h_{\mathrm{MBS}}[\mathbf{x}] = \delta[\mathbf{x}] - 2c\exp(-\|\mathbf{x}\|^2/2r_0^2)\sum_{i=1}^{6}\cos(2\pi\mathbf{x}\cdot\mathbf{u}_{m,i})$$
$$-\,2(c/g)\exp(-\|\mathbf{x}\|^2/2r_0^2)\sum_{i=7}^{10}\cos(2\pi\mathbf{x}\cdot\mathbf{u}_{m,i}). \tag{10.21}$$

Perspective views of the power density spectra of the input halftone image and the output of the multi-band-stop filter are shown in Figure 10.11. We see that the components on the halftone lattice have been effectively removed. The output image is shown

Figure 10.12 Output of the multi-band-stop filter with stop bands at the 10 frequencies within the unit cell.

in Figure 10.12. We observe for example that the halftone pattern is completely removed in flat areas such as the car and the forehead. □

10.5 Frequency-Domain Design of Multidimensional FIR Filters

FIR filter design involves the determination of the unit sample response of a filter such that its frequency response approximates a desired response as closely as possible. Many methods have been developed over the years. We present two of these methods here, namely the *window method* and the *least pth optimization method*. A detailed study of two-dimensional filter design can be found in Lu and Antoniou (1992).

Frequency-domain design methods require the specification of an ideal frequency response, denoted $H_I(\mathbf{u})$, within one unit cell of the reciprocal lattice. This ideal response may be determined from theoretical considerations, such as to remove a certain band of frequencies, or from analytical minimization of some error measure as in Wiener filtering. It may also be determined by experiments, for example with simulations using Fourier transforms.

A common type of filter is a pass-stop filter, where certain frequency bands are to pass through essentially unchanged, and others are to be suppressed, as in some examples already seen. For example, an ideal circularly symmetric low-pass filter would have the ideal response

$$H_I(\mathbf{u}) = \begin{cases} 1 & \|\mathbf{u}\| \leq u_c; \\ 0 & \|\mathbf{u}\| > u_c. \end{cases} \tag{10.22}$$

However, such a sharp transition in the frequency domain leads to large oscillations in the unit-sample response, and thus in the step response. This causes an effect called *ringing* near sharp edges in the image. This can be alleviated by using a transition

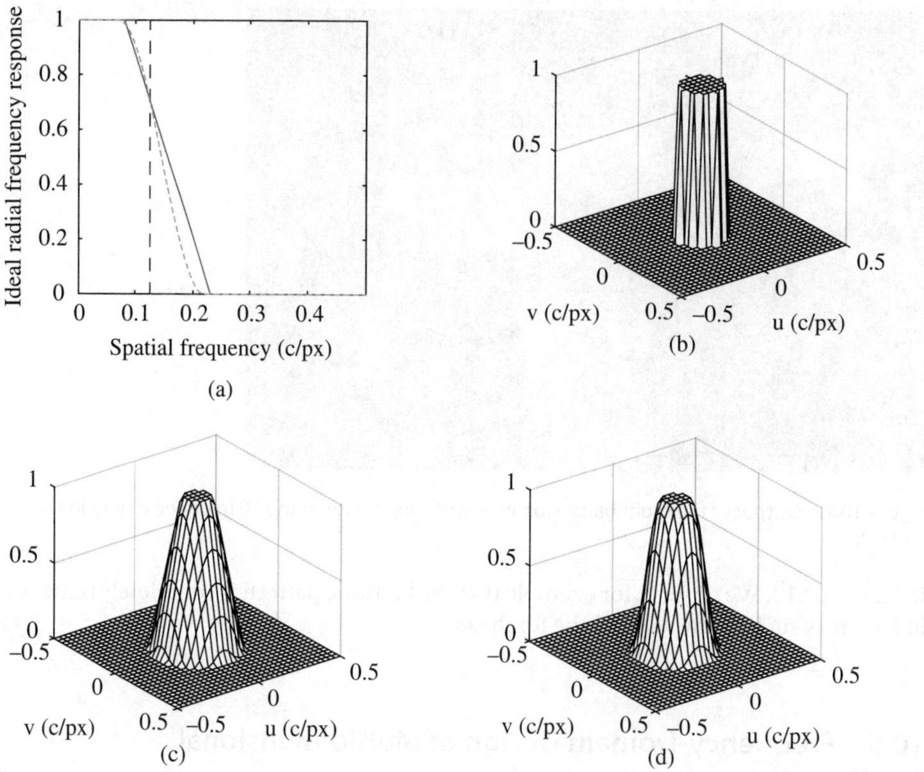

Figure 10.13 (a) Radial profile of three circularly symmetric ideal responses with 3 dB radial frequency cutoff of 0.125 c/px: step, linear, raised cosine. Perspective view of ideal response with (b) step transition; (c) linear transition; (d) raised-cosine transition.

band, allowing the ideal response to smoothly transition from 1 to 0. For example, this transition could be linear or a raise-cosine function. If we consider the case of Example 10.1, we desire a low-pass filter with a response of 0.707 at radial frequency 0.125 c/px. Thus, the ideal radial response could have the form of one of the curves in Figure 10.13. Perspective views of the ideal response with a step transition, linear transition, and raised-cosine transition are shown in Figures 10.13(b)–(d) respectively.

It is often convenient to delineate the passband and stopband regions with polygons, since they are often related to the unit cell of a reciprocal lattice, for which we generally use polygons, for example the Voronoi cell. In this case, we can use a planar function for the transition band. These bands generally have some form of symmetry, with zero-phase or quadrantal symmetry. Figure 10.14(a) shows an example of the boundaries of an ideal response with polygonal pass and stop band regions and quadrantal symmetry; Figure 10.14(b) shows a perspective view of the ideal response with a planar transition band.

10.5.1 FIR Filter Design Using Windows

Let $H_I(\mathbf{u})$ be an ideal frequency response that will be approximated by the FIR filter with unit sample response $h[\mathbf{x}]$ defined on lattice Λ, confined to the finite set of sample

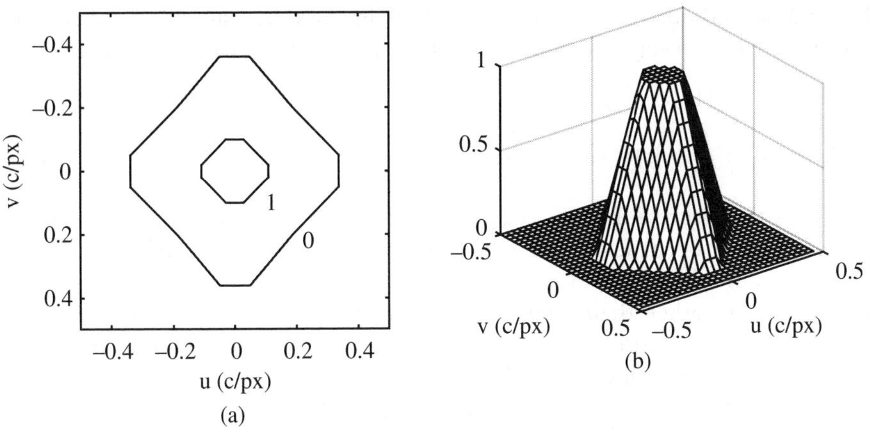

Figure 10.14 (a) Boundaries of polygonal pass and stop bands of an example ideal low-pass frequency response. (b) Perspective view of the ideal frequency response with planar transition band.

locations $\mathcal{A}$. The ideal response is specified on a unit cell $\mathcal{P}^*$ of Λ^*. We can determine the unit sample response corresponding to the ideal frequency response by taking the inverse Fourier transform to obtain

$$h_I[\mathbf{x}] = \mathrm{d}(\Lambda) \int_{\mathcal{P}^*} H_I(\mathbf{u}) \exp(\mathrm{j}2\pi \mathbf{u} \cdot \mathbf{x})\, \mathrm{d}\mathbf{u}. \tag{10.23}$$

However, this will in general result in a unit sample response that is not confined to $\mathcal{A}$, and is most likely of infinite extent. It can simply be truncated to $\mathcal{A}$ by multiplying by the zero-one function $p_{\mathcal{A}}[\mathbf{x}]$:

$$h[\mathbf{x}] = h_I[\mathbf{x}] p_{\mathcal{A}}[\mathbf{x}], \tag{10.24}$$

where

$$p_{\mathcal{A}}[\mathbf{x}] = \begin{cases} 1, & \mathbf{x} \in \mathcal{A} \\ 0, & \mathbf{x} \in \Lambda \backslash \mathcal{A}. \end{cases} \tag{10.25}$$

By Property 3.5 of the Fourier transform

$$H(\mathbf{u}) = \mathrm{d}(\Lambda) \int_{\mathcal{P}^*} H_I(\mathbf{r}) P_{\mathcal{A}}(\mathbf{u} - \mathbf{r})\, \mathrm{d}\mathbf{r} \tag{10.26}$$

so that $H(\mathbf{u})$ may end up being quite different from $H_I(\mathbf{u})$ after convolving with $P_{\mathcal{A}}(\mathbf{u})$ in the frequency domain. Typically $\mathcal{A}$ is a rectangular region, so that $p_{\mathcal{A}}$ is a discrete rect function whose Fourier transform can be computed in a straightforward fashion. In fact, we have already done it in computing the frequency response of the moving average filter; the only difference is that we don't have the $1/(2L+1)^2$ term. Thus for a $(2L+1) \times (2L+1)$ filter,

$$P_{\mathcal{A}}(\mathbf{u}) = \frac{\sin(\pi u(2L+1)X) \sin(\pi v(2L+1)Y)}{\sin(\pi uX) \sin(\pi vY)}. \tag{10.27}$$

This function $p_{\mathcal{A}}$ is sometimes referred to as a "boxcar window." Convolving the ideal response with the Fourier transform of the window will smear out the response and introduce ripples due to the sidelobes of $P_{\mathcal{A}}(\mathbf{u})$ (see Figure 10.1).

In the window method, we obtain $h[\mathbf{x}]$ by multiplying $h_I[\mathbf{x}]$ by a suitable *window function* $w[\mathbf{x}]$ with region of support confined to $\mathcal{A}$, i.e. $w[\mathbf{x}] = 0$ for $\mathbf{x} \in \Lambda \backslash \mathcal{A}$. Then

$$h[\mathbf{x}] = h_I[\mathbf{x}]w[\mathbf{x}], \tag{10.28}$$

$$H(\mathbf{u}) = \mathrm{d}(\Lambda) \int_{\mathcal{P}^*} H_I(\mathbf{r})W(\mathbf{u} - \mathbf{r})\, \mathrm{d}\mathbf{r}, \tag{10.29}$$

where $W(\mathbf{u})$ is the discrete-domain Fourier transform of $w[\mathbf{x}]$ on the lattice Λ. Many windows $w[\mathbf{x}]$ have been developed that reduce the level of the sidelobes without widening the main lobe by much [Harris (1978)]. These windows taper off to a value of zero outside $\mathcal{A}$ rather than simply truncating h_I. Windows have been studied extensively in one-dimensional signal processing. Many of these windows have a continuous analytical specification $w(t)$ where $w(t) = 0$ for $|t| > 0.5$. They can be extended to two-dimensional windows either separably or by forming a circularly symmetric window [Huang (1972)]. Examples of 1D windows are Hamming, Hanning, Kaiser, Blackman and others. For example, the Hamming window is defined as

$$w_H(t) = \begin{cases} 0.54 + 0.46\cos(2\pi t), & \text{if } |t| \le 0.5; \\ 0, & \text{otherwise,} \end{cases} \tag{10.30}$$

and the boxcar window is given by $w_B(t) = \mathrm{rect}(t)$.

A two-dimensional window with an approximately rectangular region of support

$$\mathcal{A} = \{(x, y) \in \Lambda \mid |x| \le Q_1, |y| \le Q_2\} \tag{10.31}$$

can be defined from a one-dimensional window $w_1(t)$ using a separable construction

$$w_{\mathrm{R}}[x, y] = w_1\left(\frac{x}{2Q_1}\right) w_1\left(\frac{y}{2Q_2}\right), \qquad (x, y) \in \Lambda. \tag{10.32}$$

Similarly, a two-dimensional window with an approximately circular region of support

$$\mathcal{A} = \{(x, y) \in \Lambda \mid \sqrt{x^2 + y^2} \le Q\}$$

can be obtained using

$$w_{\mathrm{C}}[x, y] = w_1\left(\frac{\sqrt{x^2 + y^2}}{2Q}\right), \qquad (x, y) \in \Lambda. \tag{10.33}$$

This construction could also be used with a rectangular region of support and applied to nonrectangular sampling structures. If w_1 is a good one-dimensional window, in general w_{R} and w_{C} will be good two-dimensional windows [Huang (1972)]. The window method extends in a straightforward way to more than two dimensions.

To implement the window method, we must first compute $h_I(\mathbf{x})$ for $\mathbf{x} \in \mathcal{A}$. For ideal zero–one responses with an elliptical or polygonal band edge, this can be done analytically using the methods of Chapters 2 and 3. Otherwise, this will typically be done by sampling $H_I(\mathbf{u})$ within a unit cell of Λ^* and using the inverse DFT (i.e. inverse discrete-domain periodic Fourier transform). As seen in Chapter 6, sampling of $H_I(\mathbf{u})$ corresponds to periodization of $h_I[\mathbf{x}]$. We need the sampling of $H_I(\mathbf{u})$ to be sufficiently dense that the periodization effect within $\mathcal{A}$ is negligible.

To illustrate, suppose that we wish to approximate the ideal circularly symmetric low-pass filter response of Figure 10.13(b) with a $(2L + 1) \times (2L + 1)$ FIR filter using the window method. Denote the radial cutoff frequency by W, where $W = 0.125$ c/px in this example. Then, from Problem 7 of Chapter 3 with $X = Y = 1$,

$$h_I[x, y] = \frac{W}{\sqrt{x^2 + y^2}} J_1(2\pi W \sqrt{x^2 + y^2}), \quad (x, y) \in \mathbb{Z}^2, \tag{10.34}$$

Where J_1 is a Bessel function of the first kind of order 1. We then obtain $h[x, y]$ by multiplying by a suitable window using Equation 10.28. Note that since the window frequency response is approximately symmetric about the midpoint of the transition, the response at W is closer to 0.5 than to 0.707. If we want a response of 0.707 at radial frequency 0.125 c/px, we should use a value of W somewhat larger than 0.125.

Figure 10.15 shows the contour and perspective plots of the frequency response designed with the window method for three sizes: 13×13 (the same as the Gaussian filter of Figure 10.3), 21×21 and 41×41. In each case, W has been chosen empirically to give a response of about 0.707 at radial frequency 0.125 c/px. Also, the response has been normalized to ensure a DC gain of 1.0.

Figure 10.16 shows the Barbara image filtered with the 13×13 and the 41×41 window-designed filters. We note that the 13×13 filter has a sharper transition than the same-size Gaussian filter. As a result, some details (stripes in the clothes) retained in Figure 10.4(b) are removed in Figure 10.16(a). The 41×41 window-designed filter has a very narrow transition band, resulting in ringing around the sharp transitions in the image. See for example the table leg.

10.5.2 FIR Filter Design Using Least-*p*th Optimization

Gaussian and window-designed filters may be sufficient to achieve desired results in many applications. However, if we require a finer control of the frequency response in pass and stop bands, or we wish to apply specific constraints to the response, we can apply optimization methods to the filter design process. The set of unknown unit sample response coefficients serves as the optimization variable of the problem and constraints on the frequency response are typically linear equality constraints. Symmetry constraints can be used to reduce the size of the optimization variable. The objective function measures the deviation between the ideal response $H_I(\mathbf{u})$ and the obtained response $H(\mathbf{u})$ over the unit cell of the reciprocal lattice. Thus, a typical optimization problem takes the form

$$\begin{aligned} &\text{minimize} && \mathcal{E}(H_I, H) \\ &\text{subject to} && H(\mathbf{u}_{c,i}) = b_i, \quad i = 1, \dots, m. \end{aligned} \tag{10.35}$$

Typically, the b_i will be 0 (where we desire perfect attenuation) or 1 (where we want no attenuation).

There are many possibilities for the error function $\mathcal{E}(H_I, H)$. The *minimax* error has the form

$$\mathcal{E}_m(H_I, H) = \max_{\mathbf{u} \in \mathcal{P}^*} |H_I(\mathbf{u}) - H(\mathbf{u})|. \tag{10.36}$$

In this case, the goal is to minimize the maximum deviation between H_I and H over $\mathcal{P}^*$, hence the name minimax. Another useful criterion is the p-error, or the least-pth

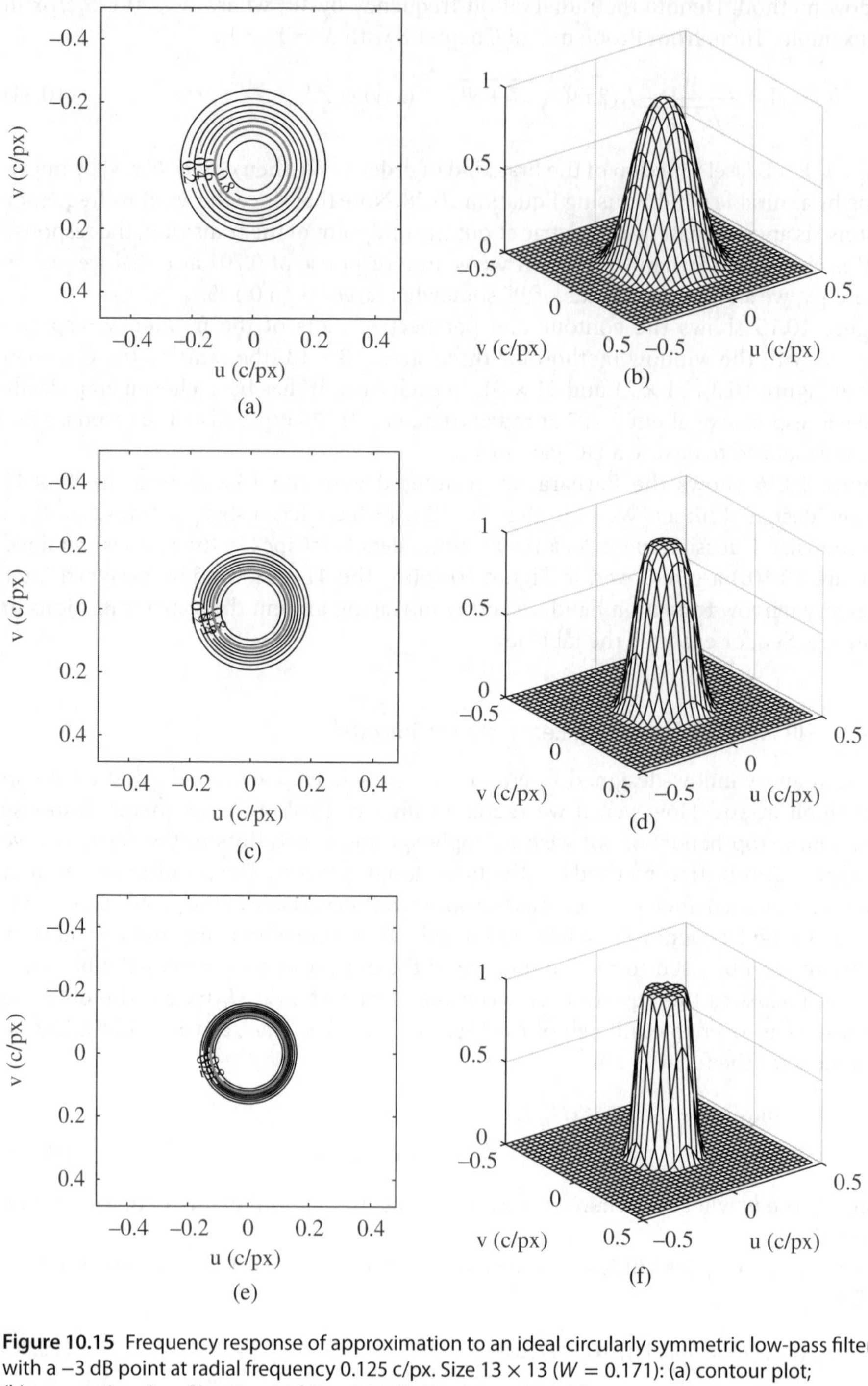

Figure 10.15 Frequency response of approximation to an ideal circularly symmetric low-pass filter with a −3 dB point at radial frequency 0.125 c/px. Size 13 × 13 ($W = 0.171$): (a) contour plot; (b) perspective view. Size 21 × 21 ($W = 0.1498$) (c) contour plot; (d) perspective view. Size 41 × 41 ($W = 0.1366$): (e) contour plot; (f) perspective view.

(a)

(b)

Figure 10.16 Result of filtering the Barbara image with the window-designed filters of Figure 10.15. (a) Size 13 × 13. (b) Size 41 × 41.

criterion, defined by

$$\mathcal{E}_p(H_I, H) = \int_{\mathcal{P}^*} |H_I(\mathbf{u}) - H(\mathbf{u})|^p \, d\mathbf{u}, \tag{10.37}$$

where p is a positive even integer. For computation, the integral can be well approximated by a sum over a sufficiently dense set of points in $\mathcal{P}^*$. For large enough p, this puts most of the weight on the largest errors and thus gives a result similar to the minimax criterion. Explicitly, the least-pth problem with no symmetry constraints becomes

$$\begin{aligned} &\underset{\mathbf{h}}{\text{minimize}} && \sum_{k=1}^{N} |H_I(\mathbf{u}_k) - \sum_{\mathbf{x}\in\mathcal{A}} h[\mathbf{x}] \exp(-\mathrm{j}2\pi \mathbf{u}_k \cdot \mathbf{x})|^p \\ &\text{subject to} && \sum_{\mathbf{x}\in\mathcal{A}} h[\mathbf{x}] \exp(-\mathrm{j}2\pi \mathbf{u}_{c,i} \cdot \mathbf{x}) = b_i, \quad i = 1, \dots, m, \end{aligned} \tag{10.38}$$

where $\mathbf{u}_k, k = 1, \dots, N$ is a sufficiently dense set of frequencies in $\mathcal{P}^*$ and $\mathbf{u}_{c,i}, i = 1, \dots, m$ are frequencies at which we impose equality constraints, e.g., $H(\mathbf{0}) = 1$ for unity DC gain.

We can often reduce the dimensionality of the problem by imposing symmetry constraints such as zero-phase symmetry, quadrantal symmetry, octal symmetry, or higher order symmetry. Consider the case of zero-phase symmetry, where $h[-\mathbf{x}] = h[\mathbf{x}]$. In this case, the frequency response is real. Define

$$\mathcal{A}_+ = \{(x_1, x_2) \in \mathcal{A} \mid x_2 > 0\} \cup \{(x_1, 0) \in \mathcal{A} \mid x_1 > 0\}. \tag{10.39}$$

Then, we see that $\mathcal{A} = \{(0, 0)\} \cup \mathcal{A}_+ \cup (-\mathcal{A}_+)$, where these three sets are disjoint and any point of $\mathcal{A}$ belongs to one of them. We can express the frequency response as

$$\begin{aligned} H(\mathbf{u}) &= h[\mathbf{0}] + \sum_{\mathbf{x}\in\mathcal{A}_+} h[\mathbf{x}] \exp(-\mathrm{j}2\pi \mathbf{u} \cdot \mathbf{x}) + \sum_{\mathbf{x}\in\mathcal{A}_+} h[-\mathbf{x}] \exp(-\mathrm{j}2\pi \mathbf{u} \cdot (-\mathbf{x})) \\ &= h[\mathbf{0}] + 2 \sum_{\mathbf{x}\in\mathcal{A}_+} h[\mathbf{x}] \cos(2\pi \mathbf{u} \cdot \mathbf{x}) \end{aligned} \tag{10.40}$$

and use as optimization variable $[h[\mathbf{0}], h[\mathbf{x}], \mathbf{x} \in \mathcal{A}_+]$. Similar approaches can be taken for quadrantal symmetry or higher order symmetry to reduce the dimensionality of the problem when H_I possesses such types of symmetry.

The optimization problem of Equation 10.38, or equivalent versions with symmetry constraints, can be solved using general purpose optimization routines (for example, those available in MATLAB). The objective function is known to be convex (Lampropoulos and Fahmy (1985)) so these routines should normally converge to a global minimum. Such routines can conveniently impose equality constraints as well. Since these are iterative solution methods, they require a starting point. The window method can be used to generate such a starting point. Since the initial solution may have a relatively large error at certain frequencies, a large value of p may cause the objective function to be very large. An approach to address this is to start with $p = 2$, and upon convergence, increase p by 2. This can be continued until there is no further reduction of the minimax error.

Example 10.4
Design a low-pass filter with quadrantal symmetry to approximate the ideal response with polygonal pass and stop band boundaries shown in Figure 10.14.

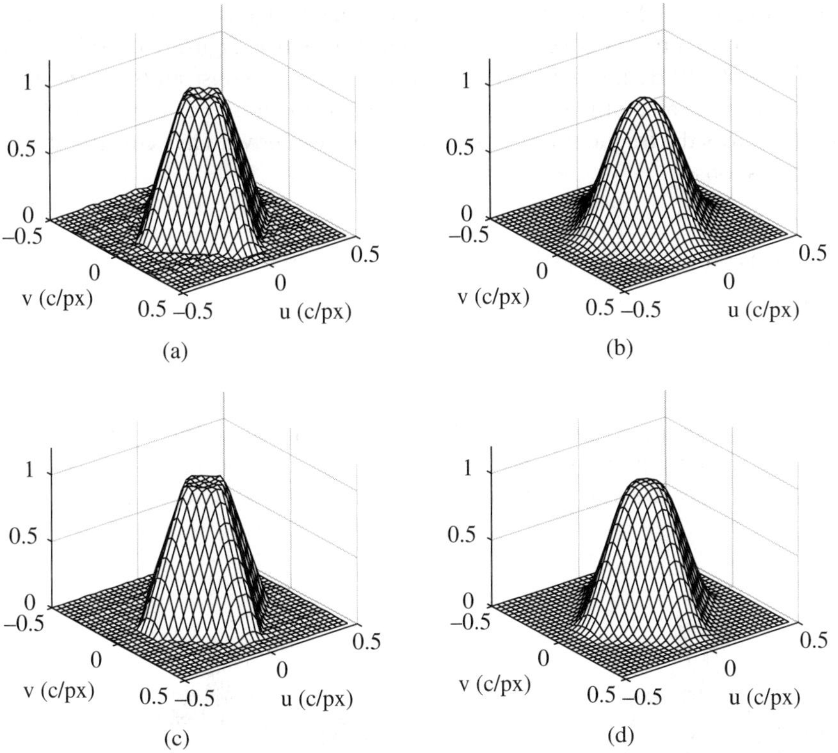

Figure 10.17 Least pth design for low-pass filter with polygonal specifications shown in Figure 10.14. (a) 13×13 least-pth filter (minimax error is 0.0395). (b) 13×13 filter designed with the window method. (c) 21×21 least-pth filter (minimax error is 0.0247). (d) 21×21 filter designed with the window method.

Solution:

We use the window method to find a starting solution. Then we apply the MATLAB optimization routine `fminunc` to find the solution. Filter coefficients in the positive quadrant of $\mathcal{A}$ are used as the optimization variable. Figure 10.17 shows perspective plots of the filters obtained for sizes 13×13 and 21×21. In both cases, the final value of p was 6. The window design for these same sizes is shown for comparison. □

Example 10.5

Design a zero-phase low pass filter to retain the baseband component of the halftone image in Figure 10.8, as seen in the spectrum estimate of Figure 10.9. Investigate the effect of constraining the response to be zero at the four closest frequency pairs of the halftone grid reciprocal lattice.

Solution:

This is an open-ended question and we just give a possible solution, obtained after some empirical tests. We specify an ideal low-pass filter with passband and stopband boundaries of diamond shape, similar to the diamond with vertices $\pm(\mathbf{v}_1 \pm \mathbf{v}_2)$. We take the passband edge at 0.4 of these values and the stopband edge at 0.6 of these values. Figure 10.18(a) shows a perspective view of the ideal response. Figure 10.18(b) shows the result of a least-pth design for filter size 21×21 with no constraints on the frequency response. The value of p at convergence is 6 and the minimax error is 0.0809. Figure 10.18(c) shows the frequency response of a 21×21 filter with five constraints on the frequency response.

$$\begin{aligned} H(\mathbf{0}) &= 1 \\ H(\mathbf{v}_1) &= 0 \\ H(\mathbf{v}_2) &= 0 \\ H(\mathbf{v}_1 + \mathbf{v}_2) &= 0 \\ H(\mathbf{v}_1 - \mathbf{v}_2) &= 0 \end{aligned} \tag{10.41}$$

In this case, we have used the MATLAB routine `fmincon`. This routine converged at $p = 4$ with a minimax error of 0.0931. Although the responses visually look very similar, they are not identical and the constraints have a clear impact

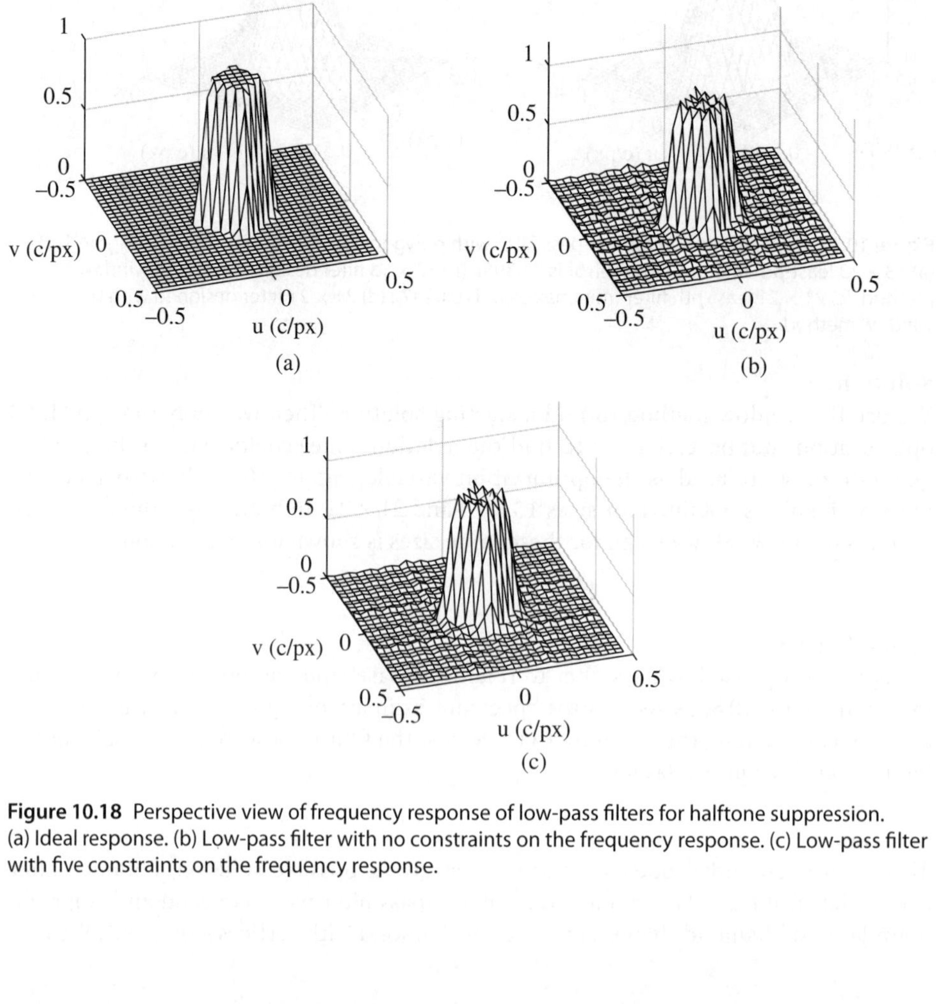

Figure 10.18 Perspective view of frequency response of low-pass filters for halftone suppression. (a) Ideal response. (b) Low-pass filter with no constraints on the frequency response. (c) Low-pass filter with five constraints on the frequency response.

(a)

(b)

Figure 10.19 Halftone image of Figure 10.8 filtered (a) without constraints on the frequency response using the filter of Figure 10.18(b); (b) with five constraints on the frequency response using the filter of Figure 10.18(c).

on the filter output. Figure 10.19(a) shows the image of Figure 10.8 filtered with the filter of Figure 10.18(b) with no frequency response constraints; the halftone pattern is still quite present (zooming of the figure may be required to see this). The result with the constrained filter is shown in Figure 10.19(b), where the halftone pattern can no longer be seen. The sharpness of the images is quite similar. Note that this example is meant to illustrate filter design and is not put forward as the optimal method for restoration of halftone images, where non-linear restoration methods may be more successful (see Stevenson (1997) for an example). □

Problems

1 Derive in detail the expression for the frequency response of a moving average filter on a rectangular lattice, with a rectangular region of support, as given in Equation 10.8.

2 Determine the frequency response of a moving average filter for a hexagonal lattice with a diamond-shaped region of support.

3 A square image (pw = ph) is sampled on the hexagonal lattice Λ generated by the sampling matrix

$$\mathbf{V} = \begin{bmatrix} X & X/2 \\ 0 & \sqrt{3}X/2 \end{bmatrix}$$

where $X = 1/512$ ph. Design a Gaussian FIR filter with unit sample response $h[\mathbf{x}] = c\exp(-\|\mathbf{x}\|^2/2r^2)$ for $\mathbf{x} \in \mathcal{A}$ having a 3 dB bandwidth of $0.2/X$ c/ph. The region of support of the FIR filter is $\mathcal{A} = \{\mathbf{x} \in \Lambda \mid \|\mathbf{x}\| \leq 3X\}$. Having determined the correct values of c and r, give the coefficients of the filter. Give an analytical approximation for the frequency response of the filter. Make a contour plot and a perspective plot of the frequency response of the filter over the frequency range $-2/X \leq u \leq 2/X$, $-4/(\sqrt{3}X) \leq v \leq 4/(\sqrt{3}X)$. Sketch (by hand if you wish) on the contour plot the points of the reciprocal lattice Λ^* and a Voronoi unit cell of Λ^*, and comment on the periodicity of the frequency response. Recall that the Voronoi unit cell consists of all points in Λ^* closer to the origin than to any other point of Λ^*.

11

Changing the Sampling Structure of an Image

11.1 Introduction

There are two main applications for changing the sampling structure of an image: resizing and format conversion (or standards conversion). Resizing generally involves displaying the image on the same display but with a larger or smaller size, and so with a different number of samples per picture height. Format conversion involves switching between different formats such as European and North American television standards, which may be displayed at the same size, but again with a different number of samples per picture height and perhaps with different temporal sampling characteristics. This application is one where the picture height is a particularly convenient choice for the unit of length.

An image $f[\mathbf{x}]$ is initially sampled on a lattice Λ, and it is required to produce a new image $g[\mathbf{x}]$ sampled on a different lattice Γ. If Γ is less dense than Λ, we are *downsampling*; conversely if Γ is denser than Λ, we are *upsampling*. The problem we are addressing is to change the sampling structure of an image with minimal loss of information and minimal introduction of artifacts.

We will consider three cases:

1) Upsampling to a superlattice
2) Downsampling to a sublattice
3) Arbitrary lattice conversion.

Before addressing these topics, we will review some additional results on lattices, specifically related to sublattices.

11.2 Sublattices

In order to study upsampling and downsampling, we review the notion of a sublattice and recall certain properties. Sublattices were introduced in Chapter 4 in the context of discrete-domain periodic signals, and many details and proofs are provided in Chapter 13.

Multidimensional Signal and Color Image Processing Using Lattices, First Edition. Eric Dubois.

Companion website: www.wiley.com/go/Dubois/multiSP

Definition 11.1 Λ is a *sublattice* of Γ if both Λ and Γ are lattices in $\mathbb{R}^D$, and every point of Λ is also a point of Γ. We write $\Lambda \subset \Gamma$. If Λ is a sublattice of Γ, then Γ is a *superlattice* of Λ.

It is generally easy to observe if a lattice is a sublattice of another lattice by sketching the points of the two lattices on a common set of axes. However, the following shows how to determine this numerically without having to sketch the lattices, which is particularly helpful in three or more dimensions.

Theorem 11.1 The lattice $\Lambda = \mathrm{LAT}(\mathbf{V}_\Lambda)$ is a sublattice of $\Gamma = \mathrm{LAT}(\mathbf{V}_\Gamma)$ if and only if $(\mathbf{V}_\Gamma)^{-1}\mathbf{V}_\Lambda$ is an integer matrix.

Proof: This theorem is the same as Theorem 13.4 where the proof is given. □

Corollary 11.1 If $\Lambda \subset \Gamma$, then $\mathrm{d}(\Lambda)$ is an integer multiple of $\mathrm{d}(\Gamma)$.

Proof: See proof of corollary to Theorem 13.4. □

Thus, by examining $\mathrm{d}(\Lambda)$ and $\mathrm{d}(\Gamma)$, we can immediately see which sublattice relation is possible, if any.

Example 11.1 Let

$$\mathbf{V}_\Lambda = \begin{bmatrix} 2X & 0 \\ 0 & 2Y \end{bmatrix} \quad \text{and} \quad \mathbf{V}_\Gamma = \begin{bmatrix} 2X & X \\ 0 & Y \end{bmatrix}.$$

Then, $\mathrm{d}(\Lambda) = 4XY$ and $\mathrm{d}(\Gamma) = 2XY$, so that $\mathrm{d}(\Lambda) = 2\mathrm{d}(\Gamma)$. Thus, $\Lambda \subset \Gamma$ is possible. Checking,

$$\mathbf{V}_\Gamma^{-1}\mathbf{V}_\Lambda = \begin{bmatrix} \frac{1}{2X} & -\frac{1}{2Y} \\ 0 & \frac{1}{Y} \end{bmatrix} \begin{bmatrix} 2X & 0 \\ 0 & 2Y \end{bmatrix} = \begin{bmatrix} 1 & -1 \\ 0 & 2 \end{bmatrix}$$

which is an integer matrix, and thus $\Lambda \subset \Gamma$ is confirmed. These ideas are illustrated in Figure 11.1, where the fact that $\Lambda \subset \Gamma$ is clear from the illustration. For this example, we arbitrarily take $Y = 0.8X$.

Now, let's see what happens in the frequency domain.

Theorem 11.2 If $\Lambda \subset \Gamma$, then $\Gamma^* \subset \Lambda^*$.

Proof: This theorem is the same as Theorem 13.5 where the proof is given. □

Example 11.1 **(continued).** A sampling matrix for Λ^* is

$$\mathbf{V}_\Lambda^{-T} = \begin{bmatrix} \frac{1}{2X} & 0 \\ 0 & \frac{1}{2Y} \end{bmatrix} = \mathbf{V}_{\Lambda^*}$$

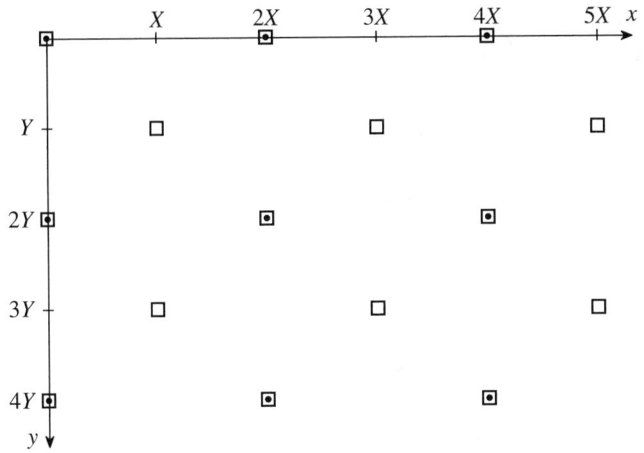

Figure 11.1 Illustration of lattices Λ (•) and Γ (□) of Example 11.1, where $\Lambda \subset \Gamma$.

and a sampling matrix for Γ^* is

$$\mathbf{V}_{\Gamma}^{-T} = \begin{bmatrix} \frac{1}{2X} & 0 \\ -\frac{1}{2Y} & \frac{1}{Y} \end{bmatrix} = \mathbf{V}_{\Gamma^*}.$$

We see that

$$(\mathbf{V}_{\Lambda^*})^{-1}\mathbf{V}_{\Gamma^*} = \begin{bmatrix} 2X & 0 \\ 0 & 2Y \end{bmatrix} \begin{bmatrix} \frac{1}{2X} & 0 \\ -\frac{1}{2Y} & \frac{1}{Y} \end{bmatrix} = \begin{bmatrix} 1 & 0 \\ -1 & 2 \end{bmatrix}$$

which is an integer matrix, confirming that $\Gamma^* \subset \Lambda^*$. Note that $\mathrm{d}(\Gamma^*) = \frac{1}{2XY}$ and $\mathrm{d}(\Lambda^*) = \frac{1}{4XY}$, and so $\mathrm{d}(\Gamma^*) = 2\mathrm{d}(\Lambda^*)$. In fact, in general,

$$\frac{\mathrm{d}(\Lambda)}{\mathrm{d}(\Gamma)} = \frac{\mathrm{d}(\Gamma^*)}{\mathrm{d}(\Lambda^*)}. \tag{11.1}$$

11.3 Upsampling

Suppose that we want to upsample an image $f[\mathbf{x}]$ defined on the lattice Λ in Figure 11.1 to the superlattice Γ, a 2:1 upsampling. If $f[\mathbf{x}]$ represents the sampled version of a continuous-domain signal, we could reconstruct that continuous-domain signal using the methods of Chapter 6 and then resample on Γ. However, we prefer to do this entirely with discrete-domain processing. In the frequency domain, $F(\mathbf{u})$ is periodic, with periodicity given by the reciprocal lattice Λ^*, as indicated by all the circles in Figure 11.2, where we assume for the purpose of this illustration that $F(\mathbf{u})$ in one unit cell of Λ^* is confined to a circular region. The desired upsampled signal $g[\mathbf{x}]$ defined on Γ has a Fourier transform $G(\mathbf{u})$ that has periodicity given by Γ^*. Both $f[\mathbf{x}]$ and $g[\mathbf{x}]$ will correspond to the same continuous-space image if we obtain $G(\mathbf{u})$ from $F(\mathbf{u})$ by setting the dashed replicas in Figure 11.2 to zero and apply an appropriate scaling. How can this be done?

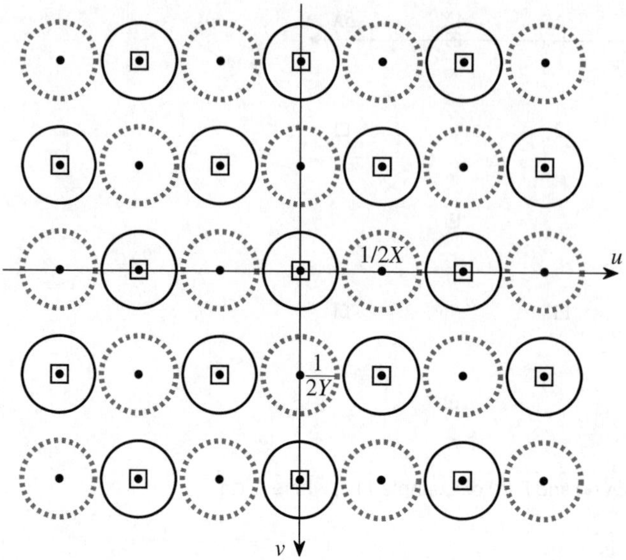

Figure 11.2 Illustration of reciprocal lattices Λ^* (•) and Γ^* (□) of Example 11.1, where $\Gamma^* \subset \Lambda^*$. Solid circles represent spectral repeats on the points of Γ^*. Dashed circles represent spectral repeats on the points of $\Lambda^* \backslash \Gamma^*$ that are to be removed in the upsampling process. (*See color plate section for the color representation of this figure.*)

It can be done in a two-step process. Let $\mathcal{S}_\Lambda$ and $\mathcal{S}_\Gamma$ be suitable spaces of signals defined on lattices Λ and Γ respectively. First $f[\mathbf{x}]$ is upsampled to Γ by inserting zeros at all the points in Γ that do not belong to Λ. We define the zero-insertion upsampler $\mathcal{U}_{\Lambda\uparrow\Gamma} : \mathcal{S}_\Lambda \to \mathcal{S}_\Gamma$ by

$$q = \mathcal{U}_{\Lambda\uparrow\Gamma} f \; : \; q[\mathbf{x}] = \begin{cases} f[\mathbf{x}], & \mathbf{x} \in \Lambda; \\ 0, & \mathbf{x} \in \Gamma \backslash \Lambda. \end{cases} \tag{11.2}$$

What does this do in the frequency domain?

$$\begin{aligned} Q(\mathbf{u}) &= \sum_{\mathbf{x}\in\Gamma} q[\mathbf{x}] \exp(-\mathrm{j}2\pi \mathbf{u}\cdot\mathbf{x}) \\ &= \sum_{\mathbf{x}\in\Lambda} f[\mathbf{x}] \exp(-\mathrm{j}2\pi \mathbf{u}\cdot\mathbf{x}) \\ &= F(\mathbf{u}). \end{aligned} \tag{11.3}$$

Thus $q[\mathbf{x}]$ and $f[\mathbf{x}]$ have identical Fourier transforms; however, $q[\mathbf{x}]$ is defined on Γ and so it can now be processed by a filter defined on Γ. If this filter is $h[\mathbf{x}]$, $\mathbf{x} \in \Gamma$, we know that its frequency response is specified by its values over one unit cell of Γ^*. In the ideal case, we want $H(\mathbf{u})$ to be equal to a constant c on the unit cell of Λ^* and zero elsewhere in a unit cell of Γ^*,

$$H(\mathbf{u}) = \begin{cases} c & \mathbf{u} \in \mathcal{P}_{\Lambda^*}; \\ 0 & \mathbf{u} \in \mathcal{P}_{\Gamma^*} \backslash \mathcal{P}_{\Lambda^*}, \end{cases} \tag{11.4}$$

where $\mathcal{P}_{\Lambda^*}$ and $\mathcal{P}_{\Gamma^*}$ are unit cells of Λ^* and Γ^* respectively. This is illustrated in Figure 11.3 for our example.

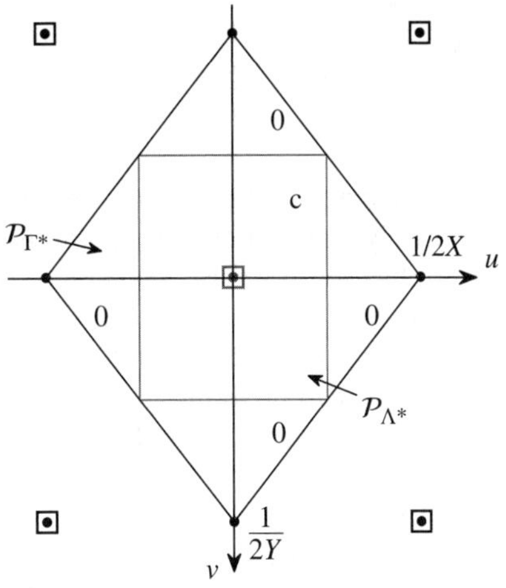

Figure 11.3 Illustration of unit cells of reciprocal lattices Λ^* (•) and Γ^* (□) of Example 11.1. $\mathcal{P}_{\Lambda^*}$ is the rectangular region and $\mathcal{P}_{\Gamma^*}$ is the diamond-shaped region. The ideal upsampling filter has a gain of c in $\mathcal{P}_{\Lambda^*}$ and a gain of 0 in $\mathcal{P}_{\Gamma^*}\backslash\mathcal{P}_{\Lambda^*}$.

What should the constant c be? We want $f[\mathbf{x}]$ and $g[\mathbf{x}]$ to correspond to the same continuous-domain signal $f_c(\mathbf{x})$. In this case,

$$F(\mathbf{u}) = \frac{1}{d(\Lambda)} F_c(\mathbf{u}) \qquad \mathbf{u} \in \mathcal{P}_{\Lambda^*} \tag{11.5}$$

$$\text{and} \quad G(\mathbf{u}) = \frac{1}{d(\Gamma)} F_c(\mathbf{u}) \qquad \mathbf{u} \in \mathcal{P}_{\Gamma^*}, \tag{11.6}$$

where we assume that $F_c(\mathbf{u}) = 0$ for $\mathbf{u} \in \mathbb{R}^D\backslash\mathcal{P}_{\Lambda^*}$. This leads us to conclude that

$$G(\mathbf{u}) = \begin{cases} \frac{d(\Lambda)}{d(\Gamma)} F(\mathbf{u}), & \mathbf{u} \in \mathcal{P}_{\Lambda^*}; \\ 0, & \mathbf{u} \in \mathcal{P}_{\Gamma^*}\backslash\mathcal{P}_{\Lambda^*}, \end{cases} \tag{11.7}$$

and so $c = d(\Lambda)/d(\Gamma)$, which is the interpolation ratio. The preceding analysis is general and applies to the upsampling of any signal defined on a lattice Λ to a superlattice Γ.

An upsampler can thus be implemented in principle using the arrangement shown in Figure 11.4, where the zero-insertion operator $\mathcal{U}_{\Lambda\uparrow\Gamma}$ is defined in Equation (11.2) and $\mathcal{H}$ is the LSI filter defined in Equation (11.4) with DC gain $c = d(\Lambda)/d(\Gamma)$. The overall upsampler is $\mathcal{H}_u = \mathcal{H}\mathcal{U}_{\Lambda\uparrow\Gamma}$, which can easily be seen to be a linear system. It is also shift invariant, in the sense that $\mathcal{H}_u\mathcal{T}_\mathbf{d}f = \mathcal{T}_\mathbf{d}\mathcal{H}_u f$ for any $\mathbf{d} \in \Lambda$. Because $\Lambda \subset \Gamma$, the shift system $\mathcal{T}_\mathbf{d}$ is well defined on both $\mathcal{S}_\Lambda$ and $\mathcal{S}_\Gamma$ for $\mathbf{d} \in \Lambda$. Of course in practice $H(\mathbf{u})$ will not be an ideal low-pass filter, but it will be an approximation designed in some way, for example using the methods of Chapter 10.

Note that an image usually has significant energy concentrated at DC, due to a nonzero average brightness. Thus, there is a great deal of energy concentrated at points of Λ^*, and consequently we want

$$H(\mathbf{u}) = 0 \quad \text{for} \quad \mathbf{u} \in \Lambda^*\backslash\Gamma^* \tag{11.8}$$

(referring to Figure 11.3, the four vertices of the diamond in this example). This should be enforced as a constraint in the filter design process along with a constraint setting the DC gain to be $d(\Lambda)/d(\Gamma)$. In addition, if we desire original input samples to be unchanged, we can impose the constraint that $h[\mathbf{0}] = 1$ and $h[\mathbf{x}] = 0$ for $\mathbf{x} \in \Lambda\backslash\mathbf{0}$.

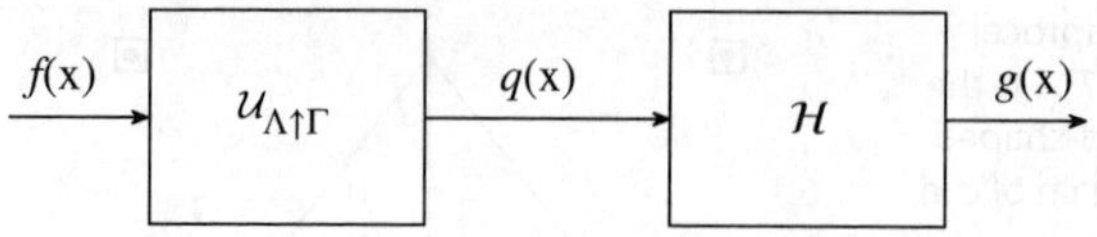

Figure 11.4 Block diagram of upsampler from Λ to superlattice Γ.

The structure of Figure 11.4 would not be used in practice to implement the upsampler, since only one sample of $q[\mathbf{x}]$ out of each $d(\Lambda)/d(\Gamma)$ is not zero. To avoid these needless multiplications by zero, a polyphase structure could be used, where the components of $g[\mathbf{x}]$ on each coset of Λ in Γ are computed separately, using only coefficients of h in a corresponding coset of Λ in Γ. The details of such implementations in the multidimensional case can be found in various sources including Vaidyanathan (1993), Coulombe and Dubois (1999), Suter (1998) and Cariolaro (2011).

Example 11.2
Design and test a system to upsample an image defined on a square lattice $\Lambda = \text{LAT}(\text{diag}(X, X))$ by a factor of four in each dimension to obtain a new image defined on the square lattice $\Gamma = \text{LAT}(\text{diag}(X/4, X/4))$.

Solution:
We have $d(\Lambda) = X^2$ and $d(\Gamma) = X^2/16$, so that the interpolation ratio is $\frac{d(\Lambda)}{d(\Gamma)} = 16$. Thus, the ideal interpolation filter will be

$$H_I(\mathbf{u}) = \begin{cases} 16 & \mathbf{u} \in \mathcal{P}_{\Lambda^*} \\ 0 & \mathbf{u} \in \mathcal{P}_{\Gamma^*} \backslash \mathcal{P}_{\Lambda^*} \end{cases}$$

For the purpose of this example, we choose the sample spacing in the upsampled image to be the unit of length (i.e. $X/4 = 1$) which we can call 1 opx for output pixel, so that

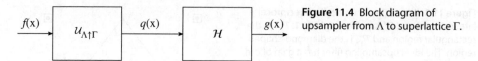

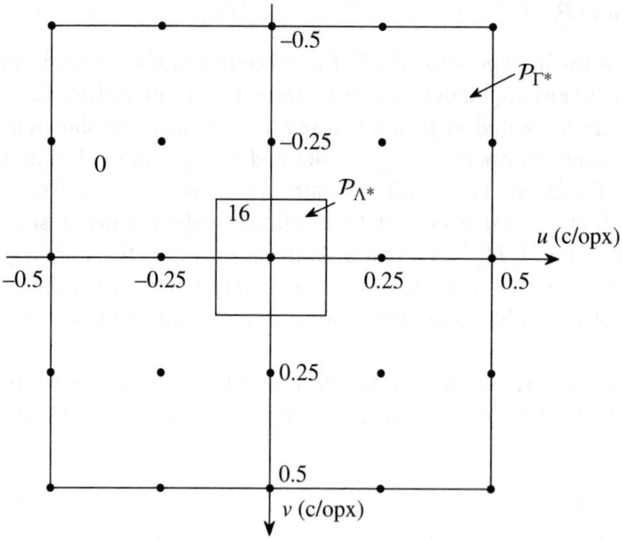

Figure 11.5 Upsampling of rectangularly sampled signal by a factor of 4 in each dimension: frequency domain view. Points of Λ^* are denoted by • and unit cells of Γ^* and Λ^* are shown. The ideal interpolation filter is $H_I(\mathbf{u}) = 16$ for $\mathbf{u} \in \mathcal{P}_{\Lambda^*}$ and $H_I(\mathbf{u}) = 0$ for $\mathbf{u} \in \mathcal{P}_{\Gamma^*} \backslash \mathcal{P}_{\Lambda^*}$. A practical filter should satisfy the constraints $H_I(\mathbf{0}) = 16$ and $H_I(\mathbf{u}) = 0$ for points $\mathbf{u} \in \Lambda^* \backslash \mathbf{0}$ within $\mathcal{P}_{\Gamma^*}$.

$\Lambda = (4\mathbb{Z})^2$ and $\Gamma = \mathbb{Z}^2$. The situation is depicted in the frequency domain in Figure 11.5 within one unit cell of Γ^*.

Since the passband of the interpolation filter is rectangular in shape and both lattices are rectangular, we could use a separable filter. The most widely used interpolation filters in practice are bilinear and bicubic separable filters. These will be discussed in more detail in Section 11.5.2. For 4:1 upsampling in each dimension, these two filters have order 7×7 and 15×15 respectively, with frequency responses shown in Figures 11.6(a) and (b). They both have a DC gain of 16 and satisfy the constraints of Equation (11.8). We can use the least-pth filter design method of Section 10.5.2 to better approximate the ideal response of the interpolation filter while satisfying the constraints. Figure 11.6(d) shows the frequency response of a 29×29 filter obtained using this method. The ideal response used for the design is shown in Figure 11.6(c); it has a passband edge at $u = 0.8 * 0.125$ and a stopband edge at $1.2 * 0.125$ with a planar transition band.

Figure 11.7 shows the result of applying these interpolation filters to a portion of the motorcyle image. The original portion is shown in (a), and the results of upsampling with bilinear, bicubic and the 29×29 interpolator obtained with the least-pth method are shown in (b)–(d) respectively. The successive improvements can be clearly seen in

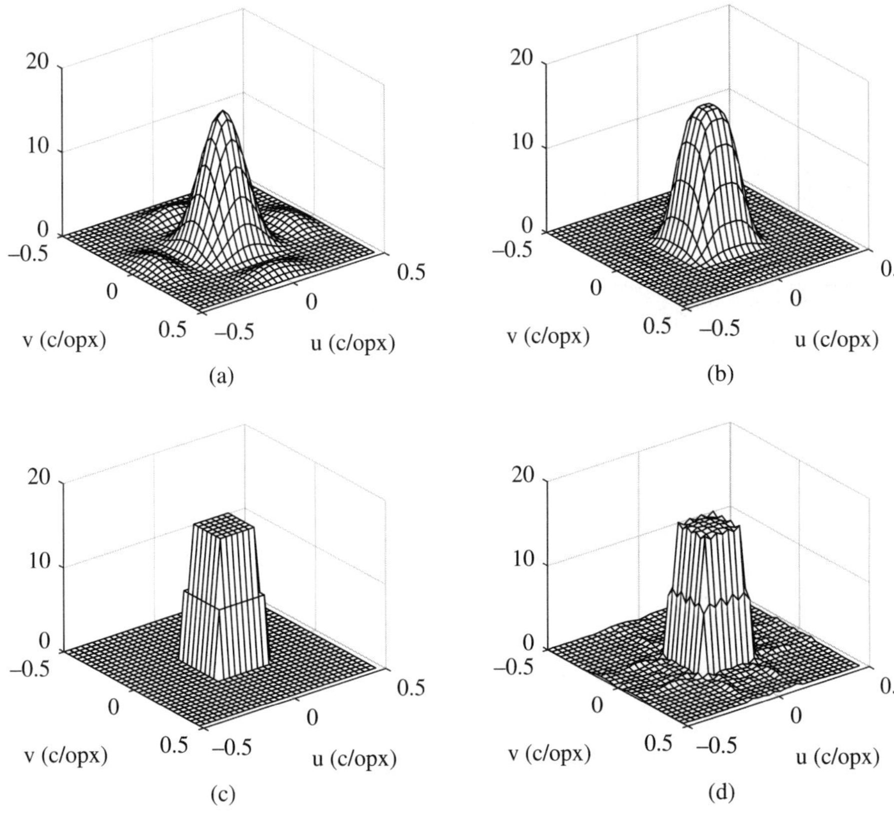

Figure 11.6 Perspective view of the frequency response of interpolation filters for 16:1 interpolation in Example 11.2. (a) Bilinear interpolator. (b) Bicubic interpolator. (c) Ideal response for least-pth design. (d) 29×29 interpolator obtained with least-pth constrained optimization and quadrantal symmetry.

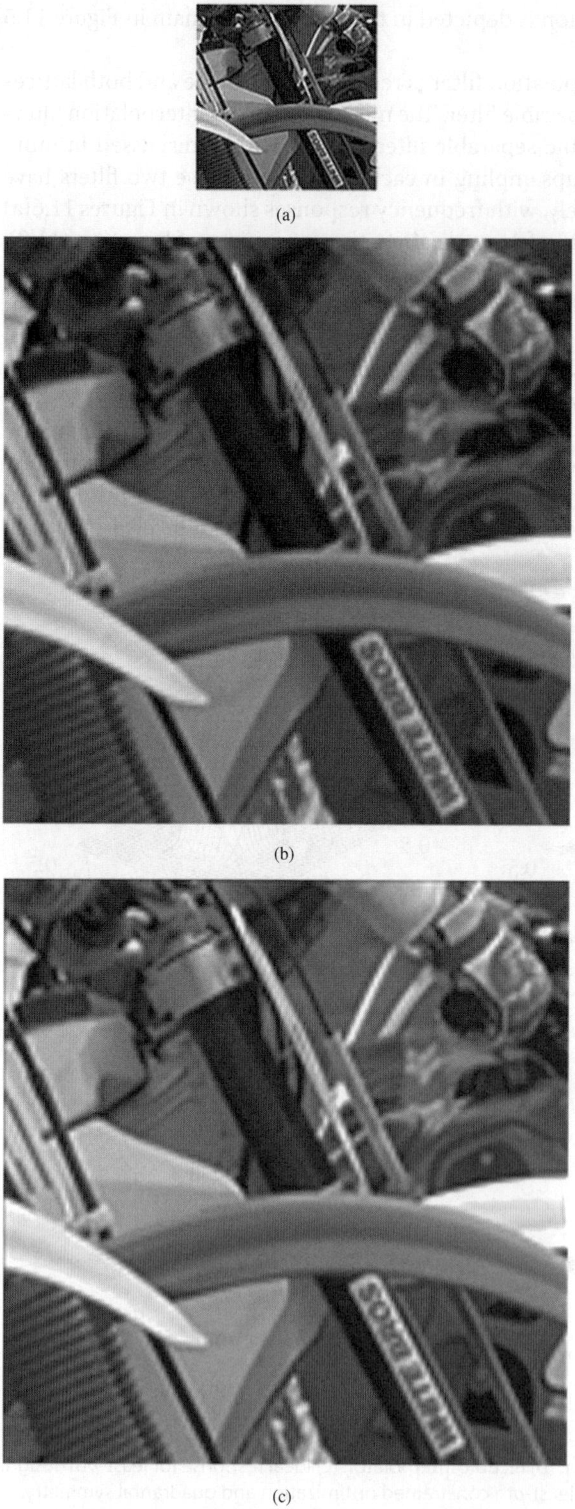

Figure 11.7 Results of interpolation the standard Kodak motorcycle image (a) with the (b) bilinear interpolator; (c) bicubic interpolator; (d) 29 × 29 interpolator obtained with the least-*p*th method.

(d)

Figure 11.7 *(Continued)*

these figures, particularly in the zipper-like artifacts along high-contrast sloping edges. Increasing the order beyond 29 × 29 had no visible effect on this particular image.

11.4 Downsampling

The reverse operation to upsampling is downsampling. Assume that f is sampled on a lattice Γ and that $\Lambda \subset \Gamma$ is a sublattice of Γ. The downsampling operator $\mathcal{D}_{\Gamma\downarrow\Lambda}$ is given by

$$g = \mathcal{D}_{\Gamma\downarrow\Lambda} f \; : \; g[\mathbf{x}] = f[\mathbf{x}], \quad \mathbf{x} \in \Lambda \tag{11.9}$$

where the samples on $\Gamma\backslash\Lambda$ are discarded. Referring to the example in Figure 11.2, it is clear that the subsampling will introduce extra spectral replicas on the points of $\Lambda^*\backslash\Gamma^*$ (the dashed replicas). In order to avoid aliasing, we need to prefilter the image so that its Fourier transform is confined to a unit cell of Λ^* before subsampling. Thus, a downsampling system has the structure shown in Figure 11.8.

We have shown that we can write

$$\Lambda^* = \bigcup_{k=0}^{K-1} (\mathbf{d}_k + \Gamma^*) \tag{11.10}$$

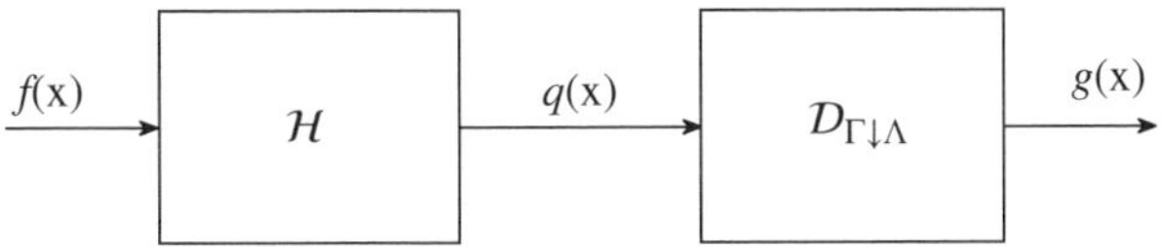

Figure 11.8 Block diagram of downsampling system from lattice Γ to sublattice Λ.

where $K = \mathrm{d}(\Lambda)/\mathrm{d}(\Gamma)$, for a suitably chosen set of $\mathbf{d}_k \in \Lambda^*$, coset representatives of Γ^* in Λ^*. This was used extensively in Chapter 4 on discrete-domain periodic signals, and is demonstrated in Chapter 13. To illustrate, in Example 11.1, $K = 2$ and we can choose we $\mathbf{d}_0 = [0 \quad 0]^T$ and $\mathbf{d}_1 = [\frac{1}{2X} \quad 0]^T$ (see Figure 11.2).

To analyze the effect of the downsampler, referring to Figure 11.8, we observe that

$$\mathcal{D}_{\Gamma\downarrow\Lambda} q = \mathcal{D}_{\Gamma\downarrow\Lambda}(q \cdot \delta_{\Gamma/\Lambda}) \tag{11.11}$$

where we recall from Chapter 4 that the periodic unit sample is

$$\delta_{\Gamma/\Lambda}[\mathbf{x}] = \begin{cases} 1 & \mathbf{x} \in \Lambda; \\ 0 & \mathbf{x} \in \Gamma\backslash\Lambda. \end{cases} \tag{11.12}$$

If we denote $r = q \cdot \delta_{\Gamma/\Lambda}$, then $g = \mathcal{D}_{\Gamma\downarrow\Lambda} r$. Also, since r is only nonzero on Λ, it follows that $r = \mathcal{U}_{\Lambda\uparrow\Gamma} g$, and so from Equation (11.3), $G(\mathbf{u}) = R(\mathbf{u})$.

The signal $\delta_{\Gamma/\Lambda}$ is a Λ-periodic signal on Γ and its discrete-domain periodic Fourier transform is easily seen to be the constant 1. Thus we have the Fourier transform synthesis Equation

$$\delta_{\Gamma/\Lambda}[\mathbf{x}] = \frac{1}{K}\sum_{k=0}^{K-1} \exp(\mathrm{j}2\pi \mathbf{d}_k \cdot \mathbf{x}), \qquad \mathbf{x} \in \Gamma, \tag{11.13}$$

so that

$$r[\mathbf{x}] = \frac{1}{K}\sum_{k=0}^{K-1} q[\mathbf{x}] \exp(\mathrm{j}2\pi \mathbf{d}_k \cdot \mathbf{x}). \tag{11.14}$$

Using the modulation property of the discrete-domain Fourier transform,

$$G(\mathbf{u}) = R(\mathbf{u}) = \frac{1}{K}\sum_{k=0}^{K-1} Q(\mathbf{u} - \mathbf{d}_k). \tag{11.15}$$

Thus, the downsampling adds $K - 1$ spectral replicas at each of the coset representatives $\mathbf{d}_k$, $k = 1, \dots, K - 1$, and these will in general overlap, causing aliasing. To eliminate such aliasing, we prefilter f with the anti-aliasing low-pass prefilter h. With such prefiltering, the overall output of the downsampling system has Fourier transform

$$G(\mathbf{u}) = \frac{1}{K}\sum_{k=0}^{K-1} H(\mathbf{u} - \mathbf{d}_k) F(\mathbf{u} - \mathbf{d}_k). \tag{11.16}$$

The "ideal" anti-aliasing prefilter has frequency response

$$H(\mathbf{u}) = \begin{cases} 1 & \mathbf{u} \in \mathcal{P}_{\Lambda^*} \\ 0 & \mathbf{u} \in \mathcal{P}_{\Gamma^*}\backslash\mathcal{P}_{\Lambda^*}. \end{cases} \tag{11.17}$$

However, such a filter could introduce objectionable ringing so a transition band should be used to balance aliasing with loss of resolution and ringing. There is no established optimal solution to this tradeoff and so engineering judgement must be used.

Similar to the case of upsampling, we would not used the structure of Figure 11.8 directly to implement the downsampling system in practice. Rather than computing all samples on Γ and then discarding $K - 1$ out of every K samples, we would only compute the desired samples on Λ. This is possible since we only use nonrecursive FIR filters for the prefiltering.

To illustrate with the lattices of Example 11.1, suppose that $f[\mathbf{x}]$ defined on Γ has Fourier transform confined to the octagonal region shown in Figure 11.9(a), replicated on the points of Γ^* and well contained in the unit cell $\mathcal{P}_{\Gamma^*}$. However, these regions will clearly overlap significantly if replicated on the points of Λ^*. To avoid aliasing upon downsampling to Λ, the ideal anti-aliasing low-pass filter has a rectangular passband, given by the unit cell $\mathcal{P}_{\Lambda^*}$, as shown in the figure. The Fourier transform of the prefiltered and downsampled signal has support in the region shown in Figure 11.9(b). The original filtered replicas on Γ^* are shown in green and the additional replicas on points of $\Lambda^*\backslash\Gamma^*$ are shown in yellow.

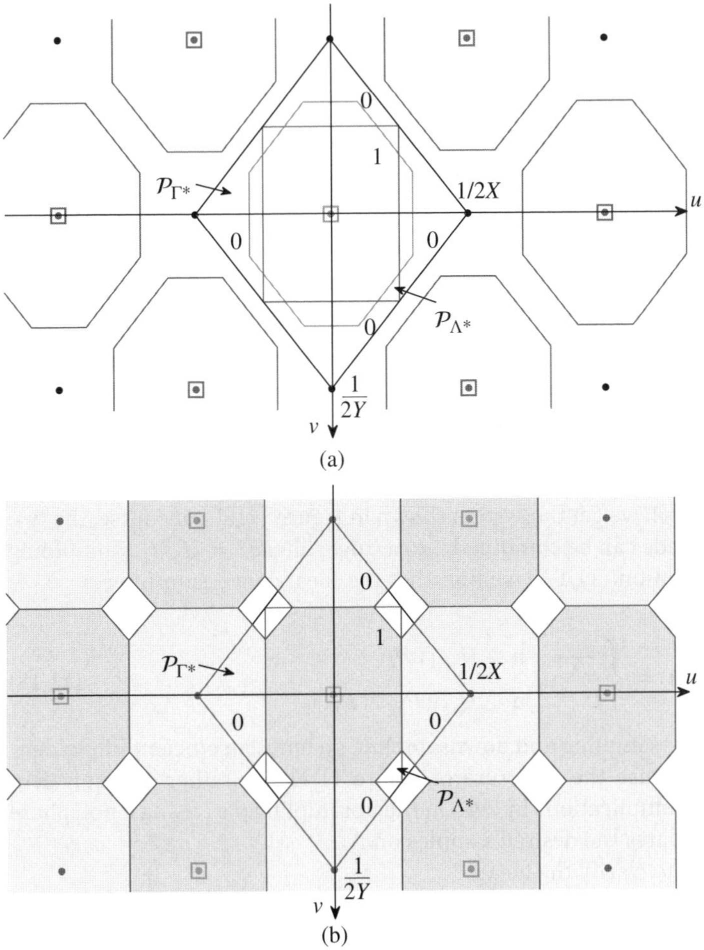

Figure 11.9 Illustration of a downsampling system using the lattices of Example 11.1. (a) Selected points of the reciprocal lattices Λ^* (•) and Γ^* (□) are shown. The Fourier transform of the original signal defined on Γ is limited to the octagonal region shown, replicated at points of Γ^*. (b) Output of the downsampling system. The original filtered replicas on Γ^* are shown in green and the additional replicas on points of $\Lambda^*\backslash\Gamma^*$ are shown in yellow. (*See color plate section for the color representation of this figure.*)

As a concrete example, consider the downsampling of the Barbara image by a factor of 2 in each direction. Let the input lattice be $\Gamma = \mathbb{Z}^2$ and the output lattice be $\Lambda = (2\mathbb{Z})^2$, i.e. the unit of length is the input pixel spacing denoted 1 ipx. The downsampled image without prefiltering is shown in Figure 11.10(a) and with a 15×15 anti-aliasing prefilter is shown in Figure 11.10(b). We also show the spectrum of the original image, the spectrum after down sampling with no prefiltering, the spectrum of the prefiltered image and the spectrum of the prefiltered image after downsampling in Figure 11.10(c)-(f) respectively. The effects of prefiltering are clearly seen.

11.5 Arbitrary Sampling Structure Conversion

What can we do if neither the input nor the output lattice is a sublattice of the other one? There are two common approaches that we present here. Suppose that the input signal is sampled on the lattice Λ_1 and the output is sampled on the lattice Λ_2, where $\Lambda_1 \not\subset \Lambda_2$ and $\Lambda_2 \not\subset \Lambda_1$. In the first approach, we try to find a third lattice Λ_3 that is a superlattice of both Λ_1 and Λ_2. We can then upsample the input signal from Λ_1 to Λ_3, and then downsample from Λ_3 to Λ_2.The second approach is to use polynomial interpolation.

11.5.1 Sampling Structure Conversion Using a Common Superlattice

Assume that we can find a common superlattice Λ_3 such that $\Lambda_1 \subset \Lambda_3$ and $\Lambda_2 \subset \Lambda_3$. Assuming that such a common superlattice exists, we want to find the least-dense common superlattice. Chapter 13 gives the conditions for which such a D-dimensional common superlattice exists and a method to find it. In the two-step procedure, the input signal is first upsampled from Λ_1 to Λ_3 using the procedure of Section 11.3. This intermediate signal is then downsampled from Λ_3 to Λ_2, with appropriate prefiltering if necessary, using the procedure of Section 11.4. Combining the block diagrams of Figure 11.8 and Figure 11.4, we get the system shown in Figure 11.11. Of course, the two filters $\mathcal{H}_1$ and $\mathcal{H}_2$ in cascade can be combined into a single filter $\mathcal{H} = \mathcal{H}_2\mathcal{H}_1$. Combining Equation (11.7) and Equation (11.17), we find that the ideal conversion filter on Λ_3 is given by

$$H(\mathbf{u}) = H_1(\mathbf{u})H_2(\mathbf{u}) = \begin{cases} \frac{d(\Lambda_1)}{d(\Lambda_3)} & \mathbf{u} \in \mathcal{P}_{\Lambda_1^*} \cap \mathcal{P}_{\Lambda_2^*} \\ 0 & \mathbf{u} \in \mathcal{P}_{\Lambda_3^*} \backslash (\mathcal{P}_{\Lambda_1^*} \cap \mathcal{P}_{\Lambda_2^*}). \end{cases} \tag{11.18}$$

Similar to the case of upsampling and downsampling systems, an efficient implementation would not directly use the structure of Figure 11.11 but rather an equivalent system that avoids the multiplications by zero in the upsampling part using a polyphase structure, and only computes the desired samples on Λ_2.

An example will serve to clarify the ideas.

Example 11.3

Design a system to convert a signal defined on the lattice $\Lambda_1 = \text{LAT}\left(\left[\begin{smallmatrix} X & 0 \\ Y & 2Y \end{smallmatrix}\right]\right) = \text{LAT}(\mathbf{V}_1)$ to a signal defined on the lattice $\Lambda_2 = \text{LAT}\left(\left[\begin{smallmatrix} 2X & X \\ 0 & 2Y \end{smallmatrix}\right]\right) = \text{LAT}(\mathbf{V}_2)$. For this example, we assume that $X = Y = 1$ px.

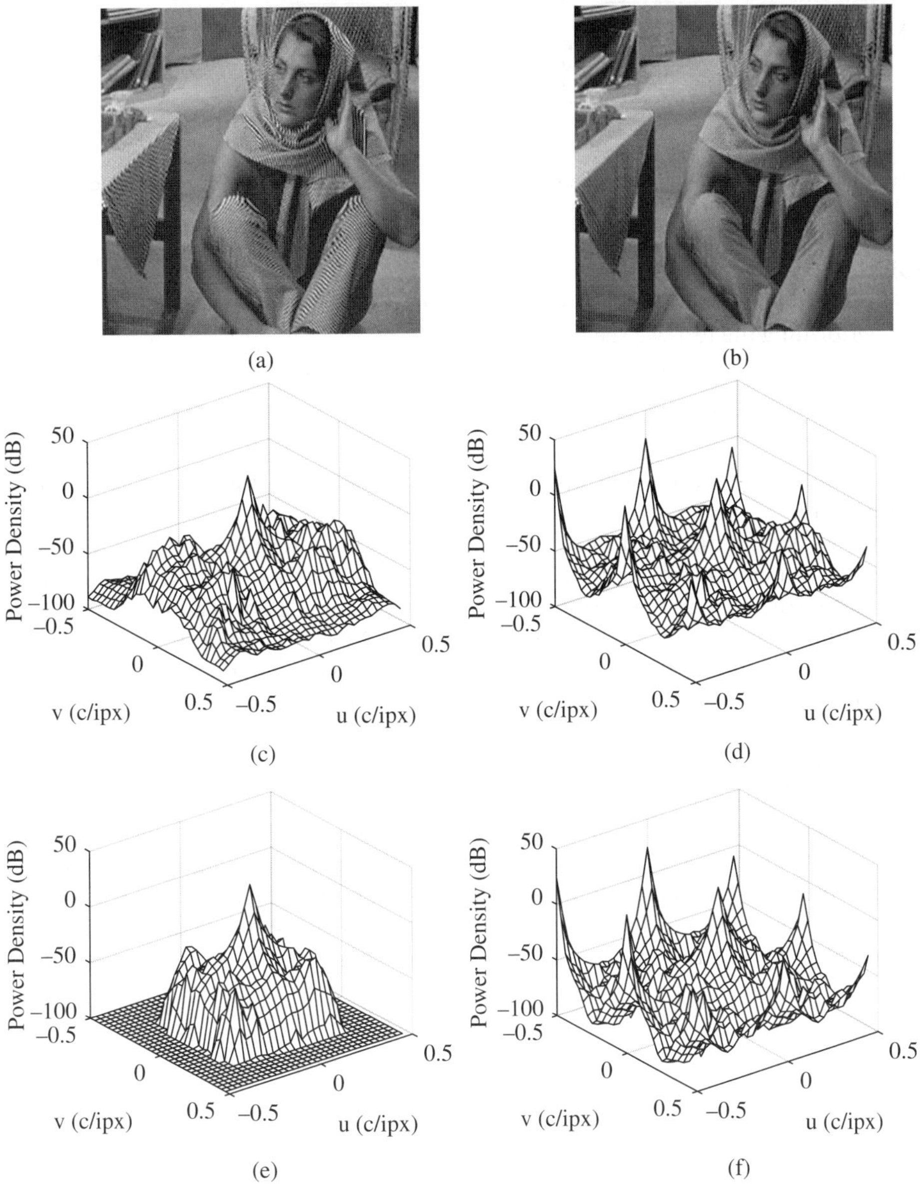

Figure 11.10 Downsampling the Barbara image by a factor of two in each dimension. (a) Without prefiltering. (b) With anti-aliasing prefilter. (c) Power density spectrum of the original Barbara image. (d) Power density spectrum of the downsampled Barbara image with no prefilter. (e) Power density spectrum of the prefiltered Barbara image. (f) Power density spectrum of the prefiltered and downsampled Barbara image.

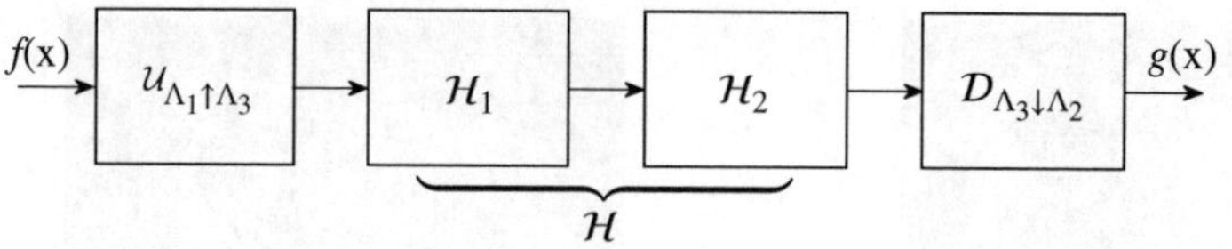

Figure 11.11 Block diagram of sampling structure conversion system using a common superlattice. The cascade of $\mathcal{H}_2$ after $\mathcal{H}_1$ is replaced with a single filter $\mathcal{H}$.

Solution:
$d(\Lambda_1) = 2$ and $d(\Lambda_2) = 4 = 2d(\Lambda_1)$ so Λ_2 has a lower sampling density than Λ_1. Is $\Lambda_2 \subset \Lambda_1$? Performing the test, we see that

$$\mathbf{V}_1^{-1}\mathbf{V}_2 = \begin{bmatrix} 1 & 0 \\ -\frac{1}{2} & \frac{1}{2} \end{bmatrix} \begin{bmatrix} 2 & 1 \\ 0 & 2 \end{bmatrix} = \begin{bmatrix} 2 & 1 \\ -1 & \frac{1}{2} \end{bmatrix}$$

which is *not* an integer matrix. Thus $\Lambda_2 \not\subset \Lambda_1$. In fact, the situation can be seen in Figure 11.12.

Since $\mathbf{V}_1^{-1}\mathbf{V}_2$ is a matrix of rational numbers, we know from Section 13.9 that there exists a two-dimensional common superlattice. Although we can use the algorithm of Section 13.9 to find it, in this case, by inspection of Figure 11.12, we can see that the least dense lattice that contains both Λ_1 and Λ_2 is $\Lambda_3 = \text{LAT}\left(\left[\begin{smallmatrix} 1 & 0 \\ 0 & 1 \end{smallmatrix}\right]\right)$. Thus, the first step is to upsample the signal from Λ_1 to Λ_3. The situation is illustrated in the frequency domain in Figure 11.13. The spectral replicas centered on points of Λ_1^* that are not in Λ_3^* must be removed by a low-pass filter defined on Λ_3. These are *not* shaded in Figure 11.13. The filter would have a gain of two in the diamond-shaped unit cell of Λ_1^* and a gain of zero in the rest of the square unit cell of Λ_3^*.

In the second step, the signal is downsampled from Λ_3 to Λ_2. The effect of this is illustrated in the frequency domain in Figure 11.14. The subsampling introduces additional copies of the spectrum of the signal at points of Λ_2^* that are not in Λ_3^*. This will clearly introduce aliasing if prefiltering is not applied. Referring to Figure 11.14, a suitable prefilter has as passband the hexagonal unit cell of Λ_2^* shown. Since in this example,

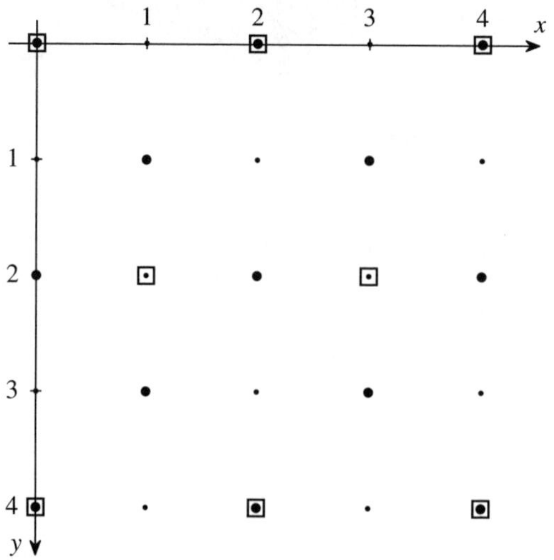

Figure 11.12 Lattices Λ_1 (•) and Λ_2 (□) for Example 11.3. Also shown is the least common superlattice Λ_3 (·).

Figure 11.13 Frequency domain situation for upconversion from Λ_1 to Λ_3 in Example 11.3. The reciprocal lattices Λ_1^* (•) and Λ_3^* (◇) and their unit cells are shown. The input signal has a Fourier transform confined to the diamond-shaped region. The interpolation filter, defined on Λ_3, should keep the shaded replicas and reject the unshaded replicas. (*See color plate section for the color representation of this figure.*)

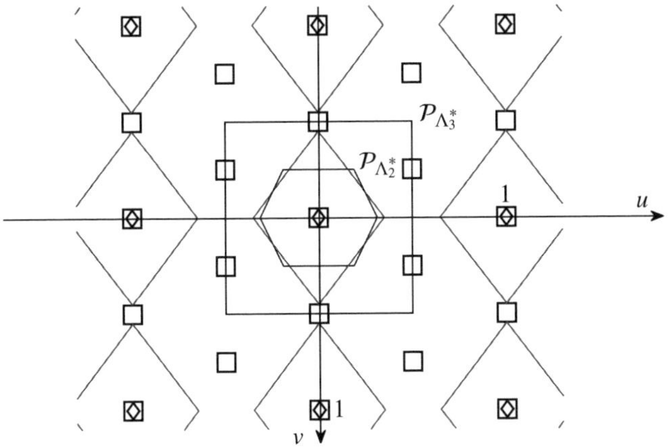

Figure 11.14 Frequency domain situation for downconversion from Λ_3 to Λ_2 in Example 11.3. The reciprocal lattices Λ_2^* (□) and Λ_3^* (◇) and their unit cells are shown. The anti-aliasing prefilter has passband equal to the hexagonal unit cell of Λ_2^*. (*See color plate section for the color representation of this figure.*)

the unit cell of Λ_2^* is entirely contained within the unit cell of Λ_1^*, the overall filter $H(u, v)$ has gain 2 in $\mathcal{P}_{\Lambda_2^*}$ and 0 in $\mathcal{P}_{\Lambda_3^*} \backslash \mathcal{P}_{\Lambda_2^*}$. □

11.5.2 Polynomial Interpolation

The most common method in practice to convert the sampling structure of a rectangularly sampled image to another lattice is separable polynomial interpolation. This can be thought of as producing a continuous version of the signal that can then be resampled at arbitrary locations. If the signal is to be downsampled, the polynomial interpolation should be preceded by an appropriate anti-aliasing prefilter adapted to the output sampling structure. This method could also be applied to a non-rectangular input lattice by operating separably along a set of basis vectors of the lattice.

The most frequently used polynomial interpolators are zero-order hold, linear (straight-line) interpolation and cubic interpolation. In the 2D case, these are usually called nearest neighbor, bilinear and bicubic interpolation. We will describe one-dimensional polynomial interpolation, which is then applied separably for multidimensional signals. The case of multivariate nonseparable polynomial interpolation is much more complex and will not be considered here. See Gasca and Sauer (2000) for a survey of some of the issues and methods.

Assume a one-dimensional signal with sample spacing X is to be interpolated. Polynomial interpolation is done piecewise: a different polynomial is used to construct each segment of length X. This is evident from the illustrations of zero-order hold and linear, straight-line interpolation shown in Figure 11.15.

We can write piecewise expressions for the interpolated signals for zero-order hold:

$$f_c^0(x) = f[kX] \quad \text{if} \quad kX - \frac{X}{2} < x \leq kX + \frac{X}{2} \tag{11.19}$$

and for linear straight-line interpolation:

$$f_c^1(x) = \left(1 - \frac{x - kX}{X}\right) f[kX] + \left(\frac{x - kX}{X}\right) f[(k+1)X] \quad \text{if} \quad kX < x \leq (k+1)X. \tag{11.20}$$

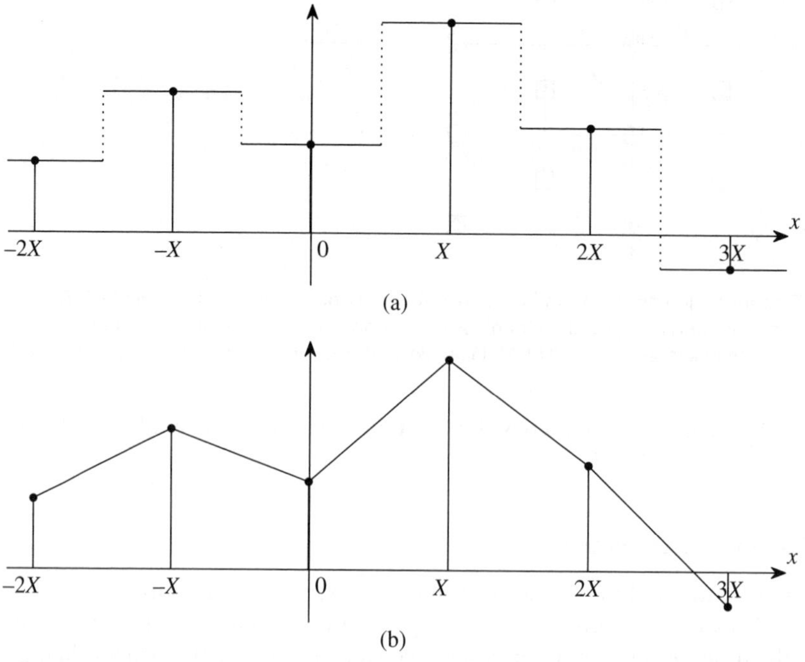

Figure 11.15 Illustration of one-dimensional (a) zero-order hold and (b) linear, straight-line interpolation.

We can write these formulas in the form of a convolution:

$$f_c^m(x) = \sum_{k=-\infty}^{\infty} f[kX]h_m(x - kX). \tag{11.21}$$

For the two cases considered above, we have

$$h_0(x) = \begin{cases} 1 & \text{if} -\frac{X}{2} < x \le \frac{X}{2} \\ 0 & \text{otherwise} \end{cases} \tag{11.22}$$

$$h_1(x) = \begin{cases} 1 - \frac{|x|}{X} & \text{if } |x| \le X \\ 0 & \text{otherwise.} \end{cases} \tag{11.23}$$

This operation is illustrated graphically in Figure 11.16.

A straightforward analysis shows that

$$F_c^m(u) = F(u)H_m(u) \tag{11.24}$$

where $H_m(u)$ has the characteristics of a low-pass filter for these polynomial interpolators. We easily find that

$$H_0(u) = \frac{\sin(\pi u X)}{\pi u} \tag{11.25}$$

$$\text{and} \qquad H_1(u) = X\frac{\sin^2(\pi u X)}{(\pi u X)^2} \tag{11.26}$$

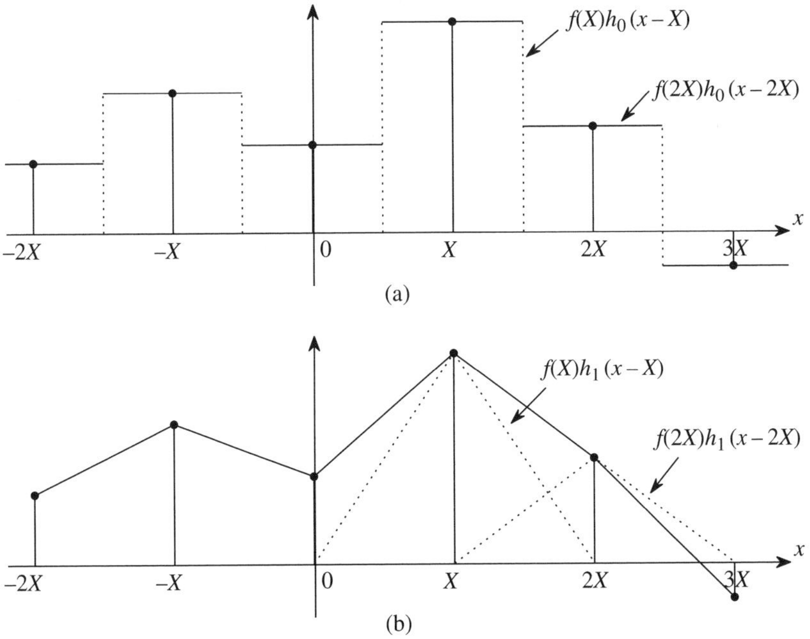

Figure 11.16 Details of interpolation illustrating the convolution form using (a) zero-order hold and (b) linear, straight-line interpolation.

where the latter is a better low-pass filter. Even better results can be obtained with cubic interpolation:

$$h_3(x) = \begin{cases} 1.5\left(\frac{|x|}{X}\right)^3 - 2.5\left(\frac{|x|}{X}\right)^2 + 1 & \text{if } 0 \leq |x| < X \\ -0.5\left(\frac{|x|}{X}\right)^3 + 2.5\left(\frac{|x|}{X}\right)^2 - 4\left(\frac{|x|}{X}\right) + 2 & \text{if } X \leq |x| < 2X \\ 0 & \text{otherwise.} \end{cases} \quad (11.27)$$

Note that $h_3(x)$ is continuous, with $h_3(0) = 1$, $h_3(\pm X) = 0$ and $h_3(\pm 2X) = 0$, so that existing samples are unchanged, as we expect from an interpolator. Also, $h_3'(x)$ is continuous, with $h_3'(0) = h_3'(\pm 2X) = 0$. Thus, the continuous-domain signal obtained with this bicubic interpolator is continuous, with continuous first derivatives, which can be very useful if we want to estimate the derivative of the image. The derivation of this cubic interpolator and its properties can be found in Keys (1981).

Two-dimensional interpolation is achieved by applying these one-dimensional interpolators separably in the x and y directions, or along the directions of the input lattice basis vectors, and similarly for higher-dimensional signal resampling. Note that the separable bilinear and bicubic interpolators were used in Example 11.2 and the frequency response of these interpolation filters was given in Figure 11.6. MATLAB implements separable polynomial interpolators using the `imresize` function, with parameters “nearest”, “bilinear” and “bicubic” respectively.

Problems

1 For each of the following pairs of lattices Λ_1 and Λ_2, state whether $\Lambda_1 \subset \Lambda_2$, $\Lambda_2 \subset \Lambda_1$ or neither. If neither, find (by inspection) the least dense lattice Λ_3 such that $\Lambda_1 \subset \Lambda_3$ and $\Lambda_2 \subset \Lambda_3$. For each lattice Λ_1, Λ_2 and Λ_3 (if required), determine and sketch the reciprocal lattice and a unit cell of the reciprocal lattice. Specify a sampling structure conversion system to transform a signal $f[\mathbf{x}]$ sampled on Λ_1 to a signal $g[\mathbf{x}]$ sampled on Λ_2. Assume that ideal low-pass filters are used where filters are required, sketch their passband in the frequency domain and indicate the gain.

(a) $V_{\Lambda_1} = \begin{bmatrix} X & 0 \\ 0 & X \end{bmatrix}$ $\quad V_{\Lambda_2} = \begin{bmatrix} 2X & 0 \\ 0 & 2X \end{bmatrix}$

(b) $V_{\Lambda_1} = \begin{bmatrix} X & 0 \\ 0 & X \end{bmatrix}$ $\quad V_{\Lambda_2} = \begin{bmatrix} 3X & X \\ 0 & X \end{bmatrix}$

(c) $V_{\Lambda_1} = \begin{bmatrix} 2X & 0 \\ 0 & 2X \end{bmatrix}$ $\quad V_{\Lambda_2} = \begin{bmatrix} X & X \\ X & -X \end{bmatrix}$

(d) $V_{\Lambda_1} = \begin{bmatrix} 1.5X & 0 \\ 0 & 1.5X \end{bmatrix}$ $\quad V_{\Lambda_2} = \begin{bmatrix} X & 0 \\ 0 & X \end{bmatrix}$

(e) $V_{\Lambda_1} = \begin{bmatrix} X & X \\ X & -X \end{bmatrix}$ $\quad V_{\Lambda_2} = \begin{bmatrix} 1.5X & 1.5X \\ 1.5X & -1.5X \end{bmatrix}$

12

Symmetry Invariant Signals and Systems

12.1 LSI Systems Invariant to a Group of Symmetries

A symmetry is a transformation that leaves a certain entity unchanged. For example, the translation of a lattice by an element of that lattice leaves the lattice unchanged (see Section 13.3). However, translations are not the only symmetries of a lattice. More generally, an isometry of $\mathbb{R}^D$ that leaves a lattice Λ invariant is called a symmetry of Λ. An isometry of $\mathbb{R}^D$ is a permutation of $\mathbb{R}^D$ that leaves the Euclidean distance between any pair of points unchanged. Besides translations, these consist of rotations and reflections. These symmetries have been well studied, especially in two or three dimensions in crystallography (e.g., Miller (1972)), and the four-dimensional case is completely covered in Brown *et al.* (1978). We first present the idea of symmetry invariance in general terms, then present some specific examples of particular interest in signal processing. An example of symmetry invariance in one dimension is inversion invariance, where the inversion operator maps $f[n]$ to $f[-n+c]$ for some $c \in \mathbb{Z}$. An LSI system is inversion invariant if the unit sample response is symmetric about the origin, i.e. $h[-n] = h[n]$, which corresponds to a zero-phase filter. The second use of symmetry is to form a signal space of symmetric (or anti-symmetric) signals and apply Fourier analysis on this signal space. The theory is presented first for discrete-domain *aperiodic* signals, but the main interest will be for discrete-domain *periodic* signals, leading to a general class of multidimensional discrete cosine transform (DCT). The examples will be mainly in two and three dimensions, but the theory is presented for general D as much as possible.

12.1.1 Symmetries of a Lattice

Symmetries are generally studied using group theory. A good treatment is given by Miller (1972) among many others, and more details on the concepts used here can be found in that book. The isometries of the Euclidean space $\mathbb{R}^D$ are given by the Euclidean group $E(D)$, which consists of translations and orthogonal transformations. We denote an element $\mathfrak{t} \in E(D)$ using the Euler Fraktur font and denote its action on an element $\mathbf{x} \in \mathbb{R}^D$ by $\mathfrak{t}\mathbf{x}$. By definition, $\mathfrak{t}$ is a one-to-one mapping of $\mathbb{R}^D$ to $\mathbb{R}^D$ such that $\|\mathfrak{t}\mathbf{x}_1 - \mathfrak{t}\mathbf{x}_2\| = \|\mathbf{x}_1 - \mathbf{x}_2\|$ for all $\mathbf{x}_1, \mathbf{x}_2 \in \mathbb{R}^D$, where $\|\cdot\|$ denotes the Euclidean distance in $\mathbb{R}^D$. Compositions of transformations in $E(D)$ are denoted as $\mathfrak{t}_i\mathfrak{t}_k$, where $\mathfrak{t}_i\mathfrak{t}_k$ means that $\mathfrak{t}_k$ is applied first, followed by $\mathfrak{t}_i$ (some authors use the opposite convention). For any $\mathbf{a} \in \mathbb{R}^D$,

Multidimensional Signal and Color Image Processing Using Lattices, First Edition. Eric Dubois.

Companion website: www.wiley.com/go/Dubois/multiSP

the translation transformation $\mathbf{t_a}$ is given by $\mathbf{t_a x} = \mathbf{a} + \mathbf{x}$, for all $\mathbf{x} \in \mathbb{R}^D$. The group $T(D)$ of all translations can be identified with $\mathbb{R}^D$ itself under addition, since

$$\mathbf{t_a t_b x} = \mathbf{t_a}(\mathbf{b} + \mathbf{x}) = \mathbf{a} + \mathbf{b} + \mathbf{x} = \mathbf{t_{a+b} x}. \tag{12.1}$$

The orthogonal group $O(D)$ consists of rotations and rotation-inversions about the origin, and can be identified with the group of $D \times D$ orthogonal matrices. Any isometry $\mathbf{t} \in E(D)$ can be written uniquely as $\mathbf{t} = \mathbf{t_a}\mathfrak{o}$ for some $\mathbf{a} \in \mathbb{R}^D$ and $\mathfrak{o} \in O(D)$, or using matrix notation, $\mathbf{tx} = \mathbf{a} + \mathbf{A}_\mathfrak{o}\mathbf{x}$ where $\mathbf{A}_\mathfrak{o}$ is the $D \times D$ orthogonal matrix for $\mathfrak{o}$. We will denote this $\mathbf{t} = \{\mathbf{a}, \mathbf{A}_\mathfrak{o}\}$.

Definition 12.1 A symmetry of a lattice Λ is an element $\mathbf{t}$ of $E(D)$ that leaves Λ invariant, i.e. $\{\mathbf{tx} \mid \mathbf{x} \in \Lambda\} = \Lambda$, or more simply $\mathbf{t}\Lambda = \Lambda$.

Definition 12.2 The *complete symmetry group* of Λ, denoted $\mathcal{G}_c(\Lambda)$ is the set of all possible symmetries of Λ:

$$\mathcal{G}_c(\Lambda) = \{\mathbf{t} \in E(D) \mid \mathbf{t}\Lambda = \Lambda\}. \tag{12.2}$$

The complete symmetry group $\mathcal{G}_c(\Lambda)$ is a discrete subgroup of $E(D)$ that also consists of translations, rotations and rotation-inversions. The translation subgroup $T(D) \cap \mathcal{G}_c(\Lambda)$ is isomorphic to Λ itself (as an additive group). Any element of $\mathcal{G}_c(\Lambda)$ can be written uniquely as $\mathbf{t} = \mathbf{t_a}\mathfrak{o}$ where $\mathbf{a} \in \Lambda$ and $\mathfrak{o}$ belongs to a finite subgroup of $O(D)$ (called a *point group* since it leaves a point, the origin, invariant). Any subgroup of $\mathcal{G}_c(\Lambda)$ is called a symmetry group of Λ, and is denoted $\mathcal{G}(\Lambda)$. $T(D) \cap \mathcal{G}(\Lambda)$ is the translation subgroup of $\mathcal{G}(\Lambda)$, and it is isomorphic to a sublattice of Λ. If it is a proper sublattice of dimension D, we would be concerned with periodically shift-variant systems. The translation subgroup could also be of dimension less than D. While these situations may be of interest in their own right when dealing with symmetry-invariant systems, in this chapter we will be concerned only with LSI systems where $T(D) \cap \mathcal{G}(\Lambda)$ is isomorphic to Λ.

Definition 12.3 The point group of $\mathcal{G}(\Lambda)$, denoted $\mathcal{G}_p(\Lambda)$ is

$$\mathcal{G}_p(\Lambda) = \{\mathfrak{o} \mid \mathbf{t_a}\mathfrak{o} \in \mathcal{G}(\Lambda) \quad \text{for any } \mathbf{a} \in \Lambda\}. \tag{12.3}$$

$\mathcal{G}_p(\Lambda)$ is a finite subgroup of $O(D)$; it is a symmetry group of Λ, although it is not necessarily a subgroup of $\mathcal{G}(\Lambda)$. We denote the point group of $\mathcal{G}_c(\Lambda)$ as $\mathcal{G}_{cp}(\Lambda)$; it is sometimes called the *holohedry* of Λ. $\mathcal{G}_p(\Lambda)$ is a subgroup of $\mathcal{G}_{cp}(\Lambda)$.

The symmetry-invariant systems we study correspond to different subgroups of $\mathcal{G}_{cp}(\Lambda)$ for a given lattice. If the subgroup consists of only the identity, we revert to the conventional class of LSI systems studied already. The number of possibilities increases rapidly with the dimension. In one dimension, there is only one point group with more than one element, consisting of reflection about the origin (inversion) and the identity. This symmetry group is a cyclic group of order 2. The inversion is an element of the complete symmetry group for any lattice, since $-\mathbf{x} \in \Lambda$ for any $\mathbf{x} \in \Lambda$. For a randomly chosen lattice with D random linearly independent vectors as basis, the inversion is likely to be the only nontrivial point symmetry. However, typical lattices used in imaging are not random and generally have considerably more symmetry than this.

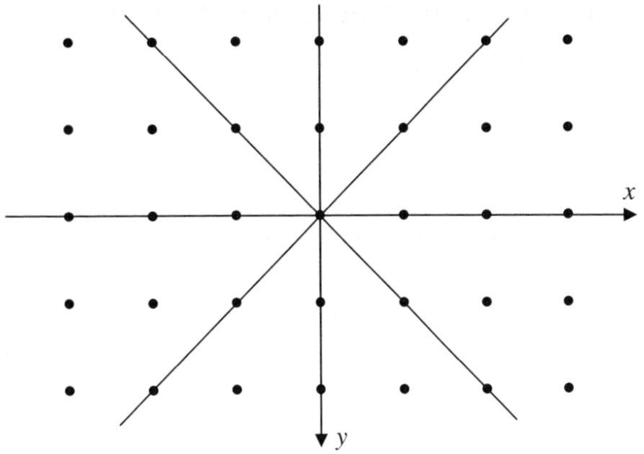

Figure 12.1 A portion of the square lattice $\mathbb{Z}^2$ and its axes of symmetry about the origin.

Table 12.1 Point symmetries of the square lattice $\mathbb{Z}^2$.

Symbol	Description	Matrix
$\mathfrak{e}$	Identity	$\begin{bmatrix} 1 & 0 \\ 0 & 1 \end{bmatrix}$
$\mathfrak{r}$	90° clockwise rotation	$\begin{bmatrix} 0 & -1 \\ 1 & 0 \end{bmatrix}$
$\mathfrak{r}^2$	180° clockwise rotation	$\begin{bmatrix} -1 & 0 \\ 0 & -1 \end{bmatrix}$
$\mathfrak{r}^3$	270° clockwise rotation	$\begin{bmatrix} 0 & 1 \\ -1 & 0 \end{bmatrix}$
$\mathfrak{h}$	Reflection about the x axis	$\begin{bmatrix} 1 & 0 \\ 0 & -1 \end{bmatrix}$
$\mathfrak{v}$	Reflection about the y axis	$\begin{bmatrix} -1 & 0 \\ 0 & 1 \end{bmatrix}$
$\mathfrak{m}$	Reflection about the line $y = x$	$\begin{bmatrix} 0 & 1 \\ 1 & 0 \end{bmatrix}$
$\mathfrak{n}$	Reflection about the line $y = -x$	$\begin{bmatrix} 0 & -1 \\ -1 & 0 \end{bmatrix}$

A very common example is the group of all symmetries of the square lattice $\mathbb{Z}^2$ in two dimensions (Figure 12.1). The set of possible point symmetries of this lattice about the origin are enumerated in Table 12.1 along with their 2×2 orthogonal matrix representations. This point group $\mathcal{G}_{cp}(\mathbb{Z}^2)$ is called the dihedral group of order 8, usually denoted D_4 (since it is the group of symmetries of the four-sided regular polygon, the square). We can consider the complete group of symmetries, or subgroups, which must be of order 2 or 4. The subgroups of order 2 are $\{\mathfrak{e}, \mathfrak{r}^2\}$, $\{\mathfrak{e}, \mathfrak{h}\}$, $\{\mathfrak{e}, \mathfrak{v}\}$, $\{\mathfrak{e}, \mathfrak{m}\}$, $\{\mathfrak{e}, \mathfrak{n}\}$. These all have the form $\{\mathfrak{e}, \mathfrak{g}\}$ where $\mathfrak{g}^2 = \mathfrak{e}$ and are thus cyclic groups of order 2. The subgroups of order 4 are $\{\mathfrak{e}, \mathfrak{r}, \mathfrak{r}^2, \mathfrak{r}^3\}$ (a cyclic group of order 4), $\{\mathfrak{e}, \mathfrak{r}^2, \mathfrak{h}, \mathfrak{v}\}$ and $\{\mathfrak{e}, \mathfrak{r}^2, \mathfrak{m}, \mathfrak{n}\}$ (these have the structure $\mathbb{Z}_2 \times \mathbb{Z}_2$).

12.1.2 Symmetry-Group Invariant Systems

Let $\mathcal{G}(\Lambda)$ be a symmetry group for a lattice Λ. An element $\mathfrak{g} \in \mathcal{G}(\Lambda)$ induces a linear system on $\mathcal{S}_\Lambda$ denoted $\mathcal{H}_\mathfrak{g}$ and defined by

$$\mathcal{H}_\mathfrak{g} : \mathcal{S}_\Lambda \to \mathcal{S}_\Lambda : (\mathcal{H}_\mathfrak{g} f)[\mathbf{x}] = f[\mathfrak{g}^{-1}\mathbf{x}]. \tag{12.4}$$

Note that this is consistent with our definition of the translation system, where $\mathfrak{t}_\mathbf{a}^{-1} = \mathfrak{t}_{-\mathbf{a}}$ and $\mathfrak{t}_{-\mathbf{a}}\mathbf{x} = \mathbf{x} - \mathbf{a}$. Compositions of lattice transformations are denoted as $\mathfrak{g}_i\mathfrak{g}_k$, where $\mathfrak{g}_i\mathfrak{g}_k$ means that $\mathfrak{g}_k$ is applied first, followed by $\mathfrak{g}_i$. Composition (i.e. cascade) of the corresponding systems can be expressed

$$(\mathcal{H}_{\mathfrak{g}_i}\mathcal{H}_{\mathfrak{g}_k} f)[\mathbf{x}] = f[\mathfrak{g}_k^{-1}\mathfrak{g}_i^{-1}\mathbf{x}] = f[(\mathfrak{g}_i\mathfrak{g}_k)^{-1}\mathbf{x}] = (\mathcal{H}_{\mathfrak{g}_i\mathfrak{g}_k} f)[\mathbf{x}] \tag{12.5}$$

so that $\mathcal{H}_{\mathfrak{g}_i}\mathcal{H}_{\mathfrak{g}_k} = \mathcal{H}_{\mathfrak{g}_i\mathfrak{g}_k}$.

Similarly to the case of shift invariant systems, we can define $\mathcal{G}(\Lambda)$-invariant systems.

Definition 12.4 A system $\mathcal{H}$ is said to be $\mathcal{G}(\Lambda)$-invariant if $\mathcal{H}\mathcal{H}_\mathfrak{g} = \mathcal{H}_\mathfrak{g}\mathcal{H}$ for all $\mathfrak{g} \in \mathcal{G}(\Lambda)$.

We will be interested in LSI systems that are also $\mathcal{G}(\Lambda)$-invariant. As the following theorem shows, we only need to be concerned with point group invariance.

Theorem 12.1 An LSI system $\mathcal{H}$ is $\mathcal{G}(\Lambda)$-invariant if and only if it is $\mathcal{G}_p(\Lambda)$-invariant.

Proof: An arbitrary element $\mathfrak{g} \in \mathcal{G}(\Lambda)$ can be written $\mathfrak{g} = \mathfrak{t}_\mathbf{a}\mathfrak{o}$ where $\mathfrak{o} \in \mathcal{G}_p(\Lambda)$ and $\mathbf{a} \in \Lambda$. Thus, $\mathcal{H}_\mathfrak{g} = \mathcal{H}_{\mathfrak{t}_\mathbf{a}}\mathcal{H}_\mathfrak{o}$. For an LSI system, $\mathcal{H}\mathcal{H}_\mathfrak{g} = \mathcal{H}\mathcal{H}_{\mathfrak{t}_\mathbf{a}}\mathcal{H}_\mathfrak{o} = \mathcal{H}_{\mathfrak{t}_\mathbf{a}}\mathcal{H}\mathcal{H}_\mathfrak{o}$. Now if $\mathcal{H}$ is $\mathcal{G}_p(\Lambda)$-invariant, $\mathcal{H}\mathcal{H}_\mathfrak{o} = \mathcal{H}_\mathfrak{o}\mathcal{H}$, and so $\mathcal{H}\mathcal{H}_\mathfrak{g} = \mathcal{H}_\mathfrak{g}\mathcal{H}$ and $\mathcal{H}$ is $\mathcal{G}(\Lambda)$-invariant. Conversely, if $\mathcal{H}$ is $\mathcal{G}(\Lambda)$-invariant, take $\mathbf{a} = 0$ and the result follows. □

The fact that an LSI system is $\mathcal{G}(\Lambda)$-invariant results in a symmetry condition on its unit sample response, as demonstrated in the following theorem.

Theorem 12.2 An LSI system is $\mathcal{G}(\Lambda)$-invariant if and only if its unit-sample response h satisfies $\mathcal{H}_\mathfrak{g} h = h$ for all $\mathfrak{g} \in \mathcal{G}_p(\Lambda)$.

Proof: Suppose that $\mathcal{H}$ is $\mathcal{G}(\Lambda)$-invariant, and thus $\mathcal{G}_p(\Lambda)$-invariant. The system $\mathcal{H}$ is characterized by its unit-sample response h. Since any element of $\mathcal{G}_p(\Lambda)$ leaves the origin invariant, $\mathcal{H}_\mathfrak{g}\delta = \delta$ for every $\mathfrak{g} \in \mathcal{G}_p(\Lambda)$. Then, $\mathcal{H}\mathcal{H}_\mathfrak{g}\delta = \mathcal{H}\delta = h$. But since $\mathcal{H}$ is $\mathcal{G}_p(\Lambda)$-invariant, $\mathcal{H}\mathcal{H}_\mathfrak{g}\delta = \mathcal{H}_\mathfrak{g}\mathcal{H}\delta = \mathcal{H}_\mathfrak{g} h$. Thus, $\mathcal{H}_\mathfrak{g} h = h$ for all $\mathfrak{g} \in \mathcal{G}_p(\Lambda)$ is a necessary condition for $\mathcal{H}$ to be $\mathcal{G}(\Lambda)$-invariant. The condition is also sufficient, since, for all $\mathfrak{g} \in \mathcal{G}_p(\Lambda)$,

$$\begin{aligned}
(\mathcal{H}\mathcal{H}_\mathfrak{g} f)[\mathbf{x}] &= \sum_{\mathbf{s}\in\Lambda} h[\mathbf{s}] f[\mathfrak{g}^{-1}(\mathbf{x} - \mathbf{s})] \\
&= \sum_{\mathbf{t}\in\Lambda} h[\mathfrak{g}t] f[\mathfrak{g}^{-1}\mathbf{x} - \mathbf{t}] \qquad (\mathbf{t} = \mathfrak{g}^{-1}\mathbf{s}) \\
&= \sum_{\mathbf{t}\in\Lambda} h[t] f[\mathfrak{g}^{-1}\mathbf{x} - \mathbf{t}] \\
&= (\mathcal{H} f)[\mathfrak{g}^{-1}\mathbf{x}] \\
&= (\mathcal{H}_\mathfrak{g}\mathcal{H} f)[\mathbf{x}].
\end{aligned} \tag{12.6}$$

□

As a simple example, suppose that $\mathfrak{g}$ is the inversion in the origin, $\mathfrak{g}\mathbf{x} = -\mathbf{x}$. The inversion system is given by $(H_{\mathfrak{g}}f)[\mathbf{x}] = f[-\mathbf{x}]$. A necessary and sufficient condition for an LSI system to be inversion invariant is $h[-\mathbf{x}] = h[\mathbf{x}]$ for all $\mathbf{x} \in \Lambda$. The inversion is a point symmetry of every lattice. The existence of other symmetries depends on the nature of the lattice. For the square lattice, each of the subgroups of the complete point symmetry group corresponds to a category of filters with certain symmetry properties, viz., zero-phase, vertical symmetry, horizontal symmetry, diagonal symmetry, quadrantal symmetry, and octagonal symmetry for the full symmetry group. These are all well-known in 2D filter design [Lu and Antoniou (1992)]. Note that if the square lattice is rotated by any angle, it is still a square lattice with the same symmetries (about rotated axes of symmetry). A common example is when the square lattice is rotated by 45° giving a diamond lattice.

In general, lattices can be classified on the basis of their complete point symmetry group. In the two-dimensional case, there are five types of lattices on this basis, as shown in Figure 12.2. They are shown with a particular orientation, but they can be oriented at any angle, as mentioned above for the square lattice. The general lattice (parallelogram) has complete point symmetry group of order two, the identity and the inversion. The square lattice we have seen has complete point symmetry group of order eight. The rectangle and rhombus classes have complete point symmetry groups of order four;

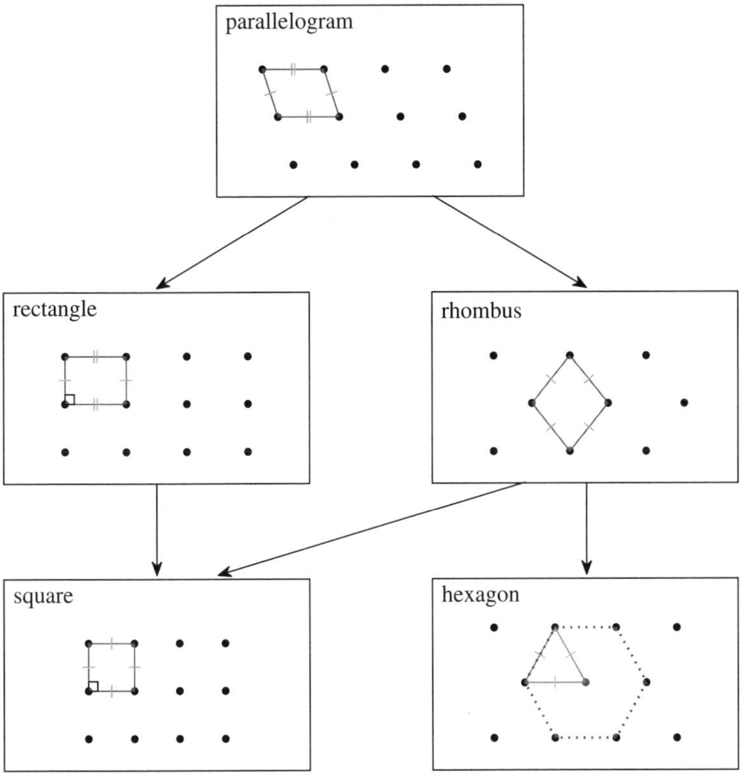

Figure 12.2 Classification of two-dimensional lattices according to the types of complete point symmetry group. Arrows point to a special case with an additional constraint as indicated by ticks on the polygon edges or the right angle symbol.

besides the identity and inversion, the rectangle class has symmetry about the major and minor sides of the rectangle, and the rhombus class has symmetry about the major and minor diagonals of the rhombus. The hexagon class has complete point symmetry group of order twelve, namely the dihedral group D_6, which is the symmetry group for the regular hexagon.

Symmetries of a lattice in the discrete (space-time) domain induce corresponding symmetries in the frequency domain. Let $\mathfrak{g}$ be a point symmetry of Λ with corresponding orthogonal matrix $\mathbf{A}_\mathfrak{g}$. Then $\mathfrak{g}^*$ is a point symmetry of Λ^*, where $\mathbf{A}_{\mathfrak{g}^*} = \mathbf{A}_\mathfrak{g}^T$. To see this, let $\mathbf{u}' = \mathbf{A}_\mathfrak{g}^T \mathbf{u}$, for $\mathbf{u} \in \Lambda^*$. Then, for all $\mathbf{x} \in \Lambda$, $\mathbf{u}' \cdot \mathbf{x} = (\mathbf{A}_\mathfrak{g}^T \mathbf{u}) \cdot \mathbf{x} = \mathbf{u} \cdot (\mathbf{A}_\mathfrak{g} \mathbf{x}) \in \mathbb{Z}$ so that $\mathbf{u}' \in \Lambda^*$. These observations lead to the following result:

Theorem 12.3 If $\mathcal{G}_p$ is a point symmetry group of a lattice Λ, defined by the orthogonal matrices $\mathbf{A}_1, \mathbf{A}_2, \ldots, \mathbf{A}_L$, then $\mathcal{G}_p^*$, defined by the orthogonal matrices $\mathbf{A}_1^{-1}, \mathbf{A}_2^{-1}, \ldots, \mathbf{A}_L^{-1}$ is a point symmetry group of Λ^* that is isomorphic to $\mathcal{G}_p$.

Proof: Since the matrices are orthogonal, $\mathbf{A}^T = \mathbf{A}^{-1}$. Since they form a group, the $\mathbf{A}_1^{-1}, \mathbf{A}_2^{-1}, \ldots, \mathbf{A}_L^{-1}$ are a rearrangement of the $\mathbf{A}_1, \mathbf{A}_2, \ldots, \mathbf{A}_L$, so the two groups are isomorphic. □

Suppose that $f \in l_1(\Lambda)$ has Fourier transform $F(\mathbf{u})$, and let $q = \mathcal{H}_\mathfrak{g} f$. Then

$$\begin{aligned} Q(\mathbf{u}) &= \sum_{\mathbf{x} \in \Lambda} f[\mathbf{A}_\mathfrak{g}^{-1} \mathbf{x}] \exp(-\mathrm{j}2\pi \mathbf{u} \cdot \mathbf{x}) \\ &= \sum_{\mathbf{t} \in \Lambda} f[\mathbf{t}] \exp(-\mathrm{j}2\pi \mathbf{u} \cdot (\mathbf{A}_\mathfrak{g} \mathbf{t})) \\ &= \sum_{\mathbf{t} \in \Lambda} f[\mathbf{t}] \exp(-\mathrm{j}2\pi (\mathbf{A}_\mathfrak{g}^T \mathbf{u}) \cdot \mathbf{t}) \\ &= F(\mathbf{A}_\mathfrak{g}^T \mathbf{u}). \end{aligned} \tag{12.7}$$

Then, if f is $\mathfrak{g}$-symmetric, $\mathcal{H}_\mathfrak{g} f = f$, it follows that $F(\mathbf{A}_\mathfrak{g}^T \mathbf{u}) = F(\mathbf{u})$, so that the Fourier transform has a corresponding symmetry. Recall that $F(\mathbf{u})$ is a function of the continuous parameter vector $\mathbf{u}$ and is Λ^*-periodic.

Referring to the case of the square lattice, as given in Table 12.1, six of the eight matrices are symmetric, so that each of these six symmetries induces the same symmetry in the frequency domain. The two that are not symmetric are $\mathfrak{r}$ and $\mathfrak{r}^3$, so that a 90° rotation in the space domain corresponds to a 270° rotation in the frequency domain, and vice versa.

As another example, let Λ be a regular hexagonal lattice with sampling matrix

$$\mathbf{V}_\Lambda = \begin{bmatrix} X & \frac{X}{2} \\ 0 & \frac{\sqrt{3}X}{2} \end{bmatrix}. \tag{12.8}$$

Suppose that h is a finite extent Gaussian filter given by

$$h[\mathbf{x}] = c \exp(-(x^2 + y^2)/2r_0^2), \quad (x, y) \in \Lambda, \quad x^2 + y^2 \leq 9X^2, \tag{12.9}$$

where $r_0 = 0.65X$ in this example and c is set to give a DC gain of 1. Since the Gaussian is circularly symmetric, it is invariant to all elements of the complete point symmetry group of the hexagonal lattice. Figure 12.3 shows a contour plot with several periods

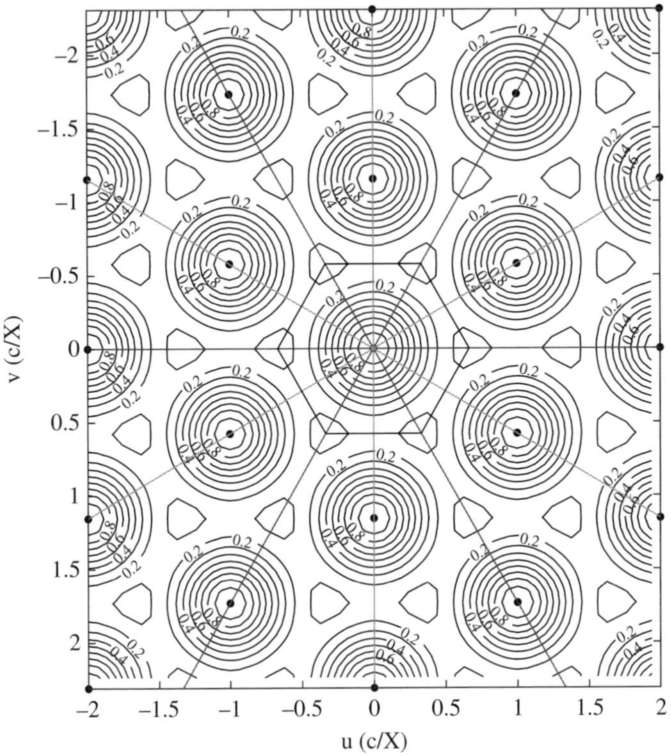

Figure 12.3 Contour plot of several periods of the frequency response of an FIR Gaussian filter on a hexagonal lattice. Indicated are points of the reciprocal lattice Λ^* (•), the hexagonal Voronoi unit cell of Λ^*, and the axes of symmetry.

of the frequency response $H(u, v)$, which also exhibits the twelve-fold symmetry. The full continuous, periodic frequency response can be reflected about six axes of the indicated hexagon (through the vertices and through the midpoints of the sides) and also be rotated by any multiple of 60° and it will remain unchanged.

12.1.3 Spaces of Symmetric Signals

Having defined the linear systems induced by the symmetries of a lattice, we can define spaces of signals that are symmetric about a certain point (which may not be the origin, or may not even be in the lattice). Thus, we assume here that the symmetry group $\mathcal{G}(\Lambda)$ has no translational component. We consider first the case of signals symmetric about the origin. In other words, we assume that $\mathcal{G}_p(\Lambda) = \mathcal{G}(\Lambda)$, i.e. $\mathcal{G}(\Lambda)$ consists of orthogonal transformations. In this section, we consider one fixed lattice Λ, so we drop the dependence on Λ in $\mathcal{G}(\Lambda)$. Let $\mathcal{G} = \{\mathfrak{g}_1, \dots, \mathfrak{g}_L\}$ where we assume that $\mathfrak{g}_1 = \mathfrak{e}$, the identity transformation.

Definition 12.5 A $\mathfrak{g}$-symmetric signal is a signal f such that $\mathcal{H}_{\mathfrak{g}} f = f$. If $\mathcal{G} = \{\mathfrak{g}_1, \dots, \mathfrak{g}_L\}$ is a group of symmetries, a signal f is $\mathcal{G}$-symmetric if $\mathcal{H}_{\mathfrak{g}_i} f = f$ for $i = 1, 2, \dots, L$.

For an arbitrary signal f, if we define

$$f_{\mathcal{G}} = \frac{1}{L}\sum_{k=1}^{L} \mathcal{H}_{\mathfrak{g}_k} f, \tag{12.10}$$

then

$$\begin{aligned} \mathcal{H}_{\mathfrak{g}_i} f_{\mathcal{G}} &= \frac{1}{L}\sum_{k=1}^{L} \mathcal{H}_{\mathfrak{g}_i}\mathcal{H}_{\mathfrak{g}_k} f \\ &= \frac{1}{L}\sum_{\ell=1}^{L} \mathcal{H}_{\mathfrak{g}_\ell} f \\ &= f_{\mathcal{G}}. \end{aligned} \tag{12.11}$$

Thus, we can create a $\mathcal{G}$-symmetric signal from an arbitrary one. Note that if f is $\mathcal{G}$-symmetric, then $f_{\mathcal{G}} = f$. As an example, suppose that f is a complex sinusoidal signal defined on a square lattice, $f[m,n] = \exp(\mathrm{j}2\pi(um + vn))$ where u and v are the horizontal and vertical frequencies. If $\mathcal{G}$ consists of the identity and the inversion (180° rotation) $\mathfrak{r}^2$, then $f_{\mathcal{G}}[m,n]$ is the real sinusoidal signal $\cos(2\pi(um + vn))$ shown in Figure 12.4(a). If $\mathcal{G}$ is the group of all eight symmetries in Table 12.1, then the resulting $f_{\mathcal{G}}$ with eight-fold symmetry shown in Figure 12.4(b) is obtained.

We define the space of signals that are $\mathcal{G}$-symmetric (about the origin) as

$$\mathcal{S}_{\mathcal{G}} = \{f \in \mathcal{S}_\Lambda \mid \mathcal{H}_{\mathfrak{g}_i} f = f, i = 1, \ldots, L\}. \tag{12.12}$$

We have already seen that if $\mathcal{H}$ is a $\mathcal{G}$-invariant system, then $h \in \mathcal{S}_{\mathcal{G}}$ (Theorem 12.2).

We can also consider a group of symmetries that leaves a point $\mathbf{a} \in \mathbb{R}^D$ different from the origin invariant. Observing Figure 12.1, we see that we can have all the symmetries in Table 12.1 about the point $\mathbf{a} = [\ 0.5\ 0.5\ ]^T$, while we can have some symmetries ($\mathfrak{e}$, $\mathfrak{r}^2$, $\mathfrak{h}$, $\mathfrak{v}$) about the point $\mathbf{a} = [\ 0.5\ 0.0\ ]^T$, but not the others. In general, there are no symmetries of this lattice for $\mathbf{a} \notin \frac{1}{2}\Lambda$. For any point $\mathbf{a} \in \Lambda$, the situation is essentially the same as for symmetry about the origin, while for points $\mathbf{a} \notin \Lambda$ the situation is essentially different. As we will see, the standard DCT-2 used in numerous compression standards involves symmetry about points not belonging to Λ, thus reinforcing the importance of this case.

Let $\mathcal{G}_{\mathbf{a}} = \{\mathfrak{g}_1, \ldots, \mathfrak{g}_L\}$ be a subgroup of $E(D)$ that leaves Λ invariant and that also leaves a point $\mathbf{a} \in \mathbb{R}^D$ fixed, where $\mathbf{a}$ does not necessarily belong to Λ. Each $\mathfrak{g}_i \in \mathcal{G}_{\mathbf{a}}$ has the form $\mathfrak{g}_i(\mathbf{x}) = \mathbf{v}_i + \mathbf{A}_i\mathbf{x}$ where $\mathbf{v}_i \in \Lambda$ and $\mathbf{A}_i \in O(D)$. The group $\mathcal{G}_{\mathbf{a}p} = \{\mathbf{A}_1, \ldots, \mathbf{A}_L\}$ is the point group of $\mathcal{G}_{\mathbf{a}}$ and it is isomorphic to $\mathcal{G}_{\mathbf{a}}$. Since $\mathcal{G}_{\mathbf{a}p}$ leaves the lattice invariant (Armstrong (1988), theorem 25.2), it must be a subgroup of the complete point symmetry group of Λ at the origin. We have already enumerated these for all possible 2D lattices.

Knowing the possible $\mathbf{A}_i$, we can identify the possible values of $\mathbf{a}$ for the different lattice classes. Since $\mathfrak{g}_i(\mathbf{a}) = \mathbf{v}_i + \mathbf{A}_i\mathbf{a} = \mathbf{a}$, we have $(\mathbf{I} - \mathbf{A}_i)\mathbf{a} = \mathbf{v}_i \in \Lambda$, which constrains the possible $\mathbf{a}$. Consider the simplest case where $\mathcal{G}_{\mathbf{a}p}$ consists of the identity and the inversion, $\mathbf{A}_1 = \mathbf{I}$ and $\mathbf{A}_2 = -\mathbf{I}$. Then $(\mathbf{I} - \mathbf{A}_1)\mathbf{a} \in \Lambda$ provides no constraint, but $(\mathbf{I} - \mathbf{A}_2)\mathbf{a} = 2\mathbf{a} \in \Lambda$ provides the constraint $\mathbf{a} \in \frac{1}{2}\Lambda$. Thus, for any point symmetry group that contains inversion about a point $\mathbf{a}$, we must have $\mathbf{a} \in \frac{1}{2}\Lambda$. Here, the number of essentially

Figure 12.4 Sinusoidal signals on square lattice ($u = 0.02$ c/px, $v = 0.05$ c/px) with (a) two-fold symmetry – invariant to the identity and inversion; (b) eight-fold symmetry – invariant to reflections about the four axes shown and to rotations by multiples of 90°.

(a)

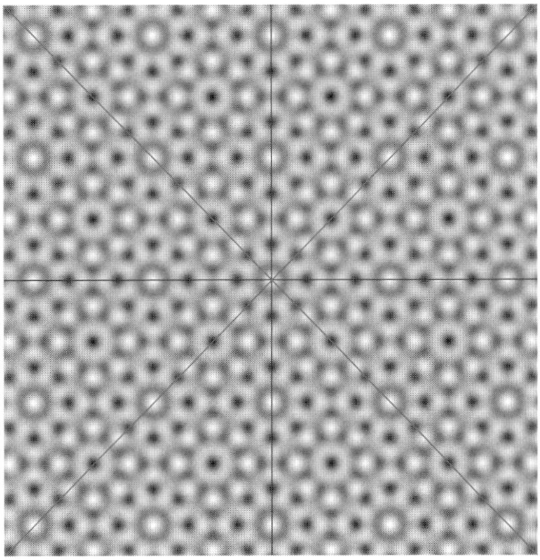

(b)

different cases would be equal to the index of Λ in $\frac{1}{2}\Lambda$, i.e. 2^D. To illustrate this, let $\mathcal{G}_{\mathbf{a}} = \{\mathfrak{g}_1, \dots, \mathfrak{g}_8\}$ be the group of eight symmetries described in Table 12.1 about the point $\mathbf{a} = [-0.5\ -0.5\ ]^T$ in the lattice $\Lambda = \mathbb{Z}^2$. These symmetries can be represented by $\mathfrak{g}_i(\mathbf{x}) = \mathbf{v}_i + \mathbf{A}_i\mathbf{x}$ or equivalently $\mathfrak{g}_i = \mathfrak{t}_{\mathbf{v}_i}\mathfrak{o}_i$ where $\mathfrak{o}_i \in D_4$. We can as above solve for the $\mathbf{v}_i$ by $\mathbf{v}_i = (\mathbf{I} - \mathbf{A}_i)\mathbf{a}$ using the $\mathbf{A}_i$ in Table 12.1. The description of the symmetry group is shown in Table 12.2.

Thus, we have the more general definition of a space of symmetric signals.

Table 12.2 Point symmetries of the square lattice $\mathbb{Z}^2$ about the point $\mathbf{a} = [-0.5 \ \ -0.5\ ]^T$.

Symbol	Description	$\mathbf{v}_i$	$\mathfrak{o}_i$	$\mathbf{A}_i$
$\mathfrak{g}_1$	Identity	$\begin{bmatrix} 0 \\ 0 \end{bmatrix}$	$\mathfrak{e}$	$\begin{bmatrix} 1 & 0 \\ 0 & 1 \end{bmatrix}$
$\mathfrak{g}_2$	90° clockwise rotation	$\begin{bmatrix} -1 \\ 0 \end{bmatrix}$	$\mathfrak{r}$	$\begin{bmatrix} 0 & -1 \\ 1 & 0 \end{bmatrix}$
$\mathfrak{g}_3$	180° clockwise rotation	$\begin{bmatrix} -1 \\ -1 \end{bmatrix}$	$\mathfrak{r}^2$	$\begin{bmatrix} -1 & 0 \\ 0 & -1 \end{bmatrix}$
$\mathfrak{g}_4$	270° clockwise rotation	$\begin{bmatrix} 0 \\ -1 \end{bmatrix}$	$\mathfrak{r}^3$	$\begin{bmatrix} 0 & 1 \\ -1 & 0 \end{bmatrix}$
$\mathfrak{g}_5$	Reflection about the line $y = 0.5$	$\begin{bmatrix} 0 \\ -1 \end{bmatrix}$	$\mathfrak{h}$	$\begin{bmatrix} 1 & 0 \\ 0 & -1 \end{bmatrix}$
$\mathfrak{g}_6$	Reflection about the line $x = 0.5$	$\begin{bmatrix} -1 \\ 0 \end{bmatrix}$	$\mathfrak{v}$	$\begin{bmatrix} -1 & 0 \\ 0 & 1 \end{bmatrix}$
$\mathfrak{g}_7$	Reflection about the line $y = x$	$\begin{bmatrix} 0 \\ 0 \end{bmatrix}$	$\mathfrak{m}$	$\begin{bmatrix} 0 & 1 \\ 1 & 0 \end{bmatrix}$
$\mathfrak{g}_8$	Reflection about the line $y = -x + 1$	$\begin{bmatrix} -1 \\ -1 \end{bmatrix}$	$\mathfrak{n}$	$\begin{bmatrix} 0 & -1 \\ -1 & 0 \end{bmatrix}$

Definition 12.6 Given a finite group of symmetries of a lattice about a point $\mathbf{a}$, we define the space of symmetric signals $S_{\mathcal{G}_\mathbf{a}}$ to be

$$S_{\mathcal{G}_\mathbf{a}} = \{f \in S_\Lambda \mid \mathcal{H}_{\mathfrak{g}_i} f = f, \text{ for all } \mathfrak{g}_i \in \mathcal{G}_\mathbf{a}\}. \tag{12.13}$$

This space is easily shown to be a subspace of the vector space S_Λ.

In the following, we assume that $\mathcal{G} = \{\mathfrak{g}_1, \dots, \mathfrak{g}_L\}$ is a group of symmetries of a lattice Λ that leave an admissible point $\mathbf{a} \in \mathbb{R}^D$ fixed. Since $\mathbf{a}$ is constant and known, we don't need to explicitly include it in the notation. For any $\mathbf{x} \in \Lambda$, $\mathcal{G}(\mathbf{x}) = \{\mathfrak{g}\mathbf{x} \mid \mathfrak{g} \in \mathcal{G}\}$ is called the $\mathcal{G}$-orbit containing $\mathbf{x}$. Each $\mathcal{G}$-orbit is an equivalence class, so the $\mathcal{G}$-orbits form a partition of Λ. The number of elements in $\mathcal{G}(\mathbf{x})$ must be a divisor of the order L of $\mathcal{G}$. Any function $f \in S_\mathcal{G}$ is constant on $\mathcal{G}$-orbits, so we only need to specify one value per $\mathcal{G}$-orbit. Let $\Lambda_\mathcal{G}$ be a set of orbit representatives chosen according to our convenience. Then

$$\Lambda = \bigcup_{\mathbf{x} \in \Lambda_\mathcal{G}} \mathcal{G}(\mathbf{x}). \tag{12.14}$$

To illustrate, consider the square lattice of Figure 12.1, and let $\mathcal{G}$ be D_4, the point group of the complete symmetry group, with elements enumerated in Table 12.1. We can identify the following orbits, which must have 1, 2, 4 or 8 elements (in fact there are none with 2 elements):

$$1: \ [0, 0]$$

$$2: \text{none}$$

$$4: \ \{[m, 0], [-m, 0], [0, m], [0, -m]\}, \{[m, m], [-m, m],$$

$$[-m, -m], [m, -m]\} \quad m \geq 1$$

$$8: \{[m, n], [n, m], [-n, m], [-m, n], [-m, -n], [-n, -m], [n, -m], [m, -n]\}$$
$$m \geq 2, \quad 1 \leq n < m.$$

If we choose (arbitrarily) the first element of each of the above sets as the orbit representatives, they lie in the shaded 45° wedge shown in Figure 12.5. The figure also shows one orbit of each type with four different symbols: $\mathcal{G}([0,0])$, $\mathcal{G}([1,0])$, $\mathcal{G}([1,1])$, $\mathcal{G}([2,1]$. The example signal in Figure 12.4(b) is also completely specified by its values in the corresponding 45° wedge.

Any signal in $\mathcal{S}_\mathcal{G}$ can be written

$$f = \sum_{\mathbf{x}\in\Lambda_\mathcal{G}} f[\mathbf{x}] \sum_{\mathbf{t}\in\mathcal{G}(\mathbf{x})} \mathcal{T}_\mathbf{t}\delta. \tag{12.15}$$

Thus, for an LSI system $\mathcal{H}$,

$$\begin{aligned} \mathcal{H}f &= \sum_{\mathbf{x}\in\Lambda_\mathcal{G}} f[\mathbf{x}] \sum_{\mathbf{t}\in\mathcal{G}(\mathbf{x})} \mathcal{H}\mathcal{T}_\mathbf{t}\delta. \\ &= \sum_{\mathbf{x}\in\Lambda_\mathcal{G}} f[\mathbf{x}] \sum_{\mathbf{t}\in\mathcal{G}(\mathbf{x})} \mathcal{T}_\mathbf{t}h. \end{aligned} \tag{12.16}$$

An LSI system is said to be $\mathcal{S}_\mathcal{G}$ invariant if $\mathcal{H}f \in \mathcal{S}_\mathcal{G}$ for all $f \in \mathcal{S}_\mathcal{G}$.

Theorem 12.4 An LSI system $\mathcal{H}$ with unit sample response h is $\mathcal{S}_\mathcal{G}$ invariant if and only if $h \in \mathcal{S}_{\mathcal{G}_p}$.

Proof: From Theorems 12.1 and 12.2, $\mathcal{H}$ is $\mathcal{G}$-invariant if and only if $h \in \mathcal{S}_{\mathcal{G}_p}$. Thus, we need to show that $\mathcal{H}$ is $\mathcal{S}_\mathcal{G}$ invariant if and only if it is $\mathcal{G}$ invariant. Suppose that $\mathcal{H}$ is $\mathcal{G}$-invariant. Then, for any $f \in \mathcal{S}_\mathcal{G}$, and for any $\mathfrak{g} \in \mathcal{G}$,

$$\mathcal{H}_\mathfrak{g}(\mathcal{H}f) = \mathcal{H}\mathcal{H}_\mathfrak{g}f = \mathcal{H}f \tag{12.17}$$

and thus $\mathcal{H}f \in \mathcal{S}_\mathcal{G}$. Conversely, if $\mathcal{H}$ is not $\mathcal{G}$-invariant, then there must be some $f \in \mathcal{S}_\mathcal{G}$ such that $\mathcal{H}\mathcal{H}_\mathfrak{g}f \neq \mathcal{H}_\mathfrak{g}\mathcal{H}f$ for some $\mathfrak{g} \in \mathcal{G}$, i.e. $\mathcal{H}f \neq \mathcal{H}_\mathfrak{g}\mathcal{H}f$, and thus $\mathcal{H}f \notin \mathcal{S}_\mathcal{G}$. □

Thus, we have established a signal space $\mathcal{S}_\mathcal{G}$ of signals on Λ that are symmetric about a specific point $\mathbf{a}$ with respect to a given group of symmetries, along with a class of LSI

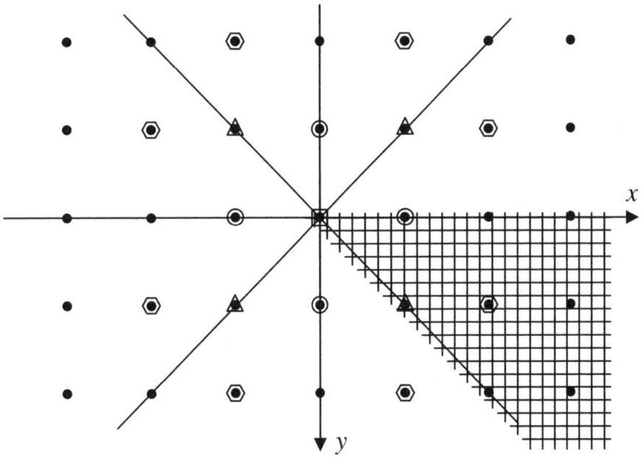

Figure 12.5 Illustration of orbits and orbit representatives for the square lattice with symmetry group D_4.

systems defined on $\mathcal{S}_{\mathcal{G}}$, where the unit sample response must belong to $\mathcal{S}_{\mathcal{G}_p}$. We see that if $\mathbf{a} \neq 0$, the signals and the unit sample response h belong to disjoint subspaces of $\mathcal{S}_\Lambda$. The unit sample response must have the form

$$h_{\rm E} = \sum_{\mathbf{x}\in\Lambda_{\mathcal{G}_p}} h_{\rm E}[\mathbf{x}] \sum_{\mathbf{t}\in\mathcal{G}_p(\mathbf{x})} \mathcal{T}_{\mathbf{t}}\delta. \tag{12.18}$$

Thus, the corresponding system $\mathcal{H}_{\rm E}$ has the form

$$\mathcal{H}_{\rm E} = \sum_{\mathbf{x}\in\Lambda_{\mathcal{G}_p}} h_{\rm E}[\mathbf{x}] \sum_{\mathbf{t}\in\mathcal{G}_p(\mathbf{x})} \mathcal{T}_{\mathbf{t}}. \tag{12.19}$$

In the case of general asymmetric aperiodic signals belonging to the signal space $\mathcal{S}_\Lambda$, we know that there are one-dimensional subspaces of $\mathcal{S}_\Lambda$ that are invariant to the action of any LSI system, namely subspaces spanned by any complex sinusoid $\phi_{\mathbf{u}}$, where $\phi_{\mathbf{u}}[\mathbf{x}] = \exp(\mathrm{j}2\pi\mathbf{u}\cdot\mathbf{x})$. Let us denote such a one dimensional subspace by $\mathcal{S}_{\mathbf{u}} = \{\alpha\phi_{\mathbf{u}} \mid \alpha \in \mathbb{C}\}$. Then for any LSI system $\mathcal{H}$, $\mathcal{H}\alpha\phi_{\mathbf{u}} = H(\mathbf{u})\alpha\phi_{\mathbf{u}} \in \mathcal{S}_{\mathbf{u}}$, where $H(\mathbf{u}) \in \mathbb{C}$ is the frequency response of $\mathcal{H}$ at frequency $\mathbf{u}$. Using Fourier analysis, an arbitrary signal is written as a superposition of complex sinusoids and the LSI system acts multiplicatively on each frequency component: if $g = \mathcal{H}f$, then $G(\mathbf{u}) = H(\mathbf{u})F(\mathbf{u})$. We wish to carry out a similar development for spaces of $\mathcal{G}$-symmetric signals.

As in the case of asymmetric aperiodic signals, we are interested in the $\mathcal{H}_{\rm E}$-invariant one-dimensional subspaces of $\mathcal{S}_{\mathcal{G}}$ in order to apply Fourier analysis. As in Equation (12.10), we can create an element of $\mathcal{S}_{\mathcal{G}}$ from a general complex sinusoid $\phi_{\mathbf{u}}$:

$$\phi_{\mathcal{G}\mathbf{u}} = \frac{1}{L}\sum_{k=1}^{L} \mathcal{H}_{\mathfrak{g}_k}\phi_{\mathbf{u}}. \tag{12.20}$$

These span the desired one-dimensional subspaces.

Theorem 12.5 The functions $\phi_{\mathcal{G}\mathbf{u}}$ defined in Equation (12.20) are eigenfunctions of any LSI $\mathcal{G}$-invariant system on the space $\mathcal{S}_{\mathcal{G}}$.

Proof: From Equation (12.19), a function ϕ is an eigenfunction of all LSI $\mathcal{G}$-invariant systems $\mathcal{H}_{\rm E}$ if and only if it is an eigenfunction of $\sum_{\mathbf{t}\in\mathcal{G}_p(\mathbf{x})}\mathcal{T}_{\mathbf{t}}$ for every $\mathbf{x} \in \Lambda$. Then, for any $\mathbf{x} \in \Lambda_{\mathcal{G}_p}$

$$\begin{aligned}\left(\sum_{\mathbf{t}\in\mathcal{G}_p(\mathbf{x})} \mathcal{T}_{\mathbf{t}}\right)\phi_{\mathcal{G}\mathbf{u}} &= \sum_{\mathbf{t}\in\mathcal{G}_p(\mathbf{x})} \frac{1}{L}\sum_{k=1}^{L} \mathcal{T}_{\mathbf{t}}\mathcal{H}_{\mathfrak{g}_k}\phi_{\mathbf{u}} \\ &= \frac{1}{L}\sum_{k=1}^{L}\sum_{\mathbf{t}\in\mathcal{G}_p(\mathbf{x})} \mathcal{T}_{\mathbf{t}}\mathcal{H}_{\mathfrak{g}_k}\phi_{\mathbf{u}}.\end{aligned} \tag{12.21}$$

Assuming that $\mathfrak{g}_k = \{\mathbf{v}_k, \mathbf{A}_k\}$, then

$$(\mathcal{H}_{\mathfrak{g}_k}\phi_{\mathbf{u}})[\mathbf{s}] = \exp(\mathrm{j}2\pi\mathbf{u}\cdot(\mathbf{A}_\ell\mathbf{s} + \mathbf{v}_\ell)), \tag{12.22}$$

where $\mathfrak{g}_k^{-1} = \mathfrak{g}_l$, and we note that the $\mathfrak{g}_l$ are a permutation of the $\mathfrak{g}_k$. Thus

$$\begin{aligned}(\mathcal{T}_{\mathbf{t}}\mathcal{H}_{\mathfrak{g}_k}\phi_{\mathbf{u}})[\mathbf{s}] &= \exp(\mathrm{j}2\pi\mathbf{u}\cdot(\mathbf{A}_\ell(\mathbf{s}-\mathbf{t}) + \mathbf{v}_\ell) \\ &= \exp(-\mathrm{j}2\pi\mathbf{u}\cdot(\mathbf{A}_\ell\mathbf{t}))\exp(\mathrm{j}2\pi\mathbf{u}\cdot(\mathbf{A}_\ell\mathbf{s} + \mathbf{v}_\ell))\end{aligned} \tag{12.23}$$

for all $\mathbf{s} \in \Lambda$. In other words

$$\mathcal{T}_{\mathbf{t}}\mathcal{H}_{\mathfrak{g}_k}\phi_{\mathbf{u}} = \exp(-\mathrm{j}2\pi\mathbf{u}\cdot(\mathbf{A}_{\ell}\mathbf{t}))\mathcal{H}_{\mathfrak{g}_k}\phi_{\mathbf{u}}. \tag{12.24}$$

Inserting this into Equation (12.21)

$$\left(\sum_{\mathbf{t}\in\mathcal{G}_p(\mathbf{x})}\mathcal{T}_{\mathbf{t}}\right)\phi_{\mathcal{G}\mathbf{u}} = \frac{1}{L}\sum_{k=1}^{L}\left(\sum_{\mathbf{t}\in\mathcal{G}_p(\mathbf{x})}\exp(-\mathrm{j}2\pi\mathbf{u}\cdot(\mathbf{A}_{\ell}\mathbf{t}))\right)\mathcal{H}_{\mathfrak{g}_k}\phi_{\mathbf{u}}. \tag{12.25}$$

Now, if $\mathbf{t}\in\mathcal{G}_p(\mathbf{x})$, then $\mathbf{A}_{\ell}\mathbf{t}\in\mathcal{G}_p(\mathbf{x})$ and the set $\{\mathbf{A}_{\ell}\mathbf{t} \mid \mathbf{t}\in\mathcal{G}_p(\mathbf{x})\}$ is a permutation of $\mathcal{G}_p(\mathbf{x})$. Thus, we can conclude that

$$\left(\sum_{\mathbf{t}\in\mathcal{G}_p(\mathbf{x})}\mathcal{T}_{\mathbf{t}}\right)\phi_{\mathcal{G}\mathbf{u}} = \left(\sum_{\mathbf{t}\in\mathcal{G}_p(\mathbf{x})}\exp(-\mathrm{j}2\pi\mathbf{u}\cdot\mathbf{t})\right)\phi_{\mathcal{G}\mathbf{u}}, \tag{12.26}$$

showing that $\phi_{\mathcal{G}\mathbf{u}}$ is an eigenfunction of $\sum_{\mathbf{t}\in\mathcal{G}_p(\mathbf{x})}\mathcal{T}_{\mathbf{t}}$, with eigenvalue $\sum_{\mathbf{t}\in\mathcal{G}_p(\mathbf{x})}\exp(-\mathrm{j}2\pi\mathbf{u}\cdot\mathbf{t})$. Applying this to a general system $\mathcal{H}_{\mathrm{E}}$ acting on $\mathcal{S}_{\mathcal{G}}$,

$$\begin{aligned}\mathcal{H}_{\mathrm{E}}\phi_{\mathcal{G}\mathbf{u}} &= \left(\sum_{\mathbf{x}\in\Lambda_{\mathcal{G}_p}} h_{\mathrm{E}}[\mathbf{x}]\sum_{\mathbf{t}\in\mathcal{G}_p(\mathbf{x})}\exp(-\mathrm{j}2\pi\mathbf{u}\cdot\mathbf{t})\right)\phi_{\mathcal{G}\mathbf{u}}\\ &= H_{\mathrm{E}}(\mathbf{u})\phi_{\mathcal{G}\mathbf{u}},\end{aligned} \tag{12.27}$$

showing that $\phi_{\mathcal{G}\mathbf{u}}$ is an eigenfunction of this system with eigenvalue $H_{\mathrm{E}}(\mathbf{u})$. □

Note that $H_{\mathrm{E}}(\mathbf{u})$ is the same as the Fourier transform of h_{E} when considered as an arbitrary element of $l_1(\Lambda)$. As mentioned in Equation (12.7), $H_{\mathrm{E}}(\mathbf{u})$ will have a corresponding symmetry, $H_{\mathrm{E}}(\mathbf{A}_{\mathfrak{g}}^T\mathbf{u}) = H_{\mathrm{E}}(\mathbf{u})$ for all $\mathfrak{g}\in\mathcal{G}_p$. Thus, we only need to specify $\mathbf{u}$ in a subset of the unit cell called a *fundamental domain*. For the square lattice with full eight-fold symmetry, this fundamental domain would be a 45° wedge comprising one-eighth of the unit cell, e.g., $\{(u,v) \mid 0 \le u \le 0.5, 0 \le v \le u\}$. Similarly for the hexagonal lattice with twelve-fold symmetry, the fundamental domain of the frequency response is a 30° sector comprising one-twelfth of the hexagonal unit cell (see Figure 12.3).

If $\mathcal{G} = \mathcal{G}_p$, i.e. the symmetries are about the origin, and $f_{\mathrm{E}}\in\mathcal{S}_{\mathcal{G}}\cap l_1(\Lambda)$, then similarly to above, the Fourier transform of f_{E} can be defined

$$F_{\mathrm{E}}(\mathbf{u}) = \sum_{\mathbf{x}\in\Lambda_{\mathcal{G}}} f_{\mathrm{E}}[\mathbf{x}]\sum_{\mathbf{t}\in\mathcal{G}(\mathbf{x})}\exp(-\mathrm{j}2\pi\mathbf{u}\cdot\mathbf{t}) \tag{12.28}$$

and if $q_{\mathrm{E}} = h_{\mathrm{E}} * f_{\mathrm{E}}$ is the output of the LSI system $\mathcal{H}_{\mathrm{E}}$, then $Q_{\mathrm{E}}(\mathbf{u}) = H_{\mathrm{E}}(\mathbf{u})F_{\mathrm{E}}(\mathbf{u})$ and all Fourier transforms have the same symmetries in the frequency domain. However, if $\mathbf{a} \neq 0$ and the symmetries of the signal are about a point different than the origin, then the situation is similar but a bit more complex. We leave this as an exercise for the reader to develop.

Example 12.1
Develop the Fourier representation for two-dimensional discrete-domain signals on the lattice $\mathbb{Z}^2$ with eight-fold symmetry about the origin.

Solution:

For this case, the symmetry group $\mathcal{G}$ is D_4 with elements shown in Table 12.1 and $\mathcal{G}_p = \mathcal{G}$. A $\mathcal{G}$-symmetric signal has independent values on the orbit representatives; we will use the ones shown in Figure 12.5. Other than the origin, points on the four axes of symmetry have orbits with four elements while all other points have orbits with eight elements. The eigenfunctions of LSI $\mathcal{G}$-invariant systems on this space have the form

$$\phi_{\mathcal{G}\mathbf{u}} = \frac{1}{8}\sum_{k=1}^{8} \mathcal{H}_{g_k}\phi_{\mathbf{u}}, \tag{12.29}$$

and expanding explicitly, we find

$$\begin{aligned}\phi_{\mathcal{G}(u,v)}[m,n] &= \frac{1}{8}(\exp(\mathrm{j}2\pi(um+vn)) + \exp(\mathrm{j}2\pi(un-vm)) \\ &\quad + \exp(\mathrm{j}2\pi(-um-vn)) + \exp(\mathrm{j}2\pi(-un+vm)) \\ &\quad + \exp(\mathrm{j}2\pi(um-vn)) + \exp(\mathrm{j}2\pi(-um+vn)) \\ &\quad + \exp(\mathrm{j}2\pi(un+vm)) + \exp(\mathrm{j}2\pi(-un-vm))) \\ &= \frac{1}{2}(\cos(2\pi um)\cos(2\pi vn) + \cos(2\pi un)\cos(2\pi vm)).\end{aligned} \tag{12.30}$$

These functions can be verified to be eigenfunctions of $\sum_{\mathbf{t}\in\mathcal{G}(\mathbf{x})}\mathcal{T}_{\mathbf{t}}\delta$ for the four types of orbits shown if Figure 12.5 (we omit the algebraic details):

i. $\mathbf{x} = [0,0]$, $\mathcal{G}(\mathbf{x}) = \{[0,0]\}$, $\mathcal{T}_0\phi_{\mathcal{G}(u,v)} = \phi_{\mathcal{G}(u,v)}$.
ii. $\mathbf{x} = [p,0]$, $\mathcal{G}(\mathbf{x}) = \{[p,0],[-p,0],[0,p],[0,-p]\}$, $(\mathcal{T}_{(p,0)} + \mathcal{T}_{(-p,0)} + \mathcal{T}_{(0,p)} + \mathcal{T}_{(0,-p)})$ $\phi_{\mathcal{G}(u,v)}[m,n] = 2(\cos(2\pi up) + \cos(2\pi vp))\phi_{\mathcal{G}(u,v)}[m,n]$.
iii. $\mathbf{x} = [p,p]$, $\mathcal{G}(\mathbf{x}) = \{[p,p],[-p,p],[p,-p],[-p,-p]\}$, $(\mathcal{T}_{(p,p)} + \mathcal{T}_{(-p,p)} + \mathcal{T}_{(p,-p)}$ $+ \mathcal{T}_{(-p,-p)})\phi_{\mathcal{G}(u,v)}[m,n] = 4\cos(2\pi up)\cos(2\pi vp)\phi_{\mathcal{G}(u,v)}[m,n]$.
iv. $\mathbf{x} = [p,q]$, $p \geq 2$, $1 \leq q < p$, $\mathcal{G}(\mathbf{x}) = \{(\pm p, \pm q), (\pm q \pm p)\}$, $(\mathcal{T}_{(p,q)} + \mathcal{T}_{(-p,q)} + \mathcal{T}_{(p,-q)}$ $+ \mathcal{T}_{(-p,-q)} + \mathcal{T}_{(q,p)} + \mathcal{T}_{(-q,p)} + \mathcal{T}_{(q,-p)} + \mathcal{T}_{(-q,-p)})\phi_{\mathcal{G}(u,v)}[m,n] = 4(\cos(2\pi up)\cos(2\pi vq)$ $+ \cos(2\pi uq)\cos(2\pi vp))\phi_{\mathcal{G}(u,v)}[m,n]$.

For an LSI $\mathcal{G}$-invariant filter of size $(2M+1)\times(2M+1)$ on this space, the frequency response is then given by

$$\begin{aligned}H_{\mathrm{E}}(u,v) &= h_{\mathrm{E}}[0,0] + 2\sum_{m=1}^{M} h_{\mathrm{E}}[m,0](\cos(2\pi um) + \cos(2\pi vm)) \\ &\quad + 4\sum_{m=1}^{M} h_{\mathrm{E}}[m,m]\cos(2\pi um)\cos(2\pi vm) \\ &\quad + 4\sum_{m=2}^{M}\sum_{n=1}^{m} h_{\mathrm{E}}[m,n](\cos(2\pi um)\cos(2\pi vn) + \cos(2\pi un)\cos(2\pi vm)).\end{aligned} \tag{12.31}$$

A $(2M+1)\times(2M+1)$ Gaussian filter would be an example of such a filter. We can also design filters with eight-fold symmetry using the methods of Chapter 10.

For an arbitrary summable signal with eight-fold symmetry, $f_E \in S_{\mathcal{G}} \cap l_1(\mathbb{Z}^2)$, the Fourier transform is

$$\begin{aligned}F_E(u,v) &= f_E[0,0] + 2\sum_{m=1}^{\infty} f_E[m,0](\cos(2\pi um)+\cos(2\pi vm))\\ &+ 4\sum_{m=1}^{\infty} f_E[m,m]\cos(2\pi um)\cos(2\pi vm)\\ &+ 4\sum_{m=2}^{\infty}\sum_{n=1}^{m} f_E[m,n](\cos(2\pi um)\cos(2\pi vn)+\cos(2\pi un)\cos(2\pi vm)).\end{aligned} \tag{12.32}$$

We note that these Fourier transforms are also invariant to all the symmetries in D_4. □

12.2 Symmetry-Invariant Discrete-Domain Periodic Signals and Systems

Now that we have presented discrete-domain aperiodic signal spaces and systems that are invariant to certain groups of symmetries of the lattice, we extend this to the case of periodic signals as treated in Chapter 4. We will show that the symmetry operators can also be applied to periodic signals under certain conditions. We also show how Fourier analysis for spaces of symmetric periodic signals leads to a general multidimensional extension of the discrete cosine transform (DCT). The main motivation for this analysis is the Fourier representation and processing of finite extent signals (image blocks), where the underlying symmetry can influence the effectiveness of processing and compression operations. To illustrate, Figure 12.6(a) shows a 64×64 block of the image 'motorcycles' repeated periodically, as implied by the discrete-domain periodic Fourier transform (i.e. multidimensional DFT) of Chapter 4. We note that the periodic signal is discontinuous at block boundaries, which is why the DFT is *not* preferred for block-based image compression. Figure 12.6(b) shows the result if the lower half of the block is reflected by an inversion ($\mathfrak{r}^2$ in Table 12.1) about the center of the block and the result is periodically replicated as above. We see that the resulting periodic signal is still discontinuous, in fact more so. Figure 12.6(c) shows what happens if the lower right quadrant of the block is reflected horizontally about the left edge, then vertically about the top edge. The resulting block has quadrantal symmetry, corresponding to symmetry group $\{\mathfrak{e}, \mathfrak{h}, \mathfrak{v}, \mathfrak{r}^2\}$ of Table 12.1 and the periodically replicated signal no longer exhibits the block discontinuity. Finally, Figure 12.6(d) shows the result of using the upper triangular portion of the lower-right quadrant of the block and replicating it with the full eight-fold symmetry group of Table 12.1. We see that these latter two periodic signals do not have the characteristic discontinuities of the first two, making them more amenable to block compression and processing. The quadrantal symmetry of Figure 12.6(c) corresponds to the popular and successful 2D separable DCT (with a certain caveat to be discussed shortly).

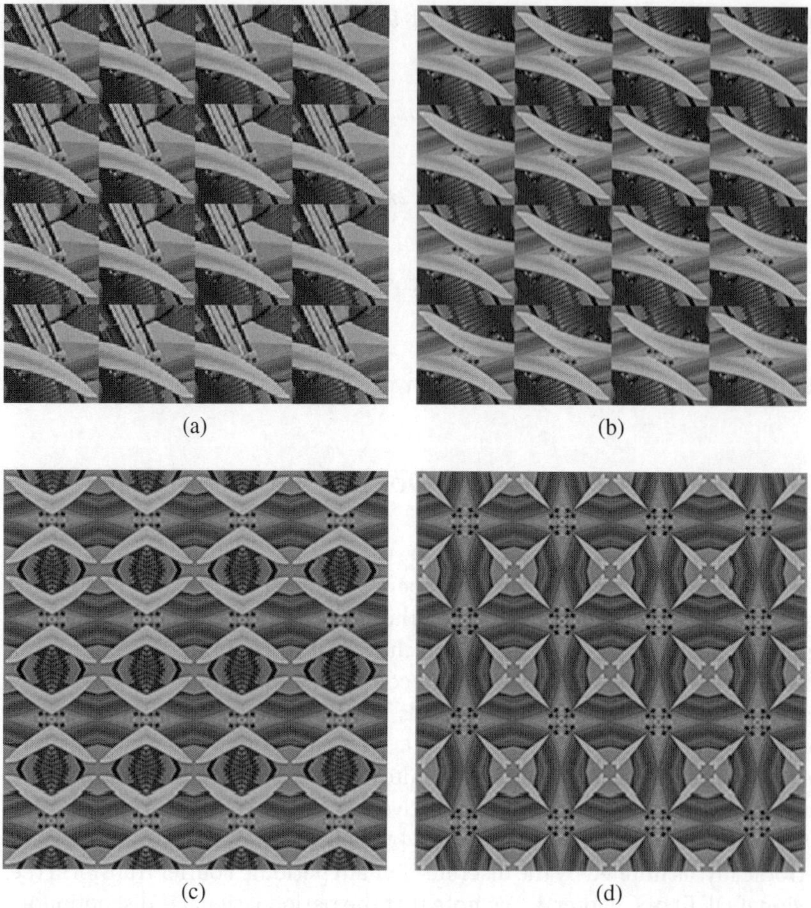

Figure 12.6 Periodic extension of image blocks with (a) no symmetry, (b) inversion symmetry, (c) quadrantal symmetry, (d) eight-fold symmetry. (*See color plate section for the color representation of this figure.*)

12.2.1 Symmetric Discrete-Domain Periodic Signals

As in Chapter 4, we assume that Λ is a lattice in D dimensions and Γ is a D-dimensional sublattice of Λ. $\mathcal{S}_{\Lambda/\Gamma}$ denotes the space of Γ-periodic signals on Λ, where $K = \mathrm{d}(\Gamma)/\mathrm{d}(\Lambda)$ is the number of samples in one period. We let $\{\mathbf{b}_0, \ldots, \mathbf{b}_{K-1}\}$ be a set of coset representatives, so that an LSI system on $\mathcal{S}_{\Lambda/\Gamma}$ can be represented as

$$\mathcal{H} = \sum_{k=0}^{K-1} h[\mathbf{b}_k]\mathcal{T}_{\mathbf{b}_k}. \tag{12.33}$$

Let $\mathcal{G}$ be a finite group of symmetries of the lattice Λ with a fixed point $\mathbf{a} \in \mathbb{R}^D$. Although it is simpler to consider only $\mathbf{a} = 0$, the ubiquitous and successful DCT-2 used widely in compression and image processing does not correspond to $\mathbf{a} = 0$, so it is necessary to treat the general case. We can apply the systems $\mathcal{H}_\mathfrak{g}$ for $\mathfrak{g} \in \mathcal{G}$ to signals in $\mathcal{S}_{\Lambda/\Gamma}$ considered as elements of $\mathcal{S}_\Lambda$ that just happen to be Γ-periodic. However, the result is not necessarily Γ-periodic, depending on the relationship between the periodicity lattice Γ and the symmetry group $\mathcal{G}$. We establish the following condition for the symmetry systems to map Γ-periodic signals to signals that are also Γ-periodic.

Theorem 12.6 Let $\mathcal{G} = \{\mathfrak{g}_1, \dots, \mathfrak{g}_L\}$ be a group of symmetries of Λ that leaves a point $\mathbf{a} \in \mathbb{R}^D$ fixed, where $\mathfrak{g}_i\mathbf{x} = \mathbf{v}_i + \mathbf{A}_i\mathbf{x}$, $\mathbf{v}_i \in \Lambda$, $\mathbf{A}_i \in O(D)$. Then, for all $f \in \mathcal{S}_{\Lambda/\Gamma}$, $\mathcal{H}_{\mathfrak{g}_i} f \in \mathcal{S}_{\Lambda/\Gamma}$ if and only if $\mathbf{A}_i^T\mathbf{d} \in \Gamma$ for all $\mathbf{d} \in \Gamma$, which means the point group $\mathcal{G}_p$ is a symmetry group of Γ.

Proof: By the periodicity condition, $f[\mathbf{x}+\mathbf{d}] = f[\mathbf{x}]$ for all $\mathbf{d} \in \Gamma$. Then, by the definition of $\mathcal{H}_{\mathfrak{g}_i}$,

$$(\mathcal{H}_{\mathfrak{g}_i} f)[\mathbf{x}] = f[\mathfrak{g}_i^{-1}\mathbf{x}] = f[\mathbf{A}_i^T(\mathbf{x}-\mathbf{v}_i)], \quad \text{and} \tag{12.34}$$

$$(\mathcal{H}_{\mathfrak{g}_i} f)[\mathbf{x}+\mathbf{d}] = f[\mathfrak{g}_i^{-1}(\mathbf{x}+\mathbf{d})] = f[\mathbf{A}_i^T(\mathbf{x}-\mathbf{v}_i) + \mathbf{A}_i^T\mathbf{d}]. \tag{12.35}$$

Thus, $(\mathcal{H}_{\mathfrak{g}_i} f)[\mathbf{x}] = (\mathcal{H}_{\mathfrak{g}_i} f)[\mathbf{x}+\mathbf{d}]$ for any $f \in \mathcal{S}_{\Lambda/\Gamma}$ and for all $\mathbf{d} \in \Gamma$ if and only if $\mathbf{A}_i^T\mathbf{d} \in \Gamma$ for all $\mathbf{d} \in \Gamma$. The symmetry defined by $\mathbf{A}_i$ is an element of $\mathcal{G}_p$ by definition, and the symmetry given by $\mathbf{A}_i^T = \mathbf{A}_i^{-1}$ is its inverse, also in $\mathcal{G}_p$. Thus, $\mathcal{G}_p$ must be a symmetry group of Γ. □

Thus, we can define a consistent space of symmetric Γ-periodic signals on Λ if $\mathcal{G}$ is a finite symmetry group of Λ and $\mathcal{G}_p$ is a symmetry group of Γ. For example, if Λ and Γ are both square lattices, any subgroup of the dihedral group shown in Table 12.1 will work. Given lattices and symmetry groups satisfying these constraints, the results of Chapter 4 and Section 12.1 are easily combined. Thus we can define the space of periodic symmetric signals as

$$\mathcal{S}_{(\Lambda/\Gamma)\mathcal{G}} = \{f \in \mathcal{S}_{\Lambda/\Gamma} \mid \mathcal{H}_{\mathfrak{g}_i} f = f \quad \text{for all } \mathfrak{g}_i \in \mathcal{G}\}. \tag{12.36}$$

One period of the signal f is defined by its values at $\mathbf{b}_0, \mathbf{b}_1, \dots, \mathbf{b}_{K-1}$. For each $\mathbf{b}_i$, there is an orbit $\mathcal{G}(\mathbf{b}_i)$ on which the symmetric signal is constant. Thus, we must choose a set of coset representatives that are also orbit representatives as the independent values of the signal. As an example, let $\Lambda = \mathbb{Z}^2$, $\Gamma = (8\mathbb{Z})^2$, and let $\mathcal{G}$ be the group of eight-fold symmetries about the point $[-0.5 \ -0.5\ ]^T$, as given in Table 12.2. In this case, $\mathcal{G}_p$ (isomorphic to D_4 as given in Table 12.1) is a symmetry group of Γ, so all conditions are satisfied. Figure 12.7 shows that of the 64 coset representatives, 10 can be chosen as orbit representatives. Thus, this signal space is isomorphic to $\mathbb{C}^{10}$. Take the set of coset representatives to be $\{(m,n) \mid 0 \le m \le 7, 0 \le n \le 7\}$. Orbits contain either 4 or 8 coset representatives; specifically, the orbits labeled A, C, F and J contain 4 elements, and those labeled B, D, E, G, H and I contain 8 elements ($4 \times 4 + 6 \times 8 = 64$ as required). In general, we denote the number of distinct elements of Λ in the $\mathcal{G}$-orbit of a point $\mathbf{x}$ by $\kappa_{\mathcal{G}}(\mathbf{x})$.

12.2.2 Discrete-Domain Periodic Symmetry-Invariant Systems

Any linear shift-invariant system acting on $\mathcal{S}_{(\Lambda/\Gamma)\mathcal{G}}$ is characterized by a unit sample response that is periodic (Chapter 4) and $\mathcal{G}_p$-invariant (Theorem 12.4), and so belongs to $\mathcal{S}_{(\Lambda/\Gamma)\mathcal{G}_p}$. Let us denote by $\mathcal{B}_{\mathcal{G}_p}$ the set of coset representatives of Γ in Λ that are

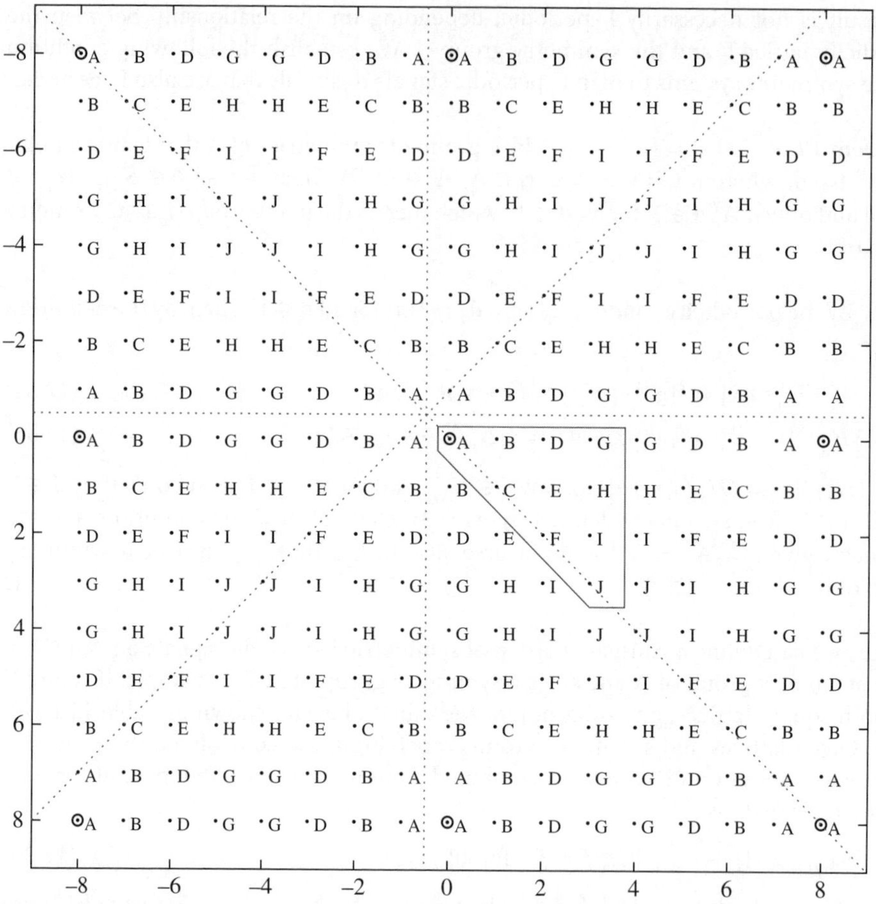

Figure 12.7 Illustration of orbits and orbit representatives for the square lattice $\Lambda = \mathbb{Z}^2$ (.) with eight-fold symmetry about the point $[-0.5\ -0.5\ ]^T$ and periodicity lattice $\Gamma = (8\mathbb{Z})^2$ (⊙). A set $\mathcal{B}_\mathcal{G}$ of orbit representatives A–J is indicated within the polygonal region. This signal is invariant to reflections about the four axes shown and to rotations of multiples of 90° about the point $[-0.5 - 0.5\ ]^T$. *(See color plate section for the color representation of this figure.)*

orbit representatives of $\mathcal{G}_p$, in analogy with $\Lambda_{\mathcal{G}_p}$ in the aperiodic case. Then, a periodic, symmetry-invariant LSI system can be expressed

$$\mathcal{H}_E = \sum_{\mathbf{x}\in\mathcal{B}_{\mathcal{G}_p}} h_E[\mathbf{x}] \sum_{\mathbf{t}\in\mathcal{G}_p(\mathbf{x})} \mathcal{T}_\mathbf{t} \tag{12.37}$$

in analogy to Equation (12.19) and Equation (4.16).

As a common example of this, when processing finite extent images, it is often assumed that they are extended symmetrically. For an image of size $M_1 \times M_2$, this corresponds to periodicity lattice $2M_1\mathbb{Z} \times 2M_2\mathbb{Z}$ and symmetry group $\mathcal{G} = \{\mathfrak{e}, \mathfrak{r}^2, \mathfrak{h}, \mathfrak{v}\}$ from Table 12.1. The filtered image maintains this symmetry if the filter is $\mathcal{G}$-invariant, i.e. h has quadrantal symmetry.

12.2.3 Discrete-Domain Symmetry-Invariant Periodic Fourier Transform

Once again, to perform Fourier analysis for this signal space and class of LSI systems, we seek the one-dimensional $\mathcal{H}_{\mathrm{E}}$-invariant subspaces of $\mathcal{S}_{(\Lambda/\Gamma)\mathcal{G}}$. As previously, we can create a $\mathcal{G}$-symmetric signal from a *periodic* complex exponential by

$$\phi_{\mathcal{G}\mathbf{u}} = \frac{1}{L}\sum_{k=1}^{L} \mathcal{H}_{\mathfrak{g}_k}\phi_{\mathbf{u}}, \tag{12.38}$$

where $\mathbf{u} \in \Gamma^*$. A function ϕ is an eigenfunction of *any* $\mathcal{H}_{\mathrm{E}}$ of the form given by Equation (12.37) if and only if it is an eigenfunction of the systems $\sum_{\mathbf{t}\in\mathcal{G}_p(\mathbf{x})}\mathcal{T}_{\mathbf{t}}$ for all $\mathbf{x} \in \mathcal{B}_{\mathcal{G}_p}$. The $\phi_{\mathcal{G}\mathbf{u}}$ satisfy this, as can be shown using exactly the same reasoning as in Theorem (12.5), the only difference being that $\mathbf{u}$ is restricted to Γ^*. Similarly, Equation (12.27) holds, with the constraint that $\mathbf{u} \in \Gamma^*$, and with $\Lambda_{\mathcal{G}_p}$ replaced with $\mathcal{B}_{\mathcal{G}_p}$:

$$H_{\mathrm{E}}(\mathbf{u}) = \sum_{\mathbf{x}\in\mathcal{B}_{\mathcal{G}_p}} h_{\mathrm{E}}[\mathbf{x}] \sum_{\mathbf{t}\in\mathcal{G}_p(\mathbf{x})} \exp(-\mathrm{j}2\pi\mathbf{u}\cdot\mathbf{t}), \qquad \mathbf{u} \in \Gamma^* \tag{12.39}$$

$$\mathcal{H}_{\mathrm{E}}\phi_{\mathcal{G}\mathbf{u}} = H_{\mathrm{E}}(\mathbf{u})\phi_{\mathcal{G}\mathbf{u}}, \qquad \mathbf{u} \in \Gamma^* \tag{12.40}$$

As before, $\mathcal{G}_p^*$ is a symmetry group for $H_{\mathrm{E}}(\mathbf{u})$ and so we have the same number of distinct values of $H_{\mathrm{E}}(\mathbf{u})$ for values of $\mathbf{u} \in \Gamma^*$ within a unit cell of Λ^* as we do of $h_{\mathrm{E}}[\mathbf{x}]$, i.e. $|\mathcal{B}_{\mathcal{G}_p}| = |\mathcal{D}_{\mathcal{G}_p}|$, where $\mathcal{D}_{\mathcal{G}_p}$ is the set of coset representatives of Λ^* in Γ^* that are orbit representatives of $\mathcal{G}_p$ acting on Γ^*.

Our signal space is isomorphic to $\mathbb{C}^{|\mathcal{B}_{\mathcal{G}}|}$ and functions $\phi_{\mathcal{G}\mathbf{u}}$ for $\mathbf{u} \in \mathcal{D}_{\mathcal{G}}$ form an orthogonal basis, with a suitable definition of inner product. We do not use the usual inner product on $\mathbb{C}^N$ but rather one that is equivalent to the inner product on $\mathcal{S}_{\Lambda/\Gamma}$ when restricted to elements of $\mathcal{S}_{(\Lambda/\Gamma)\mathcal{G}}$.

Theorem 12.7

$$\langle f_1 \mid f_2\rangle_{\mathcal{S}_{(\Lambda/\Gamma)\mathcal{G}}} = \sum_{\mathbf{x}\in\mathcal{B}_{\mathcal{G}}} f_1[\mathbf{x}]f_2^*[\mathbf{x}]\kappa_{\mathcal{G}}(\mathbf{x}) \tag{12.41}$$

is a valid inner product on $\mathcal{S}_{(\Lambda/\Gamma)\mathcal{G}}$ and is equal to the inner product of f_1 and f_2 when considered as elements of $\mathcal{S}_{\Lambda/\Gamma}$.

Proof: The properties of an inner product are easily seen to be satisfied by this definition. When f_1 and f_2 are considered to be elements of $\mathcal{S}_{\Lambda/\Gamma}$, the inner product is given by Equation (4.23),

$$\begin{aligned}\langle f_1 \mid f_2\rangle_{\mathcal{S}_{\Lambda/\Gamma}} &= \sum_{\mathbf{x}\in\mathcal{B}} f_1[\mathbf{x}]f_2^*[\mathbf{x}] \\ &= \sum_{\mathbf{x}\in\mathcal{B}_{\mathcal{G}}}\sum_{\mathbf{t}\in\mathcal{G}(\mathbf{x})} f_1[\mathbf{x}]f_2^*[\mathbf{x}] \\ &= \sum_{\mathbf{x}\in\mathcal{B}_{\mathcal{G}}} f_1[\mathbf{x}]f_2^*[\mathbf{x}]\kappa_{\mathcal{G}}(\mathbf{x})\end{aligned} \tag{12.42}$$

since f_1 and f_2 are constant on $\mathcal{G}(\mathbf{x})$ and there are $\kappa_{\mathcal{G}}(\mathbf{x})$ elements in this set. □

Theorem 12.8 The nonzero functions $\phi_{\mathcal{G}\mathbf{u}}$ defined in Equation (12.38) for $\mathbf{u}$ in a set $\mathcal{D}_I \subset \mathcal{D}_{\mathcal{G}_p}$, where $|\mathcal{D}_I| = |\mathcal{B}_{\mathcal{G}}|$, form an orthogonal basis for $\mathcal{S}_{(\Lambda/\Gamma)\mathcal{G}}$, where the inner product is defined in Theorem 12.7. Specifically,

$$\langle \phi_{\mathcal{G}\mathbf{u}_1} \mid \phi_{\mathcal{G}\mathbf{u}_2} \rangle_{\mathcal{S}_{(\Lambda/\Gamma)\mathcal{G}}} = \frac{K}{\beta(\mathbf{u}_1)} \delta[\mathbf{u}_1 - \mathbf{u}_2], \qquad \mathbf{u}_1, \mathbf{u}_2 \in \mathcal{D}_I \tag{12.43}$$

where $\beta(\mathbf{u})$ is a constant defined below.

Proof: Let $\mathbf{u}_1$ and $\mathbf{u}_2$ be two different elements of $\mathcal{D}_{\mathcal{G}_p}$. Then, by Theorem 12.7,

$$\begin{aligned}\langle \phi_{\mathcal{G}\mathbf{u}_1} \mid \phi_{\mathcal{G}\mathbf{u}_2} \rangle_{\mathcal{S}_{(\Lambda/\Gamma)\mathcal{G}}} &= \langle \phi_{\mathcal{G}\mathbf{u}_1} \mid \phi_{\mathcal{G}\mathbf{u}_2} \rangle_{\mathcal{S}_{\Lambda/\Gamma}} \\ &= \frac{1}{L^2} \left\langle \sum_{k=1}^{L} \mathcal{H}_{\mathfrak{g}_k} \phi_{\mathbf{u}_1} \,\middle|\, \sum_{\ell=1}^{L} \mathcal{H}_{\mathfrak{g}_\ell} \phi_{\mathbf{u}_2} \right\rangle_{\mathcal{S}_{\Lambda/\Gamma}}.\end{aligned} \tag{12.44}$$

With $\phi_{\mathbf{u}_1}[\mathbf{x}] = \exp(\mathrm{j}2\pi\mathbf{u}_1 \cdot \mathbf{x})$ and using Equation (12.34), we obtain for each k

$$\begin{aligned}(\mathcal{H}_{\mathfrak{g}_k} \phi_{\mathbf{u}_1})[\mathbf{x}] &= \exp(\mathrm{j}2\pi\mathbf{u}_1 \cdot \mathbf{A}_k^T(\mathbf{x} - \mathbf{v}_k)) \\ &= \exp(\mathrm{j}2\pi(\mathbf{A}_k\mathbf{u}_1) \cdot (\mathbf{x} - \mathbf{v}_k)) \\ &= \exp(-\mathrm{j}2\pi(\mathbf{A}_k\mathbf{u}_1) \cdot \mathbf{v}_k) \exp(\mathrm{j}2\pi(\mathbf{A}_k\mathbf{u}_1) \cdot \mathbf{x})\end{aligned} \tag{12.45}$$

or in other words

$$\mathcal{H}_{\mathfrak{g}_k} \phi_{\mathbf{u}_1} = \exp(-\mathrm{j}2\pi(\mathbf{A}_k\mathbf{u}_1) \cdot \mathbf{v}_k) \phi_{\mathbf{A}_k\mathbf{u}_1}, \tag{12.46}$$

with a similar equation for $\phi_{\mathbf{u}_2}$. We note that $\mathbf{A}_k\mathbf{u}_1 \in \mathcal{G}_p(\mathbf{u}_1)$, the orbit containing $\mathbf{u}_1$, and similarly $\mathbf{A}_\ell\mathbf{u}_2 \in \mathcal{G}_p(\mathbf{u}_2)$. Thus

$$\begin{aligned}\langle \phi_{\mathcal{G}\mathbf{u}_1} \mid \phi_{\mathcal{G}\mathbf{u}_2} \rangle_{\mathcal{S}_{\Lambda/\Gamma}} = \frac{1}{L^2} \sum_{k=1}^{L} \sum_{\ell=1}^{L} &\exp(-\mathrm{j}2\pi(\mathbf{A}_k\mathbf{u}_1) \cdot \mathbf{v}_k) \exp(\mathrm{j}2\pi(\mathbf{A}_\ell\mathbf{u}_2) \cdot \mathbf{v}_\ell) \\ &\langle \phi_{\mathbf{A}_k\mathbf{u}_1} \mid \phi_{\mathbf{A}_\ell\mathbf{u}_2} \rangle_{\mathcal{S}_{\Lambda/\Gamma}}.\end{aligned} \tag{12.47}$$

Since orbits of $\mathbf{u}_1$ and $\mathbf{u}_2$ under $\mathcal{G}_p$ are disjoint, $\mathbf{A}_k\mathbf{u}_1$ and $\mathbf{A}_\ell\mathbf{u}_2$ are distinct elements of $\mathcal{D}$ and thus $\langle \phi_{\mathbf{A}_k\mathbf{u}_1} \mid \phi_{\mathbf{A}_\ell\mathbf{u}_2} \rangle_{\mathcal{S}_{\Lambda/\Gamma}} = 0$ by Theorem 4.1. Thus, the $\phi_{\mathcal{G}\mathbf{u}}$ for $\mathbf{u} \in \mathcal{D}_{\mathcal{G}_p}$ are pairwise orthogonal and so the nonzero elements form an orthogonal basis for their span. This set of $\mathbf{u} \in \mathcal{D}_{\mathcal{G}_p}$ for which $\phi_{\mathcal{G}\mathbf{u}}$ is nonzero is denoted $\mathcal{D}_I$. We know from Chapter 4 that any element of $\mathcal{S}_{(\Lambda/\Gamma)\mathcal{G}}$ can be expressed as a linear combination of the $\phi_{\mathcal{G}\mathbf{u}}$, and so these must span $\mathcal{S}_{(\Lambda/\Gamma)\mathcal{G}}$. Since it is possible that $|\mathcal{D}_{\mathcal{G}_p}| > \mathcal{B}_{\mathcal{G}}$, some of the $\phi_{\mathcal{G}\mathbf{u}}$ could be identically zero, as will see in examples.

If $\mathbf{u}_1 = \mathbf{u}_2 = \mathbf{u}$ and $\phi_{\mathcal{G}\mathbf{u}}$ is not identically zero, then in Equation (12.47)

$$\langle \phi_{\mathbf{A}_k\mathbf{u}} \mid \phi_{\mathbf{A}_\ell\mathbf{u}} \rangle_{\mathcal{S}_{\Lambda/\Gamma}} = K\delta_{\Lambda^*}[\mathbf{A}_k\mathbf{u} - \mathbf{A}_\ell\mathbf{u}] \tag{12.48}$$

where for a given $\mathbf{u}$ and k, it is possible that $\mathbf{A}_\ell\mathbf{u} = \mathbf{A}_k\mathbf{u} \pmod{\Lambda^*}$ for more than one value of ℓ, depending on the size of the orbit contining $\mathbf{u}$. It follows that

$$\|\phi_{\mathcal{G}\mathbf{u}}\|^2_{\mathcal{S}_{\Lambda/\Gamma}} = \frac{K}{L^2} \sum_{k=1}^{L} \sum_{\substack{\ell=1 \\ \mathbf{A}_\ell\mathbf{u}=\mathbf{A}_k\mathbf{u}}}^{L} \exp(\mathrm{j}2\pi(\mathbf{A}_k\mathbf{u}) \cdot (\mathbf{v}_\ell - \mathbf{v}_k)). \tag{12.49}$$

To simplify this expression, recall that $\mathbf{v}_k$ satisfies $\mathbf{v}_k = (\mathbf{I} - \mathbf{A}_k)\mathbf{a}$, so that $\mathbf{v}_\ell - \mathbf{v}_k = (\mathbf{A}_k - \mathbf{A}_\ell)\mathbf{a}$. Using the identity $(\mathbf{Ab}) \cdot \mathbf{c} = \mathbf{b} \cdot (\mathbf{A}^T\mathbf{c})$ and noting that $\mathbf{A}_k^T\mathbf{A}_k = \mathbf{I}$, we observe that $(\mathbf{A}_k\mathbf{u}) \cdot (\mathbf{v}_\ell - \mathbf{v}_k) = (\mathbf{I} - \mathbf{A}_\ell^T\mathbf{A}_k)\mathbf{u} \cdot \mathbf{a}$. But the condition $\mathbf{A}_\ell\mathbf{u} = \mathbf{A}_k\mathbf{u}$ implies that $\mathbf{u} = \mathbf{A}_\ell^T\mathbf{A}_k\mathbf{u}$ so that the argument of the exp function is 0 for all ℓ such that $\mathbf{A}_\ell\mathbf{u} = \mathbf{A}_k\mathbf{u}$ and the inner summation is equal to the number of values of ℓ for which the condition holds. We define $\mathcal{G}_p^{\mathbf{u}} = \{\mathfrak{g} \in \mathcal{G}_p \mid \mathfrak{g}\mathbf{u} = \mathbf{u}\}$, known as the isotropy subgroup of $\mathcal{G}_p$ at $\mathbf{u}$. It is known that the size of the isotropy subgroup, which is the number of terms in the inner summation, can be found in terms of the size of the $\mathcal{G}_p$ orbits: $|\mathcal{G}_p^{\mathbf{u}}|\kappa_{\mathcal{G}_p}(\mathbf{u}) = |\mathcal{G}_p| = L$ [Miller (1972), theorem 1.6]. So finally, we can conclude that

$$\|\phi_{\mathcal{G}\mathbf{u}}\|^2_{\mathcal{S}_{\Lambda/\Gamma}} = \frac{K}{\kappa_{\mathcal{G}_p}(\mathbf{u})}, \tag{12.50}$$

or in other words $\beta(u) = \kappa_{\mathcal{G}_p}(\mathbf{u})$, the size of the $\mathcal{G}_p$ orbit of $\mathbf{u}$. Then we can divide the $\phi_{\mathcal{G}\mathbf{u}}$ by $\alpha\sqrt{K/\beta(\mathbf{u})}$, where α is any complex number with $|\alpha| = 1$, to get an orthonormal basis. □

Having an orthogonal basis, we can now expand arbitrary elements of $\mathcal{S}_{(\Lambda/\Gamma)\mathcal{G}}$ in terms of this basis. This is Fourier analysis for this space. Let $|\mathcal{B}_{\mathcal{G}_p}| = |\mathcal{D}_{\mathcal{G}_p}| = N$. Then, for $f \in \mathcal{S}_{(\Lambda/\Gamma)\mathcal{G}}$,

$$f = \sum_{n=0}^{N-1} F[\mathbf{u}_n]\phi_{\mathcal{G}\mathbf{u}_n} \tag{12.51}$$

where

$$F[\mathbf{u}_n] = \frac{\langle f \mid \phi_{\mathcal{G}\mathbf{u}_n}\rangle}{\|\phi_{\mathcal{G}\mathbf{u}_n}\|^2} = \frac{\langle f \mid \phi_{\mathcal{G}\mathbf{u}_n}\rangle}{K/\beta(\mathbf{u}_n)}. \tag{12.52}$$

These expressions can be further expanded using the respective definitions:

$$\phi_{\mathcal{G}\mathbf{u}}[\mathbf{x}] = \frac{1}{L}\sum_{k=1}^{L} \exp(-\mathrm{j}2\pi(\mathbf{A}_k\mathbf{u}) \cdot \mathbf{v}_k)\exp(\mathrm{j}2\pi(\mathbf{A}_k\mathbf{u}) \cdot \mathbf{x}) \tag{12.53}$$

$$\begin{aligned} F[\mathbf{u}] &= \frac{\beta(\mathbf{u})}{KL}\sum_{k=1}^{L} \exp(\mathrm{j}2\pi(\mathbf{A}_k\mathbf{u}) \cdot \mathbf{v}_k)\sum_{\mathbf{x}\in\mathcal{B}_\mathcal{G}} f[\mathbf{x}]\exp(-\mathrm{j}2\pi(\mathbf{A}_k\mathbf{u}) \cdot \mathbf{x})\kappa_\mathcal{G}(\mathbf{x}) \\ &= \frac{\beta(\mathbf{u})}{KL}\sum_{\mathbf{x}\in\mathcal{B}_\mathcal{G}} f[\mathbf{x}]\kappa_\mathcal{G}(\mathbf{x})\sum_{k=1}^{L} \exp(\mathrm{j}2\pi(\mathbf{A}_k\mathbf{u}) \cdot \mathbf{v}_k)\exp(-\mathrm{j}2\pi(\mathbf{A}_k\mathbf{u}) \cdot \mathbf{x}) \end{aligned} \tag{12.54}$$

$$f[\mathbf{x}] = \frac{1}{L}\sum_{\mathbf{u}\in\mathcal{D}_I} F[\mathbf{u}]\sum_{k=1}^{L} \exp(-\mathrm{j}2\pi(\mathbf{A}_k\mathbf{u}) \cdot \mathbf{v}_k)\exp(\mathrm{j}2\pi(\mathbf{A}_k\mathbf{u}) \cdot \mathbf{x}) \tag{12.55}$$

where in all equations, there are N distinct values of $\mathbf{x} \in \mathcal{B}_\mathcal{G}$ and of $\mathbf{u} \in \mathcal{D}_I$. These equations define the discrete-domain symmetry-invariant periodic Fourier transform. This representation is simply the conventional discrete-domain periodic Fourier transform of Chapter 4 restricted to the space of symmetry-invariant signals and expressed in terms of independent values at orbit representatives. Although this may be of interest in its own right, we will mainly consider it for the block-based representation of images.

Note that in most cases of interest, the point symmetry group $\mathcal{G}_p$ contains the inversion, so that the basis functions can be made real and built out of cosine functions, as we will see in the examples that will be considered. This allows us to define a general class of discrete cosine transforms on arbitrary lattices. We illustrate these concepts with several examples that highlight specific aspects of the theory. We start with two one-dimensional examples, which are instances of the DCT. The first exhibits symmetry about zero so that the ideas are quite straightforward. The second has symmetry about the point −0.5, which introduces some of the complexities that occur when the symmetry group is different from the point group. However, it is an important case, corresponding to the widely used DCT-2. We follow this with the two-dimensional version of the DCT-2, which is a straightforward separable extension of the one-dimensional version. Then we consider a more complex example using the case of eight-fold symmetry as shown in Figure 12.7.

In our first two examples, we consider the one-dimensional case, which leads to different types of DCT. In this case, we can take $\Lambda = \mathbb{Z}$ and $\Gamma = K\mathbb{Z}$. The corresponding reciprocal lattices are $\Lambda^* = \mathbb{Z}$ and $\Gamma^* = \frac{1}{K}\mathbb{Z}$. The only nontrivial point symmetry group consists of the identity $\mathfrak{e}$ and the inversion, denoted $\mathfrak{i}$, where $\mathfrak{i}n = -n$. Thus $L = 2$. The center of symmetry must lie in $\frac{1}{2}\Lambda = \frac{1}{2}\mathbb{Z}$. There are only two essentially distinct cases, namely $a = 0$ and $a = -0.5$. The situation is slightly different depending on whether K is even or odd. This leads to four different transforms, often called DCT-1, DCT-2, DCT-5, and DCT-6. Other types of DCT are obtained by including some odd symmetry as well. We will not consider these cases. In the previous derivations, all vectors and matrices reduce to scalars, so the expressions are considerably simpler. For example, the matrix representations of the orthogonal point symmetries are simply $A_1 = +1$ and $A_2 = -1$. If $a = 0$, then $v_1 = v_2 = 0$, and for $a = -0.5$, $v_1 = 0$ and $v_2 = -1$. We examine only the cases with K even, first with $a = 0$ then with $a = -0.5$. The cases with K odd are left as an exercise.

Example 12.2

Develop the Fourier representation for one-dimensional periodic signals with even-length period and symmetry about the origin.

Solution:

In this case, $\Lambda = \mathbb{Z}$, $\Gamma = K\mathbb{Z}$ where K is even, and the symmetry group consists of the identity $\mathfrak{e}$ and the inversion $\mathfrak{i}$.

The set of coset representatives for Λ in Γ that we take is

$$\mathcal{B} = \{0, 1, \dots, K-1\} = \left\{0, 1, \dots, \frac{K}{2} - 1, \frac{K}{2}, \frac{K}{2} + 1, \dots, K-1\right\}. \tag{12.56}$$

If we map these points with the inversion $\mathfrak{i}$ (mod K), we get

$$\mathfrak{i}(\mathcal{B}) = \left\{0, K-1, K-2, \dots, \frac{K}{2} + 1, \frac{K}{2}, \frac{K}{2} - 1, \dots, 1\right\}. \tag{12.57}$$

Thus, we see two orbits with a single element, namely $\mathcal{G}(0)$ and $\mathcal{G}(K/2)$, and all the remaining orbits have two elements. As a set of orbit representatives, we choose

$$\mathcal{B}_{\mathcal{G}} = \left\{0, 1, \dots, \frac{K}{2} - 1, \frac{K}{2}\right\} \tag{12.58}$$

with corresponding sizes of the orbits

$$\kappa_{\mathcal{G}}(\mathcal{B}_{\mathcal{G}}) = \{1, 2, 2, \ldots, 2, 2, 1\}. \tag{12.59}$$

Thus, $|\mathcal{B}_{\mathcal{G}}| = N = \frac{K}{2} + 1$, or alternatively, $K = 2(N-1)$. The basis functions are given by

$$\phi_{\mathcal{G}u} = \frac{1}{2}(\mathcal{H}_{\mathrm{e}}\phi_u + \mathcal{H}_{\mathrm{i}}\phi_u) \tag{12.60}$$

or explicitly

$$\begin{aligned}\phi_{\mathcal{G}u}[n] &= \frac{1}{2}(\phi_u[n] + \phi_u[-n]) \\ &= \frac{1}{2}(\exp(\mathrm{j}2\pi un) + \exp(-\mathrm{j}2\pi un)) \\ &= \cos(2\pi un), \quad u \in \frac{1}{K}\mathbb{Z}.\end{aligned} \tag{12.61}$$

A set of coset representatives for Λ^* in Γ^* is

$$\mathcal{D} = \left\{0, \frac{1}{K}, \ldots, \frac{K-1}{K}\right\} = \left\{0, \frac{1}{2(N-1)}, \ldots, \frac{2(N-1)-1}{2(N-1)}\right\}. \tag{12.62}$$

Since $\phi_{\mathcal{G}u} = \phi_{\mathcal{G}(-u)}$, we can take as orbit representatives in the frequency domain

$$\mathcal{D}_{\mathcal{G}_p} = \left\{0, \frac{1}{K}, \ldots, \frac{1}{2}\right\} = \left\{0, \frac{1}{2(N-1)}, \ldots, \frac{1}{2}\right\}, \tag{12.63}$$

where $|\mathcal{D}_{\mathcal{G}_p}| = |\mathcal{B}_{\mathcal{G}}|$ and so $\mathcal{D}_I = \mathcal{D}_{\mathcal{G}_p}$. To evaluate $\beta(u)$, we note that $A_1u - A_2u \in \Lambda^* = \mathbb{Z}$ (i.e. $2u \in \mathbb{Z}$) if $u = 0$ or $u = \frac{1}{2}$. Thus, evaluating Equation (12.50),

$$\|\phi_{\mathcal{G}u}\|^2_{S_{\Lambda/\Gamma}} = \begin{cases} K & u \in \left\{0, \frac{1}{2}\right\} \\ \frac{K}{2} & u \in \mathcal{D}_I \backslash \left\{0, \frac{1}{2}\right\}, \end{cases} \tag{12.64}$$

and so

$$\beta(u) = \begin{cases} 1 & u \in \left\{0, \frac{1}{2}\right\} \\ 2 & u \in \mathcal{D}_I \backslash \left\{0, \frac{1}{2}\right\}. \end{cases} \tag{12.65}$$

Then the analysis equation is

$$\begin{aligned}F\left[\frac{k}{K}\right] = F\left[\frac{k}{2(N-1)}\right] &= \frac{\beta(k/(2(N-1)))}{4(N-1)} \sum_{n=0}^{N-1} f[n]\kappa_{\mathcal{G}}(n) \cos\left(\frac{\pi kn}{N-1}\right) \\ &= \frac{\beta(k/(2(N-1)))}{4(N-1)} (f[0] + 2\sum_{n=1}^{N-2} f[n] \cos\left(\frac{\pi kn}{N-1}\right) \\ &\quad + f[N-1](-1)^k),\end{aligned} \tag{12.66}$$

and the synthesis equation is

$$f[n] = \sum_{k=0}^{N-1} F\left[\frac{k}{2(N-1)}\right] \cos\left(\frac{\pi kn}{N-1}\right). \tag{12.67}$$

This representation is equivalent to the conventional DCT-1 [Strang (1999)], other than a scale factor. □

Example 12.3
Develop the Fourier representation for one-dimensional periodic signals with even-length period and symmetry about the point −0.5.

Solution:
As mentioned above, $v_1 = 0$ and $v_2 = -1$, and so the transformation group is $\mathcal{G} = \{\mathfrak{e}, \mathcal{T}_{-1}\mathfrak{i}\}$. The set of coset representatives is still given by Equation (12.56), and if we map these with $\mathfrak{g}_2 = \mathcal{T}_{-1}\mathfrak{i}\ (\text{mod}\ K)$, we find

$$\mathfrak{g}_2(\mathcal{B}) = \{K-1, K-2, \dots, 1, 0\} \tag{12.68}$$

so that all orbits have two elements, $\kappa_{\mathcal{G}}(\mathbf{x}) = 2$ for all $\mathbf{x}$. As orbit representatives, we choose

$$\mathcal{B}_{\mathcal{G}} = \left\{0, 1, \dots, \frac{K}{2} - 1\right\} \tag{12.69}$$

so that $N = \frac{K}{2}$ or $K = 2N$. The basis functions are given by

$$\phi_{\mathcal{G}u} = \frac{1}{2}(\mathcal{H}_{\mathfrak{e}}\phi_u + \mathcal{H}_{\mathcal{T}_{-1}\mathfrak{i}}\phi_u) \tag{12.70}$$

or explicitly

$$\begin{aligned}
\phi_{\mathcal{G}u}[n] &= \frac{1}{2}(\phi_u[n] + \phi_u[-n-1]) \\
&= \frac{1}{2}(\exp(\mathrm{j}2\pi un) + \exp(-\mathrm{j}2\pi u)\exp(-\mathrm{j}2\pi un)) \\
&= \frac{1}{2}\exp(-\mathrm{j}\pi u)(\exp(\mathrm{j}\pi u)\exp(\mathrm{j}2\pi un) + \exp(-\mathrm{j}\pi u)\exp(-\mathrm{j}2\pi un)) \\
&= \exp(-\mathrm{j}\pi u)\cos\left(2\pi u\left(n + \frac{1}{2}\right)\right).
\end{aligned} \tag{12.71}$$

Since $\exp(-\mathrm{j}\pi u)$ is just a constant for each basis vector, we can divide by these constants to get real basis vectors

$$\phi'_{\mathcal{G}u}[n] = \cos\left(2\pi u\left(n + \frac{1}{2}\right)\right). \tag{12.72}$$

The point group is $\mathcal{G}_p = \{\mathfrak{e}, \mathfrak{i}\}$. As in Example 12.2, the orbit representatives are

$$\mathcal{D}_{\mathcal{G}_p} = \left\{0, \frac{1}{K}, \dots, \frac{1}{2}\right\} = \left\{0, \frac{1}{2N}, \dots, \frac{N}{2N}\right\}, \tag{12.73}$$

so that $|\mathcal{D}_{\mathcal{G}_p}| = N + 1 > |\mathcal{B}_{\mathcal{G}}|$. However, we note that $\phi'_{\mathcal{G}\frac{1}{2}}[n] = \cos(\pi(n + \frac{1}{2})) = 0$ for all n. Thus, we take

$$\mathcal{D}_I = \left\{0, \frac{1}{2N}, \dots, \frac{N-1}{2N}\right\}. \tag{12.74}$$

and so

$$\phi'_{\mathcal{G}\frac{k}{2N}}[n] = \cos\left(\frac{\pi k(n + \frac{1}{2})}{N}\right), \qquad n = 0, \dots, N-1; k = 0, \dots, N-1, \tag{12.75}$$

which are the standard non-normalized DCT-2 basis vectors. We evaluate $\beta(u)$ in a similar fashion to the DCT-1: $A_1 u - A_2 u = 2u \in \mathbb{Z}$ if $u = 0$, and so

$$\|\phi_{\mathcal{G}u}\|^2_{S_{\Lambda/\Gamma}} = \begin{cases} K & u = 0 \\ \frac{K}{2} & u \in \mathcal{D}_I \backslash 0 \end{cases} \tag{12.76}$$

and

$$\beta(u) = \begin{cases} 1 & u = 0 \\ 2 & u \in \mathcal{D}_I \backslash 0. \end{cases} \tag{12.77}$$

The analysis and synthesis equations are

$$F\left[\frac{k}{2N}\right] = \frac{\beta(k/2N)}{4N} \sum_{n=0}^{N-1} f[n] \cos\left(\frac{\pi k(n+\frac{1}{2})}{N}\right) \tag{12.78}$$

$$f[n] = \sum_{k=0}^{N-1} F\left[\frac{k}{2N}\right] \cos\left(\frac{\pi k(n+\frac{1}{2})}{N}\right). \tag{12.79}$$

□

Example 12.4
Develop the Fourier representation for two-dimensional periodic signals on $\mathbb{Z}^2$ with even-length period and quadrantal symmetry about the point $[-0.5, -0.5]$.

Solution:
When the lattices Λ and Γ are rectangular, the DCTs just seen can be applied separably, using the quadrantal symmetry group $\mathcal{G}_p = \{\mathfrak{e}, \mathfrak{h}, \mathfrak{v}, \mathfrak{r}^2\}$ in two dimensions and the corresponding shifts in Table 12.2, and octantal or hyperoctantal symmetry groups in three or more dimensions. We present the 2D case, since the extension to three or more dimensions is straightforward.

In the 2D case, separability results from the fact that $\mathfrak{r}^2 = \mathfrak{h}\mathfrak{v} = \mathfrak{v}\mathfrak{h}$. Although everything in the following applies to rectangular (not necessarily square) lattices, we carry out the example for the common case of square lattices to simplify notation. Thus, we set $\Lambda = \mathbb{Z}^2$ and $\Gamma = (M\mathbb{Z})^2$, where we assume that M is even; then $K = M^2$. We have four essentially distinct possibilities for the center of symmetry: $\mathbf{a} \in \{[\,0\;0\,]^T, [\,0\;-0.5\,]^T, [-0.5\;0\,]^T, [-0.5\;-0.5\,]^T\}$. This would correspond to applying the DCT-1 in a dimension where the offset is 0 and the DCT-2 in a dimension where the offset is -0.5. For our example, we suppose that $\mathbf{a} = [-0.5\;-0.5\,]^T$.

Although we can directly apply the result of Example 12.3 separably to rows and columns, we develop this as a two-dimensional example to illustrate the concepts in the 2D case. The set of coset representatives of Γ in Λ that we use are

$$\mathcal{B} = \{(n_1, n_2), 0 \le n_1, n_2 \le M-1\}. \tag{12.80}$$

The four elements of the symmetry group $\mathcal{G}$ are given by the matrices and shift vectors $\mathbf{A}_i, \mathbf{v}_i$ in Table 12.2 for $i = 1, 5, 6, 3$ respectively. Applying the three nonidentity transformations to $\mathcal{B}$, we find (mod M)

$$\mathfrak{g}_2(\mathcal{B}) = \{(n_1, M-1-n_2), \quad 0 \le n_1, n_2 \le M-1\}$$

$$\mathfrak{g}_3(\mathcal{B}) = \{(M-1-n_1, n_2), \quad 0 \le n_1, n_2 \le M-1\}$$
$$\mathfrak{g}_4(\mathcal{B}) = \{(M-1-n_1, M-1-n_2), \quad 0 \le n_1, n_2 \le M-1\}.$$

Of course, $\mathfrak{g}_1(\mathcal{B}) = \mathcal{B}$. For the same reason as in Example 12.3, all orbits have four elements (because $n = M-1-n$ has no integer solution for M even). As orbit representatives, we choose

$$\mathcal{B}_{\mathcal{G}} = \{(n_1, n_2), 0 \le n_1, n_2 \le \frac{M}{2} - 1\}, \tag{12.81}$$

so that $N = |\mathcal{B}_{\mathcal{G}}| = \left(\frac{M}{2}\right)^2$.

The Fourier basis functions are given by

$$\phi_{\mathcal{G}\mathbf{u}} = \frac{1}{4}\sum_{i=1}^{4} \mathcal{H}_{\mathfrak{g}_i}\phi_{\mathbf{u}}, \tag{12.82}$$

or explicitly

$$\phi_{\mathcal{G}(u,v)}[n_1, n_2] = \frac{1}{4}(\phi_{u,v}(n_1, n_2) + \phi_{u,v}(n_1, -n_2-1) + \phi_{u,v}(-n_1-1, n_2) + \phi_{u,v}(-n_1-1, -n_2-1)). \tag{12.83}$$

Since

$$\phi_{u,v}[n_1, n_2] = \exp(\mathrm{j}2\pi(un_1 + vn_2)) = \exp(\mathrm{j}2\pi un_1)\exp(\mathrm{j}2\pi vn_2) = \phi_u[n_1]\phi_v[n_2], \tag{12.84}$$

it follows that

$$\begin{aligned}\phi_{\mathcal{G}(u,v)}[n_1, n_2] &= \frac{1}{4}(\phi_u[n_1] + \phi_u[n_1-1])(\phi_v[n_2] + \phi_v[-n_2-1]) \\ &= \exp(-\mathrm{j}\pi(u+v))\cos\left(2\pi u\left(n_1 + \frac{1}{2}\right)\right)\cos\left(2\pi v\left(n_2 + \frac{1}{2}\right)\right)\end{aligned} \tag{12.85}$$

using the corresponding result in Example 12.3. As in the 1D case, we can divide each basis vector by the (unit-norm) constant $\exp(-\mathrm{j}\pi(u+v))$ to get the real basis vectors

$$\phi'_{\mathcal{G}(u,v)}[n_1, n_2] = \cos\left(2\pi u\left(n_1 + \frac{1}{2}\right)\right)\cos\left(2\pi v\left(n_2 + \frac{1}{2}\right)\right) \tag{12.86}$$

which are separable versions of the basis vectors in Example 12.3, as expected. Thus, everything else follows through separably, as in Example 12.3 and the details are omitted. However, we return to this case in the next section on the block-based representation of images. □

Example 12.5

Develop the Fourier representation for two-dimensional periodic signals on $\mathbb{Z}^2$ with even-length period and full eight-fold symmetry about the point $[-0.5, -0.5]$.

Solution:

The parameters are the same as in Example 12.4 except that now $L = 8$ and the symmetry group consists of all the symmetries given in Table 12.2. Figure 12.7 illustrates the situation for the case $M = 8$. We use same set of coset representatives as for the separable 2D DCT (Equation (12.80)). Referring to Figure 12.7, we choose as orbit representatives

$$\mathcal{B}_{\mathcal{G}} = \left\{ (n_1, n_2), 0 \leq n_1 \leq \frac{M}{2} - 1, 0 \leq n_2 \leq n_1 \right\}. \tag{12.87}$$

We note that

$$N = |\mathcal{B}_{\mathcal{G}}| = \sum_{m=0}^{\frac{M}{2}-1} (m+1) = \frac{M(M+2)}{8}. \tag{12.88}$$

In our example with $M = 8$, we have $N = 10$ as shown in Figure 12.7. Again referring to Figure 12.7, we see that points on the diagonal $k_1 = k_2$ have orbits with four points while all others have eight points, i.e.

$$\kappa_{\mathcal{G}}(n_1, n_2) = \begin{cases} 4 & (n_1, n_2) \in \mathcal{B}_{\mathcal{G}}, n_1 = n_2 \\ 8 & (n_1, n_2) \in \mathcal{B}_{\mathcal{G}}, n_1 \neq n_2. \end{cases} \tag{12.89}$$

The Fourier basis functions are given by

$$\phi_{\mathcal{G}\mathbf{u}} = \frac{1}{8} \sum_{i=1}^{8} \mathcal{H}_{\mathfrak{g}_i} \phi_{\mathbf{u}}, \tag{12.90}$$

where

$$\mathcal{H}_{\mathfrak{g}_i} \phi_{\mathbf{u}}[\mathbf{x}] = \exp(\mathrm{j}2\pi \mathbf{u} \cdot \mathbf{A}_i^T (\mathbf{x} - \mathbf{v}_i)) \tag{12.91}$$

with the $\mathbf{A}_i$ and $\mathbf{v}_i$ given in Table 12.2.

Although we can evaluate this expression explicitly, it is simpler to note that the four additional symmetries compared to Example 12.4 can all be obtained by composing the original symmetries with $\mathfrak{g}_7$ (i.e. $\mathfrak{g}_2 = \mathfrak{g}_7\mathfrak{g}_5$, $\mathfrak{g}_4 = \mathfrak{g}_7\mathfrak{g}_6$, $\mathfrak{g}_7 = \mathfrak{g}_7\mathfrak{g}_1$, $\mathfrak{g}_8 = \mathfrak{g}_7\mathfrak{g}_3$). If we denote the Fourier basis vectors of Example 12.4 by $\phi^{(4)}_{\mathcal{G}\mathbf{u}}$, then the basis vectors of this example are given by

$$\begin{aligned} \phi_{\mathcal{G}\mathbf{u}} &= \frac{1}{2}(\phi^{(4)}_{\mathcal{G}\mathbf{u}} + \mathcal{H}_{\mathfrak{g}_7}\phi^{(4)}_{\mathcal{G}\mathbf{u}}) \\ \phi_{\mathcal{G}\mathbf{u}}[n_1, n_2] &= \frac{1}{2}(\phi^{(4)}_{\mathcal{G}\mathbf{u}}[n_1, n_2] + \phi^{(4)}_{\mathcal{G}\mathbf{u}}[n_2, n_1]) \\ &= \frac{1}{2}\exp(-\mathrm{j}\pi(u+v))\left(\cos\left(2\pi u\left(n_1 + \frac{1}{2}\right)\right)\cos\left(2\pi v\left(n_2 + \frac{1}{2}\right)\right)\right. \\ &\quad \left. + \cos\left(2\pi u\left(n_2 + \frac{1}{2}\right)\right)\cos\left(2\pi v\left(n_1 + \frac{1}{2}\right)\right)\right), \quad (u, v) \in \left(\frac{1}{M}\mathbb{Z}\right)^2. \end{aligned} \tag{12.92}$$

As in the previous cases, we divide by the unit-norm constants $\exp(-\mathrm{j}2\pi(u + v))$ to get the real basis vectors

$$\begin{aligned} \phi'_{\mathcal{G}\mathbf{u}}[n_1, n_2] = \frac{1}{2}&\left(\cos\left(2\pi u\left(n_1 + \frac{1}{2}\right)\right)\cos\left(2\pi v\left(n_2 + \frac{1}{2}\right)\right)\right. \\ &\left. + \cos\left(2\pi u\left(n_2 + \frac{1}{2}\right)\right)\cos\left(2\pi v\left(n_1 + \frac{1}{2}\right)\right)\right). \end{aligned} \tag{12.93}$$

In the frequency domain, as usual we take

$$\mathcal{D} = \left\{ \left(\frac{k_1}{M}, \frac{k_2}{M} \right), 0 \le k_1, k_2 \le M-1 \right\}. \tag{12.94}$$

The point group $\mathcal{G}_p = D_4$ and we choose as orbit representatives

$$\mathcal{D}_{\mathcal{G}_p} = \left\{ \left(\frac{k_1}{M}, \frac{k_2}{M} \right), 0 \le k_1 \le \frac{M}{2}, 0 \le k_2 \le k_1 \right\}. \tag{12.95}$$

We note that $|\mathcal{D}_{\mathcal{G}_p}| = (M+2)(M+4)/8 > |\mathcal{B}_{\mathcal{G}}|$. However, all the $\phi_{\mathcal{G}(u,v)}$ are identically zero when $u = 0.5$, and so we choose

$$\mathcal{D}_I = \left\{ \left(\frac{k_1}{M}, \frac{k_2}{M} \right), 0 \le k_1 \le \frac{M}{2} - 1, 0 \le k_2 \le k_1 \right\}. \tag{12.96}$$

where now $|\mathcal{D}_I| = |\mathcal{B}_{\mathcal{G}}| = N$. Next we need to find $\beta(u,v)$ for $(u,v) \in \mathcal{D}_I$ using Equation (12.50).

$$\beta(u,v) = \begin{cases} 1 & u = v = 0 \\ 4 & v = 0, u = k_1/M, 0 \le k_1 \le \frac{M}{2} - 1 \\ 4 & v = u = k_1/M, 0 \le k_1 \le \frac{M}{2} - 1 \\ 8 & \text{otherwise.} \end{cases} \tag{12.97}$$

Given these results, we can write the analysis and synthesis expressions for an element $f \in \mathcal{S}_{(\Lambda/\Gamma)\mathcal{G}}$, using the real basis vectors as

$$\begin{aligned} f[n_1,n_2] = \frac{1}{2} \sum_{k_1=0}^{\frac{M}{2}-1} \sum_{k_2=0}^{k_1} F[k_1,k_2] \Bigg(& \cos\left(2\pi \frac{k_1}{M}\left(n_1 + \frac{1}{2}\right)\right) \cos\left(2\pi \frac{k_2}{M}\left(n_2 + \frac{1}{2}\right)\right) \\ & + \cos\left(2\pi \frac{k_1}{M}\left(n_2 + \frac{1}{2}\right)\right) \cos\left(2\pi \frac{k_2}{M}\left(n_1 + \frac{1}{2}\right)\right) \Bigg); \end{aligned} \tag{12.98}$$

$$\begin{aligned} F[k_1,k_2] = \frac{\beta\left(\frac{k_1}{M}, \frac{k_2}{M}\right)}{2M^2} \sum_{n_1=0}^{\frac{M}{2}-1} \sum_{n_2=0}^{n_1} f[n_1,n_2] \kappa_{\mathcal{G}}(n_1,n_2) \Bigg(& \cos\left(2\pi \frac{k_1}{M}\left(n_1 + \frac{1}{2}\right)\right) \\ & \times \cos\left(2\pi \frac{k_2}{M}\left(n_2 + \frac{1}{2}\right)\right) + \cos\left(2\pi \frac{k_1}{M}\left(n_2 + \frac{1}{2}\right)\right) \\ & \times \cos\left(2\pi \frac{k_2}{M}\left(n_1 + \frac{1}{2}\right)\right) \Bigg). \end{aligned} \tag{12.99}$$

□

12.3 Vector-Space Representation of Images Based on the Symmetry-Invariant Periodic Fourier Transform

In Section 4.7, we saw how the discrete-domain periodic Fourier transform can be used to represent a signal of finite extent, considered as one period of a periodic signal. This can then be used to develop a block-based representation of a larger signal. It was mentioned there that the block-based approach can be used with any orthogonal basis for finite extent blocks. We now extend this approach to the orthogonal bases we found for symmetry-invariant signals where a block is formed from the orbit representatives. This approach includes the standard block-based separable DCT-2 representation of images that is ubiquitous in image compression standards, and which has also found many other applications. However, the approach considerably generalizes the DCT to arbitrary lattices and periodicities.

Let $f[\mathbf{x}]$ be a signal, defined on a lattice Λ, with a finite region of support $\mathcal{A}$, so that $f[\mathbf{x}] = 0$ for $\mathbf{x} \in \Lambda \backslash \mathcal{A}$. Assume that the region $\mathcal{A}$ can be put into a one-to-one correspondence with a set $\mathcal{B}_{\mathcal{G}}$ of orbit representatives of a $\mathcal{G}$-invariant periodic signal with periodicity lattice Γ, for example the rectangular set of orbit representatives of a signal on a square lattice with four-fold symmetry (e.g., the DCT-2). We can apply the symmetries to get one period of the symmetric signal,

$$\tilde{f}[\mathbf{t}] = f[\mathbf{x}], \quad \mathbf{t} \in \mathcal{G}(\mathbf{x}), \mathbf{x} \in \mathcal{B}_{\mathcal{G}}, \tag{12.100}$$

which can be periodized on Γ to get the full symmetric periodic signal.

Assume that $\mathcal{A}$ consists of N samples. We have found a basis of N signals $\phi_{\mathcal{G}\mathbf{u}}$ to represent this signal. The basis is orthogonal with respect to the inner product of Theorem 12.7, which is suitable for symmetric periodic signals. However, for nonsymmetric finite-extent signals, we are more likely interested in using the usual Euclidean inner product

$$\langle f_1 \mid f_2 \rangle_{\mathbb{C}^N} = \sum_{\mathbf{x} \in \mathcal{B}_{\mathcal{G}}} f_1[\mathbf{x}] f_2^*[\mathbf{x}]. \tag{12.101}$$

To get basis functions orthogonal with respect to this inner product, we can define

$$\overline{\phi}_{\mathcal{G}\mathbf{u}}[\mathbf{x}] = \phi_{\mathcal{G}\mathbf{u}}[\mathbf{x}] \sqrt{\kappa_{\mathcal{G}}(\mathbf{x})}, \tag{12.102}$$

for then

$$\begin{aligned} \langle \overline{\phi}_{\mathcal{G}\mathbf{u}_1} \mid \overline{\phi}_{\mathcal{G}\mathbf{u}_2} \rangle_{\mathbb{C}^N} &= \sum_{\mathbf{x} \in \mathcal{B}_{\mathcal{G}}} \phi_{\mathcal{G}\mathbf{u}_1}[\mathbf{x}] \phi^*_{\mathcal{G}\mathbf{u}_2}[\mathbf{x}] \kappa_{\mathcal{G}}(\mathbf{x}) \\ &= \langle \phi_{\mathcal{G}\mathbf{u}_1} \mid \phi_{\mathcal{G}\mathbf{u}_2} \rangle_{S_{(\Lambda/\Gamma)\mathcal{G}}} \\ &= \frac{K}{\beta(\mathbf{u}_1)} \delta(\mathbf{u}_1 - \mathbf{u}_2), \quad \mathbf{u}_1, \mathbf{u}_2 \in \mathcal{D}_I. \end{aligned} \tag{12.103}$$

As before, we can divide by $\alpha\sqrt{K/\beta(\mathbf{u})}$ where α is any complex number with $|\alpha| = 1$ to get an orthonormal basis with respect to the usual Euclidean inner product.

If the shifted versions of $\mathcal{A}$ tile the image, we can use exactly the method of Section 4.7.2 to get an orthogonal (or orthonormal) local basis for the entire image. If the shifted versions cannot tile the image, we may have to use more than one block shape. For example, we cannot tile a rectangular image with the triangular set of orbit representatives shown in Figure 12.7. However, we could tile the image with two different triangular regions that could each serve as a valid set of orbit representatives.

Example 12.6
Obtain the orthonormal DCT-2 basis vectors for the block-based representation of an image and display the 8×8 basis vectors.

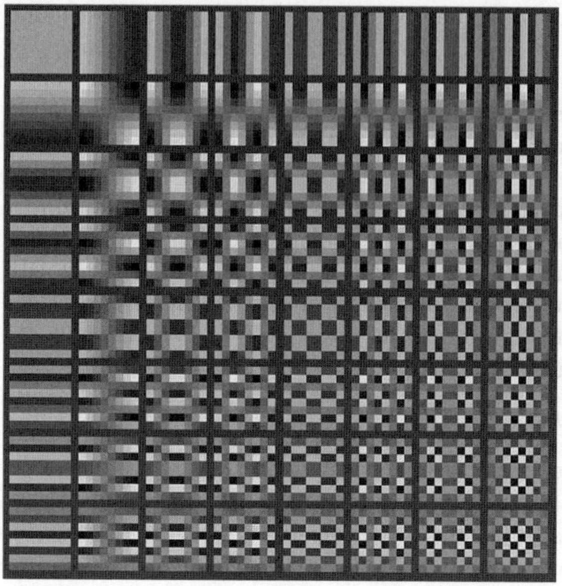

Figure 12.8 Illustration of the 64 orthonormal 8×8 2D DCT-2 basis vectors, where the basis vector $\mathbf{z}_{k_1k_2}$ is displayed in column k_1 and row k_2 of this array.

Solution:
The DCT-2 basis vectors for symmetric periodic signals with fourfold symmetry are given by Equation (12.85). All orbits have four elements, so that $\kappa_{\mathcal{G}}(\mathbf{x}) = 4$ for all $\mathbf{x} \in \mathcal{B}_{\mathcal{G}}$. Let the block size be N_1 ($N_1 = 8$ for the displayed basis vectors) so that $N = N_1^2$ and $K = 4N_1^2$. Let us denote the desired basis vectors as $\mathbf{z}_{k_1k_2}$, k_1, $k_2 = 0, \dots, N_1 - 1$. Since $\kappa_{\mathcal{G}}(\mathbf{x})$ is constant, the basis vectors of Equation (12.85) are orthogonal with respect to the Euclidean norm; to get orthornormal basis vectors on $\mathcal{B}_{\mathcal{G}}$, we divide these by $\alpha\sqrt{K/\beta(u,v)}$ for $(u,v) = \left(\frac{k_1}{2N_1}, \frac{k_2}{2N_1}\right)$, $0 \le k_1, k_2 \le N_1 - 1$, $|\alpha| = 1$. We use $\alpha = \exp(-\mathrm{j}\pi(u+v))$ to have real basis vectors, and note that $\beta\left(\frac{k_1}{2N_1}, \frac{k_2}{2N_1}\right) = \beta_1\left(\frac{k_1}{2N_1}\right)\beta_1\left(\frac{k_2}{2N_1}\right)$ where $\beta_1(u)$ is the function given by Equation (12.77), i.e.

$$\beta_1\left(\frac{k}{2N_1}\right) = \begin{cases} 1 & k = 0 \\ 2 & k = 1, \dots, N_1 - 1. \end{cases} \tag{12.104}$$

This gives the desired basis vectors

$$\begin{aligned} z_{k_1k_2}[n_1,n_2] &= \frac{\sqrt{\beta_1\left(\frac{k_1}{2N_1}\right)\beta_1\left(\frac{k_2}{2N_1}\right)}\sqrt{\kappa_{\mathcal{G}}(n_1,n_2)}}{2N_1\exp\left(-\mathrm{j}2\pi\frac{k_1+k_2}{2N_1}\right)}\phi_{\mathcal{G}\left(\frac{k_1}{2N_1},\frac{k_2}{2N_1}\right)}[n_1,n_2] \\ &= \frac{\sqrt{\beta_1\left(\frac{k_1}{2N_1}\right)\beta_1\left(\frac{k_2}{2N_1}\right)}}{N_1}\cos\left(\pi\frac{k_1}{N_1}\left(n_1+\frac{1}{2}\right)\right) \\ &\quad\times\cos\left(\pi\frac{k_2}{N_1}\left(n_2+\frac{1}{2}\right)\right). \end{aligned} \tag{12.105}$$

The 8×8 basis vectors obtained with $N_1 = 8$ are illustrated in Figure 12.8. □

Example 12.7

Compare the rate-distortion performance of DCT1 and DCT2 block-based representation for a collection of gray-scale images and a collection of Bayer CFA images.

Solution:

Image compression is outside the scope of this book. There are many excellent texts that cover this rapidly changing field. For the purpose of this example, a good proxy for the rate-distortion performance of an image basis is to apply uniform quantization to the transform coefficients and then compute the sample mean squared error and entropy over a suitable collection of images. By varying the quantizer step size, we can generate rate-distortion curves. The steps of our approach are as follows.

1) Partition the image database into a collection of R $N_1 \times N_1$ blocks $f^{(r)}[n_1, n_2]$, $r = 1, \dots, R$, $0 \le n_1, n_2 \le N_1 - 1$.
2) Expand each block in terms of the chosen orthonormal basis for $N_1 \times N_1$ blocks, denoted $\mathbf{z}_{k_1 k_2}$

$$f^{(r)}[n_1, n_2] = \sum_{k_1=0}^{N_1-1} \sum_{k_2=0}^{N_1-1} F^{(r)}[k_1, k_2] z_{k_1 k_2}[n_1, n_2], \tag{12.106}$$

where

$$F^{(r)}[k_1, k_2] = \sum_{n_1=0}^{N_1-1} \sum_{n_2=0}^{N_1-1} f^{(r)}[n_1, n_2] z_{k_1 k_2}[n_1, n_2]. \tag{12.107}$$

3) Apply a uniform quantizer encoder with step size Δ to each transform coefficient

$$Q_{\mathrm{E}}^{(r)}[k_1, k_2] = \mathrm{round}(F^{(r)}[k_1, k_2]/\Delta), \tag{12.108}$$

where $Q_{\mathrm{E}}^{(r)}[k_1, k_2] \in \mathbb{Z}$.
4) Determine the relative frequency of occurence for each quantized index $q \in \mathbb{Z}$ over the set of RN_1^2 coefficients,

$$p(q) = |\{Q_{\mathrm{E}}^{(r)}[k_1, k_2] = q\}|/RN_1^2, \qquad q \in \mathbb{Z}. \tag{12.109}$$

There are a finite number of values q for which $p(q) \neq 0$, and $\sum_{q \in \mathbb{Z}} p(q) = 1$.
5) Estimate the rate required to encode the quantizer encoder indexes by the sample entropy

$$H(\Delta) = -\sum_{q \in \mathbb{Z}} p(q) \log_2(p(q)). \tag{12.110}$$

6) Reconstruct each block with the quantized coefficients

$$\hat{f}^{(r)}[n_1, n_2] = \sum_{k_1=0}^{N_1-1} \sum_{k_2=0}^{N_1-1} \Delta Q_{\mathrm{E}}^{(r)}[k_1, k_2] z_{k_1 k_2}[n_1, n_2]. \tag{12.111}$$

7) Measure the average mean squared error distortion introduced over the entire database by the quantization

$$D(\Delta) = \frac{1}{RN_1^2} \sum_{r=1}^{R} \sum_{n_1=0}^{N_1-1} \sum_{n_2=0}^{N_2-1} (f^{(r)}[n_1, n_2] - \hat{f}^{(r)}[n_1, n_2])^2. \tag{12.112}$$

8) Plot $\mathrm{PSNR}(\Delta) = 10 \log_{10}(1/D(\Delta))$ as a function of $H(\Delta)$ over a suitable range of values of Δ.

We illustrate this approach using the suite of 24 Kodak images both in grayscale version and in CFA version. We note that if a CFA image block is reflected horizontally and vertically about the point $(-0.5, -0.5)$ as in the DCT-2, the CFA pattern is disturbed. However, if the blocks are reflected horizontally and vertically about the origin as in the DCT-1, the CFA structure is maintained. Thus we expect to reach different comparisons between DCT-1 and DCT-2 on grayscale and CFA images. Figure 12.9 shows the average PSNR as a function of sample entropy over the Kodak dataset for different block size N_1. Results are shown separately for the grayscale images and the Bayer CFA images, and for the DCT-1 and DCT-2 bases. We can make a few observations. Generally the PSNR increases with increasing blocksize, but with diminishing returns. The blocksize of 32 is best for all cases except the DCT-2 on the grayscale image where $N_1 = 16$ is best, by a small margin. The effect of blocksize is more pronounced on the Bayer CFA images

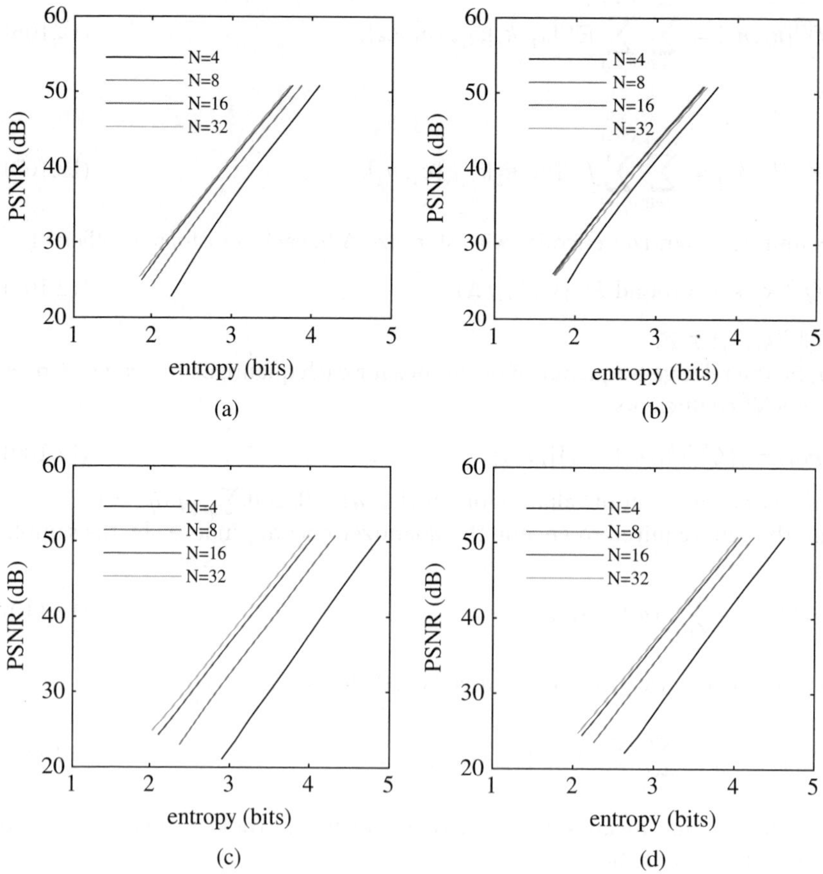

Figure 12.9 Curves of PSNR versus sample entropy over the Kodak dataset for different blocksize N_1. (a) DCT-1, grayscale images; (b) DCT-2, grayscale images; (c) DCT-1, Bayer CFA images; (d) DCT-2, Bayer CFA images. (*See color plate section for the color representation of this figure.*)

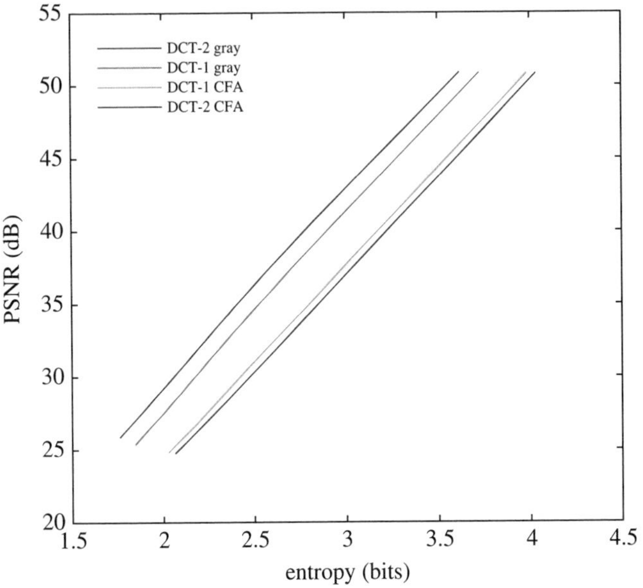

Figure 12.10 Curves of PSNR versus sample entropy over the Kodak dataset for DCT-1 and DCT-2 basis, applied to grayscale images and Bayer CFA images. The legend identifies the curves in decreasing order of PSNR. (*See color plate section for the color representation of this figure.*)

than on the grayscale images. We will use $N_1 = 32$ for the comparison between DCT-1 and DCT-2 on the two datasets.

Figure 12.10 shows a comparison of the DCT-1 and the DCT-2 with a blocksize of $N_1 = 32$ for both the grayscale images and the Bayer CFA images. As expected the PSNR is significantly higher for the grayscale images than for the Bayer CFA images for a given entropy. We do see, as alluded to above, that the DCT-2 is superior to the DCT-1 for grayscale images, but the DCT-1 is superior to the DCT-2 for CFA images. This shows that the best choice of basis can be related to the symmetry properties of the signal to be represented. □

13

Lattices

13.1 Introduction

This chapter gathers in one place the main definitions and results on lattices used throughout this book. Proofs of many results are also provided here. Lattices are used in many fields including the geometry of numbers [Cassels (1997)], crystallography [Miller (1972)], communications [Wübben *et al.* (2011)], cryptography [Micciancio and Goldwasser (2002)], vector quantization [Gersho and Gray (1992)], and in our domain of interest, the sampling of multidimensional signals. Some of these applications involve a relatively high number of dimensions, whereas in multidimensional sampling we generally deal with two–four dimensions. Thus, the choice of properties and algorithms to present here is guided by our application domain, where complexity issues in a high number of dimensions do not really concern us.

13.2 Basic Definitions

A lattice is a regular, discrete set of points in a D-dimensional Euclidean space $\mathbb{R}^D$. It is discrete in the sense that it is composed of discrete isolated points in $\mathbb{R}^D$, and it is regular in the sense that the neighborhood of every lattice point looks the same. We assume that $\mathbb{R}^D$ is equipped with an orthonormal basis, typically corresponding to x, y, z, and t axes, as needed. We seldom go above four dimensions in this work, although most results apply to arbitrary D. An element of $\mathbb{R}^D$, denoted $\mathbf{x}$, can be represented by a $D \times 1$ column matrix of its coordinates with respect to the chosen orthonormal basis, and the same symbol is used to denote this representation, e.g., $\mathbf{x} = [\, x \;\; y \;\; t \,]^T$ in $\mathbb{R}^3$.

Although the following can be derived from other defining properties, we use it here as the definition of a lattice.

Definition 13.1 A lattice Λ in D dimensions is the set of all linear combinations with *integer* coefficients of D linearly independent vectors in $\mathbb{R}^D$ (called basis vectors of the lattice),

$$\Lambda = \{n_1\mathbf{v}_1 + \cdots + n_D\mathbf{v}_D \mid n_i \in \mathbb{Z}\}, \tag{13.1}$$

where $\mathbb{Z}$ is the set of integers.

Multidimensional Signal and Color Image Processing Using Lattices, First Edition. Eric Dubois.

Companion website: www.wiley.com/go/Dubois/multiSP

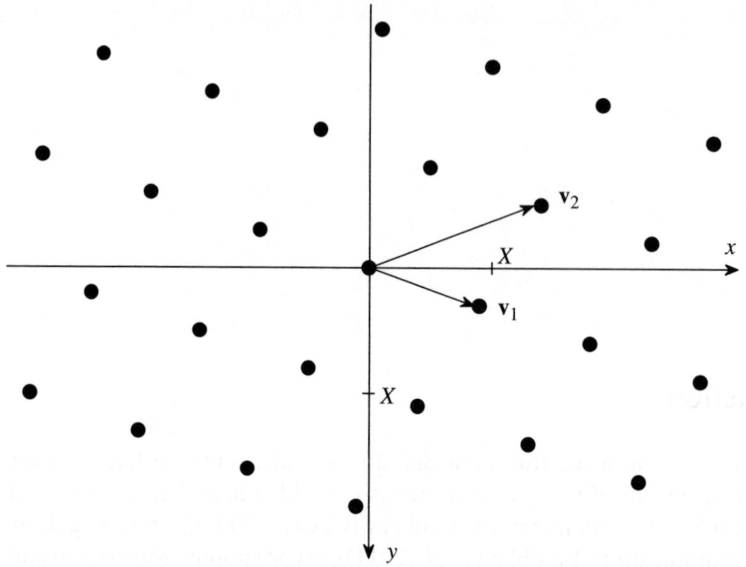

Figure 13.1 Portion of a lattice in two dimensions with basis vectors $\mathbf{v}_1$ and $\mathbf{v}_2$.

For our purposes, D will generally be one, two, three or four dimensions. Figure 13.1 shows an example of an arbitrary lattice in two dimensions, with basis vectors $\mathbf{v}_1 = [0.9X\ 0.3X]^T$ and $\mathbf{v}_2 = [1.4X\ -0.5X]^T$. As is the convention in this book, the y-axis is oriented downward. X is some arbitrary reference unit of distance.

It is possible to define a lattice in $\mathbb{R}^D$ with $K < D$ basis vectors. In this case, K is called the *rank* of the lattice, which lies within the K-dimensional subspace of $\mathbb{R}^D$ spanned by these basis vectors. This can have some applications in multidimensional sampling, for example to describe the vertical-temporal scanning structure in analog television. However, this potential application does not warrant the extra complexity of handling this case throughout, so that we will always assume unless otherwise specified that $K = D$. Since the vectors $\mathbf{v}_1, \dots, \mathbf{v}_D$ are linearly independent, they also form a basis for $\mathbb{R}^D$.

A convenient way to represent a lattice is the *sampling matrix*

$$\mathbf{V} = [\mathbf{v}_1 \mid \mathbf{v}_2 \mid \cdots \mid \mathbf{v}_D] \tag{13.2}$$

whose columns are the basis vectors $\mathbf{v}_i$ represented as column matrices. We denote the lattice Λ determined by any $D \times D$ nonsingular matrix $\mathbf{V}$ by $\Lambda = \mathrm{LAT}(\mathbf{V})$. In terms of the sampling matrix, the lattice can be expressed as

$$\Lambda = \mathrm{LAT}(\mathbf{V}) = \{\mathbf{V}\mathbf{n} \mid \mathbf{n} \in \mathbb{Z}^D\}. \tag{13.3}$$

The sampling matrix for the lattice of Figure 13.1 with respect to the given basis vectors is

$$\mathbf{V} = \begin{bmatrix} 0.9X & 1.4X \\ 0.3X & -0.5X \end{bmatrix}. \tag{13.4}$$

We can thus write

$$\Lambda = \mathrm{LAT}(\mathbf{V}) = \left\{ \begin{bmatrix} (0.9n_1 + 1.4n_2)X \\ (0.3n_1 - 0.5n_2)X \end{bmatrix} \mid n_1, n_2 \in \mathbb{Z} \right\}. \tag{13.5}$$

The basis or sampling matrix for any given lattice is not unique. For example, we can easily verify by inspection that the sampling matrix

$$\mathbf{V}_1 = \begin{bmatrix} 0.9X & 0.5X \\ 0.3X & -0.8X \end{bmatrix}$$

also generates the lattice of Figure 13.1. We will show that LAT(**V**) = LAT(**VE**) where **E** is any integer matrix with det(**E**) = ±1. In the above example, $\mathbf{V}_1 = \mathbf{VE}$ where

$$\mathbf{E} = \mathbf{V}^{-1}\mathbf{V}_1 = \begin{bmatrix} 1 & -1 \\ 0 & 1 \end{bmatrix}$$

and det(**E**) = 1.

Definition 13.2 A nonsingular integer matrix **E** is said to be *unimodular* if its inverse is also an integer matrix. Equivalently, an integer matrix **E** is unimodular if det(**E**) = ±1.

If **E** is unimodular, then $\mathbf{E}^{-1}$ is also unimodular.

Theorem 13.1 Given two sampling matrices $\mathbf{V}_1$ and $\mathbf{V}_2$, then $\text{LAT}(\mathbf{V}_1) = \text{LAT}(\mathbf{V}_2)$ if and only if $\mathbf{V}_2^{-1}\mathbf{V}_1$ is unimodular.

Proof: Suppose that $\mathbf{E} = \mathbf{V}_2^{-1}\mathbf{V}_1$ is unimodular. Thus we can write $\mathbf{V}_1 = \mathbf{V}_2\mathbf{E}$ and also $\mathbf{V}_2 = \mathbf{V}_1\mathbf{E}^{-1}$. Then

$$\begin{aligned} \text{LAT}(\mathbf{V}_1) &= \{\mathbf{V}_2\mathbf{E}\mathbf{n} \mid \mathbf{n} \in \mathbb{Z}^D\} \\ &\subset \text{LAT}(\mathbf{V}_2). \end{aligned}$$

Since **E** is an integer matrix, $\mathbf{En} \in \mathbb{Z}^D$ and thus $\mathbf{V}_2(\mathbf{En}) \in \text{LAT}(\mathbf{V}_2)$. Similarly,

$$\begin{aligned} \text{LAT}(\mathbf{V}_2) &= \{\mathbf{V}_1\mathbf{E}^{-1}\mathbf{n} \mid \mathbf{n} \in \mathbb{Z}^D\} \\ &\subset \text{LAT}(\mathbf{V}_1). \end{aligned}$$

Thus $\text{LAT}(\mathbf{V}_1) = \text{LAT}(\mathbf{V}_2)$.

Conversely, suppose that $\text{LAT}(\mathbf{V}_1) = \text{LAT}(\mathbf{V}_2)$. Thus, for any $\mathbf{n} \in \mathbb{Z}^D$, $\mathbf{V}_2\mathbf{n} \in \text{LAT}(\mathbf{V}_1)$ and so $\mathbf{V}_2\mathbf{n} = \mathbf{V}_1\mathbf{m}$ for some $\mathbf{m} \in \mathbb{Z}^D$. Thus $\mathbf{V}_1^{-1}\mathbf{V}_2\mathbf{n} \in \mathbb{Z}^D$ for all $\mathbf{n} \in \mathbb{Z}^D$. In particular, suppose that $n_k = 1$ for some k, and $n_i = 0$ for $i \neq k$. Then $\mathbf{V}_1^{-1}\mathbf{V}_2\mathbf{n}$ is the kth column of $\mathbf{V}_1^{-1}\mathbf{V}_2$. Since this is true for all $1 \leq k \leq D$, we conclude that $\mathbf{V}_1^{-1}\mathbf{V}_2$ is an integer matrix. Following the same reasoning in the other direction, we can also conclude that $\mathbf{V}_2^{-1}\mathbf{V}_1$ is an integer matrix. Since $(\mathbf{V}_2^{-1}\mathbf{V}_1)^{-1} = \mathbf{V}_1^{-1}\mathbf{V}_2$, $\mathbf{V}_2^{-1}\mathbf{V}_1$ must be unimodular. □

If $\mathbf{V}_1 = \mathbf{V}_2\mathbf{E}$, where **E** is a unimodular integer matrix, we say that $\mathbf{V}_1$ and $\mathbf{V}_2$ are right equivalent. This can easily be seen to be an equivalence relation, and so the set of all sampling matrices that define a given lattice form an equivalence class with respect to right equivalence. We can choose any member of the equivalence class to serve as sampling matrix according to our convenience.

Definition 13.3 A *unit cell* of a lattice Λ is a set $\mathcal{P} \subset \mathbb{R}^D$ such that the copies of $\mathcal{P}$ translated to each lattice point tile the whole space $\mathbb{R}^D$ without overlap: $(\mathcal{P} + \mathbf{s}_1) \cap (\mathcal{P} + \mathbf{s}_2) = \emptyset$ for $\mathbf{s}_1, \mathbf{s}_2 \in \Lambda$, $\mathbf{s}_1 \neq \mathbf{s}_2$, and $\cup_{\mathbf{s}\in\Lambda}(\mathcal{P} + \mathbf{s}) = \mathbb{R}^D$.

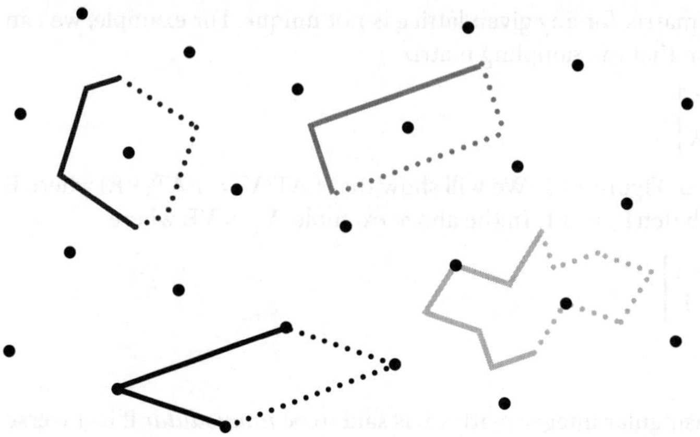

Figure 13.2 Portion of the lattice of Figure 13.1 with different possible unit cells. Clockwise from bottom left: fundamental parallelepiped, Voronoi cell, rectangular unit cell, irregular unit cell. All unit cells have the same area, $0.87X^2$. Solid borders are part of the unit cells, dashed borders are not. Axes are not shown, so that in each case the origin can be chosen as the one lattice point contained in the unit cell. (*See color plate section for the color representation of this figure.*)

For a given $\mathbf{s} \in \mathbb{R}^D$, we denote $\mathcal{P} + \mathbf{s} = \{\mathbf{x} + \mathbf{s} \mid \mathbf{x} \in \mathcal{P}\}$, a shifted copy of $\mathcal{P}$. We will use this type of notation for the shifted copy of a set regularly.

Without loss of generality, we assume that $\mathbf{0} \in \mathcal{P}$, and according to the definition, this can be the only lattice point in $\mathcal{P}$. The unit cell is not unique, as is illustrated in Figure 13.2. A region congruent to $\mathcal{P}$ with the D-dimensional volume (or D-volume) of $\mathcal{P}$ can be associated with each lattice point. Since these tile the space, the D-volume associated with each point must be independent of the specific unit cell. It is a lattice invariant, denoted $\mathrm{d}(\Lambda)$, called the *determinant of the lattice*. The reciprocal $1/\mathrm{d}(\Lambda)$ is called the sampling density in units of samples per unit D-volume. By D-volume, we mean length, area, volume and hypervolume for $D = 1, 2, 3, \geq 4$ respectively. Although unit cells can be of unusual shape, two shapes are of particular interest and utility, the fundamental parallelepiped and the Voronoi unit cell.

Definition 13.4 Given a sampling matrix $\mathbf{V}$ for a lattice $\Lambda = \mathrm{LAT}(\mathbf{V})$, the *fundamental parallelepiped* unit cell is given by

$$\mathcal{P}_P(\mathbf{V}) = \left\{ \sum_{i=1}^{D} a_i \mathbf{v}_i \mid 0 \leq a_i < 1 \right\}. \tag{13.6}$$

This unit cell depends on the choice of basis, so it is not a lattice invariant. An example for the lattice and basis of Figure 13.1 is shown in Figure 13.2.

Definition 13.5 Given a lattice Λ, the interior of the *Voronoi* unit cell consists of all points closer to the origin than to any other lattice point. Points on the boundary, equidistant from two or more lattice points, are assigned by some arbitrary rule such that half of them lie in the Voronoi cell.

The Voronoi cell is independent of the basis, so it is a lattice invariant. The assignment of boundary points to the unit cell is not usually important, unless there are singularities

(Dirac deltas) located on the boundary. The closure of the Voronoi cell is a polytope (i.e. D-dimensional polygon or polyhedron) whose boundary consists of hyperplanes (lines or planes when $D = 2$ or $D = 3$) perpendicularly bisecting lines from the origin to nearest lattice points. Figure 13.2 illustrates.

Theorem 13.2 For a lattice $\Lambda = \mathrm{LAT}(\mathbf{V})$

$$\mathrm{d}(\Lambda) = |\det(\mathbf{V})| \tag{13.7}$$

Proof: Since $\mathrm{d}(\Lambda)$ is the volume of any unit cell, we choose the fundamental parallelepiped $\mathcal{P}_P(\mathbf{V})$ as the easiest one for which we can find the D-volume, specifically

$$\mathrm{d}(\Lambda) = \int_{\mathcal{P}_P(\mathbf{V})} \mathrm{d}\mathbf{x}. \tag{13.8}$$

By defining $\mathbf{x} = \sum_{i=1}^{D} a_i \mathbf{v}_i = \mathbf{V}\mathbf{a}$, the Jacobian of the transformation is $\mathbf{J} = \det(\mathbf{V})$. Using the standard formula for change of variables in a multiple integral (e.g., section 4.8 Kaplan (1984))

$$\mathrm{d}(\Lambda) = \int_0^1 \cdots \int_0^1 |\det(\mathbf{V})|\, da_1 \cdots da_D = |\det(\mathbf{V})|. \tag{13.9}$$

□

This is independent of the choice of sampling matrix for the lattice, since any other sampling matrix has the form $\mathbf{V}_1 = \mathbf{VE}$ where $|\det(\mathbf{E})| = 1$. From the elementary properties of determinants,

$$|\det(\mathbf{V}_1)| = |\det(\mathbf{V})||\det(\mathbf{E})| = \mathrm{d}(\Lambda). \tag{13.10}$$

13.3 Properties of Lattices

A lattice has the algebraic structure of an additive Abelian group. The axioms of a group are easily seen to hold from the definition of a lattice and the properties of real numbers. For any $\mathbf{x}, \mathbf{y}, \mathbf{z} \in \Lambda$ we have:

$$\begin{aligned}
&\mathbf{x} + \mathbf{y} \in \Lambda && \text{closure} \\
&(\mathbf{x} + \mathbf{y}) + \mathbf{z} = \mathbf{x} + (\mathbf{y} + \mathbf{z}) && \text{associativity} \\
&\mathbf{x} + \mathbf{y} = \mathbf{y} + \mathbf{x} && \text{commutativity} \\
&\mathbf{0} \in \Lambda, \quad \mathbf{x} + \mathbf{0} = \mathbf{x} && \text{neutral element} \\
&-\mathbf{x} \in \Lambda, \quad (-\mathbf{x}) + \mathbf{x} = \mathbf{0} && \text{additive inverse.}
\end{aligned}$$

As a group, Λ is isomorphic to $\mathbb{Z}^D$, which can be thought of as the canonical D-dimensional lattice.

From these properties, we see that a lattice is invariant under translation by any element of the lattice:

$$\text{if } \mathbf{d} \in \Lambda, \text{then } \Lambda + \mathbf{d} = \Lambda, \text{where } \Lambda + \mathbf{d} = \{\mathbf{x} + \mathbf{d} \mid \mathbf{x} \in \Lambda\}. \tag{13.11}$$

Similarly, any lattice is invariant to inversion about the origin:

$$-\Lambda = \Lambda, \text{where } -\Lambda = \{-\mathbf{x} \mid \mathbf{x} \in \Lambda\}. \tag{13.12}$$

13.4 Reciprocal Lattice

The *reciprocal lattice* (sometimes called the dual or polar lattice) plays an important role in the frequency domain representation of signals defined on a lattice throughout this book, but it arises in many other contexts. In the case of a sinusoidal signal $\exp(j2\pi\mathbf{u}\cdot\mathbf{x})$, where $\mathbf{u} \in \mathbb{R}^D$ denotes the D-dimensional frequency vector and $\mathbf{u}\cdot\mathbf{x}$ denotes $\mathbf{u}^T\mathbf{x}$, we see that the sinusoids of frequencies $\mathbf{u}$ and $\mathbf{u}+\mathbf{r}$ are identical if $\mathbf{r}\cdot\mathbf{x}$ is an integer for all $\mathbf{x} \in \Lambda$. If we denote the set of all such points as Λ^*, the following theorem shows that Λ^* is in fact a lattice as well, and that is what we call the reciprocal lattice.

Theorem 13.3 If $\Lambda = \mathrm{LAT}(\mathbf{V})$, then

$$\Lambda^* = \{\mathbf{r} \in \mathbb{R}^D \mid \mathbf{r}\cdot\mathbf{x} \in \mathbb{Z} \quad \text{for all} \quad \mathbf{x} \in \Lambda\} = \mathrm{LAT}(\mathbf{V}^{-T}). \tag{13.13}$$

Proof: Let $\mathcal{R} = \{\mathbf{r} \mid \mathbf{r}\cdot\mathbf{x} \in \mathbb{Z} \quad \text{for all} \quad \mathbf{x} \in \Lambda\}$. Suppose that $\mathbf{r} \in \mathrm{LAT}(\mathbf{V}^{-T})$. Then $\mathbf{r} = \mathbf{V}^{-T}\mathbf{k}$ for some integer vector $\mathbf{k} \in \mathbb{Z}^D$. Thus,

$$\mathbf{r}\cdot\mathbf{x} = \mathbf{k}^T\mathbf{V}^{-1}\mathbf{V}\mathbf{n} = \mathbf{k}^T\mathbf{n} \in \mathbb{Z}$$

so that $\mathbf{r} \in \mathcal{R}$ and we can conclude that $\mathrm{LAT}(\mathbf{V}^{-T}) \subset \mathcal{R}$.

Now suppose that $\mathbf{r} \in \mathcal{R}$. Let $\mathbf{a} = \mathbf{V}^T\mathbf{r}$. Then

$$\mathbf{r}\cdot\mathbf{x} = \mathbf{a}^T\mathbf{V}^{-1}\mathbf{V}\mathbf{n} = \mathbf{a}^T\mathbf{n}.$$

This must be an integer for all possible $\mathbf{x} \in \Lambda$, and therefore for all possible $\mathbf{n} \in \mathbb{Z}^D$. In particular, if $\mathbf{n}$ is 1 in position k and zero elsewhere, then it follows that $a_k \in \mathbb{Z}$ for any k, i.e. $\mathbf{a} \in \mathbb{Z}^D$ and so $\mathbf{r} \in \mathrm{LAT}(\mathbf{V}^{-T})$. Thus $\mathcal{R} \subset \mathrm{LAT}(\mathbf{V}^{-T})$, and combining with the first result, $\mathcal{R} = \mathrm{LAT}(\mathbf{V}^{-T})$. □

Note that

$$\begin{aligned} \mathrm{d}(\Lambda^*) &= |\det(\mathbf{V}^{-T})| = \frac{1}{|\det(\mathbf{V})|} \\ &= \frac{1}{\mathrm{d}(\Lambda)} \end{aligned}$$

using standard properties of determinants. Thus, the density of the reciprocal lattice is the reciprocal of the density of the original lattice.

If we denote $\mathbf{U} = \mathbf{V}^{-T}$, then $\mathbf{U}^T\mathbf{V} = \mathbf{I}_D$. Thus, the corresponding basis vectors of the reciprocal lattice $\mathbf{u}_1, \dots, \mathbf{u}_D$ (the columns of $\mathbf{U}$) satisfy $\mathbf{u}_i^T\mathbf{v}_k = \delta_{ik}$. With respect to the basis $\mathbf{v}_1, \dots, \mathbf{v}_D$ of $\mathbb{R}^D$, the $\mathbf{u}_i$ form the *reciprocal basis* in the usual sense of vector spaces.

Continuing with the example of Figure 13.1, the reciprocal lattice is defined by

$$\mathbf{V}^{-T} = \frac{1}{X}\begin{bmatrix} 0.5747 & 0.3448 \\ 1.6092 & -1.0345 \end{bmatrix} \tag{13.14}$$

and $\mathrm{d}(\Lambda^*) = 1/\mathrm{d}(\Lambda) = 1.1494/X^2$. This reciprocal lattice is illustrated in Figure 13.3.

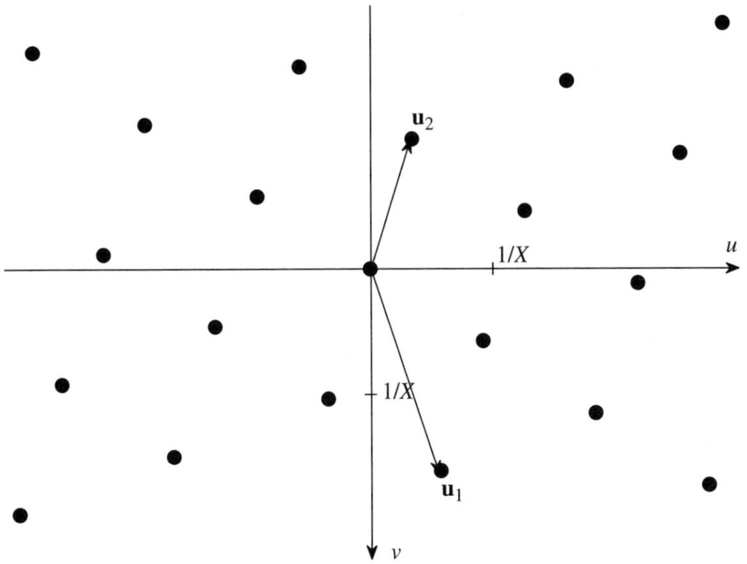

Figure 13.3 Portion of the reciprocal lattice for the lattice in Figure 13.1, with the reciprocal basis vectors.

13.5 Sublattices

Definition 13.6 Γ is a *sublattice* of Λ if both Λ and Γ are lattices in $\mathbb{R}^D$, and every point of Γ is also a point of Λ. We write $\Gamma \subset \Lambda$. If Γ is a sublattice of Λ, then Λ is a *superlattice* of Γ.

It is generally easy to observe if a lattice is a sublattice of another lattice by sketching the points of the two lattices on a common set of axes. However, the following shows how to determine this numerically without having to sketch the lattices, which is particularly helpful in three or more dimensions.

Theorem 13.4 The lattice $\Gamma = \text{LAT}(\mathbf{V}_\Gamma)$ is a sublattice of $\Lambda = \text{LAT}(\mathbf{V}_\Lambda)$ if and only if $\mathbf{V}_\Lambda^{-1}\mathbf{V}_\Gamma$ is an integer matrix.

Proof: Assume that $\Gamma \subset \Lambda$. We can write $\Gamma = \{\mathbf{V}_\Gamma \mathbf{n} \mid \mathbf{n} \in \mathbb{Z}^D\}$. Then for any $\mathbf{n} \in \mathbb{Z}^D$, $\mathbf{V}_\Gamma \mathbf{n} = \mathbf{V}_\Lambda \mathbf{m}$ for some integer vector $\mathbf{m} \in \mathbb{Z}^D$. Thus $\mathbf{V}_\Lambda^{-1}\mathbf{V}_\Gamma \mathbf{n}$ is an integer vector for any $\mathbf{n} \in \mathbb{Z}^D$. In particular, if $\mathbf{n}$ consists of 1 in position k and 0 elsewhere, then $\mathbf{V}_\Lambda^{-1}\mathbf{V}_\Gamma \mathbf{n}$ is the kth column of $\mathbf{V}_\Lambda^{-1}\mathbf{V}_\Gamma$ which must be an integer vector. Since this holds for all k from 1 to D, all columns of $\mathbf{V}_\Lambda^{-1}\mathbf{V}_\Gamma$ are integer, and so $\mathbf{V}_\Lambda^{-1}\mathbf{V}_\Gamma$ is an integer matrix.

Conversely, if $\mathbf{V}_\Lambda^{-1}\mathbf{V}_\Gamma$ is an integer matrix $\mathbf{M}$, then $\mathbf{V}_\Gamma = \mathbf{V}_\Lambda \mathbf{M}$. It follows that $\mathbf{V}_\Gamma \mathbf{n} = \mathbf{V}_\Lambda \mathbf{M}\mathbf{n} = \mathbf{V}_\Lambda \mathbf{m} \in \Lambda$ for any integer vector $\mathbf{n}$. Thus $\Gamma \subset \Lambda$. □

Corollary 13.1 If $\Gamma \subset \Lambda$, then $d(\Gamma)$ is an integer multiple of $d(\Lambda)$.

Proof: If $\Gamma \subset \Lambda$, then $\mathbf{V}_\Gamma = \mathbf{V}_\Lambda \mathbf{M}$ for some integer matrix $\mathbf{M}$. Then

$$\begin{aligned} d(\Gamma) &= |\det(\mathbf{V}_\Gamma)| = |\det(\mathbf{V}_\Lambda \mathbf{M})| \\ &= |\det(\mathbf{V}_\Lambda)||\det(\mathbf{M})| = |\det(\mathbf{M})| d(\Lambda) \end{aligned} \tag{13.15}$$

and the determinant of any integer matrix is an integer. □

The ratio $d(\Gamma)/d(\Lambda)$ is called the index of Γ in Λ. For any two lattices Λ and Γ, by examining $d(\Lambda)$ and $d(\Gamma)$, we can immediately see which sublattice relation is possible, if any.

Note the similarity of this proof with that of Theorem 13.1. In fact, Theorem 13.1 follows simply from Theorem 13.4, since $\Lambda = \Gamma$ if and only if $\Lambda \subset \Gamma$ and $\Gamma \subset \Lambda$. In this case, both $\mathbf{M} = (\mathbf{V}_\Lambda)^{-1}\mathbf{V}_\Gamma$ and $(\mathbf{V}_\Gamma)^{-1}\mathbf{V}_\Lambda$ are integer matrices. But then $\mathbf{M}^{-1} = (\mathbf{V}_\Gamma)^{-1}\mathbf{V}_\Lambda$ is an integer matrix, so $\mathbf{M}$ must be unimodular, and $|\det(\mathbf{M})| = 1$.

Theorem 13.5 If $\Gamma \subset \Lambda$, then $\Lambda^* \subset \Gamma^*$.

Proof: Suppose that $\mathbf{k} \in \Lambda^*$. Then, by definition, $\mathbf{k} \cdot \mathbf{x} \in \mathbb{Z}$ for all $\mathbf{x} \in \Lambda$. Since, $\Gamma \subset \Lambda$, it follows that $\mathbf{k} \cdot \mathbf{x} \in \mathbb{Z}$ for all $\mathbf{x} \in \Gamma$. Thus $\mathbf{k} \in \Gamma^*$ and so $\Lambda^* \subset \Gamma^*$. □

Alternatively, if $\mathbf{V}_\Gamma = \mathbf{V}_\Lambda \mathbf{M}$ for integer matrix $\mathbf{M}$, then $(\mathbf{V}_\Lambda)^{-T} = (\mathbf{V}_\Gamma)^{-T}\mathbf{M}^T$, where $\mathbf{M}^T$ is also an integer matrix, and the result follows from Theorem 13.4.

Suppose that Λ is the lattice of Figure 13.1 with $\mathbf{V}_\Lambda = \begin{bmatrix} 0.9X & 1.4X \\ 0.3X & -0.5X \end{bmatrix}$, and consider the lattice Γ with sampling matrix

$$\mathbf{V}_\Gamma = \begin{bmatrix} 0.4X & 1.4X \\ 1.1X & -0.5X \end{bmatrix}. \tag{13.16}$$

We note that $d(\Gamma) = 1.74X^2 = 2d(\Lambda)$, so it is possible for Γ to be a sublattice of Λ. Applying the test of Theorem 13.4,

$$\mathbf{V}_\Lambda^{-1}\mathbf{V}_\Gamma = \begin{bmatrix} 2 & 0 \\ -1 & 1 \end{bmatrix} \tag{13.17}$$

is an integer matrix, so Γ is indeed a sublattice of Λ. The two lattices are illustrated in Figure 13.4.

13.6 Cosets and the Quotient Group

Let $\Gamma \subset \Lambda$, with $d(\Gamma)/d(\Lambda) = K$. Essentially, there are K elements of Λ for every element of Γ, and we want to provide some structure for these. This will be crucial in the study of periodic signals and sampling structure conversion, and the structure of this set is fundamental to the fast Fourier transform algorithm.

Definition 13.7 Let $\mathbf{c}$ be any element of Λ, and let Γ be a sublattice of Λ. The set

$$\mathbf{c} + \Gamma = \{\mathbf{c} + \mathbf{x} \mid \mathbf{x} \in \Gamma\} \tag{13.18}$$

is called a *coset* of Γ in Λ.

The term coset comes from group theory, and the alternate term *class* is sometimes used. We define the following equivalence relation on Λ: $\mathbf{x} \sim \mathbf{y}$ if $\mathbf{x} - \mathbf{y} \in \Gamma$. We see

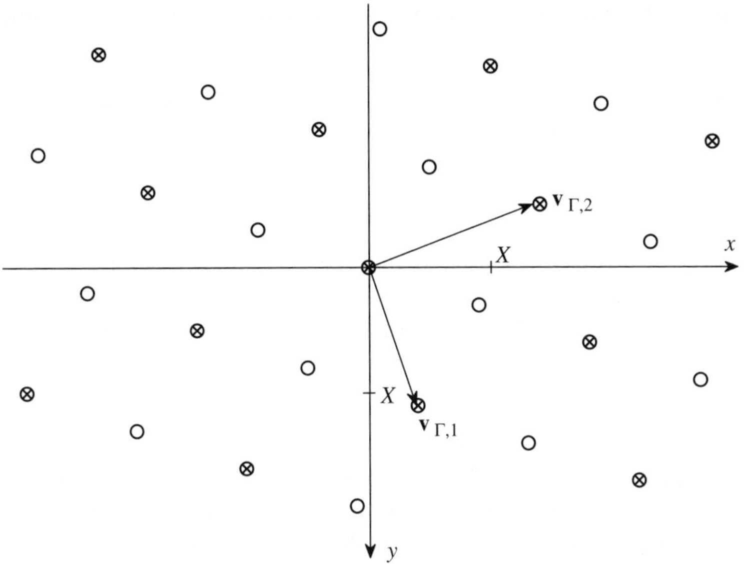

Figure 13.4 Portion of the lattices Λ (O) and $\Gamma(\times)$, showing that $\Gamma \subset \Lambda$.

that any two elements of Λ are equivalent if and only if they belong to the same coset. The cosets are the equivalence classes of the equivalence relation, and they partition Λ into disjoint cosets. Each coset can be specified by an arbitrary element called a coset representative. This section is mainly concerned with how we can enumerate the coset representatives, and the algebraic structure of the set of cosets. The following theorem shows one way to enumerate the coset representatives, and shows that there are indeed K cosets.

Theorem 13.6 Let Λ be a lattice and let $\Gamma = \mathrm{LAT}(\mathbf{V}_\Gamma)$ be a sublattice. A set of representatives for the cosets of Γ in Λ is the set of elements of Λ contained in a fundamental parallelepiped unit cell of Γ, and the number of cosets is $K = \mathrm{d}(\Gamma)/\mathrm{d}(\Lambda)$.

Proof: Let $\mathcal{P}_P(\mathbf{V}_\Gamma)$ be the unit cell of Γ. By definition of the unit cell, no two points of Λ within $\mathcal{P}_P(\mathbf{V}_\Gamma)$ can belong to the same coset. Let $\mathbf{x}$ be any point of Λ and express it as a linear combination of the basis vectors of Γ, which we can do since they also form a basis for $\mathbb{R}^D$:

$$\mathbf{x} = \sum_{i=1}^{D} a_i \mathbf{v}_{i\Gamma}. \tag{13.19}$$

Since $\mathbf{x}$ is not necessarily in Γ, the a_i are not necessarily integers. Let

$$\lfloor \mathbf{x} \rfloor = \sum_{i=1}^{D} \lfloor a_i \rfloor \mathbf{v}_{i\Gamma}, \tag{13.20}$$

where $\lfloor a_i \rfloor$ is the largest integer less than or equal to a_i. Then $\lfloor \mathbf{x} \rfloor \in \Gamma$. Also $\mathbf{x} - \lfloor \mathbf{x} \rfloor \in \mathcal{P}_P(\mathbf{V}_\Gamma) \cap \Lambda$, and is in the same coset as $\mathbf{x}$. Thus every $\mathbf{x} \in \Lambda$ can be put into one-to-one correspondence with a point of $\mathcal{P}_P(\mathbf{V}_\Gamma) \cap \Lambda$ belonging to the same coset, so these can form a set of coset representatives. Clearly the number is finite. In fact the number must be the ratio of the volume of $\mathcal{P}_P(\mathbf{V}_\Gamma)$ to the volume of any unit cell of Λ, i.e. $d(\Gamma)/d(\Lambda) = K$. □

Let us denote a set of coset representatives as $\mathbf{b}_0, \mathbf{b}_1, \dots, \mathbf{b}_{K-1}$. Then, the expression of Λ as a disjoint union of the cosets of Γ in Λ is

$$\Lambda = \bigcup_{k=0}^{K-1} (\mathbf{b}_k + \Gamma). \tag{13.21}$$

To simplify the notation, for any $\mathbf{x} \in \Lambda$, we denote the coset containing $\mathbf{x}$ as $\mathbf{x} + \Gamma = [\mathbf{x}]_\Gamma$. Further, when it clear which sublattice is concerned, we write simply $[\mathbf{x}]$.

The set of cosets has the structure of an Abelian group; it is called the quotient group and denoted Λ/Γ. The addition operation is defined by

$$[\mathbf{x}]_\Gamma + [\mathbf{y}]_\Gamma = [\mathbf{x} + \mathbf{y}]_\Gamma, \tag{13.22}$$

which is well defined and independent of the specific choices of $\mathbf{x}$ and $\mathbf{y}$ within the respective cosets. The neutral element is $[0]_\Gamma = \Gamma$ and the additive inverse of $[\mathbf{x}]_\Gamma$ is $[-\mathbf{x}]_\Gamma$. This group is the domain for Γ-periodic signals on Λ, and it is the structure of the group that leads to many of the fast Fourier transform algorithms. In this chapter, we will derive the precise structure of this quotient group.

As an example, let

$$\Lambda = \mathrm{LAT}\left(\begin{bmatrix} 2 & -1 \\ 0 & 1 \end{bmatrix}\right) \qquad \Gamma = \mathrm{LAT}\left(\begin{bmatrix} 10 & 0 \\ 0 & 6 \end{bmatrix}\right). \tag{13.23}$$

We see that

$$\mathbf{M} = \mathbf{V}_\Lambda^{-1}\mathbf{V}_\Gamma = \begin{bmatrix} 5 & 3 \\ 0 & 6 \end{bmatrix} \tag{13.24}$$

is an integer matrix, and thus $\Gamma \subset \Lambda$. In this example, $d(\Lambda) = 2$, $d(\Gamma) = 60$ and so $K = 30$, and we need to identify 30 coset representatives. Figure 13.5 shows the two lattices and the fundamental parallelepiped unit cell of Γ. The 30 coset representatives within the unit cell are identified arbitrarily from 0 to 29, in row-by-row order.

Note that we could also identify the coset representatives with two indexes as follows

$$\mathbf{b}_{i\ell} = \begin{bmatrix} 2i + (\ell \bmod 2) \\ \ell \end{bmatrix}, \qquad 0 \le i \le 4, 0 \le \ell \le 5. \tag{13.25}$$

With this set of coset representatives, the group structure of Λ/Γ is not at all evident. We will show how to use basis transformations to remedy this situation.

13.7 Basis Transformations

Problems involving sublattices such as down- or upsampling and representation of periodic signals are greatly simplified when the corresponding bases satisfy specific relationships. These are found mainly by using the Hermite normal form and the Smith normal form. Both are constructed systematically by using elementary column and row operations, such that the matrix $\mathbf{M}$ relating the two sampling matrices is in upper triangular or diagonal form respectively.

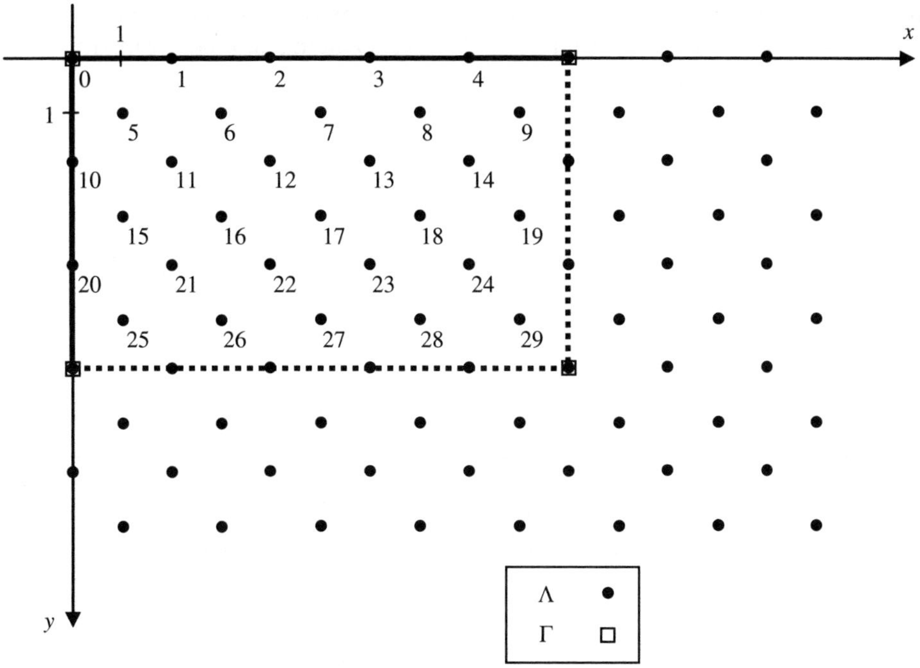

Figure 13.5 Portion of lattices Λ and Γ showing a rectangular unit cell of Γ and the corresponding coset representatives. The solid edges are included in the unit cell, while the dashed edges are not.

13.7.1 Elementary Column Operations

Let $\mathbf{V} = [\,\mathbf{v}_1 \;\cdots\; \mathbf{v}_D\,]$ be the sampling matrix for a lattice $\Lambda = \text{LAT}(\mathbf{V})$. *Elementary column operations* effect a change of basis corresponding to certain elementary unimodular integer matrices. Specifically, these are

(i) Exchange of any two columns i and k of $\mathbf{V}$: $\mathbf{v}'_i = \mathbf{v}_k$, $\mathbf{v}'_k = \mathbf{v}_i$, $\mathbf{v}'_l = \mathbf{v}_l$ for $l \neq i, k$. This is just a reordering of the basis vectors. In matrix notation, we denote the corresponding unimodular matrix transformation by $\mathbf{V}' = \mathbf{V}\mathbf{E}_{i\leftrightarrow k}$ where

$$[\mathbf{E}_{i\leftrightarrow k}]_{lm} = \begin{cases} 1 & l = m, l \neq i, l \neq k \\ 1 & l = i, m = k \\ 1 & l = k, m = i \\ 0 & \text{otherwise} \end{cases} . \tag{13.26}$$

(ii) Multiplication of any column k by -1: $\mathbf{v}'_k = -\mathbf{v}_k$, $\mathbf{v}'_l = \mathbf{v}_l$ for $l \neq k$. The corresponding unimodular matrix is

$$[\mathbf{E}_{-k}]_{lm} = \begin{cases} 1 & l = m, l \neq k \\ -1 & l = m = k \\ 0 & \text{otherwise} \end{cases} . \tag{13.27}$$

(iii) Addition of a times column k to column i, for any two columns i and k, and for $a \in \mathbb{Z}$: $\mathbf{v}_i' = \mathbf{v}_i + a\mathbf{v}_k$, $\mathbf{v}_l' = \mathbf{v}_l$ for $l \neq i$. The corresponding unimodular matrix is

$$[\mathbf{E}_{i+ak}]_{lm} = \begin{cases} 1 & l = m \\ a & l = k, m = i \, . \\ 0 & \text{otherwise} \end{cases} \tag{13.28}$$

The inverse of an elementary column operation is again an elementary column operation. In fact, the first two types are their own inverses, whereas $\mathbf{E}_{i+ak}^{-1} = \mathbf{E}_{i+(-a)k}$.

13.7.2 Hermite Normal Form

Suppose that $\text{LAT}(\mathbf{V}_\Gamma) \subset \text{LAT}(\mathbf{V}_\Lambda)$, so that $\mathbf{V}_\Gamma = \mathbf{V}_\Lambda \mathbf{M}$ for some nonsingular integer matrix $\mathbf{M}$. If we apply a change of basis to Γ via some unimodular matrix $\mathbf{E}$, then $\mathbf{V}_\Gamma' = \mathbf{V}_\Gamma \mathbf{E}$. Then $\mathbf{V}_\Gamma' = \mathbf{V}_\Lambda(\mathbf{ME}) = \mathbf{V}_\Lambda \mathbf{M}'$, so that the integer matrix $\mathbf{M}$ undergoes the same unimodular transformation. This allows us to find a new basis for Γ such that $\mathbf{M}$ is in a canonical upper triangular form known as the Hermite normal form.

Definition 13.8 An integer matrix $\mathbf{M} = [m_{ij}]$ is in *Hermite normal form* if

(i) $\mathbf{M}$ is upper triangular, $m_{ij} = 0$ if $i > j$.
(ii) Diagonal elements are positive, $m_{ii} > 0$ for $i = 1, 2, \ldots, D$.
(iii) Elements to the right of the diagonal in each row satisfy $0 \leq m_{ij} < m_{ii}$ for $j > i$.

Such a matrix must be nonsingular, since its determinant is equal to the product of the positive diagonal elements $\det(\mathbf{M}) = \prod_{i=1}^{D} m_{ii} > 0$. Note that more general definitions exist for singular or nonsquare matrices, but we are only concerned here with nonsingular matrices. An equivalent definition also exists for the lower-triangular form of the Hermite normal form.

Theorem 13.7 Every nonsingular integer matrix $\mathbf{M}$ is right equivalent to a *unique* matrix in Hermite normal form.

This is a standard result. A proof can be found in Newman (1972), and proofs in the form of algorithms can be found in Cohen (1993) and Holt *et al.* (2005).

In our previous example of Figure 13.4, $\mathbf{M} = \left[\begin{smallmatrix} 2 & 0 \\ -1 & 1 \end{smallmatrix}\right]$ is not in Hermite normal form. By choosing

$$\mathbf{V}_\Gamma' = \begin{bmatrix} 1.8X & 1.4X \\ 0.6X & -0.5X \end{bmatrix} = \mathbf{V}_\Gamma \begin{bmatrix} 1 & 0 \\ 1 & 1 \end{bmatrix} \tag{13.29}$$

we find

$$\mathbf{M}' = \mathbf{ME} = \begin{bmatrix} 2 & 0 \\ -1 & 1 \end{bmatrix} \begin{bmatrix} 1 & 0 \\ 1 & 1 \end{bmatrix} = \begin{bmatrix} 2 & 0 \\ 0 & 1 \end{bmatrix} \tag{13.30}$$

which is in Hermite normal form. In general the transformed matrix would be upper triangular and not diagonal as in the example. The above conversion could be obtained using algorithm 2.4.4 of Cohen (1993), resulting in the following sequence of elementary column operations on $\mathbf{M}$:

$$\mathbf{M}' = \mathbf{M}\mathbf{E}_{1\leftrightarrow 2}\mathbf{E}_{-2}\mathbf{E}_{1+(-1)2}\mathbf{E}_{2+(1)1}$$
$$= \mathbf{M}\begin{bmatrix} 0 & 1 \\ 1 & 0 \end{bmatrix}\begin{bmatrix} 1 & 0 \\ 0 & -1 \end{bmatrix}\begin{bmatrix} 1 & 0 \\ -1 & 1 \end{bmatrix}\begin{bmatrix} 1 & 1 \\ 0 & 1 \end{bmatrix}. \tag{13.31}$$

For the example of Figure 13.5, the matrix $\mathbf{M} = \left[\begin{smallmatrix} 5 & 3 \\ 0 & 6 \end{smallmatrix}\right]$ is already in Hermite normal form, so no further transformation is needed.

We now present two specific applications of the Hermite normal form. The first gives an explicit method to find the coset representative corresponding to any element of Λ and thus a way to implement the operations in Λ/Γ. The second gives a method to systematically enumerate all sublattices Γ of a given lattice Λ with a given index $K = \mathrm{d}(\Gamma)/\mathrm{d}(\Lambda)$.

Suppose that $\Gamma = \mathrm{LAT}(\mathbf{V}_\Gamma)$ is a sublattice of $\Lambda = \mathrm{LAT}(\mathbf{V}_\Lambda)$ and that the bases have been chosen so that $\mathbf{M} = \mathbf{V}_\Lambda^{-1}\mathbf{V}_\Gamma$ is in Hermite normal form. Thus the basis vectors have the form

$$\begin{aligned}
\mathbf{v}_{\Gamma 1} &= m_{11}\mathbf{v}_{\Lambda 1} \\
\mathbf{v}_{\Gamma 2} &= m_{12}\mathbf{v}_{\Lambda 1} + m_{22}\mathbf{v}_{\Lambda 2} \\
&\vdots \\
\mathbf{v}_{\Gamma D} &= m_{1D}\mathbf{v}_{\Lambda 1} + \cdots + m_{DD}\mathbf{v}_{\Lambda D}.
\end{aligned} \tag{13.32}$$

From this, we observe that in $\mathbb{R}^D$,

$$\mathrm{span}(\mathbf{v}_{\Gamma 1}, \dots, \mathbf{v}_{\Gamma k}) = \mathrm{span}(\mathbf{v}_{\Lambda 1}, \dots, \mathbf{v}_{\Lambda k}), \qquad k = 1, 2, \dots, D. \tag{13.33}$$

Because of the upper triangular form, if $\mathbf{x} \in \Lambda$ is expressed

$$\mathbf{x} = \sum_{i=1}^{D} n_i \mathbf{v}_{\Lambda i}, \qquad n_i \in \mathbb{Z}, \tag{13.34}$$

then in terms of the basis of Γ

$$\mathbf{x} = \sum_{i=1}^{D} \frac{n_i}{m_{ii}} \mathbf{v}_{\Gamma i}. \tag{13.35}$$

Our problem is given any $\mathbf{x} \in \Lambda$, to find $\mathbf{y} \in \Lambda \cap \mathcal{P}_P(\mathbf{V}_\Gamma)$ such that $\mathbf{x} - \mathbf{y} \in \Gamma$, or in other words, the unique coset representative of $[\mathbf{x}]_\Gamma$ lying in $\mathcal{P}_P(\mathbf{V}_\Gamma)$ as stated in Theorem 13.6. We do this by successively subtracting a multiple of $\mathbf{v}_{\Gamma k}$ as k runs backward from D to 1 such that the coefficients of $\mathbf{y}$ in the basis $\mathbf{v}_{\Gamma 1}, \dots, \mathbf{v}_{\Gamma D}$ of $\mathbb{R}^D$ are between 0 and 1 for $k \le l \le D$.

Algorithm 13.1 Given $\mathbf{x} \in \Lambda = \mathrm{LAT}(\mathbf{V}_\Lambda)$ and sublattice $\Gamma = \mathrm{LAT}(\mathbf{V}_\Gamma)$ where integer matrix $\mathbf{M} = \mathbf{V}_\Lambda^{-1}\mathbf{V}_\Gamma$ is in Hermite normal form, output the unique $\mathbf{y} \in \Lambda \cap \mathcal{P}_P(\mathbf{V}_\Gamma)$ such that $\mathbf{x} - \mathbf{y} \in \Gamma$.

1) [Initialize] Set $\mathbf{y} \leftarrow \mathbf{x}$ and express $\mathbf{y} = \sum_{i=1}^{D} a_i \mathbf{v}_{\Lambda i}$ for $a_i \in \mathbb{Z}$.
2) [Loop] For $k = D, D-1, \dots, 1$
   $c = \lfloor a_k/m_{kk} \rfloor$
   $\mathbf{y} \leftarrow \mathbf{y} - c\mathbf{v}_{\Gamma k}$.
3) [Finish] Return $\mathbf{y}$.

Since at each stage, we subtract an element of Γ, $\mathbf{y} \in [\mathbf{x}]_\Gamma$ at all times. At iteration k, when we subtract an integer multiple of $\mathbf{v}_{\Gamma k}$, the coefficients $a_{k+1}, \ldots, a_D$ are unchanged. After iteration k, we have $0 \le a_k/m_{kk} < 1$. Thus, on termination,

$$\mathbf{y} = \sum_{i=1}^{D} \frac{a_i}{m_{ii}} \mathbf{v}_{\Gamma i} \in \mathcal{P}_P(\mathbf{V}_\Gamma) \tag{13.36}$$

as desired. This algorithm is adapted from the algorithm of figure 8.1 in Micciancio and Goldwasser (2002). Thus, given elements of $\mathbf{y} \in \Lambda \cap \mathcal{P}_P(\mathbf{V}_\Gamma)$, they can be added, subtracted or negated in Λ and then reduced back to $\mathbf{y} \in \Lambda \cap \mathcal{P}_P(\mathbf{V}_\Gamma)$ using Algorithm 13.1, allowing us to implement all operations in Λ/Γ. However, this still does not reveal the structure of Λ/Γ; this will come with the Smith normal form.

The second application of the Hermite normal form is to enumerate all distinct sublattices Γ of a given lattice $\Lambda = \mathrm{LAT}(\mathbf{V}_\Lambda)$ with index $K = \mathrm{d}(\Gamma)/\mathrm{d}(\Lambda)$. The solution to this problem was presented by Cortelazzo and Manduchi (1993). As we have seen, $\Gamma = \mathrm{LAT}(\mathbf{V}_\Gamma)$ is a sublattice of Λ if and only if $\mathbf{V}_\Gamma = \mathbf{V}_\Lambda \mathbf{M}$ for some nonsingular integer matrix $\mathbf{M}$. Also, $\mathbf{M}_1$ and $\mathbf{M}_2$ describe the same sublattices if they are right associates. Thus, to get distinct sublattices, we want to identify a single representative of each equivalence class. Since every integer matrix is right equivalent to a unique matrix in Hermite normal form, we can use these as the class representatives. We can enumerate all distinct sublattices of index K by enumerating all $D \times D$ integer matrices in Hermite normal form with determinant K. The determinant is the product of the diagonal entries, so we need to find all distinct factorizations of K into D terms placed on the diagonal, and then enumerate all ways to place the entries to the right of the diagonal that satisfy the definition of the Hermite normal form. For example, if $D = 2$ and $K = 4$, we seek all 2×2 integer matrices in Hermite normal form with determinant 4. These are given in Table I of Cortelazzo and Manduchi (1993), and are:

$$\mathbf{M} \in \left\{ \begin{bmatrix} 4 & 0 \\ 0 & 1 \end{bmatrix}, \begin{bmatrix} 4 & 1 \\ 0 & 1 \end{bmatrix}, \begin{bmatrix} 4 & 2 \\ 0 & 1 \end{bmatrix}, \begin{bmatrix} 4 & 3 \\ 0 & 1 \end{bmatrix}, \begin{bmatrix} 2 & 0 \\ 0 & 2 \end{bmatrix}, \begin{bmatrix} 2 & 1 \\ 0 & 2 \end{bmatrix}, \begin{bmatrix} 1 & 0 \\ 0 & 4 \end{bmatrix} \right\}. \tag{13.37}$$

13.8 Smith Normal Form

Although very useful, the Hermite normal form does not fully expose the structure of the quotient group. This is accomplished using the Smith normal form; Newman calls this representation theorem "quite possibly the most important theorem in all of elementary matrix theory" [Newman (1972)]. This representation allows us to find bases for both Λ and Γ such that the corresponding basis vectors are collinear, making operations in Λ/Γ very straightforward.

Definition 13.9 A $D \times D$ integer matrix $\mathbf{D} = [d_{ij}]$ is in *Smith normal form* if

(i) It is diagonal, $d_{ij} = 0$ if $i \ne j$.
(ii) Diagonal elements are positive, $d_{ii} > 0$ for $i = 1, \ldots, D$.
(iii) $d_{i+1,i+1}$ divides d_{ii}, for $i = 1, 2, \ldots, D-1$.

Such a matrix is nonsingular, with $\det(\mathbf{D}) = \prod_{i=1}^{D} d_{ii} > 0$. More general definitions exist for singular or nonsquare matrices, but we are only concerned here with nonsingular square matrices.

Theorem 13.8 For every nonsingular $D \times D$ integer matrix $\mathbf{M}$, there exist unimodular matrices $\mathbf{E}_1$ and $\mathbf{E}_2$ such that $\mathbf{E}_1\mathbf{M}\mathbf{E}_2$ is in Smith normal form.

A proof of this well-known result can be found in Newman (1972), and algorithms can be found in Cohen (1993) and Holt *et al.* (2005). Note that this decomposition is not unique.

Suppose that Γ is a sublattice of Λ, where $\Lambda = \mathrm{LAT}(\mathbf{V}_\Lambda)$, $\Gamma = \mathrm{LAT}(\mathbf{V}_\Gamma)$ and $\mathbf{M} = \mathbf{V}_\Lambda^{-1}\mathbf{V}_\Gamma$ is an integer matrix. Express $\mathbf{M}$ in Smith normal form, so that $\mathbf{E}_1\mathbf{M}\mathbf{E}_2 = \mathbf{D}$, and define new bases for Λ and Γ as follows:

$$\mathbf{V}'_\Lambda = \mathbf{V}_\Lambda\mathbf{E}_1^{-1} \tag{13.38}$$

$$\mathbf{V}'_\Gamma = \mathbf{V}_\Gamma\mathbf{E}_2. \tag{13.39}$$

Then

$$\mathbf{M}' = \mathbf{E}_1\mathbf{V}_\Lambda^{-1}\mathbf{V}_\Gamma\mathbf{E}_2 = \mathbf{D} \tag{13.40}$$

or equivalently $\mathbf{V}'_\Gamma = \mathbf{V}'_\Lambda\mathbf{D}$. In terms of the basis vectors,

$$\mathbf{v}'_{\Gamma i} = d_{ii}\mathbf{v}'_{\Lambda i} \tag{13.41}$$

showing that the corresponding basis vectors are collinear. The coset representatives in a fundamental parallelepiped of $\mathcal{P}_P(\mathbf{V}'_\Gamma)$ are then

$$\mathbf{b}_{k_1,\dots,k_D} = \sum_{l=1}^{D} k_l\mathbf{v}'_{\Lambda l}, \qquad 0 \le k_l \le d_{ll} - 1. \tag{13.42}$$

Finding a coset representative amounts to reducing the coefficient of $\mathbf{v}'_{\Lambda l}$ modulo d_{ll}.

The structure for Λ/Γ that corresponds to this decomposition is

$$\Lambda/\Gamma \approx \mathbb{Z}_{d_{11}} \oplus \mathbb{Z}_{d_{22}} \oplus \cdots \oplus \mathbb{Z}_{d_{DD}} \tag{13.43}$$

where $\mathbb{Z}_d$ is the cyclic group of order d. If any of the $d_{ll} = 1$, the corresponding terms and subscripts are not needed, as in the example to follow. These bases are not necessarily "nice" or intuitive, but can be very useful in establishing certain results, or for computational purposes.

Continuing with the example of Figure 13.5, using the Smith normal form algorithm, we find that

$$\mathbf{M} = \mathbf{V}_\Lambda^{-1}\mathbf{V}_\Gamma = \begin{bmatrix} 5 & 3 \\ 0 & 6 \end{bmatrix} \tag{13.44}$$

$$= \mathbf{E}_1^{-1}\mathbf{D}\mathbf{E}_2^{-1} = \begin{bmatrix} 18 & 19 \\ 17 & 18 \end{bmatrix} \begin{bmatrix} 30 & 0 \\ 0 & 1 \end{bmatrix} \begin{bmatrix} 3 & -2 \\ -85 & 57 \end{bmatrix}. \tag{13.45}$$

The new sampling matrices are

$$\mathbf{V}'_\Lambda = \mathbf{V}_\Lambda\mathbf{E}_1^{-1} = \begin{bmatrix} 19 & 20 \\ 17 & 18 \end{bmatrix} \tag{13.46}$$

$$\mathbf{V}'_\Gamma = \mathbf{V}_\Gamma\mathbf{E}_2 = \begin{bmatrix} 570 & 20 \\ 510 & 18 \end{bmatrix}. \tag{13.47}$$

Since $d_{22} = 1$, we see this is essentially a one-dimensional down-sampling along the basis vector $\mathbf{v}'_{\Lambda 1}$, and the coset representatives can be chosen to be $\mathbf{b}_k = k\mathbf{v}'_{\Lambda 1}$, $k = 0, 1, \dots, 29$. The structure of Λ/Γ is that of $\mathbb{Z}_{30}$.

The situation in the reciprocal lattice domain is the dual of that in the lattice domain. The corresponding sampling matrices for the reciprocal lattices are

$$\mathbf{V}'_{\Lambda^*} = (\mathbf{V}'_{\Lambda})^{-T} = \mathbf{V}_{\Lambda}^{-T}\mathbf{E}_1^T \tag{13.48}$$

$$\mathbf{V}'_{\Gamma^*} = (\mathbf{V}'_{\Gamma})^{-T} = \mathbf{V}_{\Gamma}^{-T}\mathbf{E}_2^{-T} \tag{13.49}$$

and it follows directly that

$$\mathbf{V}'_{\Lambda^*} = \mathbf{V}'_{\Gamma^*}\mathbf{D}. \tag{13.50}$$

Thus, we can choose the coset representatives of Λ^* in Γ^* to be

$$\mathbf{d}_{i_1,i_2,\dots,i_D} = \sum_{l=1}^{D} i_{ll}\mathbf{v}'_{\Gamma^* l}, \quad 0 \leq i_{ll} \leq d_{ll} - 1 \tag{13.51}$$

and the structure of Γ^*/Λ^* is identical to that of Λ/Γ.

13.9 Intersection and Sum of Lattices

In many situations of interest, we deal with signals defined on two lattices, neither of which is a sublattice of the other. In this case, we will need the related concepts of intersection and sum of lattice.

Definition 13.10 The intersection of two lattices Λ_1 and Λ_2, denoted $\Lambda_1 \cap \Lambda_2$, is the set of all points in $\mathbb{R}^D$ that belong to both Λ_1 and Λ_2.

The intersection is a sublattice of both Λ_1 and Λ_2, and so is called a *common sublattice.* Since it contains all common sublattices, it is called the *greatest common sublattice.* It is possible for the intersection of two lattices to consist of only the origin, or to be a lattice of dimension less than D. We are interested in the case where the intersection is a full D-dimensional lattice. Before establishing the condition for this to hold, we introduce an adaptation of the Smith normal form to rational matrices (by which we mean matrices of rational numbers). This adaptation is sometimes called the Smith–MacMillan normal form [Vidyasagar (2011)] although the Smith normal form can be defined for rational matrices [Newman (1972)].

Definition 13.11 A $D \times D$ rational matrix $\mathbf{C} = [c_{ij}]$ is in Smith–MacMillan normal form if

(i) It is diagonal, $c_{ij} = 0$ if $i \neq j$.
(ii) Diagonal elements are positive rational numbers, $c_{ii} = \frac{\alpha_i}{\beta_i} > 0$ for $i = 1, \dots, D$, where α_i and β_i are coprime integers (no common factors).
(iii) α_{i+1} divides α_i and β_i divides β_{i+1}.

The corresponding Smith–MacMillan form decomposition is given by the following theorem.

Theorem 13.9 For every nonsingular $D \times D$ rational matrix $\mathbf{M}$, there exist unimodular integer matrices $\mathbf{E}_1$ and $\mathbf{E}_2$ such that $\mathbf{E}_1\mathbf{M}\mathbf{E}_2$ is in Smith–MacMillan normal form.

Proof: Reduce all elements of $\mathbf{M}$ so that the numerators and denominators are coprime. Let $p > 0$ be the least common multiple of all the denominators. Then $p\mathbf{M}$ is an integer matrix. Choose $\mathbf{E}_1$ and $\mathbf{E}_2$ so that $\mathbf{E}_1(p\mathbf{M})\mathbf{E}_2 = p\mathbf{E}_1\mathbf{M}\mathbf{E}_2 = \mathbf{D}$ is in Smith normal form, as per Theorem 13.8. Then, $\mathbf{C} = \frac{1}{p}\mathbf{D}$ satisfies all the conditions of the definition of the Smith–MacMillan normal form. Conditions (i) and (ii) are straightforward. For condition (iii), with $d_{ii}/p = \alpha_i/\beta_i$, we can write

$$p = \frac{d_{ii}\beta_i}{\alpha_i} = \frac{d_{i+1,i+1}\beta_{i+1}}{\alpha_{i+1}}. \tag{13.52}$$

Since $d_{i+1,i+1}$ divides d_{ii} from the definition of the Smith normal form, $\frac{\alpha_i\beta_{i+1}}{\beta_i\alpha_{i+1}}$ is an integer. Since α_i and β_i are coprime, β_i must divide β_{i+1}. Similarly, since α_{i+1} and β_{i+1} are coprime, α_{i+1} must divide α_i. □

If we denote $\mathbf{A}_\alpha = \text{diag}(\alpha_1, \dots, \alpha_D)$ and $\mathbf{A}_\beta = \text{diag}(\beta_1, \dots, \beta_D)$, then the decomposition can be written $\mathbf{E}_1\mathbf{M}\mathbf{E}_2 = \mathbf{A}_\alpha\mathbf{A}_\beta^{-1}$.

Theorem 13.10 Let $\Lambda_1 = \text{LAT}(\mathbf{V}_1)$ and $\Lambda_2 = \text{LAT}(\mathbf{V}_2)$ be D-dimensional lattices. Then $\Lambda_1 \cap \Lambda_2$ is a D-dimensional lattice if and only if $\mathbf{V}_1^{-1}\mathbf{V}_2$ is a matrix of rational numbers.

Proof: Suppose that $\Lambda_1 \cap \Lambda_2$ is a D-dimensional lattice with sampling matrix $\mathbf{V}_{12}$. Then $\mathbf{M}_1 = \mathbf{V}_1^{-1}\mathbf{V}_{12}$ and $\mathbf{M}_2 = \mathbf{V}_2^{-1}\mathbf{V}_{12}$ are both integer matrices. It follows that $\mathbf{V}_1\mathbf{M}_1 = \mathbf{V}_2\mathbf{M}_2$, and so $\mathbf{V}_1^{-1}\mathbf{V}_2 = \mathbf{M}_1\mathbf{M}_2^{-1}$, which is clearly a matrix of rational numbers.

Conversely, suppose that $\mathbf{H} = \mathbf{V}_1^{-1}\mathbf{V}_2$ is a matrix of rational numbers. By Theorem 13.9, we find $\mathbf{E}_1$ and $\mathbf{E}_2$ such that $\mathbf{E}_1\mathbf{V}_1^{-1}\mathbf{V}_2\mathbf{E}_2 = \mathbf{A}_\alpha\mathbf{A}_\beta^{-1}$ is in Smith–MacMillan normal form. Rearranging, we find

$$\mathbf{V}_2\mathbf{E}_2\mathbf{A}_\beta = \mathbf{V}_1\mathbf{E}_1^{-1}\mathbf{A}_\alpha = \mathbf{V}_3. \tag{13.53}$$

Since $\mathbf{A}_\beta$ and $\mathbf{A}_\alpha$ are integer matrices,

$$\text{LAT}(\mathbf{V}_3) \subset \text{LAT}(\mathbf{V}_2\mathbf{E}_2) = \Lambda_2 \tag{13.54}$$

$$\text{LAT}(\mathbf{V}_3) \subset \text{LAT}(\mathbf{V}_1\mathbf{E}_1^{-1}) = \Lambda_1 \tag{13.55}$$

so that $\text{LAT}(\mathbf{V}_3) \subset \Lambda_1 \cap \Lambda_2$. Since $\text{LAT}(\mathbf{V}_3)$ is clearly a D-dimensional lattice, $\Lambda_1 \cap \Lambda_2$ must be a D-dimensional lattice as well. □

Given these preliminaries, we can now show that the lattice $\text{LAT}(\mathbf{V}_3)$ given in the proof of Theorem 13.10 is in fact the intersection of Λ_1 and Λ_2, and so we have two possible forms for the sampling matrix.

Theorem 13.11 Let $\Lambda_1 = \text{LAT}(\mathbf{V}_1)$ and $\Lambda_2 = \text{LAT}(\mathbf{V}_2)$ be D-dimensional lattices such that $\mathbf{V}_1^{-1}\mathbf{V}_2$ is a matrix of rational numbers. Let $\mathbf{E}_1$ and $\mathbf{E}_2$ be unimodular integer matrices such that $\mathbf{E}_1\mathbf{V}_1^{-1}\mathbf{V}_2\mathbf{E}_2 = \mathbf{A}_\alpha\mathbf{A}_\beta^{-1}$ as defined above. Then $\Lambda_1 \cap \Lambda_2 = \text{LAT}(\mathbf{V}_2\mathbf{E}_2\mathbf{A}_\beta) = \text{LAT}(\mathbf{V}_1\mathbf{E}_1^{-1}\mathbf{A}_\alpha)$.

Proof: We have already shown in Theorem 13.10 that $\mathrm{LAT}(\mathbf{V}_2\mathbf{E}_2\mathbf{A}_\beta) \subset \Lambda_1 \cap \Lambda_2$. To prove the converse, suppose that $\mathbf{x} \in \Lambda_1 \cap \Lambda_2$. It follows that $\mathbf{x} = (\mathbf{V}_1\mathbf{E}_1^{-1})\mathbf{n}_1 = (\mathbf{V}_2\mathbf{E}_2)\mathbf{n}_2$ for some integer vectors $\mathbf{n}_1$ and $\mathbf{n}_2$. Then

$$\mathbf{n}_1 = \mathbf{E}_1\mathbf{V}_1^{-1}\mathbf{V}_2\mathbf{E}_2\mathbf{n}_2 = \mathbf{A}_\alpha\mathbf{A}_\beta^{-1}\mathbf{n}_2, \tag{13.56}$$

so that $n_{1i} = (\alpha_i/\beta_i)n_{2i}$ for $i = 1, \ldots, D$. Since α_i and β_i are coprime integers, β_i divides n_{2i} for $i = 1, \ldots, D$, and thus $\mathbf{n}_2 = \mathbf{A}_\beta\mathbf{m}$ for some integer vector $\mathbf{m}$. It follows that

$$\mathbf{x} = \mathbf{V}_2\mathbf{E}_2\mathbf{A}_\beta\mathbf{m} \in \mathrm{LAT}(\mathbf{V}_2\mathbf{E}_2\mathbf{A}_\beta), \tag{13.57}$$

so that $\Lambda_1 \cap \Lambda_2 \subset \mathrm{LAT}(\mathbf{V}_2\mathbf{E}_2\mathbf{A}_\beta)$. Thus $\Lambda_1 \cap \Lambda_2 = \mathrm{LAT}(\mathbf{V}_2\mathbf{E}_2\mathbf{A}_\beta)$, which is equal to $\mathrm{LAT}(\mathbf{V}_1\mathbf{E}_1^{-1}\mathbf{A}_\alpha)$ by Theorem 13.10. □

Corollary 13.2 $\mathrm{d}(\Lambda_1 \cap \Lambda_2) = \mathrm{d}(\Lambda_2)\det(\mathbf{A}_\beta) = \mathrm{d}(\Lambda_1)\det(\mathbf{A}_\alpha)$.

A dual concept to the intersection of lattices is the sum of lattices.

Definition 13.12 The sum of two lattices Λ_1 and Λ_2, denoted $\Lambda_1 + \Lambda_2$, is defined as $\{\mathbf{x} + \mathbf{y} \mid \mathbf{x} \in \Lambda_1, \mathbf{y} \in \Lambda_2\}$.

It is clear from the definition that $\Lambda_1 + \Lambda_2$ is a superlattice of Λ_1 and Λ_2, and that it must be a sublattice of any common superlattice. Thus $\Lambda_1 + \Lambda_2$ is the *least common superlattice* of Λ_1 and Λ_2. The relation between $\Lambda_1 \cap \Lambda_2$ and $\Lambda_1 + \Lambda_2$ follows from standard results in group theory. Specifically, by the second isomorphism theorem for groups (e.g., Armstrong (1988), theorem (16.4)),

$$(\Lambda_1 + \Lambda_2)/\Lambda_1 \approx \Lambda_2/(\Lambda_1 \cap \Lambda_2) \tag{13.58}$$

$$(\Lambda_1 + \Lambda_2)/\Lambda_2 \approx \Lambda_1/(\Lambda_1 \cap \Lambda_2). \tag{13.59}$$

Thus, if $\Lambda_1 \cap \Lambda_2$ is a lattice of dimension D, then so is $\Lambda_1 + \Lambda_2$, and $\mathrm{d}(\Lambda_1)/\mathrm{d}(\Lambda_1 + \Lambda_2) = \mathrm{d}(\Lambda_1 \cap \Lambda_2)/\mathrm{d}(\Lambda_2)$. The construction of a lattice matrix for the sum of two lattices follows the same lines as for the intersection.

Theorem 13.12 Let $\Lambda_1 = \mathrm{LAT}(\mathbf{V}_1)$ and $\Lambda_2 = \mathrm{LAT}(\mathbf{V}_2)$ be D-dimensional lattices such that $\mathbf{V}_1^{-1}\mathbf{V}_2$ is a matrix of rational numbers. Let $\mathbf{E}_1$ and $\mathbf{E}_2$ be unimodular integer matrices such that $\mathbf{E}_1\mathbf{V}_1^{-1}\mathbf{V}_2\mathbf{E}_2 = \mathbf{A}_\alpha\mathbf{A}_\beta^{-1}$ as defined above. Then $\Lambda_1 + \Lambda_2 = \mathrm{LAT}(\mathbf{V}_2\mathbf{E}_2\mathbf{A}_\alpha^{-1}) = \mathrm{LAT}(\mathbf{V}_1\mathbf{E}_1^{-1}\mathbf{A}_\beta^{-1})$.

Proof: Rearranging the expression for the Smith–MacMillan decomposition of $\mathbf{V}_1^{-1}\mathbf{V}_2$, we have

$$\mathbf{V}_2\mathbf{E}_2\mathbf{A}_\alpha^{-1} = \mathbf{V}_1\mathbf{E}_1^{-1}\mathbf{A}_\beta^{-1} = \mathbf{V}_4. \tag{13.60}$$

Since $\mathbf{A}_\beta$ and $\mathbf{A}_\alpha$ are integer matrices, we conclude that Λ_1 and Λ_2 are both sublattices of $\mathrm{LAT}(\mathbf{V}_4)$, and thus $\Lambda_1 + \Lambda_2 \subset \mathrm{LAT}(\mathbf{V}_4)$. From the definition, $d(\mathrm{LAT}(\mathbf{V}_4)) = \mathrm{d}(\Lambda_2)/\det(\mathbf{A}_\alpha)$. However, we have above that $\mathrm{d}(\Lambda_1 + \Lambda_2) = \mathrm{d}(\Lambda_2)\mathrm{d}(\Lambda_1)/\mathrm{d}(\Lambda_1 \cap \Lambda_2) = \mathrm{d}(\Lambda_2)/\det(\mathbf{A}_\alpha)$, and thus $\Lambda_1 + \Lambda_2 = \mathrm{LAT}(\mathbf{V}_4)$. □

To illustrate these ideas, we present the following example. Let Λ_1 and Λ_2 be given by the lattice matrices

$$\mathbf{V}_1 = \begin{bmatrix} 2 & 0 \\ 0 & 2 \end{bmatrix} \qquad \mathbf{V}_2 = \begin{bmatrix} 2 & 1 \\ 2 & -1 \end{bmatrix}. \tag{13.61}$$

Note that $d(\Lambda_1) = d(\Lambda_2) = 4$, so the two lattices have the same sampling density. Computing $\mathbf{V}_1^{-1}\mathbf{V}_2$ we find

$$\mathbf{V}_1^{-1}\mathbf{V}_2 = \begin{bmatrix} 1 & \frac{1}{2} \\ 1 & -\frac{1}{2} \end{bmatrix} \tag{13.62}$$

which is not an integer matrix, so the two lattices are distinct. However, it is a rational matrix, so the sum and intersection will be two-dimensional lattices. We can apply the Smith–MacMillan decomposition to find unimodular $\mathbf{E}_1$ and $\mathbf{E}_2$ so that $\mathbf{E}_1\mathbf{V}_1^{-1}\mathbf{V}_2\mathbf{E}_2 = \mathbf{A}_\alpha\mathbf{A}_\beta^{-1}$:

$$\begin{bmatrix} 1 & 1 \\ 0 & -1 \end{bmatrix} \begin{bmatrix} 1 & \frac{1}{2} \\ 1 & -\frac{1}{2} \end{bmatrix} \begin{bmatrix} 1 & 0 \\ 2 & 1 \end{bmatrix} = \begin{bmatrix} 2 & 0 \\ 0 & \frac{1}{2} \end{bmatrix} = \begin{bmatrix} 2 & 0 \\ 0 & 1 \end{bmatrix} \begin{bmatrix} 1 & 0 \\ 0 & 2 \end{bmatrix}^{-1}. \tag{13.63}$$

With this decomposition, we can compute lattice matrices $\mathbf{V}_3$ for $\Lambda_1 \cap \Lambda_2$ and $\mathbf{V}_4$ for $\Lambda_1 + \Lambda_2$ using the formulas given in Theorem 13.11 and Theorem 13.12 respectively:

$$\mathbf{V}_3 = \mathbf{V}_1\mathbf{E}_1^{-1}\mathbf{A}_\alpha = \begin{bmatrix} 4 & 2 \\ 0 & 2 \end{bmatrix} \tag{13.64}$$

$$\mathbf{V}_4 = \mathbf{V}_1\mathbf{E}_1^{-1}\mathbf{A}_\beta^{-1} = \begin{bmatrix} 2 & 1 \\ 0 & 1 \end{bmatrix}. \tag{13.65}$$

The lattices involved are illustrated in Figure 13.6.

As a simple example where the sum and intersection are not a D-dimensional lattice, consider the case where Λ_1 is the rectangular integer lattice $\mathbb{Z}^2$ and Λ_2 is a regular hexagonal lattice, with lattice matrix

$$\mathbf{V}_2 = \begin{bmatrix} 1 & 0.5 \\ 0 & \frac{\sqrt{3}}{2} \end{bmatrix}. \tag{13.66}$$

Thus, $\mathbf{V}_1^{-1}\mathbf{V}_2 = \mathbf{V}_2$ which is *not* a rational matrix. The intersection of these two lattices is a one-dimensional lattice along the x-axis ($\mathbb{Z}$).

There is an interesting relationship between sum and intersection in the lattice domain and the reciprocal lattice domain.

Theorem 13.13 $\Lambda_1 + \Lambda_2 = (\Lambda_1^* \cap \Lambda_2^*)^*$.

Proof: We show that $(\Lambda_1 + \Lambda_2)^* = \Lambda_1^* \cap \Lambda_2^*$. Suppose that $\mathbf{r} \in (\Lambda_1 + \Lambda_2)^*$. Then $\mathbf{r} \cdot \mathbf{x} \in \mathbb{Z}$ for all $\mathbf{x} \in \Lambda_1 + \Lambda_2$. Thus, $\mathbf{r} \cdot \mathbf{x} \in \mathbb{Z}$ for all $\mathbf{x} \in \Lambda_1$ so $\mathbf{r} \in \Lambda_1^*$, and $\mathbf{r} \cdot \mathbf{x} \in \mathbb{Z}$ for all $\mathbf{x} \in \Lambda_2$, so $\mathbf{r} \in \Lambda_2^*$. It follows that $\mathbf{r} \in \Lambda_1^* \cap \Lambda_2^*$, so that $(\Lambda_1 + \Lambda_2)^* \subset \Lambda_1^* \cap \Lambda_2^*$.

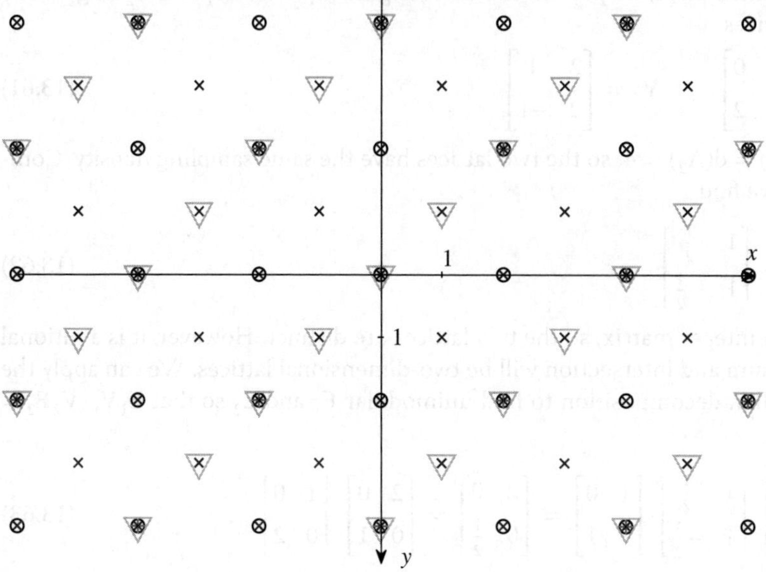

Figure 13.6 Intersection and sum of lattices. Λ_1: O; Λ_2: ∇; $\Lambda_1 \cap \Lambda_2$: +; $\Lambda_1 + \Lambda_2$: ×. *(See color plate section for the color representation of this figure.)*

Conversely, suppose that $\mathbf{r} \in \Lambda_1^* \cap \Lambda_2^*$, so that $\mathbf{r} \cdot \mathbf{x} \in \mathbb{Z}$ for all $\mathbf{x} \in \Lambda_1$ and for all $\mathbf{x} \in \Lambda_2$. Any $\mathbf{z} \in \Lambda_1 + \Lambda_2$ can be written $\mathbf{z} = \mathbf{x}_1 + \mathbf{x}_2$ where $\mathbf{x}_1 \in \Lambda_1$ and $\mathbf{x}_2 \in \Lambda_2$. Then $\mathbf{r} \cdot \mathbf{z} = \mathbf{r} \cdot \mathbf{x}_1 + \mathbf{r} \cdot \mathbf{x}_2 \in \mathbb{Z}$ and thus $\mathbf{r} \in (\Lambda_1 + \Lambda_2)^*$. Thus $\Lambda_1^* \cap \Lambda_2^* \subset (\Lambda_1 + \Lambda_2)^*$, establishing that $(\Lambda_1 + \Lambda_2)^* = \Lambda_1^* \cap \Lambda_2^*$. □

The Smith–MacMillan form allows us to find bases for Λ_1 and Λ_2 that are collinear. This allows us to explicitly find bases for $\Lambda_1 \cap \Lambda_2$ and $\Lambda_1 + \Lambda_2$ by operating independently along each of these basis vectors. However, these are not the only collinear bases, and may not be the most convenient ones. To illustrate these ideas with more a complex example, consider the lattices that arise in digital video with 525 line/60 Hz and 625 line/50 Hz interlaced signals with 720 samples per line and 4:3 aspect ratio, as per ITU-R Recommendation BT.601. The corresponding sampling matrices are respectively (note that 720 samples per line corresponds to 540 samples per picture height)

$$\mathbf{V}_1 = \begin{bmatrix} \frac{1}{540} & 0 & 0 \\ 0 & \frac{2}{525} & \frac{1}{525} \\ 0 & 0 & \frac{1}{60} \end{bmatrix} \qquad \mathbf{V}_2 = \begin{bmatrix} \frac{1}{540} & 0 & 0 \\ 0 & \frac{2}{625} & \frac{1}{625} \\ 0 & 0 & \frac{1}{50} \end{bmatrix}. \tag{13.67}$$

Spatial dimensions x and y are in picture height (ph), and temporal dimension t is in seconds. The rational matrix $\mathbf{V}_1^{-1}\mathbf{V}_2$ is

$$\mathbf{V}_1^{-1}\mathbf{V}_2 = \begin{bmatrix} 1 & 0 & 0 \\ 0 & \frac{21}{25} & -\frac{9}{50} \\ 0 & 0 & \frac{6}{5} \end{bmatrix}. \tag{13.68}$$

The Smith–MacMillan normal form of this matrix is found to be

$$\mathbf{A}_\alpha\mathbf{A}_\beta^{-1} = \frac{1}{50}\begin{bmatrix} 4200 & 0 & 0 \\ 0 & 30 & 0 \\ 0 & 0 & 1 \end{bmatrix} = \begin{bmatrix} 84 & 0 & 0 \\ 0 & 3 & 0 \\ 0 & 0 & 1 \end{bmatrix}\begin{bmatrix} 1 & 0 & 0 \\ 0 & \frac{1}{5} & 0 \\ 0 & 0 & \frac{1}{50} \end{bmatrix}. \tag{13.69}$$

The corresponding unimodular matrices such that $\mathbf{E}_1\mathbf{V}_1^{-1}\mathbf{V}_2\mathbf{E}_2 = \mathbf{A}_\alpha\mathbf{A}_\beta^{-1}$ are:

$$\mathbf{E}_1 = \begin{bmatrix} 84 & 9700 & 1385 \\ -33 & -3810 & -544 \\ -1 & -119 & -17 \end{bmatrix} \qquad \mathbf{E}_2 = \begin{bmatrix} 2856 & 51 & 1 \\ 55 & 1 & 0 \\ 8190 & 148 & 1 \end{bmatrix}. \tag{13.70}$$

The resulting collinear bases for Λ_1 and Λ_2 are

$$\mathbf{V}_1' = \mathbf{V}_1\mathbf{E}_1^{-1} = \begin{bmatrix} \frac{34}{540} & \frac{85}{540} & \frac{50}{540} \\ \frac{83}{525} & \frac{210}{525} & \frac{42}{525} \\ \frac{117}{60} & \frac{296}{60} & \frac{60}{60} \end{bmatrix} \tag{13.71}$$

$$\mathbf{V}_2' = \mathbf{V}_2\mathbf{E}_2 = \begin{bmatrix} \frac{2856}{540} & \frac{51}{540} & \frac{1}{540} \\ \frac{8300}{625} & \frac{150}{625} & \frac{1}{625} \\ \frac{8190}{50} & \frac{148}{50} & \frac{1}{50} \end{bmatrix} \tag{13.72}$$

where the proportionality factors are 84, 3/5 and 1/50 respectively.

The corresponding bases for $\Lambda_1 \cap \Lambda_2$ and $\Lambda_1 + \Lambda_2$ are respectively

$$\mathbf{V}_3 = \mathbf{V}_1'\mathbf{A}_\alpha = \begin{bmatrix} \frac{238}{45} & \frac{17}{36} & \frac{5}{54} \\ \frac{332}{25} & \frac{6}{5} & \frac{2}{25} \\ \frac{819}{5} & \frac{74}{5} & 1 \end{bmatrix} \tag{13.73}$$

$$\mathbf{V}_4 = \mathbf{V}_1'\mathbf{A}_\beta^{-1} = \begin{bmatrix} \frac{17}{270} & \frac{17}{540} & \frac{1}{540} \\ \frac{83}{525} & \frac{2}{25} & \frac{1}{625} \\ \frac{39}{20} & \frac{74}{75} & \frac{1}{50} \end{bmatrix}. \tag{13.74}$$

Rather than directly using the Smith–MacMillan normal form, we can observe that the first basis vector is the same in both original bases for Λ_1 and Λ_2. Thus, we can restrict our search for collinear basis vectors to the $y - t$ plane. If we do this, the corresponding collinear bases for Λ_1 and Λ_2 are

$$\tilde{\mathbf{V}}_1 = \begin{bmatrix} \frac{1}{540} & 0 & 0 \\ 0 & \frac{1}{105} & \frac{2}{75} \\ 0 & \frac{7}{60} & \frac{1}{3} \end{bmatrix} \qquad \tilde{\mathbf{V}}_2 = \begin{bmatrix} \frac{1}{540} & 0 & 0 \\ 0 & \frac{4}{25} & \frac{1}{625} \\ 0 & \frac{49}{25} & \frac{1}{50} \end{bmatrix} \tag{13.75}$$

with

$$\tilde{\mathbf{V}}_1^{-1}\tilde{\mathbf{V}}_2 = \begin{bmatrix} 1 & 0 & 0 \\ 0 & \frac{84}{5} & 0 \\ 0 & 0 & \frac{3}{5} \end{bmatrix} \tag{13.76}$$

which are simpler forms that those obtained directly using the 3×3 Smith–MacMillan normal form. The corresponding bases for $\Lambda_1 \cap \Lambda_2$ and $\Lambda_1 + \Lambda_2$ are respectively

$$\tilde{\mathbf{V}}_3 = \begin{bmatrix} \frac{1}{540} & 0 & 0 \\ 0 & \frac{4}{5} & \frac{2}{25} \\ 0 & \frac{49}{5} & 1 \end{bmatrix} \qquad \tilde{\mathbf{V}}_4 = \begin{bmatrix} \frac{1}{540} & 0 & 0 \\ 0 & \frac{1}{525} & \frac{1}{1875} \\ 0 & \frac{7}{300} & \frac{1}{150} \end{bmatrix}. \tag{13.77}$$

These can be verified to be right equivalent to $\mathbf{V}_3$ and $\mathbf{V}_4$, showing that they describe the same lattices.

Appendix A

Equivalence Relations

Equivalence relations and equivalence classes arise in many contexts in this book. For example, the elements of a color space are defined as equivalence classes of light spectral densities that appear identical to a given viewer (an equivalence relation). We provide here the basic definition and properties. More details can be found in any textbook of abstract algebra, e.g., section 10, chapter II of Warner (1965) or section 0.1 of Dummit and Foote (2004).

1. Let $\mathcal{A}$ be a set. A relation R is a subset of $\mathcal{A} \times \mathcal{A}$. If $(x_1, x_2) \in R$, we write $x_1 \, R \, x_2$. Examples of relations on $\mathbb{R}$ are $x_1 = x_2$ and $x_1 < x_2$.
2. A relation is said to be an *equivalence relation* if it satisfies the following three properties:

$$\begin{aligned} &\text{Reflexivity:} && x \, R \, x \\ &\text{Symmetry:} && x_1 \, R \, x_2 \Rightarrow x_2 \, R \, x_1 \\ &\text{Transitivity:} && x_1 \, R \, x_2 \text{ and } x_2 \, R \, x_3 \Rightarrow x_1 \, R \, x_3. \end{aligned}$$

   Of the above examples, $x_1 = x_2$ is an equivalence relation and $x_1 < x_2$ is not.
3. Let $\sim$ be an equivalence relation on a set $\mathcal{A}$, and define $(z) = \{x \in \mathcal{A} \mid x \sim z\}$ for any $z \in \mathcal{A}$. This set is called the equivalence class containing z. Let (z) and (y) be two equivalence classes. Then either $(z) = (y)$ or $(z) \cap (y) = \emptyset$. A consequence of this is that if y is any element of (z), then $(y) = (z)$, i.e. any element of the equivalence class can be used as the class representative. A second consequence is that every element $x \in \mathcal{A}$ belongs to one and only one equivalence class. Thus, the set of equivalence classes forms a partition of $\mathcal{A}$.

Multidimensional Signal and Color Image Processing Using Lattices, First Edition. Eric Dubois.

Companion website: www.wiley.com/go/Dubois/multiSP

Appendix B

Groups

The group is the basic algebraic structure with one operation. Groups appear in many contexts in this book. For example, all the signal domains introduced in Chapters 2–5 have the structure of an additive commutative group. Also, sets of transformations of domains and signals have the structure of a noncommutative group. This is used extensively in Chapter 12. This appendix provides basic definitions and properties of groups, but any standard text on abstract algebra should be consulted for a more detailed treatment, e.g, part I of Dummit and Foote (2004) or chapter 1 of Miller (1972).

1. A *binary operation* Δ in a set $\mathcal{A}$ is a function from $\mathcal{A} \times \mathcal{A}$ into $\mathcal{A}$. We write $x\Delta y$ for $x, y \in \mathcal{A}$.
   For example, addition $+$ is a binary operation on the set $\mathbb{Z}$ of integers.
2. A binary operation Δ on a set $\mathcal{A}$ is *associative* if $(x\Delta y)\Delta z = x\Delta(y\Delta z)$ for all $x, y, z \in \mathcal{A}$.
   For an associative operation, the parentheses are not required, and we can write simply $x\Delta y\Delta z$. Addition on $\mathbb{Z}$ is associative.
3. A binary operation Δ on a set $\mathcal{A}$ is *commutative* if $x\Delta y = y\Delta x$ for all $x, y \in \mathcal{A}$.
   Addition on $\mathbb{Z}$ is also commutative.
4. A *semigroup* is a set $\mathcal{A}$ with one associative binary operation Δ, and is denoted $(\mathcal{A}, \Delta)$. A semigroup is *commutative* if its binary operation is commutative.
   The set of strictly positive integers $\mathbb{Z}_+^*$ is a commutative semigroup.
5. Let Δ be a binary operation on a set $\mathcal{A}$. An element $e \in \mathcal{A}$ is a *neutral element* if $e\Delta x = x\Delta e = x$ for all $x \in \mathcal{A}$. If a neutral element exists, it is unique.
   The set of strictly positive integers $\mathbb{Z}_+^*$ under addition has no neutral element while the set of non-negative integers $\mathbb{Z}_+$ has the neutral element 0.
6. Let $(\mathcal{A}, \Delta)$ be a semigroup with a neutral element e. An element $x \in \mathcal{A}$ is *invertible* if there exists $y \in \mathcal{A}$ such that $x\Delta y = y\Delta x = e$. Such an element y is called the inverse of x, and it is unique if it exists. The inverse is denoted x^{-1} in general, or $-x$ when the operation is addition.
7. A *group* is a semigroup $(\mathcal{A}, \Delta)$ with a neutral element such that every element of $\mathcal{A}$ is invertible. The group $(\mathcal{A}, \Delta)$ is called Abelian (or cummutative) if the operation Δ is commutative. We will generally write that $\mathcal{A}$ is a group under the operation Δ or simply $\mathcal{A}$ when the operation is clear from context.
   All of the domains studied in Chapters 2–5 are examples of additive Abelian groups.
8. Let $\mathcal{A}$ be a group. The nonempty subset $\mathcal{B}$ of $\mathcal{A}$ is a subgroup of $\mathcal{A}$ if $\mathcal{B}$ is closed under the group operation and under inverses. If $x, y \in \mathcal{B}$, then $x\Delta y \in \mathcal{B}$ and $x^{-1} \in \mathcal{B}$. If $\mathcal{A}$

Multidimensional Signal and Color Image Processing Using Lattices, First Edition. Eric Dubois.

Companion website: www.wiley.com/go/Dubois/multiSP

is a finite group, then the number of elements in any subgroup divides the number of elements in the group (Lagrange's Theorem).
The set $\{nK \mid n \in \mathbb{Z}\}$ for any $K \in \mathbb{Z}_+^*$ is a subgroup of the additive Abelian group $\mathbb{Z}$. We denote this subgroup $K\mathbb{Z}$.

9. Let $\mathcal{B}$ be a subgroup of $\mathcal{A}$. For any $x \in \mathcal{A}$, let

$$x\mathcal{B} = \{x\Delta y \mid y \in \mathcal{B}\} \qquad \mathcal{B}x = \{y\Delta x \mid y \in \mathcal{B}\}. \tag{B.1}$$

These are called respectively a left coset of $\mathcal{B} \in \mathcal{A}$ and a right coset of $\mathcal{B} \in \mathcal{A}$. If $\mathcal{A}$ is an Abelian group, the left and right cosets are the same and we write $x + \mathcal{B}$ for a coset of $\mathcal{B}$ in $\mathcal{A}$.
The set $m + K\mathbb{Z}$ is a coset of $K\mathbb{Z}$ in $\mathbb{Z}$ for any m.

10. Let $\mathcal{A}$ be an Abelian group and let $\mathcal{B}$ be a subgroup. For any $x_1, x_2 \in \mathcal{A}$, we say that $x_1 \sim x_2$ if $x_1 - x_2 \in \mathcal{B}$. Then $\sim$ is an equivalence relation, and the equivalence classes are the cosets of $\mathcal{B}$ in $\mathcal{A}$. Any element of a coset can be used as a coset representative. The cosets of $\mathcal{B}$ in $\mathcal{A}$ form a partition of $\mathcal{A}$.
There are K cosets of $K\mathbb{Z}$ in $\mathbb{Z}$. We can use the elements $0, 1, \ldots, K-1$ as coset representatives.

11. Let $(\mathcal{A}, +)$ be an Abelian group and $\mathcal{B}$ a subgroup. Let $\mathcal{A}/\mathcal{B}$ denote the set of cosets of $\mathcal{B}$ in $\mathcal{A}$. We define a binary operation $\oplus$ on $\mathcal{A}/\mathcal{B}$ to be $(x + \mathcal{B}) \oplus (y + \mathcal{B}) = (x + y + \mathcal{B})$. This operation is well defined and $(\mathcal{A}/\mathcal{B}, \oplus)$ is an Abelian group called the quotient group.
Note that these last two concepts can be defined on nonAbelian groups as well, but we do not use them explicitly in this book.

Appendix C

Vector Spaces

This appendix briefly summarizes some of the key results from linear algebra that are used in this book. A great more detail can be found in standard texts, such as Hoffman and Kunze (1971), Shaw (1982), and Olver and Shakiban (2006).

1. A general vector space is defined with respect to a field K. In this book, we will only be concerned with real and complex vector spaces defined with respect to the fields $\mathbb{R}$ and $\mathbb{C}$ of real and complex numbers. The development is carried out for complex vector spaces, since it is then easy to restrict everything to be real. In this appendix, elements of a vector space are denoted in boldface with an arrow over them, e.g., $\vec{\mathbf{x}}$. Several different types of vector spaces appear in the main text, and they don't generally use this notation.
2. A complex vector space $\mathcal{S}$ is a set on which two operations are defined,
   vector addition $+ : \mathcal{S} \times \mathcal{S} \to \mathcal{S}$,
   multiplication by a scalar $\cdot : \mathbb{C} \times \mathcal{S} \to \mathcal{S}$
   that satisfy the following properties:
   A. For each $\vec{\mathbf{x}}, \vec{\mathbf{y}}, \vec{\mathbf{z}} \in \mathcal{S}$
      (a) $\vec{\mathbf{x}} + \vec{\mathbf{y}} = \vec{\mathbf{y}} + \vec{\mathbf{x}}$ — commutativity
      (b) $\vec{\mathbf{x}} + (\vec{\mathbf{y}} + \vec{\mathbf{z}}) = (\vec{\mathbf{x}} + \vec{\mathbf{y}}) + \vec{\mathbf{z}}$ — associativity
      (c) There is a unique $\vec{\mathbf{0}}$ such that $\vec{\mathbf{x}} + \vec{\mathbf{0}} = \vec{\mathbf{x}}, \quad \forall \vec{\mathbf{x}}$ — zero vector
      (d) For each $\vec{\mathbf{x}}$ there exists $(-\vec{\mathbf{x}})$ such that $\vec{\mathbf{x}} + (-\vec{\mathbf{x}}) = \vec{\mathbf{0}}$ — negative.
   B. For each $\alpha, \beta \in \mathbb{C}, \vec{\mathbf{x}}, \vec{\mathbf{y}} \in \mathcal{S}$
      (a) $\alpha \cdot (\beta \cdot \vec{\mathbf{x}}) = (\alpha\beta) \cdot \vec{\mathbf{x}}$
      (b) $1 \cdot \vec{\mathbf{x}} = \vec{\mathbf{x}}, \quad \forall \vec{\mathbf{x}}$
      (c) $\alpha \cdot (\vec{\mathbf{x}} + \vec{\mathbf{y}}) = \alpha \cdot \vec{\mathbf{x}} + \alpha \cdot \vec{\mathbf{y}}$
      (d) $(\alpha + \beta) \cdot \vec{\mathbf{x}} = \alpha \cdot \vec{\mathbf{x}} + \beta\vec{\mathbf{x}}$.
      Note that $(\mathcal{S}, +)$ is a group. The symbol "$\cdot$" for multiplication by a scalar can usually be omitted, writing $\alpha\vec{\mathbf{x}}$ in place of $\alpha \cdot \vec{\mathbf{x}}$.
3. Let $\vec{\mathbf{x}}_1, \vec{\mathbf{x}}_2, \dots, \vec{\mathbf{x}}_p \in \mathcal{S}$. An expression of the form

$$\alpha_1\vec{\mathbf{x}}_1 + \alpha_2\vec{\mathbf{x}}_2 + \cdots + \alpha_p\vec{\mathbf{x}}_p = \sum_{i=1}^{p} \alpha_i\vec{\mathbf{x}}_i$$

   is called a linear combination. The set of all linear combinations of $\vec{\mathbf{x}}_1, \vec{\mathbf{x}}_2, \dots, \vec{\mathbf{x}}_p$ is denoted $\text{span}(\vec{\mathbf{x}}_1, \vec{\mathbf{x}}_2, \dots, \vec{\mathbf{x}}_p)$.

Multidimensional Signal and Color Image Processing Using Lattices, First Edition. Eric Dubois.

Companion website: www.wiley.com/go/Dubois/multiSP

4. The vectors $\vec{x}_1, \dots, \vec{x}_p$ are said to be linearly independent if and only if

$$\sum_{i=1}^{p} \alpha_i \vec{x}_i = \vec{0} \qquad \text{implies } \alpha_i = 0, i = 1, \dots, p.$$

Otherwise, $\vec{x}_1, \dots, \vec{x}_p$ are said to be linearly dependent. If $\vec{x}_1, \dots, \vec{x}_p$ are linearly independent, then

$$\sum_{i=1}^{p} \alpha_i \vec{x}_i = \sum_{i=1}^{p} \beta_i \vec{x}_i \quad \text{implies } \alpha_i = \beta_i, i = 1, \dots, p,$$

i.e. a vector can be expressed in only one way as a linear combination of p linearly independent vectors.

5. If there exist p linearly independent vectors $\vec{b}_1, \dots, \vec{b}_p$ for some finite integer p such that *any* element of S can be written as a (unique) linear combination of $\vec{b}_1, \dots, \vec{b}_p$, then S is a finite-dimensional vector space of dimension p, and $\{\vec{b}_1, \dots, \vec{b}_p\}$ is called a basis. Otherwise, S is an infinite-dimensional vector space. Let $\mathcal{B} = \{\vec{b}_1, \dots, \vec{b}_p\}$ be a basis and let $\vec{x} \in S$ be expressed $\vec{x} = x_1\vec{b}_1 + \dots + x_p\vec{b}_p$. Then we can conveniently represent $\vec{x}$ as a $p \times 1$ column matrix with respect to the basis $\mathcal{B}$ as

$$\mathbf{x}_{\mathcal{B}} = \begin{bmatrix} x_1 \\ \vdots \\ x_p \end{bmatrix}.$$

6. A *subspace* of S is a subset $\mathcal{R}$ of S, which is itself a vector space. In particular, it is closed under addition and scalar multiplication. We denote this $\mathcal{R} \subset S$. For any $\{\vec{x}_1, \dots, \vec{x}_p \mid \vec{x}_i \in S\}$, $\text{span}(\vec{x}_1, \vec{x}_2, \dots, \vec{x}_p)$ is a subspace of S.

7. If $\mathcal{R}, \mathcal{T} \subset S$, we define the subspaces

$$\mathcal{R} + \mathcal{T} = \{\vec{r} + \vec{t} \mid \vec{r} \in \mathcal{R}, \vec{t} \in \mathcal{T}\},$$
$$\mathcal{R} \cap \mathcal{T} = \{\vec{x} \mid \vec{x} \in \mathcal{R} \text{ and } \vec{x} \in \mathcal{T}\}.$$

8. Let $\mathcal{R}_1, \dots, \mathcal{R}_K$ be subspaces of S. We say that $\mathcal{R}_1, \dots, \mathcal{R}_K$ are independent if $\vec{r}_1 + \dots + \vec{r}_K = \vec{0}$ for $\vec{r}_i \in \mathcal{R}_i$ implies $\vec{r}_i = \vec{0}$ for $i = 1, \dots, K$. If $S = \mathcal{R}_1 + \dots + \mathcal{R}_K$, it follows that any $\vec{x} \in S$ can be written in a *unique* fashion as

$$\vec{x} = \vec{r}_1 + \dots + \vec{r}_K, \quad \vec{r}_i \in \mathcal{R}_i.$$

In this situation, we write

$$S = \mathcal{R}_1 \oplus \mathcal{R}_2 \oplus \dots \oplus \mathcal{R}_K$$

and say that S is the *direct sum* of $\mathcal{R}_1, \dots, \mathcal{R}_K$.

9. Let $\mathcal{R}$ and S be vector spaces. A linear transformation from $\mathcal{R}$ to S is a function $\mathcal{T} : \mathcal{R} \to S$ such that $\mathcal{T}(\alpha\vec{x} + \vec{y}) = \alpha\mathcal{T}\vec{x} + \mathcal{T}\vec{y}$ for all $\vec{x}, \vec{y} \in \mathcal{R}$ and all $\alpha \in \mathbb{C}$. Let $\mathcal{B} = \{\vec{b}_1, \dots, \vec{b}_p\}$ be a basis for $\mathcal{R}$ and $\mathcal{D} = \{\vec{d}_1, \dots, \vec{d}_q\}$ be a basis for S. Then, a linear transformation is completely specified numerically by its action on the p basis vectors of $\mathcal{R}$, expressed with respect to the q basis vectors of S. Let

$$\mathcal{T}(\vec{b}_i) = \sum_{k=1}^{q} t_{ki}\vec{d}_k. \tag{C.1}$$

Then, if $\vec{y} = \mathcal{T}(\vec{x})$, we have

$$\begin{aligned}\vec{y} &= \mathcal{T}\left(\sum_{i=1}^{p} x_i \vec{b}_i\right)\\ &= \sum_{i=1}^{p} x_i \sum_{k=1}^{q} t_{ki}\vec{d}_k\\ &= \sum_{k=1}^{q}\left(\sum_{i=1}^{p} t_{ki}x_i\right)\vec{d}_k\\ &= \sum_{k=1}^{q} y_k\vec{d}_k.\end{aligned}$$

Thus, $\mathbf{y}_D = \mathbf{T}_{B\to D}\mathbf{x}_B$, where $\mathbf{T}_{B\to D}$ is the $q \times p$ matrix $[t_{ki}]$ defined by Equation (C.1).

The *kernel* (or null space) of $\mathcal{T}$ is the subspace

$$\ker \mathcal{T} = \{\vec{x} \in \mathcal{R} \mid \mathcal{T}\vec{x} = \vec{0}\} \subset \mathcal{R}.$$

10. Let $\mathcal{R} \subset \mathcal{S}$. We say that $\vec{x}, \vec{y} \in \mathcal{S}$ are equivalent mod $\mathcal{R}$ if $\vec{x} - \vec{y} \in \mathcal{R}$. This relation satisfies all the properties of an equivalence relation. The equivalence classes are called cosets of the subspace $\mathcal{R}$:

$$\vec{x} + \mathcal{R} = \{\vec{y} \mid \vec{y} - \vec{x} \in \mathcal{R}\}.$$

As usual, two equivalence classes are either identical or disjoint and the set of all equivalence classes forms a partition of $\mathcal{S}$.

11. We define the *quotient space* (or factor space) $\mathcal{S}/\mathcal{R}$ as the set of all equivalence classes,

$$\mathcal{S}/\mathcal{R} = \{\vec{x} + \mathcal{R} \mid \vec{x} \in \mathcal{S}\}.$$

The set of equivalence classes forms a vector space under the operations

$$\begin{aligned}(\vec{x} + \mathcal{R}) + (\vec{y} + \mathcal{R}) &= (\vec{x} + \vec{y}) + \mathcal{R},\\ \alpha(\vec{x} + \mathcal{R}) &= (\alpha\vec{x}) + \mathcal{R}.\end{aligned}$$

It is straightforward and standard to show that these operations are well defined and unambiguous. The function $\mathcal{Q} : \vec{x} \mapsto \vec{x} + \mathcal{R}$ from $\mathcal{S}$ to $\mathcal{S}/\mathcal{R}$ is a linear transformation called the quotient transformation or the canonical projection of $\mathcal{S}$ on $\mathcal{S}/\mathcal{R}$.

12. An inner product on a complex vector space is a mapping from $\mathcal{S} \times \mathcal{S}$ to $\mathbb{C}$, denoted $\langle \vec{x} \mid \vec{y}\rangle$, that satisfies the following properties.
    (a) $\langle \vec{x} \mid \vec{y}\rangle = (\langle \vec{y} \mid \vec{x}\rangle)^*$
    (b) $\langle \alpha_1\vec{x}_1 + \alpha_2\vec{x}_2 \mid \vec{y}\rangle = \alpha_1\langle \vec{x}_1 \mid \vec{y}\rangle + \alpha_2\langle \vec{x}_2 \mid \vec{y}\rangle$
    (c) $\langle \vec{x} \mid \vec{x}\rangle \geq 0$ and $\langle \vec{x} \mid \vec{x}\rangle = 0$ if and only if $\vec{x} = \vec{0}$.
13. If $\langle \vec{x} \mid \vec{y}\rangle = 0$, we say that $\vec{x}$ is orthogonal to $\vec{y}$. A basis $\vec{b}_1, \ldots, \vec{b}_p$ such that

$$\langle \vec{b}_i \mid \vec{b}_j\rangle = 0 \quad \text{if } i \neq j$$

is called an orthogonal basis.

14. For any $\vec{x} \in S$, we denote $\|\vec{x}\|^2 = \langle \vec{x} \mid \vec{x} \rangle$, which satisfies the properties of a *norm*. An orthogonal basis satisfying the additional property that $\|\vec{b}_i\|^2 = 1$, $i = 1, \dots, p$, is called an orthonormal basis.
15. If $\vec{b}_1, \dots, \vec{b}_p$ is an orthogonal basis for S, then any $\vec{x} \in S$ has a unique representation

$$\vec{x} = \sum_{i=1}^{p} \alpha_i \vec{b}_i.$$

It follows that

$$\begin{aligned} \langle \vec{x} \mid \vec{b}_j \rangle &= \sum_{i=1}^{p} \alpha_i \langle \vec{b}_i \mid \vec{b}_j \rangle \\ &= \alpha_j \langle \vec{b}_j \mid \vec{b}_j \rangle \end{aligned}$$

so that

$$\alpha_j = \frac{\langle \vec{x} \mid \vec{b}_j \rangle}{\|\vec{b}_j\|^2} \qquad j = 1, \dots, p.$$

For an orthonormal basis,

$$\alpha_j = \langle \vec{x} \mid \vec{b}_j \rangle, \quad j = 1, \dots, p.$$

Appendix D

Multidimensional Fourier Transform Properties

This appendix provides a synthesis of the Fourier transform properties presented in Chapters 2–5 in a single table. All corresponding properties appear in a single row of this table.

The duality property for the discrete-domain Fourier series is not given in the main text. In this property, $\mathbf{C}$ is an isomorphism from Λ/Γ to Γ^*/Λ^*, with inverse $\mathbf{C}^{-1}$. This property can be proved using the methods of Section 4.6.2 with the Smith normal form.

Multidimensional Signal and Color Image Processing Using Lattices, First Edition. Eric Dubois.

Companion website: www.wiley.com/go/Dubois/multiSP

Domain	**Continuous-domain, non-periodic**	**Discrete-domain (Λ), non-periodic**
Name of the transform	Continuous-domain Fourier transform (CDFT)	Discrete-domain Fourier transform (DDFT)
Signals and domains	$f_c(\mathbf{x}) \overset{\text{CDFT}}{\longleftrightarrow} F_c(\mathbf{u}) \quad \mathbf{x}, \mathbf{x}_0 \in \mathbb{R}^D$ $g_c(\mathbf{x}) \overset{\text{CDFT}}{\longleftrightarrow} G_c(\mathbf{u}) \quad \mathbf{u}, \mathbf{u}_0 \in \mathbb{R}^D$	$f[\mathbf{x}] \overset{\text{DDFT}}{\longleftrightarrow} F(\mathbf{u}) \quad \mathbf{x}, \mathbf{x}_0 \in \Lambda$ $g[\mathbf{x}] \overset{\text{DDFT}}{\longleftrightarrow} G(\mathbf{u}) \quad \mathbf{u}, \mathbf{u}_0 \in \mathbb{R}^D$
Periodicity	none	$F(\mathbf{u}+\mathbf{r}) = F(\mathbf{u}), \quad \mathbf{r} \in \Lambda^*$
Period	none	$\mathbf{u} : \mathcal{P}_{\Lambda^*}, \ \lvert\mathcal{P}_{\Lambda^*}\rvert = 1/d(\Lambda)$
Analysis	$F_c(\mathbf{u}) = \int_{\mathbb{R}^D} f_c(\mathbf{x}) \exp(-j2\pi\mathbf{u}\cdot\mathbf{x})\, d\mathbf{x}$	$F(\mathbf{u}) = \sum_{\mathbf{x}\in\Lambda} f[\mathbf{x}] \exp(-j2\pi\mathbf{u}\cdot\mathbf{x})$
Synthesis	$f_c(\mathbf{x}) = \int_{\mathbb{R}^D} F_c(\mathbf{u}) \exp(j2\pi\mathbf{u}\cdot\mathbf{x})\, d\mathbf{u}$	$f[\mathbf{x}] = d(\Lambda) \int_{\mathcal{P}_{\Lambda^*}} F(\mathbf{u}) \exp(j2\pi\mathbf{u}\cdot\mathbf{x})\, d\mathbf{u}$
Linearity	$Af_c(\mathbf{x}) + Bg_c(\mathbf{x}) \overset{\text{CDFT}}{\longleftrightarrow} AF_c(\mathbf{u}) + BG_c(\mathbf{u})$	$Af[\mathbf{x}] + Bg[\mathbf{x}] \overset{\text{DDFT}}{\longleftrightarrow} AF(\mathbf{u}) + BG(\mathbf{u})$
Shift	$f_c(\mathbf{x}-\mathbf{x}_0) \overset{\text{CDFT}}{\longleftrightarrow} F_c(\mathbf{u}) \exp(-j2\pi\mathbf{u}\cdot\mathbf{x}_0)$	$f[\mathbf{x}-\mathbf{x}_0] \overset{\text{DDFT}}{\longleftrightarrow} F(\mathbf{u}) \exp(-j2\pi\mathbf{u}\cdot\mathbf{x}_0)$
Modulation	$f_c(\mathbf{x}) \exp(j2\pi\mathbf{u}_0\cdot\mathbf{x}) \overset{\text{CDFT}}{\longleftrightarrow} F_c(\mathbf{u}-\mathbf{u}_0)$	$f[\mathbf{x}] \exp(j2\pi\mathbf{u}_0\cdot\mathbf{x}) \overset{\text{DDFT}}{\longleftrightarrow} F(\mathbf{u}-\mathbf{u}_0)$
Convolution	$\int_{\mathbb{R}^D} f_c(\mathbf{s}) g_c(\mathbf{x}-\mathbf{s})\, d\mathbf{s} \overset{\text{CDFT}}{\longleftrightarrow} F_c(\mathbf{u}) G_c(\mathbf{u})$	$\sum_{\mathbf{s}\in\Lambda} f[\mathbf{s}] g[\mathbf{x}-\mathbf{s}] \overset{\text{DDFT}}{\longleftrightarrow} F(\mathbf{u}) G(\mathbf{u})$
Multiplication	$f_c(\mathbf{x}) g_c(\mathbf{x}) \overset{\text{CDFT}}{\longleftrightarrow} \int_{\mathbb{R}^D} F_c(\mathbf{w}) G_c(\mathbf{u}-\mathbf{w})\, d\mathbf{w}$	$f[\mathbf{x}] g[\mathbf{x}] \overset{\text{DDFT}}{\longleftrightarrow} d(\Lambda) \int_{\mathcal{P}_{\Lambda^*}} F(\mathbf{w}) G(\mathbf{u}-\mathbf{w})\, d\mathbf{w}$
Automorphism of domain	$f_c(\mathbf{A}\mathbf{x}) \overset{\text{CDFT}}{\longleftrightarrow} \frac{1}{\lvert\det \mathbf{A}\rvert} F_c(\mathbf{A}^{-T}\mathbf{u})$	$f[\mathbf{A}\mathbf{x}] \overset{\text{DDFT}}{\longleftrightarrow} F(\mathbf{A}^{-T}\mathbf{u})$
Differentiation	$\nabla_{\mathbf{x}} f_c(\mathbf{x}) \overset{\text{CDFT}}{\longleftrightarrow} j2\pi\mathbf{u} F_c(\mathbf{u})$	N/A
Differentiation in frequency	$\mathbf{x} f_c(\mathbf{x}) \overset{\text{CDFT}}{\longleftrightarrow} \frac{j}{2\pi} \nabla_{\mathbf{u}} F_c(\mathbf{u})$	$\mathbf{x} f[\mathbf{x}] \overset{\text{DDFT}}{\longleftrightarrow} \frac{j}{2\pi} \nabla_{\mathbf{u}} F(\mathbf{u})$
Complex conjugation	$f_c^*(\mathbf{x}) \overset{\text{CDFT}}{\longleftrightarrow} F_c^*(-\mathbf{u})$	$f^*[\mathbf{x}] \overset{\text{DDFT}}{\longleftrightarrow} F^*(-\mathbf{u})$
Parseval	$\int_{\mathbb{R}^D} f_c(\mathbf{x}) g_c^*(\mathbf{x})\, d\mathbf{x} = \int_{\mathbb{R}^D} F_c(\mathbf{u}) G_c^*(\mathbf{u})\, d\mathbf{u}$	$\sum_{\mathbf{x}\in\Lambda} f[\mathbf{x}] g^*[\mathbf{x}] = d(\Lambda) \int_{\mathcal{P}_{\Lambda^*}} F(\mathbf{u}) G^*(\mathbf{u})\, d\mathbf{u}$
Duality	$F_c(\mathbf{x}) \overset{\text{CDFT}}{\longleftrightarrow} f_c(-\mathbf{u})$	$\tilde{F}_c[\mathbf{x}] \overset{\text{DDFT}}{\longleftrightarrow} d(\Gamma) \tilde{f}_c(-\mathbf{u})$

Domain	Continuous-domain, periodic (Γ)	Discrete-domain (Λ), periodic ($\Gamma \subset \Lambda$)
Name of the transform	Continuous-domain Fourier series (CDFS)	Discrete-domain Fourier series (DDFS)
Signals and domains	$\tilde{f}_c(\mathbf{x}) \overset{\text{CDFS}}{\longleftrightarrow} \tilde{F}_c[\mathbf{u}] \quad \mathbf{x}, \mathbf{x}_0 \in \mathbb{R}^D$ $\tilde{g}_c(\mathbf{x}) \overset{\text{CDFS}}{\longleftrightarrow} \tilde{G}_c[\mathbf{u}] \quad \mathbf{u}, \mathbf{u}_0 \in \Gamma^*$	$\tilde{f}[\mathbf{x}] \overset{\text{DDFS}}{\longleftrightarrow} \tilde{F}[\mathbf{u}] \quad \mathbf{x}, \mathbf{x}_0 \in \Lambda$ $\tilde{g}[\mathbf{x}] \overset{\text{DDFS}}{\longleftrightarrow} \tilde{G}[\mathbf{u}] \quad \mathbf{u}, \mathbf{u}_0 \in \Gamma^*$
Periodicity	$\tilde{f}_c(\mathbf{x}+\mathbf{s}) = \tilde{f}_c(\mathbf{x}), \quad \mathbf{s} \in \Gamma$	$\tilde{f}[\mathbf{x}+\mathbf{s}] = \tilde{f}[\mathbf{x}], \quad \mathbf{s} \in \Gamma$ $\tilde{F}[\mathbf{u}+\mathbf{r}] = \tilde{F}[\mathbf{u}], \quad \mathbf{r} \in \Lambda^*$
Period	$\mathbf{x} : \mathcal{P}_\Gamma,\ \lvert\mathcal{P}_\Gamma\rvert = d(\Gamma)$	$\mathbf{x} : \mathcal{B},\ \mathbf{u} : \mathcal{D},\ \lvert\mathcal{B}\rvert = \lvert\mathcal{D}\rvert = K$
Analysis	$\tilde{F}_c[\mathbf{u}] = \int_{\mathcal{P}_\Gamma} \tilde{f}_c(\mathbf{x}) \exp(-j2\pi\mathbf{u}\cdot\mathbf{x})\, d\mathbf{x}$	$\tilde{F}[\mathbf{u}] = \sum_{\mathbf{x}\in\mathcal{B}} \tilde{f}[\mathbf{x}] \exp(-j2\pi\mathbf{u}\cdot\mathbf{x})$
Synthesis	$\tilde{f}_c(\mathbf{x}) = \frac{1}{d(\Gamma)} \sum_{\mathbf{u}\in\Gamma^*} \tilde{F}_c[\mathbf{u}] \exp(j2\pi\mathbf{u}\cdot\mathbf{x})$	$\tilde{f}[\mathbf{x}] = \frac{1}{K} \sum_{\mathbf{u}\in\mathcal{D}} \tilde{F}[\mathbf{u}] \exp(j2\pi\mathbf{u}\cdot\mathbf{x})$
Linearity	$A\tilde{f}_c(\mathbf{x}) + B\tilde{g}_c(\mathbf{x}) \overset{\text{CDFS}}{\longleftrightarrow} A\tilde{F}_c[\mathbf{u}] + B\tilde{G}_c[\mathbf{u}]$	$A\tilde{f}[\mathbf{x}] + B\tilde{g}[\mathbf{x}] \overset{\text{DDFS}}{\longleftrightarrow} A\tilde{F}[\mathbf{u}] + B\tilde{G}[\mathbf{u}]$
Shift	$\tilde{f}_c(\mathbf{x}-\mathbf{x}_0) \overset{\text{CDFS}}{\longleftrightarrow} \tilde{F}_c[\mathbf{u}] \exp(-j2\pi\mathbf{u}\cdot\mathbf{x}_0)$	$\tilde{f}[\mathbf{x}-\mathbf{x}_0] \overset{\text{DDFS}}{\longleftrightarrow} \tilde{F}[\mathbf{u}] \exp(-j2\pi\mathbf{u}\cdot\mathbf{x}_0)$
Modulation	$\tilde{f}_c(\mathbf{x}) \exp(j2\pi\mathbf{u}_0\cdot\mathbf{x}) \overset{\text{CDFS}}{\longleftrightarrow} \tilde{F}_c[\mathbf{u}-\mathbf{u}_0]$	$\tilde{f}[\mathbf{x}] \exp(j2\pi\mathbf{u}_0\cdot\mathbf{x}) \overset{\text{DDFS}}{\longleftrightarrow} \tilde{F}[\mathbf{u}-\mathbf{u}_0]$
Convolution	$\int_{\mathcal{P}_\Gamma} \tilde{f}_c(\mathbf{s})\tilde{g}_c(\mathbf{x}-\mathbf{s})\, d\mathbf{s} \overset{\text{CDFS}}{\longleftrightarrow} \tilde{F}_c[\mathbf{u}]\tilde{G}_c[\mathbf{u}]$	$\sum_{\mathbf{s}\in\mathcal{B}} \tilde{f}[\mathbf{s}]\tilde{g}[\mathbf{x}-\mathbf{s}] \overset{\text{DDFS}}{\longleftrightarrow} \tilde{F}[\mathbf{u}]\tilde{G}[\mathbf{u}]$
Multiplication	$\tilde{f}_c(\mathbf{x})\tilde{g}_c(\mathbf{x}) \overset{\text{CDFS}}{\longleftrightarrow} \frac{1}{d(\Gamma)} \sum_{\mathbf{w}\in\Gamma^*} \tilde{F}_c[\mathbf{w}]\tilde{G}_c[\mathbf{u}-\mathbf{w}]$	$\tilde{f}[\mathbf{x}]\tilde{g}[\mathbf{x}] \overset{\text{DDFS}}{\longleftrightarrow} \frac{1}{K} \sum_{\mathbf{w}\in\mathcal{D}} \tilde{F}[\mathbf{w}]\tilde{G}[\mathbf{u}-\mathbf{w}]$
Automorphism of domain	$\tilde{f}_c(\mathbf{A}\mathbf{x}) \overset{\text{CDFS}}{\longleftrightarrow} \tilde{F}_c[\mathbf{A}^{-T}\mathbf{u}]$	$\tilde{f}[\mathbf{A}\mathbf{x}] \overset{\text{DDFS}}{\longleftrightarrow} \tilde{F}[\mathbf{A}^{-T}\mathbf{u}]$
Differentiation	$\nabla_{\mathbf{x}}\tilde{f}_c(\mathbf{x}) \overset{\text{CDFS}}{\longleftrightarrow} j2\pi\mathbf{u}\tilde{F}_c[\mathbf{u}]$	N/A
Differentiation in frequency	N/A	N/A
Complex conjugation	$\tilde{f}_c^*(\mathbf{x}) \overset{\text{CDFS}}{\longleftrightarrow} \tilde{F}_c^*[-\mathbf{u}]$	$\tilde{f}^*[\mathbf{x}] \overset{\text{DDFS}}{\longleftrightarrow} \tilde{F}^*[-\mathbf{u}]$
Parseval	$\int_{\mathcal{P}_\Gamma} \tilde{f}_c(\mathbf{x})\tilde{g}_c^*(\mathbf{x})\, d\mathbf{x} = \frac{1}{d(\Gamma)} \sum_{\mathbf{u}\in\Gamma^*} \tilde{F}_c[\mathbf{u}]\tilde{G}_c^*[\mathbf{u}]$	$\sum_{\mathbf{x}\in\mathcal{B}} \tilde{f}[\mathbf{x}]\tilde{g}^*[\mathbf{x}] = \frac{1}{K} \sum_{\mathbf{w}\in\mathcal{D}} \tilde{F}[\mathbf{u}]\tilde{G}^*[\mathbf{u}]$
Duality	$F(\mathbf{x}) \overset{\text{CDFS}}{\longleftrightarrow} \frac{1}{d(\Lambda)} f[-\mathbf{u}]$	$\tilde{F}[\mathbf{C}\mathbf{x}] \overset{\text{DDFS}}{\longleftrightarrow} K\tilde{f}[-\mathbf{C}^{-1}\mathbf{u}]$

References

Armstrong, M.A. (1988) *Groups and Symmetry*, Springer-Verlag, New York, NY.

Barrett, H.H. and Myers, K.J. (2004) *Foundations of Image Science*, Wiley-Interscience, Hoboken, NJ.

Bracewell, R.N. (2000) *The Fourier Transform and its Applications*, McGraw Hill, Boston, MA, 3rd edn.

Brandolini, L., Colzani, L., and Travaglini, G. (1997) Average decay of Fourier transforms and integer points in polyhedra. *Ark. Mat.*, **35**, 253–275.

Brillinger, D.R. (2001) *Time Series: Data Analysis and Theory*, SIAM, Philadelphia, PA.

Brown, H., Bülow, R., Neubüser, J., Wondratschek, H., and Zassenhaus, H. (1978) *Crystallographic Groups of Four-Dimensional Space*, Wiley, New York, NY.

Cariolaro, G. (2011) *Unified Signal Theory*, Springer, London.

Cassels, J.W.S. (1997) *An Introduction to the Geometry of Numbers*, Springer-Verlag, Berlin.

Cohen, H. (1993) *A Course in Computational Algebraic Number Theory*, Springer-Verlag, Berlin.

Cortelazzo, G. and Manduchi, R. (1993) On the determination of all the sublattices of preassigned index and its application to multidimensional sampling. *IEEE Trans. Circuits Syst. Video Technol.*, **3** (4), 318–320.

Coulombe, S. and Dubois, E. (1999) Nonuniform perfect reconsruction filter banks over lattices with application to transmultiplexers. *IEEE Trans. Image Process.*, **47** (4), 1010–1023.

Do, M.N. and Lu, Y.M. (2011) Multidimensional filter banks and multiscale geometric representations. *Foundations and Trends in Signal Processing*, **5** (3), 157–264.

Dubois, E. (1985) The sampling and reconstruction of time-varying imagery with application in video systems. *Proc. IEEE*, **73** (4), 502–522.

Dubois, E. (2005) Frequency-domain methods for demosaicking of Bayer-sampled color images. *IEEE Signal Process. Lett.*, **12** (12), 847–850.

Dubois, E. (2009) Color filter array sampling of color images: Frequency-domain analysis and associated demosaicking algorithms, in *Single Sensor Imaging: Methods and Applications for Digital Cameras* (ed. R. Lukac), CRC Press, Boca Raton, FL, chap. 7, pp. 183–212.

Dubois, E. (2010) *The Structure and Properties of Color Spaces and the Representation of Color Images*, Morgan and Claypool.

Dubois, E., Sabri, M.S., and Ouellet, J.-Y. (1982) Three-dimensional spectrum and processing of digital NTSC color signals. *SMPTE J.*, **91** (4), 372–378.

Multidimensional Signal and Color Image Processing Using Lattices, First Edition. Eric Dubois.

Companion website: www.wiley.com/go/Dubois/multiSP

Dudgeon, D.E. and Mersereau, R.M. (1984) *Multidimensional Digital Signal Processing*, Prentice-Hall, Englewood Cliffs, NJ.

Dummit, D.S. and Foote, R.M. (2004) *Abstract Algebra*, Wiley, Hoboken, NJ, 3rd edn.

Farrell, J., Ng, G., Ding, X., Larson, K., and Wandell, B. (2008) A display simulation toolbox for image quality evaluation. *J. Display Technol.*, **4** (2), 262–270.

Fieguth, P. (2011) *Statistical Image Processing and Multidimensional Modeling*, Springer, New York, NY.

Fieguth, P. and Zhang, J. (2005) Random field models, in *Handbook of Image and Video Processing* (ed. A. Bovik), Elsevier, Burlington, MA, chap. 4.3, pp. 361–375, 2nd edn.

Gasca, M. and Sauer, T. (2000) Polynomial interpolation in several variables. *Adv. Comput. Math.*, **12**, 377–410.

Gasquet, C. and Witomski, P. (1999) *Fourier Analysis and Applications: Filtering, Numerical Computation, Wavelets*, Springer, New York, NY.

Germer, T.A., Zwinkels, J.C., and Tsai, B.K. (eds) (2014) *Spectrophotometry: Accurate Measurement of Optical Properties of Materials*, Academic Press, Amsterdam.

Gersho, A. and Gray, R.M. (1992) *Vector Quantization and Signal Compression*, Kluwer, Boston, MA.

Grassmann, H.G. (1854) On the theory of compound colors. *Philosophic Magazine*, **4** (7), 254–264. Translated from original German version which appeared in Annelen der Physik und Chemie (Poggendorf), vol. **89**, pp. 69–84, 1853. Reproduced in D.L. MacAdam, *Sources of Color Science*, MIT Press, 1979, pp. 53–60.

Gray, R.M. and Goodman, J.W. (1995) *Fourier Transforms: An Introduction for Engineers*, Kluwer Academic Publishers, Boston, MA.

Gunturk, B.K., Glotzbach, J., Altunbasak, Y., Schafer, R.W., and Mersereau, R.M. (2005) Demosaicking: Color filter array interpolation. *IEEE Signal Process. Mag.*, **22** (1), 44–54.

Hanselman, D. and Littlefield, B. (2012) *Mastering MATLAB*, Pearson, Upper Saddle River, NJ.

Harris, F.J. (1978) On the use of windows for harmonic analysis with the discrete Fourier transform. *Proc. IEEE*, **66** (1), 51–83.

Hirakawa, K. and Wolfe, P.J. (2007) Fourier domain display color filter array design, in *Proc. IEEE Int. Conf. Image Processing*, San Antonio, TX, pp. III–429–III–432.

Hoffman, K. and Kunze, R. (1971) *Linear Algebra*, Prentice-Hall, Upper Saddle River, NJ, 2nd edn.

Holt, D.F., Eick, B., and O'Brien, E.A. (2005) *Handbook of Computational Group Theory*, Chapman & Hall/CRC, Boca Raton, FL.

Hu, J.V. and Rabiner, L.R. (1972) Design techniques for two-dimensional digital filters. *IEEE Trans. Audio Electroacoust.*, **AU-20** (4), 249–257.

Huang, T.S. (1972) Two-dimensional windows. *IEEE Trans. Audio Electroacoust.*, **AU-20** (1), 88–89.

Johnson, G.M. and Fairchild, M.D. (2003) A top down description of S-CIELAB and CIEDE2000. *Color Res. Applicat.*, **28** (6), 425–435.

Kalker, T. (1998) On multidimensional sampling, in *The Digital Signal Processing Handbook* (eds V.K. Madisetti and D.B. Williams), CRC Press, Boca Raton, FL, chap. 4, pp. 4–1 – 4–21.

Kammler, D.W. (2000) *A First Course in Fourier Analysis*, Prentice-Hall, Upper Saddle River, NJ.

Kaplan, W. (1984) *Advanced Calculus*, Addison-Wesley, Reading,MA, 3rd edn.

Keys, R.G. (1981) Cubic convolution interpolation for digital image processing. *IEEE Trans. Acoust. Speech Signal Process.*, **ASSP-29** (6), 1153–1160.

Klompenhouwer, M.A. and de Haan, G. (2003) Subpixel image scaling for color matrix displays. *J. Soc. Inform. Display*, **11** (1), 99–108.

Komrska, J. (1982) Simple derivation of formulas for Fraunhofer diffraction at polygonal apertures. *J. Opt. Soc. Am.*, **72** (10), 1382–1384.

Kretz, F. and Sabatier, J. (1981) Échantillonnage des images de télévision: analyse dans le domaine spatio-temporel et dans le domaine de Fourier. *Annales des Télécommunications*, **36** (3-4), 231–273.

Kutas, G., Choh, H.K., Kwak, Y., Bodrogi, P., and Czúni, L. (2006) Subpixel arrangements and color image rendering methods for multiprimary displays. *J. Electr. Imaging*, **15** (2), 023 002–1–023 002–9.

Lagendijk, R.L. and Biemond, J. (2009) Basic methods for image restoration and identification, in *The Essential Guide to Image Processing* (ed. A. Bovik), Academic Press, Burlington, MA, chap. 14, pp. 323–348.

Lampropoulos, G.A. and Fahmy, M.M. (1985) A new technique for the design of two-dimensional FIR and IIR filters. *IEEE Trans. Acoust. Speech Signal Process.*, **ASSP-33** (1), 268–280.

Leon-Garcia, A. (2008) *Probability and Random Processes for Electrical Engineering*, Prentice Hall, Upper Saddle River, NJ, 3rd edn.

Leung, B., Jeon, G., and Dubois, E. (2011) Least-squares luma-chroma demultiplexing algorithm for Bayer demosaicking. *IEEE Trans. Image Process.*, **20** (7), 1885–1894.

Lu, W.S. and Antoniou, A. (1992) *Two-Dimensional Digital Filters*, Marcel Dekker, New York, NY.

Lu, Y.M., Do, M.N., and Laugesen, R.S. (2009) A computable Fourier condition generating alias-free sampling lattices. *IEEE Trans. Signal Process.*, **57** (5), 1768–1782.

Micciancio, D. and Goldwasser, S. (2002) *Complexity of Lattice Problems: A Cryptographic Perspective*, Kluwer Academic Publishers, Boston, MA.

Miller, Jr., W. (1972) *Symmetry Groups and Their Applications*, Academic Press, New York, NY.

Newman, M. (1972) *Integral Matrices*, Academic Press, New York, NY.

Olver, P.J. and Shakiban, C. (2006) *Applied Linear Algebra*, Pearson Prentice Hall, Upper Saddle River, NJ.

Oppenheim, A.V., Schafer, R.W., and Buck, J.R. (1999) *Discrete-Time Signal Processing*, Prentice Hall, Upper Saddle River, NJ, 2nd edn.

Oppenheim, A.V. and Willsky, A.S. (1997) *Signals and Systems*, Prentice Hall, Upper Saddle River, NJ, 2nd edn.

Ouellet, J.Y. and Dubois, E. (1981) Sampling and reconstruction of NTSC video signals at twice the color subcarrier frequency. *IEEE Trans. Commun.*, **COM-29**, 1823–1832.

Papoulis, A. (1968) *Systems and Transforms with Applications in Optics*, McGraw-Hill, New York, NY.

Papoulis, A. and Unnikrishna Pillai, S. (2002) *Probability, Random Variables and Stochastic Processes*, McGraw Hill, New York, NY, 4th edn.

Petersen, D.P. and Middleton, D. (1962) Sampling and reconsruction of wave-number-limited functions in N-dimensional Euclidean spaces. *Informat. Contr.*, **5**, 279–323.

Platt, J.C. (2000) Optimal filtering for patterned displays. *IEEE Signal Process. Lett.*, **7** (7), 179–181.

Poirson, A.B. and Wandell, B.A. (1993) The appearance of colored patterns: pattern-color separability. *J. Opt. Soc. Am. A, Opt. Image Sci.*, **10** (12), 2458–2470.

Poirson, A.B. and Wandell, B.A. (1996) Pattern-color separable pathways predict sensitivity to simple colored patterns. *Vision Res.*, **36** (4), 515–526.

Poularikas, A.D. (1998) *The Handbook of Formulas and Tables for Signal Processing*, CRC Press, Boca Raton, FL.

Poynton, C. (2012) *Digital Video and HD: Algorithms and Interfaces*, Elsevier, Waltham, MA, 2nd edn.

Poynton, C. and Funt, B. (2014) Perceptual uniformity in digital image representation and display. *Color Research & Application*, **39** (1), 6–15.

Priestley, M.B. (1981) *Spectral Analysis and Time Series: Univariate Series*, vol. 1, Academic Press, London.

Rao, K.R., Kim, D.N., and Hwang, J.J. (2010) *Fast Fourier Transform: Algorithms and Applications*, Springer, Dordrecht.

Richards, I. and Youn, H. (1990) *Theory of Distributions: A Non-Technical Introduction*, Cambridge University Press, Cambridge, UK.

Rudin, W. (1962) *Fourier Analysis on Groups*, Wiley-Interscience, New York, NY.

Seviora, R.E. (1971) *Generalized Digital Filters*, Ph.D. thesis, University of Toronto.

Shaw, R. (1982) *Linear Algebra and Group Representations*, vol. 1, Academic Press, London.

Stark, H. and Woods, J.W. (2002) *Probability and Random Processes with Applications to Signal Processing*, Prentice-Hall, Englewood Cliffs, NJ, 3rd edn.

Stein, E. and Weiss, G. (1971) *Introduction to Fourier Analysis on Euclidean Spaces*, Princeton University Press, Princeton, NJ.

Stevenson, R.L. (1997) Inverse halftoning via MAP estimation. *IEEE Trans. Image Process.*, **6** (4), 574–583.

Strang, G. (1999) The discrete cosine transform. *SIAM Rev.*, **41** (1), 135–147.

Suter, B.W. (1998) *Multirate and Wavelet Signal Processing*, Academic Press, San Diego, CA.

Vaidyanathan, P.P. (1993) *Multirate Systems and Filter Banks*, Prentice Hall, Englewood Cliffs, NJ.

Vidyasagar, M. (2011) *Control System Synthesis: A Factorization Approach, Part I*, Morgan and Claypool.

Wang, Y., Ostermann, J., and Zhang, Y.Q. (2002) *Video Processing and Communications*, Prentice Hall, Upper Saddle River, NJ.

Warner, S. (1965) *Modern Algebra*, vol. 1, Prentice-Hall, Englewood Cliffs, NJ.

Witt, K. (2007) CIE color difference metrics, in *Colorimetry: Understanding the CIE System* (ed. J. Schanda), Wiley, Hoboken, NJ, chap. 4, pp. 79–100.

Wonham, W.M. (1985) *Linear Multivariable Control: A Geometric Approach*, Springer-Verlag, New York, NY, 3rd edn.

Woods, J.W. (2006) *Multidimensional Signal, Image, and Video Processing and Coding*, Academic Press, Burlington, MA.

Wright, W.D. (2007) Professor Wright's Paper from the Golden Jubilee Book: The historical and experimental background to the 1931 CIE system of colorimetry, in *Colorimetry: Understanding the CIE System* (ed. J. Schanda), Wiley, Hoboken, NJ, chap. 2, pp. 9–23.

Wübben, D., Seethaler, D., Jaldén, J., and Matz, G. (2011) Lattice reduction: a survey with applications in wireless communications. *IEEE Signal Process. Mag.*, **28** (3), 70–91.

Xu, J., Farrell, J., Matskewich, T., and Wandell, B. (2008) Prediction of preferred ClearType filters using the S-CIELAB metric, in *Proc. IEEE Int. Conf. Image Processing*, San Diego, CA, pp. 361–364.

Zhang, X. (1998) S-CIELAB: A spatial extension to the CIE L*a*b* DeltaE color difference metric. URL http://white.stanford.edu/~brian/scielab/.

Zhang, X. and Wandell, B.A. (1997) A spatial extension of CIELAB for digital color image reproduction. *J. Soc. Inform. Display*, **5** (1), 61–67.

Zhang, X. and Wandell, B.A. (1998) Color image fidelity metrics evaluated using image distortion maps. *Signal Process.*, **70** (3), 201–214.

Zheng, X. (2014) *Optimization of Sampling Structure Conversion Methods for Color Mosaic Displays*, Master's thesis, University of Ottawa.

Index

Multidimensional Signal and Color Image Processing Using Lattices, First Edition. Eric Dubois.

Companion website: www.wiley.com/go/Dubois/multiSP